PARZEN • Modern Probability Theory and Its Applications
PURI and SEN • Nonparametric Methods in General Linear Models
PURI and SEN • Nonparametric Methods in Multivariate Analysis
RANDLES and WOLFE • Introduction to the Theory of Nonparametric Statistics
RAO • Linear Statistical Inference and Its Applications, *Second Edition*
RAO • Real and Stochastic Analysis
RAO and SEDRANSK • W.G. Cochran's Impact on Statistics
ROHATGI • An Introduction to Probability Theory and Mathematical Statistics
ROHATGI • Statistical Inference
ROSS • Stochastic Processes
RUBINSTEIN • Simulation and The Monte Carlo Method
SCHEFFE • The Analysis of Variance
SEBER • Linear Regression Analysis
SEBER • Multivariate Observations
SEN • Sequential Nonparametrics: Invariance Principles and Statistical Inference
SERFLING • Approximation Theorems of Mathematical Statistics
SHORACK and WELLNER • Empirical Processes with Applications to Statistics
TJUR • Probability Based on Radon Measures
WILLIAMS • Diffusions, Markov Processes, and Martingales, Volume I: Foundations
ZACKS • Theory of Statistical Inference

Applied Probability and Statistics
ABRAHAM and LEDOLTER • Statistical Methods for Forecasting
AGRESTI • Analysis of Ordinal Categorical Data
AICKIN • Linear Statistical Analysis of Discrete Data
ANDERSON, AUQUIER, HAUCK, OAKES, VANDAELE, and WEISBERG • Statistical Methods for Comparative Studies
ARTHANARI and DODGE • Mathematical Programming in Statistics
BAILEY • The Elements of Stochastic Processes with Applications to the Natural Sciences
BAILEY • Mathematics, Statistics and Systems for Health
BARNETT • Interpreting Multivariate Data
BARNETT and LEWIS • Outliers in Statistical Data, *Second Edition*
BARTHOLOMEW • Stochastic Models for Social Processes, *Third Edition*
BARTHOLOMEW and FORBES • Statistical Techniques for Manpower Planning
BECK and ARNOLD • Parameter Estimation in Engineering and Science
BELSLEY, KUH, and WELSCH • Regression Diagnostics: Identifying Influential Data and Sources of Collinearity
BHAT • Elements of Applied Stochastic Processes, *Second Edition*
BLOOMFIELD • Fourier Analysis of Time Series: An Introduction
BOX • R. A. Fisher, The Life of a Scientist
BOX and DRAPER • Empirical Model-Building and Response Surfaces
BOX and DRAPER • Evolutionary Operation: A Statistical Method for Process Improvement
BOX, HUNTER, and HUNTER • Statistics for Experimenters: An Introduction to Design, Data Analysis, and Model Building
BROWN and HOLLANDER • Statistics: A Biomedical Introduction
BUNKE and BUNKE • Statistical Inference in Linear Models, Volume I
CHAMBERS • Computational Methods for Data Analysis
CHATTERJEE and PRICE • Regression Analysis by Example
CHOW • Econometric Analysis by Control Methods
CLARKE and DISNEY • Probability and Random Processes: A First Course with Applications, *Second Edition*
COCHRAN • Sampling Techniques, *Third Edition*
COCHRAN and COX • Experimental Designs, *Second Edition*

CONOVER • Practical Nonparametric Statistics, *Second Edition*
CONOVER and IMAN • Introduction to Modern Business Statistics
CORNELL • Experiments with Mixtures: Designs, Models and The Analysis of Mixture Data
COX • Planning of Experiments
DANIEL • Biostatistics: A Foundation for Analysis in the Health Sciences, *Third Edition*
DANIEL • Applications of Statistics to Industrial Experimentation
DANIEL and WOOD • Fitting Equations to Data: Computer Analysis of Multifactor Data, *Second Edition*
DAVID • Order Statistics, *Second Edition*
DAVISON • Multidimensional Scaling
DEGROOT, FIENBERG and KADANE • Statistics and the Law
DEMING • Sample Design in Business Research
DILLON and GOLDSTEIN • Multivariate Analysis: Methods and Applications
DODGE • Analysis of Experiments with Missing Data
DODGE and ROMIG • Sampling Inspection Tables, *Second Edition*
DOWDY and WEARDEN • Statistics for Research
DRAPER and SMITH • Applied Regression Analysis, *Second Edition*
DUNN • Basic Statistics: A Primer for the Biomedical Sciences, *Second Edition*
DUNN and CLARK • Applied Statistics: Analysis of Variance and Regression
ELANDT-JOHNSON and JOHNSON • Survival Models and Data Analysis
FLEISS • Statistical Methods for Rates and Proportions, *Second Edition*
FLEISS • The Design and Analysis of Clinical Experiments
FOX • Linear Statistical Models and Related Methods
FRANKEN, KÖNIG, ARNDT, and SCHMIDT • Queues and Point Processes
GALAMBOS • The Asymptotic Theory of Extreme Order Statistics
GIBBONS, OLKIN, and SOBEL • Selecting and Ordering Populations: A New Statistical Methodology
GNANADESIKAN • Methods for Statistical Data Analysis of Multivariate Observations
GOLDSTEIN and DILLON • Discrete Discriminant Analysis
GREENBERG and WEBSTER • Advanced Econometrics: A Bridge to the Literature
GROSS and CLARK • Survival Distributions: Reliability Applications in the Biomedical Sciences
GROSS and HARRIS • Fundamentals of Queueing Theory, *Second Edition*
GUPTA and PANCHAPAKESAN • Multiple Decision Procedures: Theory and Methodology of Selecting and Ranking Populations
GUTTMAN, WILKS, and HUNTER • Introductory Engineering Statistics, *Third Edition*
HAHN and SHAPIRO • Statistical Models in Engineering
HALD • Statistical Tables and Formulas
HALD • Statistical Theory with Engineering Applications
HAND • Discrimination and Classification
HILDEBRAND, LAING, and ROSENTHAL • Prediction Analysis of Cross Classifications
HOAGLIN, MOSTELLER and TUKEY • Exploring Data Tables, Trends and Shapes
HOAGLIN, MOSTELLER, and TUKEY • Understanding Robust and Exploratory Data Analysis
HOEL • Elementary Statistics, *Fourth Edition*

(*continued on back*)

Systems in Stochastic Equilibrium

Systems in Stochastic Equilibrium

PETER WHITTLE, FRS

Churchill Professor of the Mathematics of Operational
Research, University of Cambridge
Fellow of Churchill College, Cambridge

JOHN WILEY AND SONS

Chichester · New York · Brisbane · Toronto · Singapore

Library of Congress Cataloging-in-Publication Data:

Whittle, Peter.
   Systems in stochastic equilibrium.

   (Wiley series in probability of mathematical
statistics. Applied probability and statistics)
   Includes index.
   1. Stochastic systems. 2. Equilibrium. I. Title.
II. Series.
QA401.W625 1986   519.2   85–17923

ISBN 0 471 90887 8

British Library Cataloguing in Publication Data:

Whittle, Peter, 1927–
   Systems in stochastic equilibrium.
   1. Stochastic processes
   I. Title
   519.2   QA274

ISBN 0 471 90887 8

Printed and bound in Great Britain.

Contents

v

Preface

This work springs from a fascination over many years with the concepts of equilibrium, reversibility and of interaction and collective behaviour in systems extended over space or over a network. Since, however, such interactive systems could be said to constitute the whole material of science and provide its great unsolved problems, there is obviously call for a quick disclaimer. The integrating theme of this work is the study of a particular type of interaction termed *weak coupling*. The system is considered to be made up of *components* (e.g. subsystems, molecules, processors, queues, organisms, offices) between which move *units* (e.g. energy quanta, atoms, jobs, customers, nutrient, messages). The only direct interactions permitted are short-range ones between units and components: a unit may, for a time, be somehow incorporated in a component to which it is adjacent. The only interaction between components is that provided by the migration of units between them.

Such a coupling is what lies behind the Gibbs distribution of statistical mechanics, Jackson networks of processors or queues and the concept of an exchange economy. A glance at the chapter titles indicates the range of topics and applications unified by the concept. That the theory is so amenable mathematically must imply that it represents what is, physically, a special case. On the other hand, it contains both mathematical and physical substance, as the treatment of phase transitions for the polymerization and socioeconomic models of Chapters 13 to 17 will perhaps make plain. Moreover, the mechanism postulated seems less special in view of the fact that interactions mediated by a field are now seen, in a quantized theory, as mediated by the passage of some kind of particle.

Chapters 18 to 20 are somewhat distinct, in that they make no appeal to the weak coupling concept. However, they embody what has long been for me an associated theme, and they make a relevant point: that a spatial process is best seen as an equilibrium version of a spatio-temporal process.

Most of the work and all of the writing of this book was carried out during my tenure of the Churchill Chair, founded by the Esso Petroleum Company, to whom I am glad to acknowledge my gratitude.

Cambridge   P. WHITTLE

Order and conventions

The diagram below shows the logical dependences between the twenty chapters.
Conventions on the numbering of sections, equations, etc. are listed in section 1.8
and notation is established in sections 1.9 to 1.15.

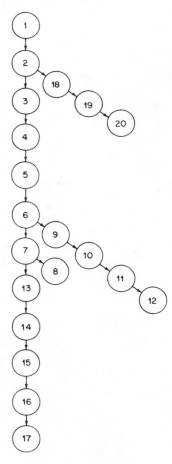

PART I

Basic material

CHAPTER 1

Equilibrium concepts; notation

1. WHY STUDY EQUILIBRIUM?

If one is studying a particular dynamic model then there are at least two reasons for taking a special interest in its equilibrium behaviour. The first is that of convenience: the study of equilibrium must necessarily be simpler than that of the more general transient case. One could put this more positively: the examination of equilibrium behaviour often seems mathematically natural, and there are whole classes of models whose equilibrium behaviour is easily characterized and whose transient behaviour is not. This view leads to the second reason, that of insight. One can argue that equilibrium behaviour reveals in summary form the intrinsic character of the model, and displays those properties which are of real significance.

So, if one studies a model of interacting molecules one expects to see in equilibrium the emergence of the possibility of several distinct states of matter. If one studies an ecological system one expects to see in equilibrium which species have survived and what mutual relations they have reached. If one studies a model of flow through a network (in any of the important contexts of traffic, communication, computation or processing generally) then one expects equilibrium behaviour to reveal the bottlenecks which limit overall capacity, and to demonstrate the typical experience of a job or customer passing through the network. In any model of economic or social structure equilibrium behaviour will be revealing, even if in such cases 'equilibrium' can be no more than a halting point on a path of indefinite evolution.

To take a simple example, consider a model of the incidence of an infectious disease in a closed community (essentially a deterministic version of the model adopted by Bartlett to represent the incidence of measles in a boarding school; see Bartlett (1949) and references to Kermack and McKendrick and to Soper quoted there). Let x_1 and x_2 represent the numbers of susceptible and infected individuals respectively, and suppose that these have rates of change

$$\dot{x}_1 = \alpha - \beta x_1 - \gamma x_1 x_2$$
$$\dot{x}_2 = \gamma x_1 x_2 - \delta x_2. \tag{1}$$

That is, susceptibles enter the community at rate α, leave it at total rate βx_1 (and so at individual rate β) and become infected at rate $\gamma x_1 x_2$ (which one might regard as proportional to the rate of encounter between susceptibles and infecteds). Infecteds are removed at individual rate δ.

Model (1) has two possible equilibria, at

$$x_1 = \frac{\alpha}{\beta}, \qquad x_2 = 0 \tag{2}$$

or

$$x_1 = \frac{\delta}{\gamma}, \qquad x_2 = \frac{\alpha\gamma - \beta\delta}{\gamma\delta}. \tag{3}$$

By the usual criteria for stability (i.e. the testing of the linearized version of the model at the equilibrium point for stability; see Exercises 4.1, 4.2 and Luenberger (1979), p. 324) we find that equilibrium (2) is stable and equilibrium (3) unstable if

$$\alpha\gamma < \beta\delta \tag{4}$$

and that the reverse is true if inequality (4) is reversed. That is, the infection is not viable if (4) holds, but is endemic in the community if (4) fails. If the inequality

$$\alpha\gamma - \beta\delta > \left(\frac{\alpha\gamma}{2\delta}\right)^2$$

holds, then, not merely is equilibrium (3) the stable one, but approach to it is oscillatory (in that the matrix of the linearized model has complex eigenvalues). That is, the endemic infection will manifest itself rather as a recurrent epidemic.

It can reasonably be claimed that the essential behaviour of the model has been revealed by this brief study of the nature of its equilibrium. Realism would, of course, require this particular model to be elaborated in a number of ways. In particular, one might allow for random effects, and formulate the model in a stochastic rather than a deterministic version. However, the points made for the simple version above will persist.

Passage to equilibrium need by no means take the form of convergence (in time) to a definite limit point such as (2) or (3). Even once transient effects have subsided, the model may show periodic, aperiodic or 'chaotic' behaviour. However, in a stochastic formulation the distinctions between these various types of behaviour become blurred, and one copes with all of them by considering the equilibrium *distribution* of the variable.

2. COVERAGE OF THE TEXT

As the title indicates, the work is concerned not merely with equilibrium, but with equilibrium of systems. That is, one is thinking of systems of interacting components, perhaps of many components. For example, we consider interaction between molecules (in the sense of competition for constituents, see Chapter 7), between and within large molecules (Chapters 13–17), between individuals in populations of various species (Chapter 8), between individuals in a role-adaptive socioeconomic model (section 17.5), between processors in a processing network (Chapters 9 and 10) and between sites in a spatial lattice (Chapters 18–20).

The study of interaction and collective effects in a system is certainly one of the most inclusive and ambitious projects that can be imagined. However, we have already set out our more modest immediate goals in the preface. We largely restrict ourselves to a form of interaction termed *weak coupling*, a form of interaction induced by the passage of units of some kind between the components of the system. The units may be quanta of energy or of matter, they may be messages, jobs, customers or packages of some commodity. Weak coupling constitutes a form of interaction both tractable and significant, and its study constitutes the integrating theme of this work.

The term 'weak coupling' does not refer to the weak interactions (the radioactive interactions) of modern physics, but rather to the weak coupling appealed to in schoolroom treatments of the perfect gas. These treatments appealed to a coupling which, apparently by its weakness, was supposed to allow gas molecules to exchange energy and yet be statistically independent. In fact, this apparently paradoxical situation is quite possible and characterizes a very substantial class of models. The conclusion is also exact, the 'weakness' of the coupling is a matter of structure rather than of degree, and no approximation is involved.

We study the general formalism of weak coupling first in Chapter 11, but begin in Chapter 5 with what is essentially a special case of it: the statistical mechanical models which lead to the Gibbs distribution. In this and related chapters (6–8, 13–17) we begin with specification of the model as a Markov model. This means that a certain amount of the usual physical discussion (classical or quantum-mechanical) is stripped away, and one begins from a model, discrete and statistical, whose relevant statistical properties are taken as axiomatic. This at least makes for economy and tidiness, and emphasizes a structure later recognizable as a special case of weak coupling.

One clear application of the models of Chapter 5 is to equilibrium chemical kinetics (Chapter 7). If molecules may be indefinitely large then one is considering polymerization processes (Chapters 13–17). Such processes provide some of the simplest examples of critical phenomena, and in considerable variety. Furthermore, their application is by no means confined to chemistry; they can be regarded as representing interaction processes in social and economic contexts, a study just beginning.

A variant of some of these models provides a not unrealistic mechanism for resource-induced ecological competition (Chapter 8). From this one deduces a stochastic version of the principle of competitive exclusion and also variational criteria to determine the species which will survive.

Chapters 9–12 concern Jackson networks and related topics. Jackson networks constitute a special class of stochastic networks, of considerable importance as models for traffic, communication, distribution, computing and processing systems. These models are more general in some respects than those of statistical mechanics, and it is at this point that a general discussion of weak coupling becomes natural.

Chapters 2–4 cover the material on Markov processes, statistical equilibrium and reversibility which is later needed. This material is largely expository, although with some novelties.

At the other end of the text, Chapters 18–20 cover some aspects of random fields. These chapters might be considered to owe their existence only to the author's interest in the subject, because there is little mention of weak coupling. However, they do reinforce two points of the work very well. For one thing, many of the puzzling features of a purely spatial process are explained when this process is seen as the equilibrium of a spatio-temporal process. For another, these chapters give new manifestations of a property by then familiar: stochastic invariance of the model under some natural group of transformations.

3. DYNAMIC MODELS; DETERMINISTIC AND STOCHASTIC PROCESSES

A dynamic model explains, at least in some degree, the variations in time of some variables of interest which we shall denote collectively by x. For example, for the infection model of section 1 we may take x as (x_1, x_2), the numbers of susceptibles and infecteds at a given time. More generally, x might be a vector function of time which obeys a system of first-order differential equations

$$\dot{x} = a(x, t). \tag{1}$$

However, equation (1) still expresses rather special dynamic assumptions. For one thing, it represents the rate of change of x as being a function of current x and of t alone, whereas x could conceivably depend upon its own past in a much more general fashion (e.g. x could obey a differential equation of higher order in t, or a dynamic relation involving time delays). For another, model (1) is *deterministic* in that the course of x is determined by initial conditions. Many applications (and those of this book may serve as examples) require models which are *stochastic*, in that the evolution of the variable is to some extent random.

Let the value of x at time t be denoted $x(t)$ and let the history of x at time t be denoted $X(t)$:

$$X(t) = \{x(\tau) : \tau \leqslant t\} \tag{2}$$

The complete x-path $X(+\infty)$ will often be denoted simply by X and termed the *realization*.

Let us say that our model is a *stochastic process* $\{x(t)\}$ if

(i) The model is stochastically complete, in that from its assumptions one can in principle calculate the value of the expectation $E[\phi(X)]$ for any scalar-valued functional $\phi(\cdot)$ in some class of interest $\mathscr{F}$.

(ii) Any quantity whose value is in principle determined at time t must be a function of $X(t)$.

If these assumptions are satisfied we shall say that x is the *process variable* and $X(t)$ the *process history* at time t.

Assumption (ii) implies a good deal. It implies that time runs forward and that the process variable provides an adequate level of description for the system one is studying. The assumption raises a number of points which we dwell upon in Exercises 1–4.

Assumption (i) implies that we can in principle calculate conditional expectations such as

$$E[\phi(x(\tau)|X(t)] \qquad (\tau \geqslant t) \tag{3}$$

for a sufficiently general choice of ϕ, and in this sense predict the forward course of the process from time t. If prediction is perfect, in that $x(\tau)$ is determined for all τ by $X(t)$ for any given t then the process is *deterministic*. It is on the whole helpful to see the deterministic case as a special case of the stochastic.

Evaluation (3) in a sense expresses the stochastic dynamics of the process, in that it indicates how past conditions future. However, it is the stochastic analogue, not merely of a stochastic equation such as (1), but also of the solution of that equation. One is interested in the minimal expression of stochastic dynamics. That is, in the least that the model need specify in order that the expectations $E[\phi(X)]$ should indeed be calculable. This would be the stochastic analogue of the specification of the 'law of motion' (1) plus its initial conditions.

The stochastic analogue of a law of motion as special as (1), a system of first-order differential equations, is a *Markov process*. For such processes stochastic dynamics are determined by specification of the *transition probabilities* or *intensities*. To define the Markov property, let us introduce the notion complementary to the history $X(t)$, the *future* at time t:

$$X^*(t) = \{x(\tau), \tau \geqslant t\}. \tag{4}$$

Suppose that $E[\phi(X^*(t))|X(t)]$ depends upon $X(t)$ only through $x(t)$ for all relevant ϕ and t. It is then plausible and in fact true (see Theorem 14.1) that one can make the identification

$$E[\phi(X^*(t))|X(t)] = E[\phi(X^*(t))|x(t)] \tag{5}$$

This condition expresses the *Markov property*, the process $\{x(t)\}$ is then a *Markov process* and the process variable x is termed a *state variable*. The term 'state variable' implies that $x(t)$ fully characterizes the state of the process at time t in the sense (5): that $x(t)$ contains all aspects of $X(t)$ relevant for prediction at time t.

In a sense a Markov formulation is the canonical form of a process: that achieved when one has chosen a process variable which gives a complete description of state. We run through the relevant Markov theory in Chapters 2 and 3, and all our temporal models are Markov.

Exercises and comments

1. Suppose x is generated by a deterministic relation

$$\dot{x} = a(x, u, t) \tag{6}$$

where $u(t)$ is not determinable from $X(t)$. Then x is not a process variable unless the model also specifies u as a function $u(t)$ of t. This is because relation (6) on its own explains the evolution of x in terms of u, but does not explain the evolution of u. If one supplements the model so that u is also explained then (x, u) would jointly be a process variable.

   In discussing a partial model such as (6) an economist would refer to x as an *endogenous* variable and u as an *exogenous* variable. For example, x might describe the economy of a country and u express factors such as weather, technological change or the economies of other countries. These are external factors in that they are not explained by a model of that country, which is then necessarily an incomplete model.

2. One could also have an incomplete stochastic model, in that an expectation such as $E[\phi(X)]$ is a function of (i.e. is *parametrized* by) an exogenous history U. One could not then even average over variation in U unless one knew its statistics, i.e. had a stochastic model for U. Causality will in general require that $E[\phi(X(t))]$ be parametrized only by $U(t)$.

3. To say that some quantity is 'in principle determined' at time t is not the same as saying that it has been observed at that time. For example, the thoughts of an experimental subject at time t_1 are in principle determined at t_1, even though they may not be accessible to the observer. On the other hand, they will not in general be determined at $t_2 < t_1$.

4. Assumption (ii) can be seen as a statement both of causality and of autonomy, in that it states that any quantity (of interest for the problem in question, be it understood) which is in principle determined at time t must be calculable from $X(t)$. Some such assumption is needed, otherwise a statement such as 'a process is deterministic if its course is fully determined by initial conditions' is meaningless. One must constrain 'initial conditions' so that they do not specify more than could be known at initial time.

5. The term 'state' is one of the all-purpose overworked drudges of the scientific vocabulary. We use it here and henceforth in the sense in which it is understood in Markov processes and control theory. Sometimes one stretches this sense in that one speaks of the 'state' of a component of a system, meaning the description of that component which would indeed be a state variable if the component were isolated. There is also the common physical sense of the term, e.g. 'states of matter', corresponding to the identification of well-defined possible regimes of a system. In the theory of spatial processes the term is sometimes used in yet another sense: that of a possible joint distribution of field variables. This usage doubtless has its motivations (e.g. as a reference to the previous usage, or to the quantum mechanical concept of state) but we shall refrain from it.

4. DETERMINISTIC AND STOCHASTIC EQUILIBRIA

A very standard type of deterministic process is the time-invariant form of (3.1); we suppose x a d-dimensional vector whose rate of change depends on current x alone:

$$\dot{x} = a(x) \tag{1}$$

Equation (1) specifies the dynamics of the process and is often termed the *process equation*. In this particular case it consists of a system of d first-order differential equations (in time) of time-invariant form.

The simple notion of equilibrium is the static one: that x no longer changes with time. An equilibrium point $\bar{x}$ for process (1) is thus a solution of

$$a(\bar{x}) = 0 \tag{2}$$

We know that such equilibrium points may be stable or unstable in varying degrees. Explicitly, let $D(\bar{x})$ be the set of $x(0)$ for which it is true that $x(t) \to \bar{x}$ as $t \to +\infty$. Then $D(\bar{x})$ is the *domain of attraction* of $\bar{x}$. The equilibrium point $\bar{x}$ will be most stable if $D(\bar{x})$ is the whole space, least stable if $D(\bar{x})$ consists only of the point $\bar{x}$, and *locally stable* if $D(\bar{x})$ contains a neighbourhood of $\bar{x}$. The usual test for local stability is to test the linearized version of the model at $\bar{x}$ for stability (see Exercise 2).

However, it is also known that the long-term behaviour of a system described by (1) can be far more complicated than that of simple convergence to a limit value $\bar{x}$. The course of $x(t)$ can ultimately become periodic or show an 'almost-recurrent' behaviour of aperiodic or chaotic character. Such behaviour deserves to be regarded as a form of equilibrium, dynamic rather than static.

In a stochastic formulation the fine structures which can be discerned in the deterministic case (e.g. the very existence of deterministic chaos) become blurred and scarcely survive. Rather than consider $x(t)$ directly one will consider its distribution

$$\pi(A, t) = P(x(t) \in A) \tag{3}$$

(The expression on the right is to be read 'the probability that $x(t)$ falls in the set A'. See section 13 for conventions on probabilistic notation.) One considers that the process is in equilibrium if $\pi(A, t)$, as a function of A and t, is independent of t. This is an assertion we shall have to expand, but for the moment it will do. A time-invariance $\pi(A, t) = \pi(A)$ is not inconsistent with equilibrium of a dynamic character, see Exercise 4 for a simple example.

A stochastic formulation even has an integrating effect on one's view of the static equilibria determined by (2). Roughly speaking, these equilibrium points are stationary points of the equilibrium distribution $\pi(x)$. The locally stable equilibria are those corresponding to local maxima of $\pi(x)$. The notions of separateness and of domains of attraction of equilibria are much weakened, as passage from one equilibrium point to another is in general possible. That is, equilibrium points mark regions of temporary rather than of permanent sojourn.

As an example, consider a model representing the behaviour of N bees in a region, x of whom have joined a swarm around a queen, and $N - x$ of whom are unattached and wandering freely. (Alternatively, there are N factory workers, x of whom have joined a union, or N molecules, x of which have become absorbed on a monolayer.) Let us assume that individual bees join and leave the swarm at rates α and $\beta e^{-\gamma x}$ respectively, so that the total rates of joining and leaving are

$$\lambda(x) = \alpha(N - x)$$

$$\mu(x) = \beta x e^{-\gamma x}$$

(4)

respectively. The term $e^{-\gamma x}$ represents the capturing effect of the swarm, a collective effect which becomes stronger as the swarm becomes larger. In a deterministic treatment the variable x then satisfies the equation

$$\dot{x} = \lambda(x) - \mu(x)$$

(5)

with possible equilibria at the roots of

$$\alpha(N - x) = \beta x e^{-\gamma x}$$

(6)

One sees graphically that equation (6) either has one root (Cases (a) and (c) of Fig. 1.4.1) or three (Case (b)). The stable solutions are those for which $\lambda'(x) - \mu'(x) < 0$, i.e. for which the graph of $\mu(x)$ crosses that of $\lambda(x)$ from below as x increases (see Exercise 2). So, the single root of Cases (a) and (c) locates a stable equilibrium, as do the outer roots of Case (b). The single root of Case (a) and the first root of Case (b) represent a regime in which the bees are largely free-living. The third root of Case (b) and the single root of Case (c) represent a regime in which the bees have formed a swarm.

Suppose now that we recognize the variable x as integer valued, and regard the rates $\lambda(x)$ and $\mu(x)$ as the probability intensities of transitions $x \rightarrow x + 1$ and $x \rightarrow x - 1$ respectively. Balance of the probability flux $x \rightleftharpoons x + 1$ in equilibrium then demands that the equilibrium distribution $\pi(x)$ satisfy

$$\lambda(x)\pi(x) = \mu(x + 1)\pi(x + 1).$$

(7)

(These matters will be given a general treatment in the next chapter, when we shall see relation (7) as the detailed balance equation satisfied by a birth and death process, see Exercise 2.7.7.) It follows then that

$$\pi(x) \propto \prod_{y \leqslant x} \frac{\lambda(y - 1)}{\mu(y)}$$

$$\propto \binom{N}{x} e^{-\delta x - \gamma x^2 / 2}$$

(8)

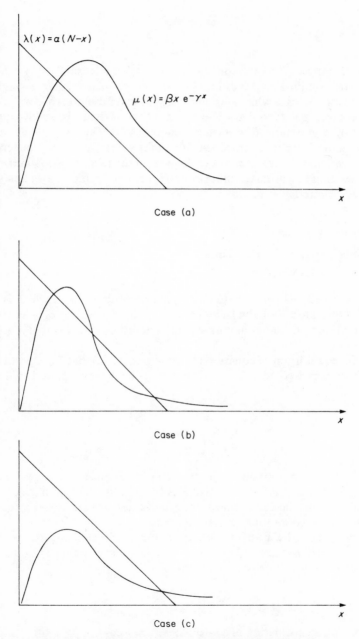

Fig. 1.4.1 Equilibria for the deterministic bee-swarming model.

where

$$e^{-\delta} = \frac{\alpha}{\beta} e^{-\gamma/2}$$

The discrete stationarity condition $\pi(x) \sim \pi(x+1)$ for π is exactly $\lambda(x) \sim \mu(x)$, the deterministic equilibrium condition (6). The case when (6) has a single root corresponds to the case when $\pi(x)$ has a single maximum; the case when (6) has three roots corresponds to that when $\pi(x)$ has two maxima (the stable equilibria) separated by a minimum (the unstable equilibrium). The two maxima vary in relative magnitude (to the point when either may vanish) as the parameters α, β, γ and N are varied. This corresponds to changes in the relative probabilities of the two regimes: that in which swarming (union membership, monolayer adsorption) scarcely occurs, and that in which it is dominant.

Exercises and comments

1. Consider the linear process equation

$$\dot{x} = Ax$$

where A is a $d \times d$ matrix. This has an equilibrium point $\bar{x} = 0$ which is locally stable (and then globally stable) if and only if all the eigenvalues of A have strictly negative real part. One often summarizes this condition by saying that A must be a stability matrix.

2. If $\bar{x}$ is an equilibrium of equation (1), set $x = \bar{x} + \xi$, so that ξ is a perturbation from $\bar{x}$. The linearized version of the model in the neighbourhood of $\bar{x}$ is then

$$\dot{\xi} = A\xi$$

where

$$a_{jk} = \left(\frac{\partial a_j(x)}{\partial x_k} \right)_{x = \bar{x}}$$

(see section 10 for conventions on matrix notation). Under regularity conditions the equilibrium at $\bar{x}$ is stable if and only if A is a stability matrix. In the case when this condition is satisfied and some of the eigenvalues of A are moreover complex, then approach to equilibrium can be oscillatory.

3. By appealing to the assertions of Exercise 2 confirm the assertions made in section 1 concerning the equilibrium of the infection model (1.1).

4. Consider the two-dimensional system

$$\dot{x}_1 = x_2$$
$$\dot{x}_2 = -x_1$$

and the polar transformation of coordinates $x_1 = r \cos \theta$, $x_2 = r \sin \theta$. Show that a possible equilibrium distribution is one in which r is arbitrarily distributed and the distribution of θ conditional on r is always uniform on $(0, 2\pi]$. Interpret. (The model is deterministic, but one can well have a probability distribution of values induced by a randomness in the specification of initial values, see section 2.8.)

5. STATIONARITY AND TIME-HOMOGENEITY

To demand that distribution (4.3) be independent of t is to demand that

$$\pi(A, t+s) = \pi(A, t) \tag{1}$$

identically in A, t, s. Any distribution satisfying this condition is termed *stationary*. However, the concept of stochastic equilibrium would seem to demand more than this, to demand that the joint statistics of $x(\cdot)$ at any number of points in time should be invariant to a bodily shift in the time axis. That is, that the statistics of the process $\{x(t)\}$ should be identical with those of the process $\{x(t+s)\}$ for any s. This is the property of *stationarity*.

To formalize this property let us recall our use of X to denote the entire realization or history $\{x(t); -\infty < t < +\infty\}$. Let us denote by $X[t]$ the value of this realization at the particular time t, so that

$$X[t] = x(t) \tag{2}$$

This could be regarded as the infinite time-axis equivalent of saying that $x(t)$ is the tth component of a vector X, the operation of extracting the tth component being denoted by $X[t]$.

Define now the time-translation operator T^s which, applied to X, shifts it bodily in time by an amount s. That is,

$$(T^s X)[t] = x(t+s) \tag{3}$$

Relation (3) is sometimes written simply as

$$T^s x(t) = x(t+s) \tag{4}$$

which is certainly more convenient and transparent, but is improper unless one is willing to make a special convention. The impropriety lies in the fact that T^s is an operator converting functions into functions, not point values into point values. (In the notation to be adopted in section 9 it is a mapping from $\mathscr{X}^R$ to $\mathscr{X}^R$, not from $\mathscr{X}$ to $\mathscr{X}$.) The value of $x(t+s)$ could not be determined from that of $x(t)$ for given t, s; it is determined from knowledge of X. The special convention would be to *understand* that $T^s x(t)$ is the value at time t of the transformed realization $T^s X$. This is a convention most of us have accepted for years, see Exercise 5.

The indicial notation T^s is appropriate, since indeed $T^{s_1} T^{s_2} = T^{s_1 + s_2}$. The operator T^s is also defined for negative values of s if $x(t)$ is assumed defined for the indefinite past ($t \to -\infty$) as well as the indefinite future ($t \to +\infty$), and T^{-s} is the operator inverse to T^s.

The formal definition is then: the stochastic process $\{x(t)\}$ is *stationary* if

$$E\phi(T^s X) = E\phi(X) \tag{5}$$

for any functional ϕ of the process for which $E\phi(X)$ is defined and for any s (but see Exercise 3 and Appendix 1). This is the most complete possible equilibrium demand. The property is sometimes referred to as *time-invariance*, meaning more exactly invariance of the statistics of the process under an arbitrary time-translation.

Time-invariance is not to be confused with the weaker property of *time-homogeneity*, which merely asserts that the dynamics of the process do not change with time. Formally, time-homogeneity requires that

$$E[\phi(T^s X)|X(t+s)] = E(\phi(X)|X(t)] \tag{6}$$

for all s and for any ϕ for which either side exists. Stationarity requires both time-homogeneity *and* equilibrium.

Exercises and comments

1. Sometimes a process is defined only on the half-axis $t \geqslant 0$, say. One can still make the stationarity demand (5) for $s \geqslant 0$. However, T^s is undefined for negative s. In forming T^s for positive s one loses $\{x(t); 0 \leqslant t < s\}$ and T^s does not have an inverse.

2. Note that the deterministic and time-varying function $x(t) = \cos(\omega t + \theta)$ is stationary if frequency ω is fixed and phase θ regarded as uniformly distributed over $(-\pi, \pi]$.

3. The conditions (5) for distinct s are not independent. Fulfilment of the condition for $s = s_0$ implies its fulfilment for s any multiple of s_0. See Appendix 1.

4. Suppose a process time-homogeneous Markov. Then, plausibly, it is stationary if and only if (1) holds, i.e. if the one-point distribution is stationary.

5. Note that the familiar derivative $\mathrm{d}x(t)/\mathrm{d}t$ is properly a functional of the whole function $x(\cdot)$, i.e. of X, rather than of the point value $x(t)$ for given t. In fact, one can make the formal identification

$$\frac{\mathrm{d}}{\mathrm{d}t} = \lim_{s \to 0} \frac{T^s - 1}{s}$$

6. STOCHASTIC INVARIANCE AND EQUIVALENCE

The idea that the statistics of a process are invariant under some transformation of interest is one that we shall meet again and again.

So, consider a transformation SX of the realization X on to the same space of realizations. Then we shall say that $\{x(t)\}$ is *stochastically invariant under S*, written $X \backsim SX$, if

$$E\phi(SX) = E\phi(X) \tag{1}$$

for any functional ϕ for which the right-hand member is defined. We shall say that the two processes are *stochastically equivalent*, written $X \sim SX$, if (1) holds for all ϕ such that either side of the relation is defined. Stochastic equivalence thus implies that the statistics of X can be determined from those of SX, and so that to this degree S possesses an inverse.

In the case of stationarity the operation which induced the transformation S was that of time translation. In fact, there was a whole set of such transformations T^s, corresponding to translation by a variable amount s. We shall later find it natural to consider stochastic invariance of random functions of space as well as of time, and under a whole range of operations: time reversal (Chapter 4), state conjugation (section 4.6), transformations of state space (section 3.6), spatial translation (sections 18.5, 19.2, 19.3), spatial rotation (section 20.3) and spatial dilation (section 20.4). Each of these operations and their iterates defines a group, and even at a superficial level one begins to see a good deal of structure, which we summarize in Appendix 1.

7. STOCHASTIC MODELS AND QUANTUM MECHANICS

The fact that some of our models (e.g. the statistical mechanical models of Chapters 5 and 6) have some basis in physical theory and yet are formulated with discrete-valued variables may prompt the reader to wonder whether an appeal is being made to quantum mechanical ideas at any point. The answer is 'no'. The passage from a classical model to a quantum mechanical version corresponds to a very particular way of stochasticizing a deterministic model, employing a very distinctive probability calculus. We are not considering models with the special structure that leads to such a calculus, but consider standard Markov models using the standard probability calculus. Nevertheless, it is true that potential connections with a quantum formalism do suggest themselves from time to time. In the study of reversibility it turns out to be natural to represent the equilibrium distribution $\pi(x)$ in the form $w(x)^2$, where the factor $w(x)$ has a significance (Theorem 4.5.1). This recalls the fact that, in quantum theory, a probability amplitude is the squared modulus of a wave function.

Again, our central concept of weak coupling, specifying that components interact only by the passage of quanta of some kind, is also reminiscent. In quantum theory, if components (e.g. particles) interact via a field, then quantization of that field represents the field by a sea of quanta, and interaction of a component with the field corresponds, in a first approximation, to absorption or emission of a field quantum by the component. However, again, the quantum field versions seem to have too distinctive a structure that one can usefully relate them to our 'classical' weak coupling mechanism.

8. ORGANIZATION OF THE TEXT; STYLE; NOTATION

The logical dependences between chapters are summarized in the diagram on p. ix.

The argument in the text is largely given in theorem/proof form. This form should be regarded neither as forbidding nor as pretentious but, in the author's view, simply as the clearest way of summarizing and punctuating a discussion. Conclusions expressed as theorems are not always either deep or novel; the point is that a definite conclusion is stated. The theorem is the string that ties up the package of a discussion.

'Exercises' are used in an almost complementary fashion: not as the 'problems' of a student text (which would be inappropriate), but often as opportunities to make miscellaneous points. These are points which, although important or interesting in themselves, would have confused discussion if they had been incorporated in the main text.

A rather wide-ranging text calls for a wide-ranging and often overworked notation. We try to clarify conventions systematically in the following sections, in the hope that a tread perhaps excessively deliberate at this point may make the walking lighter later.

A reference to section 2 is to section 2 of the current chapter; a reference to section 4.2 is to section 2 of Chapter 4. Likewise, references to equations (1), (2.1) and (4.2.1) are references respectively to equation (1) of the current section, equation (1) of section 2 of the current chapter, and equation (1) of section 4.2. Theorems and figures always carry section identification as well. Therefore Theorem 2.1 refers to Theorem 1 of section 2 of the current chapter, Theorem 4.2.1 refers to Theorem 2.1 of Chapter 4. A similar convention is followed for figures.

9. NOTATION FOR SETS AND SPACES

Spaces and whole sets are generally indicated in upper case script, subsets in upper case italic. Thus, $\mathcal{X}$ is the set with elements x. The statement that x belongs to $\mathcal{X}$ is written $x \in \mathcal{X}$, the statement that it does not is written $x \notin \mathcal{X}$. The statement that A is a subset of $\mathcal{X}$ is written $A \subset \mathcal{X}$. In a context where $\mathcal{X}$ is understood as the 'whole set' the complement of A in $\mathcal{X}$ is written $\bar{A}$. The quantities $A \cup B$, $A \cap B$ and $A \setminus B$ are respectively the union, intersection and difference of sets A and B (the difference being A less $A \cap B$).

We use $\{x; (\cdot)\}$ for the set of elements x with characterization $(\cdot)$. So, the direct product $A \otimes B$ of two sets is $\{x, y; x \in A, y \in B\}$. If f is a function defined on A then $f(A)$ is the set $\{f(x); x \in A\}$. If we replace the idea of a function by that of a transformation σ taking x into σx then σA is the set $\{\sigma x; x \in A\}$.

If we write $f: \mathcal{X} \to \mathcal{F}$ then it is understood that f is a mapping from $\mathcal{X}$ to $\mathcal{F}$, a function of x in $\mathcal{X}$ taking values in $\mathcal{F}$. The set of all such possible functions is written $\mathcal{F}^{\mathcal{X}}$.

There are some standard sets. In a departure from the script notation the real axis will be denoted $\mathbb{R}$ and the set of signed integers $\{\ldots, -2, -1, 0, 1, 2, \ldots\}$ denoted $\mathbb{Z}$. Thus $\mathbb{R}$ is the set of real numbers and $\mathbb{R}^d$ the set of d-tuples of such numbers (d-dimensional Euclidean space). An element x of $\mathbb{R}^d$ is a d-dimensional vector with elements x_j ($j = 1, 2, \ldots, d$) and length

$$|x| = \left(\sum_j x_j^2 \right)^{1/2}$$

$\mathbb{R}_+$ is the positive real axis and $\mathbb{R}_+^d$ the positive orthant in $\mathbb{R}^d$; the set $x \geq 0$. $\mathbb{Z}^d$ is the set of d-tuples of signed integers; the 'cubic lattice' in $\mathbb{R}^d$. $\mathbb{Z}_+$ is the set of non-negative integers $\{0, 1, 2, \ldots\}$ and $\mathbb{Z}_+^d$ the set of d-tuples of such numbers.

10. NOTATION FOR MATRICES, VECTORS AND VECTOR DERIVATIVES

If we write of a matrix A that

$$A = (a_{jk})$$

then we mean that a_{jk} is the jk th element of A, the element in the j th row and k th column. In Markov contexts we sometimes use rather a notation such as

$$A = (a(x, x'))$$

where x, x' are values of state, labelling again row and column respectively.

The identity matrix $I = (\delta_{jk})$ is necessarily square, with a dimension depending upon the context. We use $\mathrm{diag}\,(w_j)$ (or $\mathrm{diag}\,(w(x))$) to denote a diagonal matrix with elements w_j (or $w(x)$) down its diagonal.

The *trace* of a square matrix A, written $\mathrm{tr}\,A$, is the sum $\sum_j a_{jj}$ of its diagonal elements.

The matrix with jk th element a_{kj} is the *transpose* of A, written A^T. More generally, Λ^T is used to denote the adjoint of a linear operator Λ (see (2.1.13)) and I is used to denote the identity operator.

The complex conjugate of a scalar a is denoted $\bar{a}$, and the matrix with jk th element $\bar{a}_{kj}$ is the *transpose conjugate* of A, denoted $A^\dagger$. If A is square then its determinant is denoted $|A|$. Its *adjugate* is denoted $\mathrm{adj}\,(A)$, so that its inverse A^{-1} is $\mathrm{adj}\,(A)/|A|$.

A vector $x = (x_j)$ is generally taken to be a column vector with elements x_j. If it is an n-vector it is then an $n \times 1$ matrix. The corresponding row vectors x^T or $x^\dagger$ are then $1 \times n$ matrices.

A real matrix A which is symmetric ($A^\mathsf{T} = A$) is non-negative definite if

$$x'Ax \geq 0 \tag{1}$$

for all real vectors x (of appropriate dimension); positive definite if equality can hold only if $x = 0$. One then writes $A \geq 0$ or $A > 0$ respectively, if the context is such that it is understood that positive-definiteness of difference supplies the ordering. For complex matrices symmetry is strengthened to the Hermitian property $(A^\dagger = A)$ and inequality (1) to

$$x^\dagger A x \geqslant 0$$

for all complex vectors of appropriate dimension x. We shall often write the column vector which has 1 in all its entries as **1**.

If A is a matrix then we shall sometimes use A_{CD} to denote the submatrix of A in which row label j is restricted to the subset C and column label k is restricted to the subset D of possible values. The ordering of rows and columns within C and D is generally immaterial so long as a consistent convention is observed. The vector $\mathbf{1}_C$ is correspondingly the vector with elements corresponding to $j \in C$ and all equal to unity. If C is a subset of rows then $\bar{C}$ is the complementary subset, correspondingly for columns. A *partitioning*

$$A = \begin{pmatrix} A_{CD} & A_{C\bar{D}} \\ A_{\bar{C}D} & A_{\bar{C}\bar{D}} \end{pmatrix}$$

of A thus in general implies a reordering of rows and columns, which is immaterial providing the reordering is observed consistently.

If a is scalar and x a column n-vector then the notations $\partial a / \partial x$ and (frequently) a_x are used to denote the *row* vector of first derivatives $\partial a / \partial x_k$. If a is a column m-vector then these quantities denote the $m \times n$ matrix of derivatives

$$a_x = \left(\frac{\partial a_j}{\partial x_k} \right).$$

If a is a scalar and x a row vector then these quantities are used to denote the *column* vector of derivatives $\partial a / \partial x_j$. If a is scalar and x a column vector then we use the notations

$$\frac{\partial^2 a}{\partial x^2} = a_{xx} = \left(\frac{\partial^2 a}{\partial x_j \partial x_k} \right)$$

for the *Hessian* of A; the square matrix of second derivatives. More generally, if a is a scalar and x, y are vectors then

$$a_{xy} = \left(\frac{\partial^2 a}{\partial x_j \partial y_k} \right)$$

The *null-space* $\mathcal{N}$ of a matrix A is the set of x such that

$$Ax = 0.$$

The *range* of A is the orthogonal complement of $\mathcal{N}$; i.e. the set of y such that

$$y^\mathsf{T} x = 0$$

for all x in $\mathcal{N}$.

11. SOME MISCELLANEOUS NOTATION

We use the symbol $:=$ to denote a defining equality, so that

$$n! := n(n-1)(n-2)\ldots 2.1 \qquad (n = 1, 2, \ldots)$$

constitutes a definition of $n!$

We use δ_j to denote the Kronecker delta function, so that

$$\delta_j = \begin{cases} 1 & j = 0 \\ 0 & j \neq 0 \end{cases}$$

Correspondingly

$$\delta_{jk} := \delta_{j-k} = \begin{cases} 1 & j = k \\ 0 & j \neq k \end{cases}$$

for integral j, k. Sometimes it is better to use a functional rather than a subscript notation, so that

$$\delta(x, x') = \begin{cases} 1 & x = x' \\ 0 & x \neq x' \end{cases}$$

where x, x' take values in some discrete set.

For x in $\mathbb{R}^d$ the quantity $\delta(x)$ is the Dirac delta function, satisfying

$$\int \theta(x)\,\delta(x)\,\mathrm{d}x = \theta(0)$$

for suitably regular $\theta(\cdot)$, with $\mathrm{d}x := \prod_j \mathrm{d}x_j$.

The quantity $n^{(s)}$ for non-negative integral n, s is the factorial power

$$n^{(s)} := n(n-1)(n-2)\ldots(n-s+1) = \frac{n!}{(n-s)!}$$

Correspondingly

$$\binom{n}{s} := \frac{n!}{(n-s)!\,s!}$$

is the binomial coefficient.

We sometimes use $\{z_j\}$ to denote the array of quantities $z_1, z_2, \ldots$. Thus, we might write the generating function $\Pi(z)$ of section 13 as $\Pi(\{z_j\})$ if there were a reason for exhibiting the components of the argument z explicitly.

12. NOTATION FOR TIME AND THINGS TEMPORAL

We denote time by t. It is considered to take values in a set $\mathbb{T}$, which will usually be either $\mathbb{R}$ or $\mathbb{Z}$, according as one is working in continuous or discrete time.

We often consider a function of time $x(t)$, this will be a function $\mathbb{T} \to \mathscr{X}$ if x is a variable taking values in $\mathscr{X}$. If $\mathscr{X}$ is a vector space (so that linear combinations of elements of $\mathscr{X}$ are defined and belong to $\mathscr{X}$) then one can formulate the notion of a time derivative. This, if it exists, will be written

$$\dot{x} = \frac{dx}{dt}$$

with the time argument indicated (i.e. $\dot{x}(t)$) if clarity requires it.

In fact, we shall mostly work with random functions of time. This is often dealt with formally by writing $x(t)$ rather as $x(\omega, t)$, where ω is a parameter whose random value indicates the particular function of time that will turn up: the *realization*. We shall not adopt this notation, but just keep in mind that $x(t)$ is indeed random, and that its value for a given value of t will not in general be known until it is observed. The course of values $\{x(\tau), \tau \leqslant t\}$ up to a given time t is the *history* or *partial realization* at time t, denoted $X(t)$. The *future* course $\{x(\tau); \tau \geqslant t\}$ is correspondingly denoted $X^*(t)$. The full realization $X(+\infty)$ will sometimes be denoted by X. Specification of X is equivalent to specification of ω, if one uses the $x(\omega, t)$ notation. The realization X takes values in $\mathscr{X}^{\mathsf{T}}$.

In considering transformations of the model we often consider transformations S of realizations to realizations, i.e. mappings $S: \mathscr{X}^{\mathsf{T}} \to \mathscr{X}^{\mathsf{T}}$. The value of the transformed realization SX at time t will be denoted $SX[t]$ if we wish to be exact (so that $X[t] = x(t)$). However, we shall often write it simply as $Sx(t)$; this is either a laxity or a convention which we have already discussed in section 5.

Another convention which may be loose but is certainly useful is to speak of 'the process $\{x(t)\}$'. By this one means not simply the realization X but also the statistics of realizations, from which required distributions or expectations can in principle be derived; see section 3.

13. NOTATION FOR PROBABILITIES AND THINGS ALEATORY; GENERATING FUNCTIONS

The notation $P(\cdot)$ is used for 'probability of $(\cdot)$'. That is, $P(A)$ is the probability of the event A; $P(x = x_0)$ is the probability that the random variable x takes the value x_0, etc. Note that $P(\cdot)$ is not a functional notation; if one wishes to express the dependence of $P(x(t) = x_0)$ upon variables occurring then one must do so explicitly, e.g.

$$P(x(t) = x_0) = \pi(x_0, t)$$

The symbol P on its own is also used to denote the transition probability matrix of a Markov chain, or, more generally, the forward conditioning operator of a Markov process; see section 2.1.

We shall usually not distinguish between random variables and particular values of them. For example, we shall write $P(x)$ for the probability that the *random variable* x takes the *generic value* x if the context is such that confusion is unlikely.

The distribution of a random variable x can be described by $P(x)$ only if the space $\mathscr{X}$ of values adopted by x is discrete. More generally, the probability distribution of x would have to be described by either a *measure* $m(\cdot)$ on subsets A of $\mathscr{X}$,

$$m(A) = P(x \in A) \tag{1}$$

or by a *density* $f(\cdot)$ with respect to a measure $\mu(\cdot)$

$$P(x \in A) = \int_A f(x)\mu(\mathrm{d}x) \tag{2}$$

In this latter case, if the distribution depends upon parameters such as time then the dependence is usually included in f, and μ is supposed to be a 'universal' measure, independent of such parameters. The density approach is a useful one in that it encompasses almost all cases. In the discrete case $P(x)$ can itself be regarded as the density with respect to counting measure: the measure such that $\mu(A)$ is the number of elements of $\mathscr{X}$ in A.

Just as P is used as a generic notation for a probability, so f is used as a generic notation for a probability density, relative to an understood measure.

E is the *expectation operator*, so that, for functions $\phi(x)$ of the random variable x,

$$E\phi(x) = \int \phi(x)m(\mathrm{d}x) \tag{3}$$

or

$$E\phi(x) = \int \phi(x)f(x)\mu(\mathrm{d}x) \tag{4}$$

in Cases (1) and (2) respectively. The relation inverse to (3) is

$$m(A) = EI_A(x) \tag{5}$$

where $I_A(\ )$ is the *indicator function* of the set A. That is

$$I_A(x) = \begin{cases} 1 & x \in A \\ 0 & x \notin A \end{cases} \tag{6}$$

More generally, if a parameter ω labels the 'elementary outcomes' in a model (i.e. the atoms of probability space) then one replaces x by ω in (3)–(6) to derive relations between expectations and probability distributions for that model.

It is plain that we are neglecting the niceties of measurability, σ-fields, etc. dear to the professional probabilist. We shall be dealing with physical contexts where these are indeed niceties which have a clear resolution and are not the points of interest.

Suppose x is a random vector with elements x_j. Then $E(x)$ is the vector with elements $E(x_j)$. The *covariance matrix* $\operatorname{cov}(x, y)$ between two such vectors is the matrix whose jk^{th} element is

$$\operatorname{cov}(x_j, y_k) = E[(x_j - E(x_j))(y_k - E(y_k))]$$
$$= E(x_j y_k) - E(x_j)E(y_k)$$

Thus

$$\operatorname{cov}(x, y) = E(xy^{\mathsf{T}}) - E(x)E(y)^{\mathsf{T}}$$

We shall write $\operatorname{cov}(x, x)$ simply as $\operatorname{cov}(x)$, the covariance matrix of the vector x. In the case when x is a scalar then $\operatorname{cov}(x)$ is just the variance of x, written $\operatorname{var}(x)$.

If x is a random vector with elements x_j then we shall often use the generating function

$$\Phi(\theta) = E(e^{\theta^{\mathsf{T}}x}) = E\left[\exp\left(\sum_j \theta_j x_j\right)\right]$$

If the x_j are integer-valued we may work rather in terms of

$$\Pi(z) = E\left(\prod_j z_j^{x_j}\right).$$

$\Phi(\theta)$ is a *moment-generating function* and $\Pi(z)$ a *probability generating function* (routinely abbreviated 'p.g.f.'). We refer to either θ_j or z_j as a *marker variable* for the random variable x_j.

The conventions for transition probabilities, densities and intensities will be explained in Chapters 2 and 3. As a rule, the notations $\pi(\cdot)$ and $\Pi(\cdot)$ indicate equilibrium distributions and p.g.f.s respectively.

14. CONDITIONING

The subject of conditioning is always important and, even in an unrigorous treatment, there are points to be clarified and emphasized.

We use $P(\cdot|B)$, $f(\cdot|B)$ and $E(\cdot|B)$ for probability, probability density and expectation respectively, conditional upon an event B. If x and y are two random variables then one can consider the conditional expectation $E(y|x = x_0)$, having the evaluation

$$E(y|x = x_0) = \frac{E(y\delta(x, x_0))}{E(\delta(x, x_0))} \tag{1}$$

if x is discrete-valued and the denominator positive. This is a function of x_0. We shall replace x_0 by the generic value x and write the conditional expectation simply as $E(y|x)$. That is, as formerly, x is used to denote both the random variable and a generic value of that variable.

Generally, the conditional expectation $E(y|x)$ is defined as a function of x satisfying the identity

$$E[\phi(x)y] = E[\phi(x)E(y|x)] \tag{2}$$

for any $\phi(x)$ (with reservations, see below). This implies the elementary evaluation (1) in the discrete-valued case. The determination of $E(y|x)$ by characterization (2) is unique to the extent that, if $\Delta(x)$ is the difference of any two such evaluations, then

$$E[\Delta(x)\phi(x)] = 0$$

for all admissible ϕ.

In the more careful approach to these matters adopted by the professional probabilist the events $x \in A$ for an assembly of sets A both rich and regular enough constitute a σ-field $\mathscr{F}$, and the idea that a random variable is a function of x is replaced by the concept that it is $\mathscr{F}$-measurable. The quantity $E(y|x)$ is then denoted rather $E^{\mathscr{F}}(y)$ and, as in (2), is defined as an $\mathscr{F}$-measurable random variable satisfying

$$E(zy) = E[zE^{\mathscr{F}}(y)] \tag{3}$$

for all $\mathscr{F}$-measurable z.

One writes $\mathscr{F}_1 \subset \mathscr{F}_2$ if $\mathscr{F}_1$-measurability implies $\mathscr{F}_2$-measurability. That is, $\mathscr{F}_2$ provides at least as detailed a description as does $\mathscr{F}_1$.

To return to the naïve notation, suppose it turns out that a conditional expectation $E(y|x_1, x_2, x_3)$ is a function of x_1 alone. That is, the requirement that a function $\chi(x_1, x_2, x_3)$ should satisfy

$$E[\phi(x_1, x_2, x_3)y] = E[\phi(x_1, x_2, x_3)\chi(x_1, x_2, x_3)] \tag{4}$$

identically in ϕ can be satisfied by $\chi = \chi_1(x_1)$, say. By choosing ϕ in (4) as a function of x_1 alone or of (x_1, x_2) alone we see that $\chi_1(x_1)$ equally provides an evaluation of $E(y|x_1)$ and of $E(y|x_1, x_2)$. We can express this in the more general conception.

Theorem 14.1 *Consider two σ-fields $\mathscr{F}_1 \subset \mathscr{F}_2$ and suppose that $E^{\mathscr{F}_2}(y)$ is $\mathscr{F}_1$-measurable. Then $E^{\mathscr{F}_2}(y)$ can be identified as $E^{\mathscr{F}}(y)$ for any $\mathscr{F}$ such that $\mathscr{F}_1 \subset \mathscr{F} \subset \mathscr{F}_2$.*

It often turns out that the distribution of a random variable y depends upon some parameter θ. We shall refer to θ as a *parametrizing variable*, and if we wish, for instance, to emphasize the fact that $E(y)$ depends upon θ, we shall write the expectation as $E(y|\theta)$. That is, we shall treat the parametrizing variable as though it were a conditioning variable. There are cases where one should make a distinction between the two concepts. In fact, if the value of θ is allowed to depend upon those of the random variables of the system ('feedback') then expectations parametrized by θ and conditioned by θ will in general differ in value. We shall not encounter such cases, however; parameters have their values prescribed exogenously.

15. DEFINITIONS FOR GRAPHS

A graph consists of a set of *nodes* connected by some pattern of *arcs*.

The arcs can be *directed* ○⟶○ or *undirected* ○——○

Multiple arcs ⬭ and loops ⬭ may be permitted.

Two graphs are *mutually unconnected* if they share neither nodes nor arcs. A graph is *connected* if it cannot be seen as a combination of mutually unconnected graphs. The connected graphs of which a graph is composed are its *components*.

A *tree* is a connected graph with the minimal number of arcs, in that the graph is no longer connected if any arc is removed.

The simplest tree is a *linear graph* ○——○——○— – – – –○

A connected graph which is not a tree then necessarily contains at least one *cycle*.

A division of the nodes of a graph into two complementary subsets is called a *cut*. A tree thus has exactly one arc crossing any cut for which the subsets are each connected.

A graph is said to be *coloured* if a variable α is defined at each of its nodes, taking values $\alpha = 1, 2, \ldots, p$, say.

CHAPTER 2

Markov processes: standard material

Markov structure is the expression of two linked and physically significant concepts: the possibility that dynamics might be cast in a simply recursive form and the idea of a dynamically sufficient description ('state').

The standard material on Markov processes is now so very standard that we shall limit ourselves to a sketchy (although self-contained) treatment of the material that will later be needed. For the reader who wishes something more extensive there are many classic texts, and, of course, specialist texts. Among recent texts, that by Grimmett and Stirzaker (1982) combines penetration with simplicity in a particularly happy fashion, and that by Ross (1980) is very accessible.

Chapter 3 treats a few topics which are less standard but will prove relevant.

1. MARKOV FORMALISM

We have already defined the Markov property in (1.3.5). A special case of this defining relation would be

$$E[\phi(x(t_1)|X(t_0))] = E[\phi(x(t_1)|x(t_0))] \qquad (t_0 \le t_1) \tag{1}$$

for all appropriate t_0, t_1 and all $\phi: \mathscr{X} \to \mathbb{R}$ in some class which we shall denote by $\mathscr{F}$.

Now, the expression in the right-hand member of (1) is itself a scalar function of $x(t_0)$, and can be regarded as a transformed version of ϕ, $P(t_0, t_1)\phi$ say, with argument $x(t_0)$. That is, the transformation $\phi \to P(t_0, t_1)\phi$ is defined by

$$(P(t_0, t_1)\phi)(x) := E[\phi(x(t_1)|x(t_0) = x]. \tag{2}$$

Otherwise expressed, $P(t_0, t_1)$ is a conditional expectation operator, taking expectations over $x(t_1)$ conditional on the value of $x(t_0)$. We shall term it the *transition operator*. It was defined as an operation upon ϕ, and we shall assume $\mathscr{F}$ large enough that $P(t_0, t_1)\phi$ belongs to $\mathscr{F}$ if ϕ does. That is, $P(t_0, t_1)$ is a mapping from $\mathscr{F}$ to $\mathscr{F}$.

One could have defined the transition operator by (2) whether the process was Markov or not. However, for a Markov process the operator compounds in a particular way and determines the stochastic dynamics of the process completely. The first assertion is made explicit in

Theorem 1.1 *If the process is Markov then the transition operator P* $(\cdot, \cdot)$ *satisfies the relations*

$$P\,(t, t) = I \tag{3}$$

$$P\,(t_0, t_2) = P\,(t_0, t_1)\,P\,(t_1, t_2) \qquad (t_0 \leqslant t_1 \leqslant t_2) \tag{4}$$

Proof Relation (3) is an immediate consequence of definition (2). For $t_0 \leqslant t_1 \leqslant t_2$ we have, quite generally,

$$E[\phi(x(t_2)|X(t_0))] = E[E[\phi(x(t_2)|X(t_1))]|X(t_0)].$$

In view of the Markov property (1.3.5) this reduces to

$$E[\phi(x(t_2)|x(t_0)] = E[E[\phi(x(t_2)|x(t_1))]|x(t_0)] \tag{5}$$

which is equivalent to (4). ∎

Relations (3), (4) express a semi-group property for the operators $P\,(\cdot, \cdot)$. For example, suppose that the process is time-homogeneous, so that $P\,(t_0, t_1)$ must be a function of the time-difference $s = t_1 - t_0$ alone, $P\,(s)$, say. Then relations (3), (4) would imply that

$$P\,(rs) = P\,(s)^r \qquad (s \geqslant 0, r = 0, 1, 2, \ldots) \tag{6}$$

Relation (4) is equivalent to the classic Chapman–Kolmogorov relation. To see this, and to perceive the distributional consequences of the Markov property generally, we note from (2) that $P\,(\cdot, \cdot)$ is in fact a linear operator with the explicit representation

$$(P\,(t_0, t_1)\phi)\,(x(t_0)) = \int \phi(x(t_1)f(x(t_1)|x(t_0))\mu(\mathrm{d}x(t_0)) \tag{7}$$

Here $f\,(x(t_1)|x(t_0))$ is the probability density of $x(t_1)$ conditional on $x(t_0)$ relative to a time-independent measure μ. If we express the transition density $f(\cdot|\cdot)$ in explicit functional form

$$f\,(x(t_1) = x_1\,|x(t_0) = x_0) = f\,(x_0, t_0; x_1, t_1) \qquad (t_0 \leqslant t_1) \tag{8}$$

then relation (7) has the more explicit form

$$(P\,(t_0, t_1)\phi)\,(x_0) = \int \phi(x_1)f(x_0, t_0; x_1, t_1)\mu(\mathrm{d}x_1) \tag{9}$$

We see from (9) that the identity (5) in ϕ, or its equivalent expression (4), can be written

$$f\,(x_0, t_0; x_2, t_2) = \int f(x_0, t_0; x_1, t_1)f(x_1, t_1; x_2, t_2)\mu(\mathrm{d}x_1) \tag{10}$$

which is just the Chapman–Kolmogorov identity.

Suppose that $x(t)$ has probability density $\pi(x, t)$ relative to μ. It is then true, Markov structure or not, that

$$\pi(x_1, t_1) = \int \pi(x_0, t_0) f(x_0, t_0; x_1, t_1) \mu(dx_0) \qquad (t_0 \leqslant t_1) \tag{11}$$

Theorem 1.2 *The transformation $\pi(\cdot, t_0) \to \pi(\cdot, t_1)$ between functions of x given by (11) can be written*

$$\pi(\cdot, t_1) = P(t_0, t_1)^\mathsf{T} \pi(\cdot, t_0) \tag{12}$$

where the adjoint P^T of the operator P is defined by the identity

$$\int \pi(x)\,(P\phi)\,(x)\,\mu(dx) = \int \phi(x)\,(P^\mathsf{T}\pi)\,(x)\mu(dx) \tag{13}$$

Proof The assertion is an immediate consequence of (9), (11) and the definition (13) of adjointness. ∎

We must now show that $P(\cdot, \cdot)$ contains all the information needed to determine the stochastic dynamics of a Markov process. Note from (1) a first consequence of Markov structure.

$$f(x(t_1)\,|\,X(t_0)) = f(x(t_1)\,|\,x(t_0)) \qquad (t_0 \leqslant t_1) \tag{14}$$

A second is

Theorem 1.3 *If $\{x(t)\}$ is Markov then the process $\{x(t_j)\}$ on a subsequence $\{t_j\}$ of Π is also Markov.*

Proof By characterizing $\{t_j\}$ as a subsequence we indicate also that it is ordered: $t_j < t_{j+1}$. Denote histories and futures for the sub-process by the subscript s:

$$X_s(t_j) = \{x(t_k), k \leqslant j\}$$
$$X_s^*(t_j) = \{x(t_k), k \geqslant j\}.$$

It then follows from (1.3.5) that

$$E[\phi(X_s^*(t_j)\,|\,X(t_j)] = E[\phi(X_s^*(t_j)\,|\,x(t_j)] \tag{15}$$

which implies, by Theorem 1.14.1, that

$$E[\phi(X_s^*(t_j)\,|\,X_s(t_j)] = E[\phi(X_s^*(t_j)\,|\,x(t_j)] \tag{16}$$

Relation (16) expresses the Markov property for the sub-process. ∎

Theorem 1.4 *If $\{x(t)\}$ is Markov then for a subsequence $\{t_j\}$ of Π*

$$f(x(t_1), x(t_2), x(t_3), \dots, x(t_n)\,|\,X(t_0)) = \prod_{j=1}^{n} f(x(t_j)\,|\,x(t_{j-1})) \tag{17}$$

and in this sense the transition operator $P(\cdot, \cdot)$ determines the full process dynamics.

Proof The multiple density is understood to be with respect to a product measure $\mu \times \mu \times \ldots \times \mu$. By appeal to the Markov property at various points we have

$$f(x(t_1), \ldots, x(t_n) | X(t_0)) = f(x(t_1), \ldots, x(t_n) | X_s(t_0))$$

$$= \prod_j f(x(t_j) | X_s(t_{j-1}))$$

$$= \prod_j f(x(t_j) | x(t_{j-1}))$$

which establishes (17). Now, specification of the transition operator $P(\cdot, \cdot)$ implies specification of the transition density $f(\cdot | \cdot)$ and we see from (17) that this implies specification of any finite-dimensional distribution conditional on the past. ∎

The Markov property can be given a slightly different characterization which is for some purposes the natural one.

Theorem 1.5 *The Markov property* (1.3.5) *can be equivalently characterized: at any time the past and the future are independent conditional upon the present.*

Proof Let us for simplicity write the descriptions $X(t)$, $x(t)$ and $X^*(t)$ of past, present and future at time t simply as X, x and X^*. Note that X and X^* both determine x.

Multiplying relation (1.3.5) by $\psi(X)$ and taking expectations conditional on x we deduce that, for any ϕ, ψ,

$$E[\phi(X^*)\psi(X)|x] = E[\phi(X^*)|x]E[\psi(X)|x] \tag{18}$$

which is the conditional independence property stated in the theorem. Conversely, if (18) holds then

$$\theta(x) := E[\phi(X^*)|x] \tag{19}$$

can be characterized as the function of x (and so of X) which satisfies

$$E[\phi(X^*)\psi(X)] = E[\theta(x)\psi(X)]$$

for any $\psi(X)$. We can thus make the identification

$$\theta(x) = E[\phi(X^*)|X] \tag{20}$$

and (19), (20) between them imply the original defining property (1.3.5). ∎

Characterization (18) is pleasing in its symmetric treatment of past and future, a symmetry less evident in condition (1.3.5).

Exercises and comments

1. Suppose that $\mathscr{X}$ is discrete and μ is counting measure. Then relation (9), for example, becomes

$$(P(t_0, t_1)\phi)(x_0) = \sum_{x_1} \phi(x_1) f(x_0, t_0; x_1, t_1)$$

In this case $P(t_0, t_1)$ can be regarded as a matrix with $(x_0, x_1)^{\text{th}}$ element

$$f(x_0, t_0; x_1, t_1) = P(x(t_1) = x_1 | x(t_0) = x_0)$$

(the transition probability). It operates on the vector ϕ with x^{th} element $\phi(x)$.

2. THE KOLMOGOROV EQUATIONS IN DISCRETE TIME

We shall restrict ourselves to time-homogeneous processes from now on. This is for notational convenience rather than any formal necessity, but it is also true that all the physical models we shall consider will be time-homogeneous. The first notational gain is that $P(t_0, t_1)$ can be written simply as $P(t_1 - t_0)$, $(t_0 \leqslant t_1)$.

Let us for a given ψ define the function of x

$$\phi(x, t) = E[\psi(x(t_1)|x(t) = x] \tag{1}$$

and, as in section 1, denote the probability density of $x(t)$ relative to μ by $\pi(x, t)$. We have then, by the definition (1.2) of $P(\cdot, \cdot)$ and by Theorem 1.2

$$\phi(\cdot, t) = P(t_1 - t)\phi(\cdot, t_1) = P(t_1 - t)\psi \tag{2}$$

$$\pi(\cdot, t) = P(t - t_0)^{\mathsf{T}}\pi(\cdot, t_0) \qquad (t_0 \leqslant t \leqslant t_1) \tag{3}$$

Suppose now that we restrict ourselves to discrete time, so that t takes only integer values (i.e. $\Pi = \mathbb{Z}$). Let us also denote $P(1)$ simply by P, which we shall term the *one-step transition operator*. Then, by Theorem 1.1,

$$P(s) = P^s \qquad (s = 0, 1, 2, \dots) \tag{4}$$

and we can write (2), (3) as

$$\phi(\cdot, t) = P^{t_1 - t}\phi(\cdot, t_1) \tag{5}$$

$$\pi(\cdot, t) = (P^{\mathsf{T}})^{t - t_0}\pi(\cdot, t_0) \qquad (t_0 \leqslant t \leqslant t_1) \tag{6}$$

The special cases of (5), (6)

$$\phi(\cdot, t) = P\phi(\cdot, t + 1) \tag{7}$$

$$\pi(\cdot, t) = P^{\mathsf{T}}\pi(\cdot, t - 1) \tag{8}$$

are termed the *Kolmogorov backward* and *forward* equations respectively. The backward equation (7) gives a backward recursion for the expectation of a future random variable; the forward equation (8) gives a forward recursion for the distribution of state. Written out in full the relations are

$$\phi(x, t) = \int f(x, x')\phi(x', t + 1)\mu(dx') \tag{7}$$

$$\pi(x, t) = \int \pi(x', t - 1)f(x', x)\mu(dx') \tag{8}$$

where $f(x, x')$ is the *transition density*, characterized by

$$P(x(t+1) \in A \mid x(t) = x) = \int_A f(x, x') \mu(\mathrm{d}x') \tag{9}$$

In the case of discrete $\mathscr{X}$ we take μ as counting measure, and the Kolmogorov equations become

$$\phi(x, t) = \sum_{x'} f(x, x') \phi(x', t+1) \tag{10}$$

$$\pi(x, t) = \sum_{x'} \pi(x', t-1) f(x', x) \tag{11}$$

where $f(x, x')$ is now the *transition probability*

$$f(x, x') = P(x(t+1) = x' \mid x(t) = x) \tag{12}$$

In this case P is a matrix, the *transition matrix*, with $(xx')^{\text{th}}$ element $f(x, x')$. The adjoint operator P^T is just the transpose of P.

The Kolmogorov forward equation (8) is the 'master equation' of statistical mechanics, so called because it determines the stochastic evolution in time of a full description of the process. Reduced or approximate versions of it are considered as one tries to work with reduced descriptions in some sense.

Exercises and comments

1. Special cases of the Kolmogorov equations are obtained by taking $\psi(x) = \delta(x, x_1)$ or $\pi(x, t_0) = \delta(x, x_0)$ in (5), (6). One can then make the identifications $\phi(x, t) = f(x, x_1; t_1 - t)$ and $\pi(x, t) = f(x_0, x; t - t_0)$ for $t_0 \leqslant t \leqslant t_1$, where $f(x, x'; s)$ is the transition density over a time lapse s. Equations (7), (8) then amount simply to the relations

$$P(s+1) = PP(s)$$
$$P(s+1) = P(s)P$$

evident from Theorem 1.1.

2. A particular case of an expectation over the future is an absorption probability. Suppose that $\psi(x) = \phi(x, t_1)$ is chosen as the indicator function of a set A. Then $\alpha(x, s) := \phi(x, t_1 - s)$ is the probability that a process beginning in state x at a given time is in the set of states A after a time-lapse s. If the process is so modified that A cannot be left once it is entered then $\alpha(x, s)$ becomes the probability that a process beginning at x at time $t = 0$ (say) has entered A by time $t = s$. This entry probability will obey the backward equation

$$\alpha(x, s) = \int f(x, x') \alpha(x', s-1) \mu(\mathrm{d}x') \qquad (x \in \bar{A}, \ s > 0) \tag{13}$$

with boundary condition

$$\alpha(x, s) = 1 \qquad (x \in A) \tag{14}$$

and end condition

$$\alpha(x, 0) = I_A(x). \tag{15}$$

The boundary condition expresses the modification that A may not be left when once entered.

3. Show that the function $\alpha(x, s)$ determined by (13)–(15) is non-decreasing in s for a given x, so that

$$\alpha(x) := \lim_{s \to \infty} \alpha(x, s)$$

exists. For the modified process the quantity $\alpha(x)$ is to be interpreted as an absorption probability: the probability that a path starting at x ultimately enters A. For the unmodified process it is to be taken as the probability that the path *sometime* enters A. Show that $\alpha(x)$ is the smallest non-negative solution of

$$\alpha(x) = \int f(x, x') \alpha(x') \mu(dx') \qquad (x \in \bar{A})$$

consistent with

$$\alpha(x) \geqslant I_A(x)$$

3. STATIONARY DISTRIBUTIONS; BALANCE EQUATIONS

A consequence of passage to equilibrium would be the convergence of the state distribution $\pi(x, t)$ to a time-independent form $\pi(x)$ satisfying the equilibrium form of (2.8)

$$\int \pi(x') f(x', x) \mu(dx') = \pi(x) \tag{1}$$

In the case of discrete $\mathscr{X}$ this *balance equation* takes the form

$$\sum_{x'} \pi(x') f(x', x) = \pi(x) \tag{2}$$

Either of these equations can be written $P^T \pi = \pi$. In the discrete case we can regard P and π as matrix and vector respectively and write (2) as

$$\pi^T P = \pi^T \tag{3}$$

A distribution π satisfying the static balance equation (1) is termed a *stationary distribution* or an *invariant measure* of the process. It has yet to be established whether such a distribution exists, whether it is unique and whether it can be said to be an equilibrium distribution in that $\pi(x, t) \to \pi(x)$ as $t \to +\infty$ for arbitrary initial conditions.

It is useful to define the quantity

$$[A, B](\pi f) = \left[\int_A \int_B - \int_B \int_A \right] \pi(x) f(x, x') \mu(dx) \mu(dx') \tag{4}$$

where in each case the first integral is with respect to x and the second with respect to x'. This can be interpreted as the *net probability flux* from state-set A to state-set B; i.e. the expected number of passages $A \to B$ in unit time less the expected number $B \to A$. It is a function of the state distribution π and the transition density f. When the values of these are understood from the context we shall write the net flux simply as $[A, B]$. Note that

$$[A, A](\pi f) = 0 \tag{5}$$

The balance equation (1), expressing the fact that π is a stationary distribution, can be written

$$[\mathscr{X}, x](\pi f) = 0. \tag{6}$$

Theorem 3.1 *Let π be a stationary distribution and $(A, \bar{A})$ an arbitrary cut of $\mathscr{X}$. Then*

$$[A, \bar{A}](\pi f) = 0 \tag{7}$$

Proof Integrating (4) with respect to x over A we deduce that

$$[\mathscr{X}, A](\pi f) = 0$$

Subtracting (5) from this equation we deduce (7). ∎

Relation (7) states the obvious: that in a stationary regime the probability fluxes across a cut in the two directions must balance. However, it is as well to have a formal proof.

For the next two sections we shall restrict ourselves to the case of discrete $\mathscr{X}$. This is not because the results of these sections are not valid more generally, but because advanced techniques are required to prove them so.

Exercises and comments

1. The matrix $P = (f(x, x'))$ of the discrete case is a *probability transition matrix*. Such a matrix has exactly two characterizing properties: its elements are non-negative and its row-sums are unity. A matrix with these properties is often termed *stochastic*. The row-sum property can be expressed

$$P\mathbf{1} = \mathbf{1}$$

That is, P has a unit eigenvalue with $\mathbf{1}$ as corresponding right eigenvector. Relation (3) exhibits a stationary distribution π as a corresponding left eigenvector.

2. *Birth and death processes* Suppose that $\mathcal{X}$ is just the interval of integers $a \leqslant x \leqslant b$, and that transitions out of x can only be to nearest neighbours. We thus have

$$f(x, x') = \begin{cases} p(x) & x' = x + 1 \\ q(x) & x' = x \\ r(x) & x' = x - 1 \end{cases}$$

as the only non-zero elements of P, with $r(a)$ and $p(b)$ both zero.

Such a process is termed a birth and death process, since the model can be seen as representing the size x of a population which can change only by single values. The balance equation (2) is then effectively a second-order difference equation in $\pi(x)$. However, if one places a cut between the states x and $x + 1$ (i.e. $A = \{a, a+1, \ldots, x\}$) then the cut-balance relation (7) yields the first-order difference equation

$$p(x)\pi(x) = r(x+1)\pi(x+1) \tag{8}$$

The simplification is because the transitions $A \rightleftharpoons \bar{A}$ can be achieved only by transitions $x \rightleftharpoons x + 1$, and the fluxes of these transitions must balance. We deduce from (8) the solution

$$\pi(x) \propto \prod_{y \leqslant x} \frac{p(y-1)}{r(y)} \tag{9}$$

3. Solution (9) (of (8)) is unique (when normalized) if all states communicate, in that passage (in general indirect) is possible between any pair of states. Suppose that $p(c-1) = r(c) = 0$ for some state c, so that the states $x < c$ cannot communicate with the states $x \geqslant c$. What solutions does (8) then have?

4. RECURRENCE, CONVERGENCE TO EQUILIBRIUM

In this and the next section we assume $\{x(t)\}$ a time-homogeneous Markov process in discrete time with discrete state-space. Proofs of standard material are relegated to the exercises if lengthy. This makes it possible to summarize the relevant standard theory effectively, and yet provide a self-sufficient account.

Let us define $\tau(x)$ as the smallest positive t such that $x(t) = x$ if $x(0) = x$. This then the *recurrence time* to x, a random quantity in general.

The following definitions are standard. State x has *period* d if d is the largest integer for which $\tau(x)$ is necessarily a multiple of d. The state is *aperiodic* if $d = 1$. State x is *recurrent* (or *persistent*) if recurrence is certain, i.e. $P(\tau(x) < \infty) = 1$. Otherwise it is *transient*. A recurrent state is *positive recurrent* if

$$m(x) := E(\tau(x)) < \infty$$

Otherwise it is *null recurrent*. The mutually exclusive possibilities of transience, null recurrence and positive recurrence constitute three recurrence types.

State x is said to *communicate* with state x' if passage from x to x' is possible, i.e. if $P(x(t) = x' | x(0) = x)$ is positive for some positive t. Two states *intercommunicate* if each communicates with the other.

Theorem 4.1   (Solidarity)   *Intercommunicating states are of the same period and the same recurrence type.*

The proof is sketched in Exercise 3.

Theorem 4.2   *Suppose that all states of the process intercommunicate. Then the states of the process are positive recurrent if and only if there is a stationary distribution. If such a distribution exists then it is unique and given by*

$$\pi(x) = m(x)^{-1} \tag{1}$$

If the process is also aperiodic then $\pi(x, t) \to \pi(x)$ as $t \to +\infty$.

The statement that there is a stationary distribution means that one can find a solution π of (3.2) which is non-negative and satisfies $\Sigma \pi(x) = 1$. This last condition is more than a simple normalization; it excludes the possibilities that, even for an unnormalized distribution, the sum $\Sigma \pi(x)$ could be either zero or infinite. That is, $\pi(x)$ must not be identically zero and must be summable.

The theorem is crucial for our purposes. Its most important combined implication is that, if the process is intercommunicating and aperiodic and one has found a stationary distribution π then π *is* the unique equilibrium distribution in that $\pi(x, t)$ converges to $\pi(x)$ with increasing t.

We show in Exercise 4 that positive recurrence implies existence of an equilibrium distribution, identified with (1). In Exercise 5 we show that existence of a stationary distribution implies positive recurrence and uniqueness of the evaluation (1). The demonstration of passage to equilibrium, $\pi(x, t) \to \pi(x)$, involves ideas of independent interest which we defer to sections 3.1 and 3.2. However, Exercises 4 and 5 demonstrate the weaker conclusion,

$$\lim_{\theta \uparrow 1} (1 - \theta) \sum_{t=0}^{\infty} \pi(x, t) \theta^t = \pi(x) \tag{2}$$

valid even if the process is periodic.

A process whose states all intercommunicate is termed *irreducible*.

Exercises and comments

1. Define the quantities

$$u_n(x, x') = P(x(n) = x' | x(0) = x)$$
$$v_n(x, x') = P(x(n) = x', x(t) \neq x' \text{ for } 0 < t < n | x(0) = x)$$

for $n = 1, 2, \ldots$ with the conventions $u_0(x, x') = \delta(x, x')$ and $v_0(x, x') = 0$. The quantity $u_n(x, x')$ is then just the n-step transition probability $f(x, x', n)$ of Exercise 2.1, and $v_n(x, x')$ is the probability that the first passage (if $x' \neq x$) or recurrence (if $x' = x$) to x' takes place after n steps. Define the generating functions

$$U(x, x', \theta) = \sum_{n=0}^{\infty} u_n(x, x')\theta^n$$

$$V(x, x', \theta) = \sum_{n=0}^{\infty} v_n(x, x')\theta^n$$

certainly convergent for $|\theta| < 1$. Show that

$$u_n(x, x) = \sum_{r=1}^{n} v_r(x, x)u_{n-r}(x, x)$$

and hence that

$$U(x, x, \theta) = [1 - V(x, x, \theta)]^{-1}. \tag{3}$$

Show correspondingly that, for $x \neq x'$,

$$U(x, x', \theta) = V(x, x', \theta)U(x', x', \theta) \tag{4}$$

2. The assumption of recurrence is equivalent to

$$\lim_{\theta \uparrow 1} V(x, x, \theta) = 1 \tag{5}$$

and so, by (3), to

$$\lim_{\theta \uparrow 1} U(x, x, \theta) = +\infty \tag{6}$$

If x is recurrent then the assumption of positive recurrence implies finiteness of the limit

$$\lim_{\theta \uparrow 1} \frac{1 - V(x, x, \theta)}{1 - \theta} = m(x) \tag{7}$$

From (3), (7) we deduce that

$$\lim_{\theta \uparrow 1} (1 - \theta)U(x, x, \theta) = m(x)^{-1} \tag{8}$$

We thus see that a recurrent state x is null recurrent if and only if

$$\lim_{\theta \uparrow 1} (1 - \theta)U(x, x, \theta) = 0 \tag{9}$$

3. The relation

$$u_{n+r+s}(x, x) \geqslant u_r(x, x')u_n(x', x')u_s(x', x)$$

holds, and if x, x' intercommunicate then one can find, r, s such that $u_r(x, x')$ and $u_s(x', x)$ are positive. Use this inequality and the characterizations (6), (9) of recurrence and null recurrence to establish Theorem 4.1.

4. We can now prove the direct part of Theorem 4.2, and show that an irreducible positive recurrent process has stationary distribution (1).

Show that if x is recurrent and communicates with x' then

$$\lim_{\theta \uparrow 1} V(x, x', \theta) = 1 \tag{10}$$

It then follows from (4), (8) and (10) that

$$\lim_{\theta \uparrow 1} (1 - \theta) U(x, x', \theta) = m(x')^{-1} \tag{11}$$

The definition of U implies the identity

$$U(y, x, \theta) = \delta(y, x) + \theta \sum_{x'} U(y, x', \theta) f(x', x) \tag{12}$$

It follows then from (11), (12) that $\pi(x) = m(x)^{-1}$ is a solution of the balance equation (3.2). It indeed constitutes a normalized stationary solution, because the identity $\sum_{x'} u_n(x, x') = 1$ implies that the sum over x' of expression (11) is unity. Relation (11) implies (2).

5. We can now prove the converse part of Theorem 4.2. Suppose that there is a stationary distribution $\pi(x)$. We then have

$$\pi(x) = \sum_{x'} \pi(x') u_n(x', x)$$

Since $\pi(x') > 0$ for some x' it thus follows that $\pi(x) > 0$ for all x. Forming a generating function over n we deduce that

$$\pi(x) = \sum_{x'} \pi(x')(1 - \theta) U(x', x, \theta) \tag{13}$$

This relation implies that $\lim_{\theta \uparrow 1} U(x', x, \theta) = \infty$ for some x', which implies (6), by (4). That is, that x is recurrent. From (11), (13) we deduce (1). That is, that the stationary distribution is unique. Also, since $m(x) = \pi(x)^{-1} < \infty$, all states are positive recurrent.

6. Revert to the case of a general state space and suppose $\pi(x)$ the density function of a stationary distribution. Consider passage (recurrence) from x into a set of states A. Suppose that for any x this passage (recurrence) is certain with finite expected passage (recurrence) time $m(x, A)$. Then m must obey the backward equation

$$m(x, A) = 1 + \int_A f(x, x') m(x', A) \mu(dx')$$

Integration over x with respect to the stationary distribution yields

$$\int_A m(x, A) \pi(x) \mu(dx) = 1 \tag{14}$$

This is the general version of relation (1), due to Kac (1947).

5. DECOMPOSITION OF STATE SPACE, ERGODIC CLASSES

A set C of states is *closed* if passage out of it is impossible, so that $f(x, x') = 0$ for $x \in C$, $x' \in \bar{C}$. It is *irreducible* if it is closed and its states intercommunicate. Two such irreducible sets are plainly disjoint, so we have

Theorem 5.1 *The state space $\mathscr{X}$ can be decomposed into disjoint classes: closed irreducible sets $\mathscr{X}(1)$, $\mathscr{X}(2)$, ... and a set of transient states $\mathscr{Y}$.*

The closed irreducible sets $\mathscr{X}(b)$ ($b = 1, 2, \ldots$) are often termed the *ergodic classes* of the process. The process on each such set can be regarded as constituting a Markov process on its own, with state space $\mathscr{X}(b)$ irreducible, and so not permitting further decomposition. The states of the residuum $\mathscr{Y}$ left when all the ergodic classes have been subtracted are certainly transient, because they communicate with some states of the ergodic classes but not conversely. Ultimate and irreversible passage out of $\mathscr{Y}$ is thus certain.

An ergodic class can be transient or null recurrent only if it is infinite, since it is only by passage of x to 'remote state space' that uncertain or infinitely-delayed recurrence can be achieved.

Theorem 5.2 *If an irreducible process has finite state space then all its states are positive recurrent.*

Proof All states are of the same recurrence type, by Theorem 4.1. If they are transient then, by (4.4), (4.6)

$$\lim_{\theta \uparrow 1} U(x, x', \theta) < \infty$$

for all x, x'. If they are null recurrent then, by (4.4), (4.9),

$$\lim_{\theta \uparrow 1} (1 - \theta)U(x, x', \theta) = 0$$

for all x, x'. But both of these relations are incompatible with

$$\sum_{x'} u_n(x, x') = 1$$

because this implies that

$$\sum_{x'} U(x, x', \theta) = (1 - \theta)^{-1} \quad \blacksquare$$

The proof appeals to finiteness of $\mathscr{X}$ in that it assumes the validity of interchanging summation over $\mathscr{X}$ and passage to the limit $\theta \uparrow 1$.

Corollary 5.1 *If $\mathscr{X}$ is finite then all recurrent states are positive recurrent, and there is at least one such state.*

Exercises and comments

1. A random walk on the non-negative integers provides the standard example of an irreducible process which may show any of the three types of recurrence behaviour. Assume the transition probabilities $f(x, x+1) = p$ $(x = 0, 1, 2, \ldots)$, $f(x, x-1) = q = 1 - p$ $(x = 1, 2, 3, \ldots)$ with the boundary condition of impenetrability at $x = 0$: $f(0, 0) = q$. The process is positive recurrent, null recurrent or transient according as p is less than, equal to or greater than q respectively.

In fact, the general solution to the balance equations is

$$\pi(x) \propto (p/q)^x$$

which is summable if and only if $p < q$.

2. Suppose all the states of a process positive recurrent. Denote the ergodic classes by $\mathscr{X}(b)$ $(b = 1, 2, \ldots)$ and let $b(x)$ denote the class of state x. Then $b(x)$ is a maximal invariant of the motion in that it is a function of x whose value cannot change under any possible transition, and any such invariant is a function of $b(x)$.

Let $\pi(x) = \Phi(x)$ be a strictly positive solution of the balance equation

$$[\mathscr{X}, x](\pi f) = 0 \tag{1}$$

Note that this also satisfies the 'partial' balance equations

$$[\mathscr{X}(b), x](\pi f) = 0 \qquad (b = 1, 2, \ldots) \tag{2}$$

and that any non-negative summable solution of (1) can be written

$$\pi(x) = h(b(x))\Phi(x) \tag{3}$$

for some $h(\cdot)$. This can be regarded as a solution corresponding to a general initial distribution of x over ergodic classes.

6. THE KOLMOGOROV EQUATIONS IN CONTINUOUS TIME

The material of section 1 and of section 2 up to equation (2.3) is valid for a process on a general ordered time set Π. We now consider the continuous time case $\Pi = \mathbb{R}$, which is the natural choice for most physical models. The passage from discrete to continuous time introduces the possibility of another degree of infinity, since it is then conceivable that there could be infinitely many transitions in state in a finite time. This can lead to pathologies. In this section we state the formalism: the results that one would expect to hold in pathology-free cases. In the next section we indicate conditions that would validate the formalism.

Let us rewrite the Chapman–Kolmogorov relations (1.4) with slightly modified arguments.

$$P(t_0, t_1) = P(t_0, t)P(t, t_1) \qquad (t_0 \leqslant t \leqslant t_1) \tag{1}$$

The Kolmogorov backward and forward equations in discrete time are essentially a pair of recursions for $P(\cdot, \cdot)$ obtained by letting t approach t_0 or t_1 respectively in

(1). In continuous time one may expect that this recursion would take a differential form. Explicitly, one would expect that

$$\lim_{t \uparrow t_0} P(t_0, t) = I \tag{2}$$

and to let t tend to t_0 in (1) says no more than

$$P(t_0, t_1) = \lim_{t \downarrow t_0} P(t, t_1) \tag{3}$$

Define the *infinitesimal transition operator* $\Lambda(t)$ by

$$\Lambda(t) = \lim_{s \downarrow 0} \frac{1}{s} \left[P(t, t+s) - I \right] \tag{4}$$

That is

$$(\Lambda(t)\phi)(x) = \lim_{s \downarrow 0} \frac{1}{s} E[\phi(x(t+s)) - \phi(x(t)) | x(t) = x] \tag{5}$$

for the admissible functions ϕ of $\mathcal{F}$. The Kolmogorov backward and forward equations, derived formally by letting t approach t_0 and t_1 respectively in (1), are then

$$-\frac{\partial}{\partial t_0} P(t_0, t_1) = \Lambda(t_0) P(t_0, t_1) \tag{6}$$

$$\frac{\partial}{\partial t_1} P(t_0, t_1) = P(t_0, t_1) \Lambda(t_1) \tag{7}$$

For the time-homogeneous case, when $\Lambda(t)$ and $P(t_0, t_1)$ can be written Λ and $P(t_1 - t_0)$ respectively, these relations become rather

$$\frac{\partial}{\partial s} P(s) = \Lambda P(s) \qquad (s > 0) \tag{8}$$

$$\frac{\partial}{\partial s} P(s) = P(s) \Lambda \qquad (s > 0) \tag{9}$$

with initial condition

$$P(0) = I \tag{10}$$

These relations have the formal solution, analogous to (2.4).

$$P(s) = e^{s\Lambda} = \sum_{j=0}^{\infty} \frac{(s\Lambda)^j}{j!} \qquad (s \geq 0) \tag{11}$$

As would be sometimes expressed, Λ is the *infinitesimal generator* of the semigroup of transformations $\{P(s), s \geq 0\}$.

The more familiar form of the Kolmogorov equations is a differential equation for expectations or distributions. We give the continuous time analogue of relations (2.7), (2.8).

Theorem 6.1 *Define* $\phi(x, t) = E[\psi(x(t_1))|x(t) = x]$ *and* $\pi(\cdot, t)$, *the probability density of* $x(t)$ *for prescribed* $\pi(\cdot, t_0)$ $(t_0 \leqslant t \leqslant t_1)$. *Then these obey respectively the backward and forward equation*

$$-\frac{\partial}{\partial t}\phi(\cdot, t) = \Lambda(t)\phi(\cdot, t) \qquad (t < t_1) \tag{12}$$

$$\frac{\partial}{\partial t}\pi(\cdot, t) = \Lambda(t)^\mathsf{T}\pi(\cdot, t) \qquad (t > t_0) \tag{13}$$

Here, as in (1.13), the adjoint Λ^T of Λ is defined by the identity

$$\int \pi(x)(\Lambda\phi)(x)\mu(dx) = \int \phi(x)(\Lambda^\mathsf{T}\pi)(x)\mu(dx) \tag{14}$$

The explicit form of $\Lambda(t)$ is given by

$$(\Lambda(t)\phi)(x) = \lim_{s\downarrow 0}\frac{1}{s}\int f(x, t, x'; t+s)[\phi(x') - \phi(x)]\mu(dx') \tag{15}$$

What form this limit may take is best considered separately in the various standard cases.

7. CONTINUOUS TIME, DISCRETE STATE SPACE

Consider the time-homogeneous discrete-state case. Define the *probability transition intensity*

$$\lambda(x, x') = \lim_{s\downarrow 0}\frac{1}{s}P(x(t+s) = x'|x(t) = x) \qquad (x \neq x') \tag{1}$$

$\lambda(x, x')$ itself being the intensity for the transition $x \to x'$. The assumption that these limits exist is close to the assumption that a transition from the currently-occupied state x to x' within a small time interval s has probability $\lambda(x, x')s + o(s)$, multiple transitions having a total probability of smaller order than s. The probability of no transition in this time interval must then be $1 - \lambda(x)s + o(s)$, where

$$\lambda(x) := \sum_{x' \neq x}\lambda(x, x') \tag{2}$$

That is,

$$\lim_{s\downarrow 0}\frac{1}{s}[P(x(t+s) = x|x(t) = x) - 1] = -\lambda(x) \tag{3}$$

Assuming then the existence of the limits (1), (3) we see that the operator Λ of (6.15) must in this case have the evaluation

$$(\Lambda\phi)(x) = \sum_{x'}\lambda(x, x')[\phi(x') - \phi(x)] \tag{4}$$

at least if the two components of the sum are absolutely convergent. The value assigned to $\lambda(x, x)$ in (4) is immaterial, since it has a zero coefficient.

The backward and forward Kolmogorov equations (6.12), (6.13) then take the forms

$$-\frac{\partial}{\partial t}\phi(x, t) = \sum_{x'} \lambda(x, x')[\phi(x', t) - \phi(x, t)] \tag{5}$$

$$\frac{\partial}{\partial t}\pi(x, t) = \sum_{x'} [\pi(x', t)\lambda(x', x) - \pi(x, t)\lambda(x, x')] \tag{6}$$

the evaluation of Λ^{T} implied in (6) being readily verified. In the case of finite $\mathscr{X}$ we can interpret Λ as a matrix

$$\Lambda = (\lambda(x, x') - \lambda(x)\delta(x, x')) \tag{7}$$

the *transition intensity matrix* and ϕ, π as time-dependent vectors obeying

$$-\dot{\phi} = \Lambda\phi$$
$$\dot{\pi} = \Lambda^{\mathsf{T}}\pi$$

All formal calculations are correct if the following conditions are fulfilled, which we shall label: (Λ) The intensity matrix Λ exists and the formal solution (6.11) for $P(s)$ satisfies

$$\pi^{\mathsf{T}}\left[\sum_{j=0}^{\infty} \frac{(s\Lambda)^j}{j!}\right]\phi = \sum_{j=0}^{\infty} \frac{s^j}{j!} (\pi^{\mathsf{T}}\Lambda^j\phi) \tag{8}$$

for any non-negative s, any distribution vector π and any vector ϕ with uniformly bounded components.

In other words, the matrix multiplications and the summation over powers of Λ must commute. For an example of a case when (Λ) is not fulfilled, see Exercise 2. A sufficient but by no means necessary condition for (Λ) is that Λ should exist and the $\lambda(x)$ be uniformly bounded.

The forward equation (6) can be seen as a mass-conserving flow equation for a flow of 'probability mass' on $\mathscr{X}$. One interprets $\pi(x, t)$ as the amount of probability mass at site x, the intensity $\lambda(x, x')$ as the 'conductance' of the direct path x to x', and $\pi(x, t)\lambda(x, x')$ as the probability flux along this path. The right-hand member of (6) is then just the difference between total fluxes into and out of x. The equilibrium version of this relation is

$$\sum_{x'} [\pi(x')\lambda(x', x) - \pi(x)\lambda(x, x')] = 0 \tag{9}$$

and any distribution $\pi(x)$ satisfying (9) is a stationary distribution for the continuous time process.

As in section 3 we find it convenient to introduce the notation

$$[A, B](\pi\lambda) = \left[\sum_A \sum_B - \sum_B \sum_A \right] \pi(x)\lambda(x, x') \qquad (10)$$

where the first and second summations are in each case with respect to x and x' respectively. This again represents the net probability flux from A to B for A, B sets of $\mathscr{X}$. Definition (10) of course differs from the discrete time definition (3.4) in that λ replaces f, but it still has the interpretation of the expected *net* number of transitions $A \rightarrow B$ per unit time under a state distribution π. The forward equation (6) can be written

$$\dot{\pi}(x) = [\mathscr{X}, x](\pi\lambda)$$

and the stationarity condition (9) can be written

$$[\mathscr{X}, x](\pi\lambda) = 0 \qquad (9')$$

If the process is in state x then the time elapsing up till exit from the state is exponentially distributed with expectation $\lambda(x)^{-1}$; see Exercise 1. This implies that periodic behaviour is impossible in the continuous-time discrete-state case, since recurrence times can take all possible values.

Because a state, once occupied, is occupied for a positive time, one modifies the definition of recurrence time in continuous time processes to 'time until first *return* to x' (i.e. departure and then return). With this modification one can again define the three recurrence types of state: transient, null recurrent and positive recurrent.

Theorem 7.1 *Suppose condition* (Λ) *satisfied. Then the process is positive recurrent if and only if there is a distribution* $\pi(x)$ *satisfying the balance equation (9). This stationary distribution is unique and is the equilibrium distribution in that* $\pi(x, t) \rightarrow \pi(x)$ *as* $t \rightarrow + \infty$.

Proof Assume existence of a π satisfying (9), or

$$\pi^T\Lambda = 0 \qquad (11)$$

It then satisfies also $\pi^T\Lambda^j = 0$, and so

$$\pi^T P(s) = \pi^T \qquad (12)$$

for any $s \geqslant 0$, by (6.11) and (8). It follows from Theorem 4.2 that the embedded discrete-time Markov process $\{x(ks); k \in \mathbb{Z}\}$ for any given s is positive recurrent. So then is the continuous time process.

Conversely, suppose the process positive recurrent. It follows then from Theorem 4.2 that, for any given s, there is a π satisfying (12) and that indeed $\pi(x, ks + \Delta) \rightarrow \pi$ as k increases through the integers. Since π is determined uniquely by the s-step transition probabilities it cannot depend upon the time translation Δ, and so cannot indeed depend upon s. There is thus a unique stationary distribution π which is also the equilibrium distribution for all initial conditions. ∎

The latter part of the argument is made more explicit in Exercise 11, for those who would like to see it expanded.

Exercises and comments

1. Let $G(s)$ be the probability that a sojourn in state x lasts for at least time s. Show that
$$\dot{G} = -\lambda(x)G$$
and hence that $G(s) = \exp[-\lambda(x)s]$.

2. Consider a pure birth process, for which $\mathscr{X} = \mathbb{Z}$ and the only possible transitions are single upward jumps, $x \to x+1$, with intensity $\lambda(x)$. The process is then a stochastic analogue of the deterministic process
$$\dot{x} = \lambda(x) \tag{13}$$
Equation (13) has integral
$$\int_{x(0)}^{x(t)} \lambda(x)^{-1}\,dx = t$$
so x will become infinite in a finite time if
$$\int_{x(0)}^{+\infty} \lambda(x)^{-1}\,dx < \infty$$

This example indicates the type of behaviour which would invalidate the formalism of this and the preceding section. For the stochastic process, the time needed to move from $x(0)$ to state y is $\sum_{x=x(0)}^{y-1} \sigma(x)$, where the $\sigma(x)$ are independent exponential variables with $E\sigma(x) = \lambda(x)^{-1}$. Relatively straightforward arguments (see the next exercise) then show that the time for passage to infinite x will be finite with probability one if
$$\sum_{x(0)}^{+\infty} \lambda(x)^{-1} < \infty \tag{14}$$
and infinite with probability one if
$$\sum_{x(0)}^{+\infty} \lambda(x)^{-1} = \infty \tag{15}$$

In the first case, (14), we must then have $\sum_x \pi(x,t) < 1$ for some t, or $\pi^T P(t)\mathbf{1} < 1$, where π is the initial distribution vector. Since $\pi^T \Lambda^j \mathbf{1} = 0$ for any finite j, at least for any π with support on bounded x, then we see that relation (8) is not satisfied in this case.

3. Let τ be the passage time from $x = x(0)$ to $x = +\infty$ for the stochastic example of the last exercise. By the characterization given this has moment-generating function

$$\Phi(\theta) = E(e^{-\theta\tau}) = \prod_x (1 - \theta/\lambda(x))^{-1}$$

where the product is from $x(0)$ to $+\infty$. In the case (15) this product is divergent in that

$$\Phi(\theta) = \begin{cases} 0 & \theta < 0 \\ +\infty & \theta > 0 \end{cases}$$

In the case (14) it is convergent and defines an increasing function of real θ in $\theta < \inf_x \lambda(x)$.

Consider case (14). An appeal to Markov's inequality (see e.g. Grimmett and Stirzaker (1982), p. 181) shows that

$$P(\tau \geqslant T) < e^{-\theta T} \Phi(\theta) \qquad (\theta \geqslant 0)$$

Since we can find a positive θ such $\Phi(\theta)$ is finite we see that τ is finite with probability one, and that $x(t)$ escapes to infinity in finite time.

An appeal to Markov's inequality likewise shows that

$$P(\tau \leqslant T) < e^{-\theta T} \Phi(\theta) \qquad (\theta \leqslant 0)$$

Choosing θ negative we see that this bound is zero for any T in case (15). That is, τ is infinite with probability one, and $x(t)$ remains finite for finite t.

4. Note that the eigenvalue 0 of Λ corresponds to the eigenvalue 1 of P; it again has $\mathbf{1}$ and π as corresponding right and left eigenvectors. Just as all eigenvalues of P lie within the unit circle, so all eigenvalues of Λ have non-positive real part.

5. *Independent migration.* Suppose that individuals can move between a number of sites labelled $j = 1, 2, \ldots, m$. Suppose further that a single individual thus migrating follows a Markov process in which the site j that he occupies constitutes his state and the intensity of transition from j to k is λ_{jk}.

Consider now a process consisting of N such individuals moving independently. This is certainly also Markov if one retains the full description: an identification of all individuals and a listing of their positions. Suppose, however, that one does not distinguish between individuals but describes the process simply by the vector of occupation numbers $n = \{n_j\}$, where n_j is the number of individuals at site j. Show that this reduced process is also Markov, with transition intensity $\lambda_{jk} n_j$ for a migration from j to k. (See section 3.7.)

Show that, if the single individual process has unique equilibrium distribution π_j then the N-individual process has unique equilibrium distribution

$$\pi(n) = N! \prod_j \frac{\pi_j^n}{n_j!} \qquad \left(\sum_j n_j = N\right)$$

6. Note the continuous-time version of Theorem 3.1, that

$$[A, \bar{A}](\pi\lambda) = 0 \tag{16}$$

if π is a stationary distribution and $(A, \bar{A})$ an arbitrary cut of $\mathscr{X}$.

7. *Birth and death processes.* These are the continuous-time versions of the processes defined in Exercise 3.2, with $\mathscr{X} = \mathbb{Z}$ and the only possible transitions out of x being to $x+1$ or $x-1$. The intensities $\lambda(x, x+1)$ and $\lambda(x, x-1)$ are interpreted as total 'birth' and 'death' rates respectively. As for the discrete-time case, the cut-balance relation (16) has the consequence

$$\pi(x)\lambda(x, x+1) = \pi(x+1)\lambda(x+1, x) \tag{17}$$

whence

$$\pi(x) \propto \prod_{y \leqslant x} \frac{\lambda(y-1, y)}{\pi(y, y-1)} \tag{18}$$

This expression gives the equilibrium distribution over any given irreducible set if summable over that set.

8. *Cropping of a resource.* Let us denote the total birth and death rates $\lambda(x, x+1)$ and $\lambda(x, x-1)$ by $\alpha(x)$ and $\beta(x)$ respectively. The deterministic analogue of the process of Exercise 7 is

$$\dot{x} = \alpha(x) - \beta(x).$$

Suppose that x represents the size of a fish population, which is fished at a constant rate γ so long as fish are there. The question is: what is the largest sustainable value of γ? Suppose then that

$$\alpha(x) = \alpha_0 + \alpha_1 x$$

$$\beta(x) = \begin{cases} \beta_1 x + \beta_2 x^2 + \gamma & x > 0 \\ 0 & x = 0 \end{cases}$$

where α_0 is an immigration rate to the area and α_1 and β_1 are individual birth and death rates. The $\beta_2 x^2$ term represents an increase in individual death rate at high population density caused by depletion of resources. The deterministic model thus becomes

$$\dot{x} = \begin{cases} \alpha_0 & x = 0 \\ \alpha_0 + (\alpha_1 - \beta_1)x - \beta_2 x^2 - \gamma & x > 0 \end{cases}$$

Suppose that $\alpha_1 - \beta_1 > 0$, i.e. that the population is intrinsically viable, and define

$$\gamma_0 = \max_x (\alpha_0 + (\alpha_1 - \beta_1)x - \beta_2 x^2) = \alpha_0 + \frac{(\alpha_1 - \beta_1)^2}{4\beta_2}$$

Then for $\gamma > \gamma_0$ only the equilibrium at $x = 0$ is possible (a 'chattering' equilibrium, in which immigrants are caught as soon as they appear). For $\alpha_0 < \gamma < \gamma_0$ the population has possible equilibria at $x = 0$ and at the roots x', x'' (where $x' < x''$) of

$$\alpha_0 + (\alpha_1 - \beta_1)x - \beta_2 x^2 - \gamma = 0 \tag{19}$$

These are respectively stable, unstable and stable. It appears then that the population can be fished up to a rate γ_0, in that there is a stable positive population size up to this value.

The argument is fallacious, and its use is said to have led to collapse of the Peruvian anchovy stocks. Return to the stochastic model, and suppose that $\alpha_0 > 0$, so that actual extinction of the population is impossible. For $\alpha_0 < \gamma < \gamma_0$ the equilibrium distribution $\pi(x)$ given by (18) has the form illustrated in Fig. 2.7.1. The deterministic equilibria at $x = 0$, x' and x'' have as analogues a peak of π at $x = 0$, a minimum near x' and a maximum near x''. As γ increases the maximum at x'' weakens; it vanishes altogether when $\gamma = \gamma_0$. The 'stable equilibrium' for the deterministic model at x'' for γ slightly less than γ_0 is thus seen to correspond to no more than a slight local maximum on a distribution whose mass is at small x values.

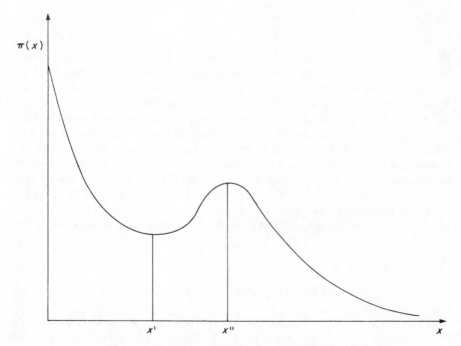

Fig. 2.7.1 The equilibrium distribution of a harvested fish population.

Note the importance of the α_0 and $\beta_2 x^2$ terms in this model (immigration and environmental limitation). They prevent the states $x = 0$, $x = +\infty$ respectively from being absorbing.

9. *Optimal cropping.* The methods of dynamic programming (see e.g. Whittle, 1982c) show that, if one wishes to crop the population in such a way as to maximize the average yield, then the cropping rate should be infinite or zero according as the population size exceeds some critical value ξ or not. The distribution of population size is then

$$\pi(x) = \frac{a(x)}{\displaystyle\sum_0^\xi a(y)} \qquad (0 \leqslant x \leqslant \xi)$$

where

$$a(x) = \prod_{y=1}^x \frac{\alpha(y-1)}{\beta(y)}$$

and $\alpha(x)$, $\beta(x)$ are the total birth and death rates for the uncropped population. Since all excess over ξ is immediately removed the average yield is $\alpha(\xi)\pi(\xi)$, and ξ should be chosen to maximize this. An approximate stationarity condition yields the relation

$$\frac{a(\xi)}{\displaystyle\sum_0^\xi a(x)} \sim 1 - \frac{\beta(\xi)}{\alpha(\xi)}$$

10. Suppose that $\pi(x)$ is a stationary distribution for the process with transition intensity $\lambda(x, x')$. Show that $\pi(x) \propto \pi(x)\lambda(x)$ is the stationary distribution of x at *the moments of transition*. That is, it is the stationary distribution of the *embedded Markov chain*, the discrete time process defined by the transition points.

11. Consider the following alternative treatment of the final point in the proof of Theorem 7.1. Let π^s be the unique stationary distribution associated with the transition matrix $P(s)$; it is desired to show that π^s is in fact independent of s. Consider values r, s whose ratio is rational, so that $r/s = m/n$, say. Show from the relation $\pi(x, jnr) = \pi(x, jms)$ that $\pi^r = \pi^s$. Thus π^s has a common value for all rational s; denote this by π. Show from (6.11) and (12) that π satisfies (11), and from this again that π satisfies (12) for all $s \geqslant 0$.

12. Let $\tau(x, x')$ be the time of first passage (recurrence) to x' from x, and define

$$V(x, x', \theta) = E(e^{-\theta \tau(x, x')})$$

Show that V satisfies

$$(\lambda(x) + \theta)V(x, x', \theta) = \lambda(x, x') + \sum_{y \neq x'} \lambda(x, y)V(y, x', \theta) \qquad (20)$$

if we take the convention $\lambda(x, x) = 0$. The relation for $x \neq x'$ comes fairly easily; the relation for $x = x'$ does appeal to the fact that recurrence is 'first return', and its derivation differs.

13. Deduce from equation (20) a set of equations for the quantities $E(\tau(x, x'))$. Hence deduce that

$$\pi(x) = (E\tau(x, x))^{-1}$$

(c.f. (4.1)).

14. *The Poisson process*. This is a standard continuous time process in which x takes values $0, 1, 2, \ldots$ and the only possible transitions are of the form $x \to x + 1$ with intensity $\lambda(t)$, independent of x but in general time-dependent. One can think of $x(t) - x(0)$ as the number of events which have occurred in the time interval $(0, t]$. The events might be, for example, registrations on a Geiger counter or traffic accidents. We speak of the events as constituting a *Poisson stream* of rate $\lambda(t)$.

What are the Kolmogorov forward equations? Show that $n(t) = x(t) - x(0)$ is Poisson distributed with parameter $\gamma(t) = \int_0^t \lambda(\tau) d\tau$. That is

$$P(n(t) = n) = e^{-\gamma(t)} \gamma(t)^n / n!$$

When one speaks of a Poisson process one generally means the particular case with constant intensity λ, so that $\gamma(t) = \lambda t$. The instants of time t_j at which a transition occurs constitute a *Poisson sequence* $\{t_j\}$. Show that the inter-event intervals $t_j - t_{j-1}$ are independently and exponentially distributed with expectation λ^{-1}.

15. *The compound Poisson process*. A time-homogeneous compound Poisson process is one for which $\mathcal{X} = \mathbb{Z}_+$ and transitions $x \to x + j$ have intensity λ_j ($j = 1, 2, \ldots$). One could view such a transition as a 'j-fold event', e.g. the birth of a litter of j, the arrival of an ionized particle bearing j quanta of charge or the arrival of a job containing j quanta of work. The total number of such quanta arriving in a time interval of length t will then have p.g.f.

$$\Pi(z) = \exp\left[\sum_j \lambda_j t(z^j - 1)\right]$$

16. *The doubly stochastic Poisson process*. A doubly stochastic Poisson process is one for which the rate is a function $\lambda(\xi(t))$ of a quantity $\xi(t)$ which is itself a random process. We shall encounter such processes in Chapter 10 when it is natural to consider a component which is 'driven' by a Poisson stream of intensity v. However, only a fraction $\psi(x(t))$ of the stream actually enters the component, where $x(t)$ is the current state of the component. The actual input stream is then a Poisson stream of rate $v\psi(x(t))$. One can regard this as a Poisson stream of rate v 'modulated' by the state-dependent factor $\psi(x)$.

17. Consider the continuous time version of Exercise 5.3. It follows, as there, that the general solution of the balance equation (9) is of the form

$$\pi(x) = h(b(x))\Phi(x) \tag{21}$$

for some $h(\cdot)$, where $\Phi(x)$ is an arbitrary strictly positive solution.

There is now an interest in 'relaxing' the process, i.e. in introducing transitions between the $\mathcal{X}(b)$ in such a way that the irreducible process thus constructed still has an equilibrium distribution of form (21) for some $h(\cdot)$. Show that one way to achieve this is to choose a single state $x(b)$ (the *port state*) in each $\mathcal{X}(b)$, and to allow communication between the $\mathcal{X}(b)$ through the port states alone. For a reference see Whittle (1983b). The significance of the relaxation concept emerges in Chapters 10 and 11.

8. DETERMINISTIC AND DIFFUSION PROCESSES

Suppose that x takes values in $\mathbb{R}^d$, and that $\{x(t)\}$ is both continuous-time Markov and deterministic in that it obeys a first-order differential equation

$$\dot{x} = a(x, t) \tag{1}$$

This could be written out componentwise

$$\dot{x}_j = a_j(x, t) \qquad (j = 1, 2, \ldots, d)$$

The action of Λ is then determined by

$$
\begin{aligned}
(\Lambda\phi)(x) &= \lim_{s \downarrow 0} \frac{1}{s}[x(t+s) - x(t)] \\
&= \lim_{s \downarrow 0} \frac{1}{s}[x + s\phi_x a + o(s) - x] \\
&= \phi_x a
\end{aligned}
\tag{2}
$$

where x is the value of $x(t)$. Recall from section 1.10 that ϕ_x is the row vector of derivatives of ϕ, so that Λ has the explicit form

$$\Lambda(t) = \sum_j a_j(x, t)\frac{\partial}{\partial x_j} \tag{3}$$

We thus deduce

Theorem 8.1 *For the deterministic equation (1) the probability density $\pi(x, t)$ relative to Lebesgue measure obeys the Kolmogorov forward equation*

$$\frac{\partial \pi}{\partial t} + \sum_j \frac{\partial}{\partial x_j}(a_j\pi) = 0 \tag{4}$$

Proof We deduce from (3) that the adjoint of $\Lambda(t)$ relative to Lebesgue measure has the action

$$\Lambda(t)^{\mathrm{T}}\pi = -\sum_j \frac{\partial}{\partial x_j}(a_j(x, t)\pi) \tag{5}$$

The result then follows from (6.13) and (5). ∎

One can speak of a probability density even though the process is deterministic. A random starting value at $t = 0$ will mean a random position at a later time t. An important class of models are those with Hamiltonian structure, see Appendix 2.

Suppose (1) relaxed to a probabilistic diffusion

$$\dot{x} = a(x, t) + \varepsilon \tag{6}$$

where $\{\varepsilon(t)\}$ is a vector white noise process with zero mean and power density matrix $C(x, t)$. One can give various explanations of what this means. The simplest for our purposes is simply to say that Λ has the action

$$\Lambda \phi = \sum_j a_j \frac{\partial \phi}{\partial x_j} + \frac{1}{2} \sum_j \sum_k c_{jk} \frac{\partial^2 \phi}{\partial x_j \partial x_k} \tag{7}$$

(see Exercise 1). This can be more compactly expressed

$$\Lambda \phi = \phi_x a + \tfrac{1}{2} tr(C \phi_{xx}) \tag{8}$$

Theorem 8.2 *For a diffusion process* (6) *the probability density* $\pi(x, t)$ *of* $x(t)$ *relative to Lebesgue measure obeys the Kolmogorov forward equation*

$$\frac{\partial \pi}{\partial t} = -\sum_j \frac{\partial}{\partial x_j} (a_j \pi) + \frac{1}{2} \sum_j \sum_k \frac{\partial^2}{\partial x_j \partial x_k} (c_{jk} \pi) \tag{9}$$

The result follows formally from (6.13) and (7). ∎

The coefficients a_j and c_{jk} are usually termed the *drift* and *diffusion* coefficients, pertaining as they do to the deterministic and random components of the motion respectively.

Exercises and comments

1. The characteristic of the white noise process (t) described is that, if one sets $\Delta w_j = \int_t^{t+s} \varepsilon_j(\tau) d\tau$ then, for small positive s

$$E(\Delta w_j) = o(s)$$
$$E(\Delta w_j \Delta w_k) = c_{jk}(x, t)s + o(s)$$

and expectations of third degree terms, etc. are $o(s)$. Here the expectations are conditional upon the (x, ε) history at t, with $x(t) = x$.

2. The characteristic of a diffusion process (of which the deterministic differential equation (1) is a special case) is that it is the only continuous-time Markov process for which $x(t)$ is continuous in t. However, one can well have hybrid discrete/continuous transitions in state. Consider the case $x = (r, y)$, where r takes values in a discrete set and y in $\mathbb{R}^d$. The variable r labels a regime and y follows deterministic dynamics

$$\dot{y}_j = a_{rj}(y)$$

specific to that regime for given r. The transition $r \to s$ between regimes has intensity λ_{rs}. Such processes are variously referred to as piecewise-deterministic (Davis, 1984) or semi-deterministic (Branford, 1983, 1984). Determine Λ, and show that the Kolmogorov forward equation is

$$\dot{\pi} + \sum_j \frac{\partial}{\partial y_j}(a_{rj}\pi_r) = \sum_s (\pi_s \lambda_{sr} - \pi_r \lambda_{rs})$$

where $\pi_r(y)$ is the probability density of $x(t)$ relative to counting measure for r and Lebesgue measure for y. How is this relation modified if λ_{rs} is y-dependent?

3. As an example, suppose that $x = (n, y)$ where $n \in \mathbb{Z}_+$ is the number of people in a group and $y \in \mathbb{R}_+$ is a variable measuring 'unrest'. Arrivals $(n \to n + 1)$ and departures $(n \to n - 1)$ have intensities v and $\mu(y)n$ respectively. Between departures unrest subsides deterministically

$$\dot{y} = -a(y)$$

but at a departure it receives an increment u with density $f(u)$. Show that

$$\Lambda \Phi(n, y) = v\phi(n + 1, y) + \mu(y) \int \phi(n - 1, y + u) f(u)\,\mathrm{d}u$$

$$- a(y)\frac{\partial \phi(n, y)}{\partial y} - (v + \mu(y))\phi(n, y)$$

and hence determine the Kolmogorov forward equation (see Branford, 1983).

9. TRANSFORM AND GENERATING FUNCTION FORMALISM

Suppose $x \in \mathbb{R}^d$. Then calculations often work out more conveniently in terms of the moment generating function (m.g.f.)

$$\Phi(\theta, t) := E(\mathrm{e}^{\theta^{\mathsf{T}} x(t)}) \tag{1}$$

of $x(t)$. This always exists if θ is purely imaginary (when it is termed the characteristic function) but may well exist in a larger θ-set.

If $x \in \mathbb{Z}^d$ then $\Phi(\theta, t)$ is of course still defined, but there is often advantage in working in terms of a transformed version of it, the probability generating function (p.g.f.)

$$\Pi(z, t) := E\left[\prod_j z_j^{x_j(t)}\right] \tag{2}$$

With the substitution $z_j = \mathrm{e}^{\theta_j}(j = 1, 2, \ldots, d)$ the p.g.f. becomes the m.g.f., the imaginary θ_j-axis corresponding to the unit circle in the complex z_j-plane.

The key property of m.g.f.s (and so of p.g.f.s) is the fact that the m.g.f. of a sum of independent random variables is the product of the m.g.f.s of those variables. See Exercise 1 for an immediate example.

However, whether for this reason or another, the Kolmogorov forward equation often becomes both more compact and more readily solved when expressed in terms of the m.g.f. rather than the distribution. There is a formalism for this calculation, due to Bartlett (1949).

Define the *derivate cumulant function*

$$L(\theta, x, t) = \lim_{s \downarrow 0} \frac{1}{s} [E(e^{\theta^{\mathsf{T}}(x(t+s)-x(t))}|x(t) = x) - 1] \qquad (3)$$

This is a scalar-valued quantity, related to the transition operator Λ by

$$L(\theta, x, t) = e^{-\theta^{\mathsf{T}}x} \Lambda e^{\theta^{\mathsf{T}}x} \qquad (4)$$

Theorem 9.1 *Under the transformation* $\pi \to \Phi$ *from density to m.g.f. the Kolmogorov forward equation formally becomes*

$$\dot{\Phi} = L\left(\theta, \frac{\partial}{\partial \theta}, t\right)\Phi \qquad (5)$$

Here $\partial/\partial\theta$ operates on θ only where it occurs in Φ, and not in L.

Proof One has

$$\Phi(\theta, t + \Delta t) = \int \pi(x, t) E[e^{\theta^{\mathsf{T}}x(t+\Delta t)}|x(t) = x]\mu(dx) \qquad (6)$$

From (6) and the definition (3) one then derives formally

$$\dot{\Phi} = \int L(\theta, x, t)e^{\theta^{\mathsf{T}}x}\pi(x, t)\mu(dx) \qquad (7)$$

whence (5) follows by appeal to formal relations such as

$$\int x_j^r e^{\theta^{\mathsf{T}}x} m(dx) = \left(\frac{\partial}{\partial\theta_j}\right)^r \int e^{\theta^{\mathsf{T}}x} m(dx) \qquad \blacksquare$$

In the case when x has integer-valued components a p.g.f. formulation is likely to be more natural. This is because, rather than meeting a power such as x_j^r in a transition intensity, one is more likely to meet the factorial power

$$x_j^{(r)} = x_j(x_j - 1) \dots (x_j - r + 1),$$

proportional to the number of ways of selecting r objects out of x_j. We have

$$\int x_j^{(r)} \left(\prod_k z_k^{x_k}\right) m(dx) = z_j^r \left(\frac{\partial}{\partial z_j}\right)^r \int \left(\prod_k z_k^{x_k}\right) m(dx) \qquad (8)$$

and so derive

Theorem 9.2 *Suppose that $x \in \mathbb{Z}_+^d$ and that there are transitions of types $i = 1, 2, \dots$ where the ith type of transition has intensity $c_i \prod_j x_j^{(r_{ij})}$ and implies that $x_j \to x_j + s_{ij} - r_{ij}$ ($j = 1, 2, \dots, d$). Then*

$$\dot{\Pi} = \sum_i c_i \left(\prod_j z_j^{s_{ij}} - \prod_j z_j^{r_{ij}} \right) \left(\prod_j \left(\frac{\partial}{\partial z_j} \right)^{r_{ij}} \right) \Pi \qquad (9)$$

Proof Relation (7) now becomes

$$\dot{\Pi} = \sum_x L^*(z, x, t) \left(\prod_j z_j^{x_j} \right) \pi(x, t) \qquad (10)$$

where L^* is the z-analogue of L, given in this case by

$$L^*(z, x, t) = \sum_i c_i \left(\left(\prod_j z_j^{s_{ij} - r_{ij}} \right) - 1 \right) \prod_j x_j^{(r_{ij})} \qquad (11)$$

Inserting (11) into (10) and appealing to (8) we deduce (9). ∎

The ith type of transition corresponds to a 'reaction' in which r_{ij} units of type j are 'consumed' and s_{ij} produced ($j = 1, 2, \ldots, d$). The term $x_j^{(r_{ij})}$ represents the number of ways in which the required r_{ij} units can be found from the x_j available.

Exercises and comments

1. Consider the independent migration model of Exercise 7.5. The variable here is $n(t) = \{n_j(t)\}$, where $n_j(t)$ is the number of individuals at site j at time t; suppose $n(0)$ prescribed. For the case $N = 1$ one then has

$$\Pi(z, t) = \prod_j \left(\sum_k f(j, k; t) z_k \right)^{n_j(0)} \qquad (12)$$

where $f(j, k; t) = P(x(t) = k \,|\, x(0) = j)$. Of course, in this case $n_j(0)$ is either zero or unity, and is unity for only one value of j. For the case of general N one has $n(t) = \sum_a n^{(a)}(t)$, where $n^{(a)}$ is the occupancy vector for the ath individual. Because these are independent, the p.g.f. is the product of individual p.g.f.s, and the form (12) persists. If the distribution tends to an equilibrium, so that $f(j, k; t) \to \pi_k$ as $t \to +\infty$, then of course

$$\Pi(z, t) \to \left(\sum_k \pi_k z_k \right)^N$$

consistently with the final formula of Exercise 7.5.

2. Note that, for the model of Exercise 1, equation (9) becomes

$$\dot{\Pi} = \sum_j \sum_k \lambda_{jk} (z_k - z_j) \frac{\partial \pi}{\partial z_j} \qquad (13)$$

Verify that expression (12) solves (13).

3. Note the p.g.f.s of some standard distributions. For a Poisson distribution of mean λ: $\exp[\lambda(z - 1)]$. For a binomial distribution (the distribution of the number of successes in n independent trials, each having success probability p):

$(pz + q)^n$, where $q = 1 - p$. For the geometric distribution qp^n $(n = 0, 1, 2, \ldots)$: $q/(1 - pz)$.

4. Consider a version of the 'emigration/unrest' model of Exercise 8.3 in which the unrest or 'excitation' variable y takes values in a discrete set. We suppose the transitions and intensities

$$n \rightarrow n + 1 \qquad \text{with intensity } v$$
$$y \rightarrow w \qquad \text{with intensity } \sigma_{yw}$$
$$(n, y) \rightarrow (n - 1, w) \qquad \text{with intensity } \lambda_{yw}n$$

These represent respectively immigration, a spontaneous change in excitation (statistically downwards) and emigration coupled with a statistical increase in excitation. Define the partial p.g.f.

$$\Pi_y(z, t) = \sum_n z^n \pi(n, y, t)$$

Show that

$$\dot{\Pi}_y = v(z - 1)\Pi_y + \sum_w (\sigma_{wy}\Pi_w - \sigma_{yw}\Pi_y) + \sum_w (\lambda_{wy}\Pi'_w - \lambda_{yw}\Pi'_y) \qquad (14)$$

where $\Pi'_y = \partial\Pi_y/\partial z$.

Now consider the 'mixed Poisson' representation

$$\Pi_y(z) = \int e^{u(z-1)} f_y(u)\, du,$$

equivalent to assuming n Poisson distributed, but with the Poisson parameter u randomly distributed with density $f_y(\cdot)$. Define the m.g.f.

$$\phi_y(\theta) = \int e^{u\theta} f_y(u)\, du$$

of this distribution; one then has the immediate relation $\phi_y(\theta) = \Pi_y(1 + \theta)$. The dynamic equations for the ϕ_y follow immediately from (14); from this one deduces the equations for the densities f_y

$$\dot{f}_y + \frac{\partial}{\partial u}\left[\left(v - \left(\sum_w \lambda_{yw}\right)u\right)f_y\right] = \sum_w [(\lambda_{wy}u + \sigma_{wy})f_w - (\lambda_{yw}u + \sigma_{yw})f_y]$$

But this is exactly the Kolmogorov forward equation for the piecewise-deterministic model in which u obeys

$$\dot{u} = v - \left(\sum_w \lambda_{yw}\right)u$$

when the excitation is at level y, and the transition $y \rightarrow w$ has intensity $\lambda_{yw}u + \sigma_{yw}$ (cf. Exercise 8.2). The solution of the original problem is thus reduced to that of the rather simpler piecewise-deterministic problem (see Branford, 1983; 1984).

Markov processes: supplementary material

1. PARTIAL BALANCE

For convenience of later application we shall discuss this concept in continuous time, but one obtains the discrete-time analogue simply by substitution of $f(x, x')$ for $\lambda(x, x')$.

Let π be a stationary distribution, so that

$$[\mathscr{X}, x](\pi\lambda) = 0 \tag{1}$$

for all x of $\mathscr{X}$. If for a particular state x and a particular set of states A it is furthermore true that

$$[A, x](\pi\lambda) = 0 \tag{2}$$

we shall say that x *is balanced against* A. The implication is that the probability fluxes $x \rightleftharpoons A$ balance, at least if π is the state distribution in force. It follows then from (1), (2) that the fluxes $x \rightleftharpoons \bar{A}$ also balance.

$$[\bar{A}, x](\pi\lambda) = 0 \tag{3}$$

We shall say that B *is balanced against* A if (2) holds for all x in B. We shall say that the set A is *partially balanced* if it is balanced against itself, so that (2) holds for all x in A. Partial balance implies that $\pi(x)$ would remain a stationary distribution in A (at least if renormalized) if the process were restricted to A, in that transitions in and out of A were forbidden. It is a property that occurs spontaneously in applications and is of great significance, as we shall see in Part III.

Applications also require us to develop a notion of simultaneous partial balance on several sets. The natural multi-set version turns out to be the statement that sets $A_1, A_2, \ldots, A_p$ are *jointly partially balanced* if partial balance holds over the intersection of any selection of these p sets.

If E is any subset of $\{1, 2, \ldots, p\}$ and F any subset of E let us define

$$A_{EF} = \left(\bigcap_F A_i\right)\left(\bigcap_{E \setminus F} \bar{A}_i\right) \tag{4}$$

The sets A_{EF} for fixed E and varying $F \subset E$ thus constitute a decomposition of $\mathscr{X}$.

Theorem 1.1 *Suppose the sets A_i ($i \in E$) jointly partially balanced. Then A_{EE} is partially balanced against the sets A_{EF} ($F \subset E$).*

Proof The assumption of joint partial balance implies that

$$[A_{GG}, x](\pi\lambda) = 0 \qquad (x \in A_{EE}, G \subset E) \tag{5}$$

Now, if $I(A, \cdot)$ is the characteristic function of a set A of $\mathscr{X}$ then we have

$$[A, x](\pi\lambda) = \sum_y I(A, y)[y, x](\pi\lambda) \tag{6}$$

and

$$I(A_{EF}, y) = \sum_{F \subset G \subset E} (-)^{|G-F|} I(A_{GG}, y) \tag{7}$$

From (5)–(7) we deduce that

$$[A_{EF}, x](\pi\lambda) = 0 \qquad (x \in A_{EE})$$

as required. ■

The following theorem gives a sufficient condition that partial balance on each A_i should imply joint partial balance. The condition itself turns out to be significant and natural.

Theorem 1.2 *Let $A_1, A_2, \ldots, A_p$ be arbitrary subsets of $\mathscr{X}$. Suppose that partial balance holds over each A_i and that there is no transition under which more than one A_i is entered or more than one A_i is vacated. Then the sets A_i are jointly partially balanced.*

Proof Denote the intersection of $A_1, A_2, \ldots, A_s$ by B_s. If we can show that partial balance holds on B_s for $s = 1, 2, \ldots$ then we shall have proved the result. By hypothesis partial balance holds on B_1, and this gives the start of an inductive proof: suppose that partial balance holds on $B_1, B_2, \ldots, B_s$ for given s. We then have

$$[\bar{B}_s, x](\pi\lambda) = 0 \qquad (x \in B_s) \tag{8}$$

and

$$[A_{s+1}, x](\pi\lambda) = 0 \qquad (x \in A_{s+1}) \tag{9}$$

Now, the entrance/exit hypothesis of the theorem implies that

$$[\bar{B}_s, x](\pi\lambda) = [A_{s+1} \backslash B_s, x](\pi\lambda) \qquad (x \in B_{s+1}) \tag{10}$$

From (8), (10) we deduce that

$$[A_{s+1} \backslash B_s, x](\pi\lambda) = 0 \qquad (x \in B_{s+1}) \tag{11}$$

and from (9), (11) that

$$[B_{s+1}, x](\pi\lambda) = 0 \qquad (x \in B_{s+1}) \tag{12}$$

There is thus partial balance on B_{s+1} and the induction is complete. ■

The material of this section is essential for later developments (see Chapters 11 and 12); it appeared first in Whittle (1985a, 1986).

2. LIAPUNOV FUNCTIONS AND THE FORWARD EQUATION

Quite the most elegant way of proving convergence of a Markov process to equilibrium, in whatever degree this holds, is to use a Liapunov function. This technique is familiar in stability theory; its application in the present context will be self-explanatory.

We consider a time-homogeneous Markov process. For the moment we assume a discrete state space, although this is a condition that one hopes would find a natural relaxation in a natural approach. We denote the transition probability over a time interval s by

$$f(x, x'; s) = P(x(t + s) = x' | x(t) = x)$$

for all relevant s (i.e. for s in $\mathbb{Z}_+$ and $\mathbb{R}_+$ in the discrete and continuous time cases respectively).

The time-dependent occupation probability $\pi(x, t)$ satisfies

$$\pi(x, t + s) = \sum_{x'} \pi(x', t) f(x', x; s) \tag{1}$$

Assume the existence of a strictly positive stationary distribution $\pi(x)$ satisfying the equilibrium form of (1) for all relevant s. Such a distribution certainly exists under the hypotheses of Theorems 2.4.2 and 2.7.1. Define the ratio

$$\xi(x, t) = \frac{\pi(x, t)}{\pi(x)} \tag{2}$$

and the Liapunov function

$$V(t) = \sum_x \pi(x) \chi(\xi(x, t)) \tag{3}$$

where χ is a scalar-valued function, bounded above and strictly concave for non-negative argument.

Theorem 2.1 *The Liapunov function $V(t)$ is monotone non-decreasing in t, and so converges to a finite limit value. In this limit all $\xi(x', t)$ for which $f(x', x; s)$ is non-zero have the same value, possibly (x, t)-dependent.*

Proof It follows from the definition (3) that

$$V(t + s) = \sum_x \pi(x) \chi\left(\sum_{x'} \pi(x', t) f(x', x; s) / \pi(x) \right)$$

$$= \sum_x \pi(x) \chi\left(\sum_{x'} f^*(x, x'; s) \xi(x', t) \right) \tag{4}$$

where we have defined

$$f^*(x, x', s) = \frac{\pi(x')}{\pi(x)} f(x', x; s) \tag{5}$$

We can regard f^* as a transition probability $x \to x'$; it satisfies

$$\sum_{x'} f^*(x, x'; s) = 1 \tag{6}$$

$$\sum_{x'} \pi(x) f^*(x, x'; s) = \pi(x') \tag{7}$$

It is in fact the transition probability of the time-reversed process (see section 4.3).

Appealing now to Jensen's inequality and (7) we deduce from (4) that

$$\begin{aligned}
V(t+s) &\geqslant \sum_{x} \sum_{x'} \pi(x) f^*(x, x', s) \chi(\xi(x', t)) \\
&= \sum_{x'} \pi(x') \chi(\xi(x', t)) \\
&= V(t) \tag{8}
\end{aligned}$$

so that $V(t)$ is indeed non-decreasing. Being bounded above it converges to a finite limit. In this limit equality will hold in (8). Equality can hold in the inequality of (8) only if, for given values of (x, t), the quantity $\xi(x', t)$ has the same value for all x' such that $f^*(x, x'; s) > 0$. This is just the final assertion of the theorem. ■

The final assertion of the theorem stops short of making the convergence assertion we would wish: that $\xi(x, t) \to 1$ as $t \to +\infty$. This is because of the complicating possibilities of multiple ergodic classes and periodicity. These can now be clarified.

Theorem 2.2 *Suppose all states intercommunicating and aperiodic. Then* $\pi(x, t) \to \pi(x)$ *as* $t \to +\infty$.

Proof The communication hypothesis implies that $\pi(x) > 0$ for all x. Periodicity can be excluded in the continuous time case. In discrete time aperiodicity implies that there is at least one state $\bar{x}$ for which $f(\bar{x}, \bar{x}; 1) > 0$. In either case, there is at least one state $\bar{x}$ for which $f(\bar{x}, \bar{x}; s) > 0$ for all relevant s. The communication hypothesis implies that, for any x, $f(x, \bar{x}; s) > 0$ for some s. For any x there thus exists an s such that both $f(\bar{x}, \bar{x}; s)$ and $f(x, \bar{x}; s)$ are both positive and so, by Theorem 2.1, $\xi(x, t) = \xi(\bar{x}, t)$ in the limit of large t. That is, in the limit of large t

$$\pi(x, t) \propto \pi(x) \tag{9}$$

for all x. But, since both distributions sum to unity, the proportionality must be an equality. ■

One could adapt the treatment to the case of a general state space, definition (3) then becoming

$$V(t) = \int \chi(\xi(x, t)) \pi(x) \mu(\mathrm{d}x)$$

where $\pi(x)$ is again a stationary distribution, and $\xi(x, t)$ a Radon–Nikodym derivative of the distribution over states at time t relative to the stationary distribution. However, for this more general case an approach via the backward equation turns out to be more natural, as the occurrence of the reversed-time transition probability (5) might have hinted.

Theorem 2.1 is due to Moran (1961); it was used by Renyi (1970) and Penrose (1970) to demonstrate convergence to equilibrium. However, the use of Liapunov arguments in this context goes back to Boltzmann (1896) who used it to prove his celebrated H-Theorem (see section 6.2). Boltzmann used a particular function χ to define entropy, which was the quantity he proved to be increasing in time. This choice was to some extent dictated by the fact that his evolution equations were nonlinear, being in fact approximate reduced forms of the 'master equation'. However, it is also true that the entropy measure plays a particular role in this context; see the discussion of Chapter 5.

Exercises and comments

1. The case when the Liapunov function reduces to something like an entropy measure is that for which

$$\chi(\xi) = \xi - \xi \log \xi$$

or, equivalently (since ξ averages to unity),

$$\chi(\xi) = -\xi \log \xi$$

which is certainly concave and bounded above. One then has

$$V(t) = -\sum_x \pi(x, t) \log[\pi(x, t)/\pi(x)]$$

a well-known measure of discrepancy between distributions. In the case of a finite state space and a uniform stationary measure ($\pi(x)$ constant) V reduces essentially to the entropy measure

$$V(t) = -\sum_x \pi(x, t) \log(\pi(x, t))$$

whose special role will become manifest in Chapter 5.

3. LIAPUNOV FUNCTIONS AND THE BACKWARD EQUATION

The backward analogue of the material of section 2 is interestingly different. It avoids the use of the reverse transition probability and seems better adapted to the case of general $\mathscr{X}$, in that it also avoids use of a Radon–Nikodym derivative. For notational convenience we shall give the treatment for discrete $\mathscr{X}$, but the formal passage to the general case is achieved simply by replacing $\sum_x (\cdot)$ wherever it occurs by $\int (\cdot) \mu(dx)$.

Define the expectation

$$\phi(x, t) = E[\psi(x(t))|x(0) = x] \tag{1}$$

for a function $\psi \colon \mathscr{X} \to \mathbb{R}$. (This differs from the $\phi(x, t)$ of (2.2.1), which would be $\phi(x, t_1 - t)$ in this notation.) It satisfies the backward equation

$$\phi(x, t + s) = \sum_{x'} f(x, x'; s)\phi(x', t) \tag{2}$$

Define now the Liapunov function as

$$V(t) = \sum_{x} \pi(x)\chi(\phi(x, t)) \tag{3}$$

As in the previous section, $\pi(x)$ is a stationary distribution and χ a scalar-valued function, uniformly bounded above and strictly concave.

Theorem 3.1   *The Liapunov function $V(t)$ is monotone non-decreasing in t, and so converges to a finite limit. In this limit all $\phi(x', t)$ for which $\pi(x) f(x, x'; s)$ is non-zero have the same limit value, possibly (x, t)-dependent.*

Proof   We have

$$\begin{aligned}
V(t + s) &= \sum_{x} \pi(x)\chi\left(\sum_{x'} f(x, x', s)\phi(x', t)\right) \\
&\geq \sum_{x} \sum_{x'} \pi(x) f(x, x', s)\chi(\phi(x', t)) \\
&= V(t)
\end{aligned}$$

and the remaining assertions follow as in the proof of Theorem 2.1. ∎

Theorem 3.2   *Suppose all states intercommunicating and aperiodic. Then $\phi(x, t) \to \sum_{x'} \pi(x')\psi(x')$ as $t \to +\infty$. In this sense, then, $\pi(x, t) \to \pi(x)$.*

Proof   There is an implied assumption that a stationary distribution exists. This plus the assumptions of the theorem imply that π is unique and all states positive recurrent (Theorems 2.4.2 and 2.7.1).

By the argument of Theorem 2.2 we see that $\phi(x, t)$ becomes independent of x as t becomes large. Otherwise expressed, there exists a function $m(x, t)$ such that

$$E[\psi(x(t))|x(0) = x] \to \sum_{x'} m(x', t)\psi(x') \tag{4}$$

and, in virtue of positive recurrence, $m(x, t)$ must be a distribution over $\mathscr{X}$ for given t. We shall write (4) as

$$P(t)\psi \to M(t)\psi := \mathbf{1}m(t)^{\mathsf{T}}\psi \tag{5}$$

where $m(t)$ is the vector with elements $m(x, t)$. Relation (5) implies that

$$M(t + s) = P(s)M(t) = M(t)P(s) \qquad (s, t \geq 0) \tag{6}$$

The first relation of (6) implies that $m(t + s) = m(t)$, so that $m(t) = m$ independent of t. The second implies that m is a stationary distribution. By uniqueness m must be identifiable with π, and (4) then implies the assertion of the theorem. ∎

The author does not know of a reference for this material in the literature, although it can scarcely be completely novel. The point is, however, that a 'backward' Liapunov treatment appears in some ways more natural than the conventional forward treatment.

4. SOJOURN TIME DISTRIBUTIONS

We shall be interested in the time τ that a Markov process takes to escape from a set of states A, wherre τ is reckoned from the moment of first entry. This is the *sojourn time* in A, or *passage time* through A.

For example, one may be interested in the time a processor takes to complete a given task. The task may consist of several phases, some of which may be taken in parallel and some of which must be taken in order. One then regards the complete processing of the task as passage, by some route, through the set of states A.

A 'simple task' consists of passage through s single state x, when we know that τ is exponentially distributed with mean $\lambda(x)^{-1}$ (see Exercise 2.7.1). What we shall in fact do is to replace a simple task by a compound task. It is then best to introduce a notation distinct from that for the x-process. We postulate a set J of discrete states $j = 1, 2, \dots$. This is not the whole state space of a Markov process, but a subset of such a space. Suppose that J is entered at time $t = 0$; let $v(j)$ be the probability that entry takes place at j. Within J, let $\lambda(j, k)$ be the probability intensity of a transition $j \to k$ and $\mu(j)$ the probability intensity of a transition $j \to \bar{J}$, i.e. of leaving J from j. Note that $v(j)$ is not an intensity but describes a probability distribution over J. Thus

$$\sum_j v(j) = 1 \tag{1}$$

if it is understood that all sums are over J.

For the purposes of calculating the sojourn time distribution we can imagine that there is only a single passage through J, and so that $\bar{J}$ is effectively a single absorbing state. Let $p(j; t)$ denote the probability that state j is occupied at time t on these assumptions. That is, on the assumptions that

$$p(j; 0) = v(j) \qquad (j \in J) \tag{2}$$

and that the variable becomes absorbed in $\bar{J}$ on leaving J. We can then make the identification

$$p(j, t) = P(\tau > t \text{ and } j \text{ is occupied at } t)$$

and these probabilities will satisfy the Kolmogorov forward equation

$$\dot{p}(j,t) = \sum_k [p(k;t)\lambda(k,j) - p(j;t)\lambda(j,k)] - \mu(j)p(j;t) \tag{3}$$

with initial condition (2).

The distribution function $F(\cdot)$ of sojourn time τ is then determined by

$$1 - F(t) = P(\tau > t) = \sum_j p(j;t) \tag{4}$$

Let us assume that, not merely is passage out of J certain (i.e. that expression (4) tends to zero with increasing t) but that $E(\tau) < \infty$. Let us define $u(j)$ as the expected time spent in state j during a sojourn in J. This can be expressed as

$$u(j) = E \int_0^\infty I_j(t)\,dt = \int_0^\infty p(j;t)\,dt \tag{5}$$

where $I_j(t)$ is the indicator function of the event that state j is occupied at time t.

Theorem 4.1 *The quantities $u(j)$, the expected sojourn times before escape, satisfy*

$$v(j) - \mu(j)u(j) + \sum_k [u(k)\lambda(k,j) - u(j)\lambda(j,k)] = 0 \qquad (j \in J) \tag{6}$$

They thus satisfy also

$$\sum_j \mu(j)u(j) = \sum_j v(j) = 1 \tag{7}$$

and the expected sojourn time in J is equal to $\sum_j u(j)$

Proof Equation (6) follows from integration of (3) over t from 0 to $+\infty$ and appeal to (5). Note that

$$\int_0^\infty \dot{p}(j;t)\,dt = p(j;\infty) - p(j;0) = -v(j)$$

Relation (7) follows from (6) by summation over J. ∎

We shall in fact often normalize the rate of the process by assuming that we have a 'standard task' of unit expected duration, so that $E(\tau) = 1$ and

$$\sum_j u(j) = 1 \tag{8}$$

The partial process defined by specification of the set J and the parameters (λ, μ, v) subject to (1) and the constraint (8) will be termed an *auxiliary process*.

The interesting point now is that one can realize virtually any sojourn time distribution by appropriate choice of an auxiliary process. That is, the working time for a task can be arbitrarily well approximated (in distribution) if we represent processing of that task by passage through some set of states of a Markov process. Explicitly, consider a sequence of auxiliary processes labelled by $i = 1, 2, \ldots$, and let τ_i be the sojourn time for the ith such process.

Theorem 4.2   *Let τ be an arbitrary non-negative scalar random variable with unit mean. Then one can find a sequence of auxiliary processes such that τ_i converges to τ in distribution.*
That is, such that

$$E\phi(\tau_i) \to E\phi(\tau)$$

as $i \to \infty$ for any bounded continuous function ϕ.

The result as stated is proved by Barbour (1976), but the intuitive basis of the assertion, the 'method of phases' is fairly immediate. We indicate the intuitive argument in Exercises 2 and 3.

We have regarded the sojourn time τ as measuring the workload in a task. It is conceivable that this load is not worked off at a uniform rate. Let τ be the actual value of workload and suppose the task begun at time t_0. Then we shall say that the task is *worked off at rate* $\rho(t)$ if the time of completion σ is related to τ by

$$\int_{t_0}^{\sigma} \rho(t)\,\mathrm{d}t = \tau \tag{9}$$

Here the work-rate $\rho(t)$ is a function of time which may itself be random, because of randomness in the state of the processor. The moment of completion σ may then be random, even when conditional upon a known value of the actual workload τ.

There are now two views of the situation. One is that the value of τ has actually been determined (as a sample from the prescribed workload distribution) when the task enters the processor, although this value is not disclosed to the processor. The other is that the random value of τ is generated in real time by passage of the state of the task through an auxiliary process, this auxiliary process now proceeding at the time-varying rate $\rho(t)$. The two views are equivalent, under an appropriate independence assumption.

Theorem 4.3   *Suppose the work-rate $\rho(t)$ may depend upon the presence of the task in question but not upon the stage of processing that has been reached (up to completion). Then representation of the processing of the task by an auxiliary process with time-dependent rates $(\lambda\rho(t), \mu\rho(t), \nu)$ is equivalent to the supposition that the task arrives with a workload τ (of distribution equal to that of sojourn time in an auxiliary process (λ, μ, ν)) which is worked off at rate $\rho(t)$. In particular, the* *statement that the task has probability intensity $\rho(t)$ of completion is equivalent to the assumption that the task arrives with a workload τ of standard exponential distribution and that this load is worked off at rate $\rho(t)$.*

Proof We can achieve the normalization $t_0 = 0$ by a change in time origin. The distribution function F of τ is determined in terms of the auxiliary process by equations (2)–(4). Let G be the distribution function of completion time σ conditional on a given realization of the function ρ. This is the completion time when processing is represented by the time-varying auxiliary process, and so G is also determined by relations (2)–(4) with the modification that the rates λ and μ are multiplied by $\rho(t)$. (An essential point here is that the rate is independent of auxiliary process state j.) We hence deduce that there is an effective time-dilation by the time-dependent factor $\rho(t)$, and hence that G and F are related by

$$G(s) = F\left(\int_0^s \rho(t)\,dt \right) \qquad (s \geqslant 0) \tag{10}$$

since $F(0) = G(0) = 0$. Since τ is given in terms of σ by (9) (with $t_0 = 0$) then relation (10) states exactly that, conditional on ρ, the effective workload τ has distribution function F. That is, the workload implied by use of a time-varying auxiliary process has the correct distribution, whatever the course of ρ up to the time of completion. ∎

Exercises and comments

1. Suppose that J is indeed a subset of a full state space $\mathscr{J}$ with transition intensities $\lambda(j,k)$ and a unique stationary distribution $\pi(j)$ defined for all j,k of $\mathscr{J}$. Determine then the values of $\nu(j)$ and $\mu(j)$ in terms of these quantities, assuming that the process is in its stationary state.

2. Suppose that J contains n states $j = 1, 2, \ldots, n$ with $\nu(1) = 1$, $\lambda(j,j+1) = \mu(n) = n/a$ $(j = 1, 2, \ldots, n-1)$ and all other λ, μ, ν set equal to zero. Then τ is a sum of n independent exponential variables, each with expectation a/n, and so has the gamma distribution, with density

$$f(\tau) = \frac{(n/a)^n \tau^{n-1} e^{-n\tau/a}}{(n-1)!}$$

As n increases this approximates the distribution $\delta(\tau - a)$ more and more closely. In fact, $E(\tau) = a$, $\mathrm{var}(\tau) = a^2/n$.

3. One can now construct an auxiliary process whose density function $f(\tau)$ approximates $\sum_r b_r \delta(\tau - a_r)$ by having several parallel chains of the type constructed in Exercise 2, the rth giving delay a_r (in the approximate sense explained above) and being selected with initial probability b_r. One can thus approximate mixtures of delta-functions and can go on to approximate any distribution on the non-negative axis. However, one can presumably find a very much more economical realization of a given smooth distribution.

4. Note that the characteristic function of the sojourn time has the evaluation

$$\Phi(\theta) = E(e^{\theta\tau}) = -\sum_j \int_0^\infty e^{\theta t} \dot{p}(j;t)\,\mathrm{d}t$$

for θ such that the expectation is finite. Show then from relations (2), (3) that

$$\Phi(\theta) = v^{\mathsf{T}} \Lambda (\Lambda + \theta I)^{-1} \mathbf{1} \qquad (9)$$

an expression rational in θ for finite J. Here Λ is the matrix with jk^{th} element $\lambda(j,k) - \delta(j,k)\left(\sum_k \lambda(j,k) + \mu(j)\right)$ for $j,k \in J$, and $I, \mathbf{1}$ are the corresponding identity matrix and column vector of units. Show from (8) that

$$E(\tau) = -v^{\mathsf{T}} \Lambda^{-1} \mathbf{1}$$

consistently with Theorem 4.1. The assumption that τ is finite is equivalent to the assumption that Λ^{-1} exists.

5. Relation (9) is certainly valid if θ has real part smaller than κ, where $-\kappa$ (real and non-positive) is the greatest eigenvalue of Λ.

6. Note that for the suggested realization of Exercise 3

$$\Phi(\theta) = \sum_r b_r (1 - \theta a_r / n_r)^{-n_r}$$

5. SEMI-MARKOV PROCESSES

The Markov process with transition intensity $\lambda(x, x')$ can be regarded as one for which the holding time $\tau(x)$ in state x is exponentially distributed with expectation

$$\lambda(x)^{-1} = \left(\sum_{x \neq x'} \lambda(x, x')\right)^{-1} \qquad (1)$$

and, at the expiry of this random holding time, the state changes to x' with probability

$$p(x, x') = \frac{\lambda(x, x')}{\lambda(x)} \qquad (x \neq x') \qquad (2)$$

A semi-Markov process is one in which the assumptions above are relaxed in that the holding time $\tau(x)$ may have an arbitrary distribution, although one that is conditioned only by current x if one conditions on the history of the process. At the expiry of the holding time transition again takes place according to the rule (2).

See Exercise 2 for a more formal statement of these assumptions. One still tends to refer to x as 'state' despite the fact that it no longer supplies a state description in the Markov sense. An implication of the semi-Markov property is that the process $\{x(t),s(t)\}$ is Markov, where $s(t)$ is the 'age' of the current state $x(t)$, i.e. the time that has elapsed since that state was entered.

As an example, one might consider a machine that becomes faulty in the course of use, and is then repaired and restored to service. It then alternates between the states of 'faulty' and 'serviceable'. However, the times spent in these states may not be exponentially distributed if the processes of wear and of repair are in some sense compound.

The discrete-time process obtained by considering the values of x only at the moments of change is the *embedded Markov process*. This has a probability transition matrix $(p(x, x'))$ which is special in that $p(x, x) = 0$ for all x (but see Exercise 4).

Theorem 5.1 *Suppose the embedded Markov chain irreducible and positive recurrent, so that its equilibrium distribution $\omega(x)$ is the unique distribution solving the balance equations*

$$\sum_{x'} \omega(x')p(x', x) = \omega(x) \tag{3}$$

Then the state x of the semi-Markov process has the unique equilibrium distribution

$$\pi(x) \propto \omega(x)E(\tau(x)) \tag{4}$$

The result is intuitive. The equilibrium distribution $\omega(x)$ at the moments of change is weighted by $E(\tau(x))$, the expected sojourn time in state x.

Proof The line of proof is to augment the x variable to make the semi-Markov process a Markov process and then deduce the equilibrium distribution of this Markov process. One such augmentation is to consider the process $\{x(t),s(t)\}$; the line of argument is indicated in Exercise 3. We shall rather use an auxiliary process in each state to realize the distribution of holding time in the state.

We augment the variable x then to (x, j_x) where j_x is the variable of an auxiliary process with parameters λ_x, μ_x, v_x. Specifically, we suppose that (x, j_x) is the variable of a Markov process with transitions and intensities

$$
\left.
\begin{aligned}
(x,j) &\to (x, k) \quad \text{with intensity} \quad \lambda(x)\lambda_x(j, k) \\
(x,j) &\to (x', k) \quad \text{with intensity} \quad \lambda(x)\mu_x(j)p(x, x')v_{x'}(k) \\
&\qquad\qquad\qquad\qquad = \mu_x(j)\lambda(x, x')v_{x'}(k)
\end{aligned}
\right\} \tag{5}
$$

This corresponds exactly to the prescription that the time spent in state x is the time needed for the auxiliary variable j_x to work through its set J_x. However, the transitions of this auxiliary process are accelerated by a factor $\lambda(x)$, so the expected time taken is $E(\tau(x)) = \lambda(x)^{-1}$. When the auxiliary variable completes its passage the variable x undergoes a transition, the transition being to x' with probability $p(x, x')$. Simultaneously a new auxiliary process is begun, that associated with the new value x' of state.

The balance equations for the Markov process with transitions and intensities given by (5) are readily found to have a stationary solution

$$\pi(x, j) = \omega(x)\lambda(x)^{-1}u_x(j) \tag{6}$$

where $u_x(j)$ is the expected time for which $j_x = j$ under operation of the x-auxiliary process at unit working rate. The corresponding distribution for x alone is then (cf. (4.8))

$$\pi(x) = \omega(x)\lambda(x)^{-1} = \omega(x)E(\tau(x)) \tag{7}$$

in agreement with (4).

If we choose the auxiliary processes so that they are irreducible (i.e. so that the equation systems corresponding to (4.6) have a unique solution) then expression (6) is the unique stationary solution of the Markov process. It and the derived distribution (7) consequently constitute the unique equilibrium distributions. Furthermore, any semi-Markov process can be realized (in the sense of Theorem 4.2) by appropriate choice of the auxiliary processes. The theorem is thus proved. ∎

We have the immediate

Corollary 5.1 *Suppose that a Markov process with transition intensity $\lambda(x, x')$ has unique equilibrium distribution $\pi(x)$. Then any semi-Markov process with the same transition matrix (2) and the same expected holding times (1) has the same equilibrium distribution $\pi(x)$ for x.*

This follows because the equilibrium distribution (4) holds for all such processes, including the Markov process itself.

Exercises and comments

1. Suppose that a holding time τ has distribution function $F(t) := P(\tau < t)$. Define the *expiry rate* $g(t)$ by

$$P(\tau > t) = \exp\left[-\int_0^t g(u)\,du\right] \tag{8}$$

We have then indeed

$$g(t) = \lim_{s \downarrow 0} \frac{1}{s}P(t < \tau \leqslant t + s \mid \tau > t)$$

so that $g(t)$ is the intensity of the event 'expiry of holding time' at time t, given that expiry had not occurred previously. It is identical with the 'hazard rate' or 'failure rate' of contexts in which τ is regarded as the time to failure of a component.

The intensity $g(t)$ will be infinite for some t if τ does not possess a probability density. However, the integral occurring in expression (8) will still be defined.

2. A semi-Markov process can be defined as a Markov process with state variable $(x,s) \in (\mathcal{X}, \mathbb{R}_+)$ and the following transition rules. The transition $(x,s) \to (x',0)$ has intensity $g(x,s)p(x,x')$ $(x' \neq x)$. If no such transitions occur in $(t_1, t_2]$ then $x(t_2) = x(t_1)$, $s(t_2) = s(t_1) + t_2 - t_1$. The rules are thus a mixture of random discrete transitions and deterministic continuous transitions.

3. Suppose the Markov process defined in Exercise 2 has an equilibrium distribution

$$P(x,s) := P(x(t) = x, s(t) \leq s)$$

We can then define the equilibrium density (relative to counting measure × Lebesgue measure)

$$\pi(x,s) = \frac{\partial P(x,s)}{\partial s}$$

Show formally that π satisfies the balance equations

$$\frac{\partial \pi(x,s)}{\partial s} + g(x,s)\pi(x,s) = 0 \qquad (s > 0)$$

$$\pi(x,0) = \sum_{x'} \int_0^\infty \pi(x,\tau)g(x,\tau)p(x',x)\,d\tau$$

and hence that

$$\pi(x,s) = \pi(x,0)G(x,s) \tag{9}$$

$$\pi(x,0) = \sum_{x'} \pi(x',0)p(x',x) \tag{10}$$

where

$$G(x,s) := P(\tau(x) > s) = \exp\left[-\int_0^s g(x,\tau)\,d\tau \right]$$

The solution

$$\pi(x,s) \propto \omega(x)G(x,s)$$

then follows from (9), (10) and implies (4).

4. One could also define a semi-Markov process for which $p(x,x)$ was not necessarily zero, formula (2) being modified to hold for all x, x' with $\lambda(x) = \sum_{x'} \lambda(x,x')$. The intensity $\lambda(x,x)$ would be that for the event of 'restart' in state x, i.e. a departure and immediate return. The holding time $\tau(x)$ would be the time elapsing until first departure or restart.

6. NON-CONSERVATIVE FLOWS

The Kolmogorov forward equation

$$\dot{\pi}(x,t) = \sum_{x'} [\pi(x',t)\lambda(x',x) - \pi(x,t)\lambda(x,x')]$$

$$= \sum_{x'} \pi(x',t)\lambda(x',x) - \lambda(x)\pi(x,t) \tag{1}$$

can, as we observed in section 2.7, be regarded as representing a flow of probability mass through state space. The inflows and outflows represented by the two sums in the second member balance overall, so that the total mass $\sum_{x} \pi(x,t)$ is conserved over time.

Suppose that one knew the solution $m(x,t)$ of the related system of equations

$$\dot{m}(x,t) = \sum_{x'} m(x',t)\lambda(x',x) \tag{2}$$

where $m(x,t)$ represents the 'amount of mass of description x' at time t. Exercise 1 gives an example in the context of molecular replication and mutation. The square matrix $(\lambda(x,s'))$ is arbitrary, and in general obeys no conditions such as $\sum_{x'} \lambda(x,x')$ $= 0$. The total mass $\sum_{x} m(x,t)$ will then in general not be conserved, but one wonders whether the normalized mass-distribution

$$\pi^*(x,t) = \frac{m(x,t)}{\sum_{x'} m(x',t)} \tag{3}$$

deduced from the solution of (2) might not satisfy the mass-conserving flow equations (1). The answer is simple.

Theorem 6.1 *The normalized mass π^* associated with the flow equations (2) obeys the equation system*

$$\pi^*(x,t) = \sum_{x'} \pi^*(x',t)\lambda(x',x) - \lambda^*(t)\pi^*(x,t) \tag{4}$$

where

$$\lambda^*(t) := \sum_{x} \pi^*(x,t)\lambda(x) \tag{5}$$

One then has identity of π and π^* (for common initial conditions at $t = 0$) if and only if for almost all $t \geqslant 0$ the values of $\lambda(x)$ for x such that $\pi(x,t) > 0$ have a common (possibly t-dependent) value.

Proof Deduction of (4) is immediate, and we see from (1), (4) that what is required for identity of π, π^* is

$$\pi(x,t)(\lambda(x) - \lambda^*(t)) = 0$$

for all x and almost all $t \geqslant 0$. The requirement is then that $\pi(x,t) > 0$ shall imply $\lambda(x) = \lambda^*(t)$. In view of (5) (with $\pi^* = \pi$) this implies simply that these $\lambda(x)$ have a common value. ■

If the process (1) is irreducible and $\mathscr{X}$ is finite then one will have $\pi(x,t) > 0$ for all x and all positive t. The theorem then says that π, π^* can agree only if $\lambda(x)$ is constant, i.e. if the exit intensity is the same for all states. This is a very strong condition. However, there are examples of physical interest in which it can plausibly be weakened in the limit of a large state space (see section 13.7).

Exercises and comments

1. Eigen, Schuster and others (see, for example, Schuster, Sigmund and Wolff, 1980) have considered the replication and mutation of large complex molecules in their studies of prebiotic evolution. In one of the simplest cases equation (2) can represent the dynamics of molecular abundance, $m(x,t)$ being the number of molecules of species x in a liquid 'culture' and $\lambda(x, x')$ a production rate for species x' from species x. It is assumed that all necessary base materials are available from the culture medium without limit, and that 'death' and interaction can be neglected. The solution of (2) would have the asymptotic form for large t

$$m(x,t) \sim \text{const.}\,\xi(x)e^{\alpha t} \tag{6}$$

where α is the largest eigenvalue of the matrix $(\lambda(x,x'))$ and $(\xi(x))$ the corresponding left eigenvector.

In experimental practice indefinite growth cannot be allowed. One can nevertheless achieve conditions which allow passage to an equilibrium if some of the culture is continuously drained from the reaction vessel and replaced with new medium. Equations (2) then become

$$\dot{m}^*(x,t) = \sum_{x'} m^*(x',t)\lambda(x',x) - Dm^*(x,t) \tag{7}$$

where D is proportional to the dilution flux and $m^*(x,t)$ is the abundance for this modified situation. The dilution flux may depend upon m^* and t but would not depend upon x unless one could remove components selectively from the culture. Such selection, even if possible, would shift the asymptotic composition from that represented by ξ. The required condition of constancy of $\lambda(x)$ in Theorem 6.1 represents a condition of 'non-selective dilution'.

One could choose D as a linear function of the production rates:

$$D = \sum_x b(x)\left[\sum_{x'} m(x',t)\lambda(x',x)\right]$$

This is equivalent to considering the normalization

$$m^*(x, t) = \frac{m(x, t)}{\sum_{x'} b(x')m(x', t)}$$

of the solution of (2). The asymptotic solution of (7) under these circumstances is

$$m^*(x) = \frac{\xi(x)}{\sum_{x'} b(x')\xi(x')}$$

7. INVARIANT AND REDUCED MARKOV PROCESSES

Suppose $\{x(t)\}$ Markov. Consider a transformation $x \to \sigma x$ of state space and consider the process obtained by this instantaneous transformation.

$$y(t) = \sigma(x(t)) \tag{1}$$

One may ask: under what conditions is the transformed process $\{y(t)\}$ also Markov? In general, it is not (see Exercise 1). If the transformation σ is non-invertible then its application loses structure and also, in general, loses the Markov property. However, there are cases where the Markov property is not lost under the reduction (Theorem 7.1). This is typically true because of a clear symmetry (i.e. invariance property) of the parent process (Theorem 7.2). However, attempted reductions of the 'master equation' correspond to attempts to establish that some significant reduced variable is at least approximately Markov.

Theorem 7.1   *The process $\{y(t)\} = \{\sigma x(t)\}$ is Markov if, for all admissible ψ, the expectation $E[\psi(Y^*(t)|x(t)]$ depends upon $x(t)$ through $y(t)$ alone, when it can be identified with $E[\psi(Y^*(t)|y(t)].*

Proof   Recall that $Y^*(t)$ is the future of the y-process at t. Let us write $\xi = (Y^*(t))$, for simplicity. By Theorem 1.14.1 and the Markov character of $\{x(t)\}$ the postulated reduction implies that

$$E[\xi | X(t)] = E[\xi|y(t)]$$

which implies, again by Theorem 1.14.1, that

$$E[\xi|Y(t)] = E[\xi|y(t)] \quad \blacksquare$$

The condition of the theorem is sufficient but not necessary, see Exercise 2. The theorem can be rephrased

Corollary 7.1   *The derived process $\{\sigma x(t)\}$ is Markov if the transition operators P (discrete time) or Λ (continuous time) induce operators $\tilde{P}$ and $\tilde{\Lambda}$ respectively on $\sigma \mathscr{X}$.*

Proof Consider the discrete time case. The condition of Theorem 7.1 will be satisfied if we require it for ξ a function of $y(t+1)$ alone, i.e.

$$E[\psi(y(t+1))|x(t)] = E[\psi(y(t+1))|y(t)]$$

That is, if $\phi(x)$ is a function of σx alone then so is $(P\phi)(x)$. But this is what we mean by saying that P induces an operator on $\sigma \mathscr{X}$, which we then denote by $\tilde{P}$. The continuous time argument is analogous. ∎

The transformation $x \to \sigma x$ of state space induces a transformation $X \to SX$ of realizations, by

$$SX[t] = \sigma(x(t))$$

Let us suppose that there are several transformations σ which constitute a group Σ. All σ of Σ are then invertible by hypothesis, and the transformations S induced by the σ of Σ also constitute a group $\mathscr{S}$. Let us assume that the process is statistically invariant under all such transformations, so that all the transformed processes are statistically equivalent (see Appendix 1).

Let us for simplicity consider just the case of discrete time, the continuous time analogue will be obvious. Suppose the process has transition density $f(x,x')$ relative to a measure μ on sets of x'. If the process is statistically invariant under a transformation σ of state space then one can at least normalize μ so that

$$\mu(\sigma A) = \mu(A) \tag{5}$$

for any subset A of $\mathscr{X}$, and then invariance will also require that

$$f(\sigma x, \sigma x') = f(x, x') \tag{6}$$

Now suppose that

$$y = a(x)$$

is a *maximal invariant* under the σ of Σ. That is, $a(\sigma x) = a(x)$, but the values of $a(x)$ and $a(x')$ are distinct if one cannot write $x' = \sigma x$ for some σ of Σ. Otherwise expressed, suppose that x and x' are regarded as *equivalent* if one can write $x' = \sigma x$ for some σ of Σ. Then y labels the equivalence classes of $\mathscr{X}$, and labels them distinctly. A function $\psi(y)$ of y is a function $\phi(x)$ of x which is invariant (i.e. $\phi(\sigma x) = \phi(x)$ for σ of Σ) and any invariant function is a function of y.

Theorem 7.2 *Suppose the Markov process $\{x(t)\}$ statistically invariant under transformations induced by a group Σ of transformations of state space, and let $y = a(x)$ be a maximal invariant under Σ. Then*

(i) *The process $\{y(t)\}$ is Markov.*
(ii) *Any stationary distribution relative to a measure normalized to invariance as in (5) is itself invariant under Σ, and so uniform over equivalence classes of $\mathscr{X}$.*

Proof Let us denote a typical pair of successive state values $x(t), x(t+1)$ by x, x'

and the corresponding $y(t)$, $y(t+1)$ by y, y'. By Theorem 7.1, we shall have demonstrated assertion (i) if we can show that $E[\psi(y')|x]$ depends upon x only through y.

We can write $\psi(y)$ as $\phi(x)$, where ϕ is invariant, and have then

$$
\begin{aligned}
E[\phi(x')|x] &= \int f(x, x')\phi(x')\mu(dx') \\
&= \int f(\sigma x, \sigma x')\phi(\sigma x')\mu(d\sigma x') \\
&= \int f(\sigma x, u)\phi(u)\mu(du)
\end{aligned}
\tag{7}
$$

The second equality follows from the invariance of μ, f and ϕ; the final equality simply from a change of variable. By comparing the second and fourth members of (7) we see that $E[\phi(x')|x]$ defines a function of x which is also invariant under all σ of Σ, and so is a function of y. Assertion (i) is thus proved.

A stationary distribution $\pi(x)$ satisfies the balance equation

$$
\pi(x') = \int \pi(x)f(x, x')\mu(dx)
$$

A repetition of the same argument shows the integral to be a function of y alone. ∎

Expression (7) implies at least a partial definition of transition density and measure, $\tilde{f}$ and $\tilde{\mu}$, of the y-process. The easiest case (but not the general one: see Exercise 3) is that when $f(x, x')$ is itself a function of y, y' alone. If one takes this as defining $\hat{f}(y, y')$ then one sees from (7) that $\tilde{\mu}$ has the effective definition

$$
\tilde{\mu}(A) = \mu(a^{-1}A)
$$

where A is a set in $\mathscr{Y}$ and $a^{-1}A$ its inverse image in $\mathscr{X}$. That is, $a^{-1}A$ is the set in $\mathscr{X}$ which maps into A under the transformation $x \to a(x)$.

As the example we shall most commonly meet, suppose that x can be represented as having N components $x = (x_1, x_2, \ldots, x_N)$ each of which can take values $j = 1, 2, \ldots, m$. (The statement $x_i = j$ can for concreteness be interpreted as saying that the ith of N particles is at site j, of m possible sites.) We can then take μ as being a simple counting measure on the m^N possible values of x. Suppose now that the process is invariant under permutation of components (i.e. of particles). The maximal invariant under such permutations is the vector of occupation numbers $n = \{n_j\}$, where n_j is the number of components with value j (i.e. the number of particles at site j). By Theorem 7.2, $\{n(t)\}$ is Markov, and, under any stationary distribution values of x corresponding to the same value of n have the same probability.

Suppose that the transition density $f(x, x')$ is a function of n, n' alone, which we take as defining the transformed transition density $\tilde{f}(n, n')$. The transformed measure $\tilde{\mu}$ is then

$$
\tilde{\mu}(A) = \sum_{n \in A} \frac{N!}{\prod_j n_j!}
$$

the combinatorial term representing the number of x-values corresponding to a given n-value. We can either say that the former equilibrium density $\pi(x) = \varpi(n)$ remains the equilibrium density, or that $N! \Big/ \left(\prod_j n_j! \right) \varpi(n)$ is the equilibrium density relative to simple counting measure on the values of n.

The migration model of Exercise 2.7.5 provides the simplest example of this situation. We shall meet it again in section 4.4 as the 'Ehrenfest urn model'.

Exercises and comments

1. A non-invertibly transformed Markov process is in fact scarcely ever Markov. That is, 'blurring' of state space in general destroys the Markov property. For a simple example which makes the point, consider random variables $y(1), y(2), \ldots$ which are independently and identically distributed, conditional on the value of a parameter θ. (For example, results of repeated assays on a randomly chosen ore body of average yield θ.) The process $\{x(t)\}$ with $x(t) = (y(t), \theta) = (y(t), z(t))$ say, is certainly Markov. However, the joint density of $y(1), y(2), \ldots, y(n)$ has the symmetric form

$$f(y(1), y(2), \ldots, y(n)) = \int \left[\prod_{t=1}^{n} f(y(t)|\theta) \right] \rho(\mathrm{d}\theta)$$

ρ being the probability measure of θ. This is consistent with Markov behaviour of the y-process only in the degenerate case when the variables are independent, i.e. when they are independent of θ.

2. To show that the condition of Theorem 7.1 is not necessary, suppose that the x-process is in discrete time with a partitioned state variable $x(t) = (y(t), w(t))$. For notational simplicity suppose all variables discrete; the Markov character of x then implies that

$$f[y(t+1)|Y(t)] = \sum_{W(t)} f[y(t+1)|Y(t), W(t)] f[W(t)|Y(t)]$$

$$= \sum_{w(t)} f[y(t+1)|y(t), w(t)] f[w(t)|Y(t)] \qquad (8)$$

If the y-process is to be Markov then we require that expression (8) reduce to a function of $y(t), y(t+1)$ alone. This it will do under the condition of Theorem 7.1.

$$f[y(t+1)|y(t), w(t)] = f[y(t+1)|y(t)] \qquad (9)$$

but also under the condition

$$f[w(t)|Y(t)] = f[w(t)|y(t)] \qquad (10)$$

As an example for which the y-process is Markov in virtue of the second condition rather than the first, consider plausible assumptions for an animal moving and foraging. Its position at time t is $y(t)$, the amount of food available at that position is $w(t)$. Assume that the next position $y(t+1)$ is conditioned by both current position $y(t)$ and amount available for forage $w(t)$, but that food availability is stochastically dependent only upon position. The condition (9) of Theorem 7.1 is then not satisfied, but the animal's wanderings are Markov in consequence of fulfilment of condition (10).

3. The invariance condition (6) certainly does not imply that $f(x, x')$ is a function of y, y' alone. For the simplest counter-example, suppose that $x = (y, w)$, where w can take only the values 1, 2 and y can take a finite number m of values. Suppose we require invariance of the process under permutation of the two w-values. Relation (6) then implies that

$$f(y, 1; y', 1) = f(y, 2; y', 2) = c(y, y')$$
$$f(y, 1; y', 2) = f(y, 2; y', 1) = d(y, y') \tag{11}$$

where these relations define c and d. The m equivalence classes are indeed those x of given y, but we see from (11) that $f(x, x')$ will be a function of y, y' alone only if c and d are chosen equal, which they need not be. The transition matrix has the form

$$P = \begin{pmatrix} C & D \\ D & C \end{pmatrix}$$

By Theorem 7.2 the y-process is Markov, with transition matrix $C + D$. If its equilibrium distribution vector is α then that of the original process is $\frac{1}{2}\binom{\alpha}{\alpha}$. That is, the two values of w are equiprobable, whatever the value of y, consistently with Theorem 7.2.

For an example, one may think of a sleeper turning and dreaming. The component w denotes the side he lies upon (left or right, for simplicity) and y denotes the state of his dreams. Relations (11) postulate that the process is symmetric in position, w. The matrices C and D are the dream-transition matrices associated with no change or with a change in position respectively.

CHAPTER 4

Reversibility

1. THE CONCEPT OF REVERSIBILITY

A process which is statistically invariant under time reversal about an arbitrary origin is termed *time-reversible* or simply *reversible*. The property is a strong one, and reversible Markov processes have an especially simple structure. It is also a natural one, in that many fundamental processes show a degree of reversibility. The equations of classical and quantum mechanics show a special form of reversibility (*dynamic* reversibility, see below) and the consequences of this permeate statistical mechanics. Perhaps more fortuitously, there is a wide range of models of practical interest in other contexts which possess some degree of reversibility.

The term 'reversible' is used generally in a number of senses, although in the context of stochastic processes it has the clear technical definition given above. For example, a chemical reaction $A \rightleftharpoons B$ is sometimes said to be reversible, meaning no more than that reactions in both directions are possible. This we would regard as something more akin to the intercommunication hypotheses of section 2.4.

The models of classical mechanics are also said to be time-reversible, in that, if $\{p(t), q(t)\}$ represents a time-path for momenta p and coordinates q which is consistent with Newton's equations, then so does $\{-p(-t), q(-t)\}$. That is, the whole set of solutions to Newton's equations is invariant under a time-reversal *and a simultaneous reversal in the sign of momenta*. This is the property of *dynamic reversibility*, which we shall characterize more generally in section 6.

Finally, Kelly (1979 and papers quoted there) has indicated a number of useful relaxations of the reversibility concept which we shall find natural for the network models of Chapters 9 and 10. One may say that a reversible process is the same process after time-reversal (in that it has the same statistics). Suppose that one has a class $\mathscr{C}$ of stochastic processes which is invariant as a class under time-reversal. That is, a member of $\mathscr{C}$ may not transform into itself under time reversal, but at least transforms into some other member of $\mathscr{C}$. The important network models of Part III demonstrate exactly this property. The property is sometimes referred to as *quasireversibility*, although this is a term which is both used in a more specific sense (see section 10.8) and also inevitably applied to almost any weakening of the full reversibility property.

77

2. REVERSIBILITY: FORMALISM AND FIRST CONSEQUENCES

Consider the operator r_a on the time variable which reverses time about the *pivot* a:

$$r_a t = a + (a - t) = 2a - t \tag{1}$$

We shall require that

$$r_a \mathbb{T} = \mathbb{T} \tag{2}$$

if the pivot is to be admissible. So, in the continuous time case ($\mathbb{T} = \mathbb{R}$) the pivot can be any real number. In the discrete time case ($\mathbb{T} = \mathbb{Z}$) the pivot must be an integer or a half-integer.

Relation (2) also constrains $\mathbb{T}$. It cannot, for example, be satisfied for any a if $\mathbb{T} = \mathbb{R}_+$. We require a doubly infinite time axis or something similar (see Exercise 5) if reversibility is to be definable.

The time reversal r_a will induce a transformation R_a of the realization:

$$(R_a X)[t] = x(2a - t) \tag{3}$$

Following the convention described in section 1.5 we shall usually write this simply as

$$R_a x(t) = x(2a - t) \tag{3'}$$

For simplicity we shall write R_0 simply as R. This special choice of pivot is simply one of convention, for the choice of time origin is in general arbitrary.

Lemma 2.1   *The time reversal and translation operators obey the relations*

$$R^2 = I \tag{4}$$

$$R_a = T^{-a} R T^a \tag{5}$$

$$R T^s = T^{-s} R \tag{6}$$

with consequences

$$R_a = T^s R_a T^s \tag{7}$$

$$R_b R_a = T^{2(a-b)} \tag{8}$$

Relations (4)–(6) are easily verified from the definitions and consequences (7), (8) are then immediate. ■

We now define a process $x(t)$ as *time-reversible* (or simply *reversible*, unless the term becomes ambiguous) if it is statistically invariant under time-reversal about an arbitrary admissible pivot. In symbols,

$$X \sim R_a X \tag{9}$$

for all admissible a.

Theorem 2.1   *Suppose* $\{x(t)\}$ *reversible. Then*

(i) *The full stochastic equivalence*

$$X \sim R_a X \tag{10}$$

   holds for all admissible a.

(ii) $\{x(t)\}$ *is stationary.*

Proof   Relation (8) implies that $R_a^2 = I$, so that

$$X = R_a(R_a X) \sim R_a X$$

This relation together with (9) implies (10). From (8) and (10) we now infer that

$$T^{2(a-b)}X = R_b R_a X \sim R_b X \sim X$$

which implies stationarity of $\{x(t)\}$. ∎

As one might imagine, these assertions are specimens of a rather simple general assertion. The general result is, that if a process $\{x(t)\}$ is stochastically invariant under a number of operations S and if repeated applications of the S generates a set of operations $\mathscr{S}$ which is a group, then the transformations of induced $\{x(t)\}$ by any member of $\mathscr{S}$ are all stochastically equivalent (see Appendix 1).

In the case of Theorem 1.1, assertion (i) follows from the fact that (I, R_a) constitutes a group, and assertion (ii) from the fact that the R_a (for admissible a) generate a group which contains the translations T^s as members. This group could also be generated from R and the elementary translations, as we see from (5). We make the point slightly differently in Theorem 2.3 below.

Theorem 2.2   *Reversibility is conserved under time translation, time reversal and time-invariant filtrations. Separate reversibility of* $\{x(t)\}$ *and* $\{y(t)\}$ *does not imply reversibility of the joint process* $\{x(t), y(t)\}$.

Proof   The first two assertions follow immediately from the equivalences $X \sim R_a X \sim T^s X$ proved in Theorem 2.1. Explicitly,

$$T^s X \sim X \sim R_a X \sim R_a(T^s X)$$
$$R_b(R_a X) \sim R_b X \sim X \sim R_a X$$

By a 'time-invariant filtration' of $\{x(t)\}$ is meant the construction of a new process $\{w(t)\}$ by a relation

$$w(t) = \phi(T^{-t}X)$$

where $\phi(X)$ is a fixed functional of X. If $\{x(t)\}$ is reversible then so plainly is $\{w(t)\}$. The last assertion is demonstrated by a counter-example (see Exercise 1). ∎

Theorem 2.3   *For a discrete-time process* $\{x(t)\}$ *reversibility is equivalent the properties that* $x(t)$ *be stationary and show the 'blockwise' reversibility*

$$P(x(t) \in A(t); t = 0, 1, \ldots, n) = P(x(t) \in A(n-t); t = 0, 1, \ldots, n) \qquad (11)$$

for arbitrary n and sets $A(t)$.

Proof Reversibility implies stationarity, and implies (11) by invariance under time-reversal about the pivot $n/2$. Stationarity and condition (11) imply that

$$E[\phi(X)] = E[\phi(R_a X)]$$

for finite a and for ϕ a functional of the process over a finite time-span. Full time-reversibility then follows by a limiting argument. ∎

The theorem will have a continuous-time analogue under the regularity condition of *separability*: that the statistics of X are determined by the finite-dimensional distributions $P(x(t_j) \in A(t_j); \quad j = 1, 2, \ldots, n)$. Exercise 6 demonstrates the need for such a condition.

Exercises and comments

1. Suppose that $\{x(t)\}$ is reversible; define Y as the lagged version $Y = T^s X$ of X. Then $\{y(t)\}$ is also reversible, but the joint process $\{z(t)\} = \{x(t), y(t)\}$ will scarcely ever be. For example, the function $T^s X - Y$ of Z is identically zero, but the same function for RZ is $T^s RX - RY = (RT^{-s} - RT^s)X \sim (I - T^{2s})X$, which will be zero only if $x(t)$ has period $2s$.

2. Suppose that $\{x(t)\}, \{y(t)\}$ are mutually independent and separately reversible. The joint process is then indeed reversible.

3. Suppose $\{x(t)\}$ stationary with $x \in \mathbb{R}^d$. The *autocovariance matrix*

$$V(s) := \text{cov}(x(t), x(t-s))$$

satisfies

$$V(s) = V(-s)^\mathsf{T}$$

as an immediate consequence of its definition. Since the reversed process will have autocovariance matrix $V(-s)$, the process can then be reversible only if $V(s)$ is symmetric, identically in s. This will always be true if $d = 1$, but only under special circumstances if $d > 1$. This observation reinforces the point made in Exercise 1.

4. Consider the scalar stationary process $\{x(t)\}$ in continuous time generated by

$$x(t) = \sum_j b_j(t - t_j)$$

where $\{t_j\}$ is a Poisson sequence (see Exercise 2.7.12). The quantity $b_j(t - t_j)$ is the transient response at time t to an impulse at time t_j, and we consider two alternative specifications.

(i) $b_j(s) = \begin{cases} 0 & (s < 0) \\ e^{-\alpha s} & (s \geqslant 0) \end{cases}$

(ii) $b_j(s) = \begin{cases} 0 & (s < 0) \\ 1 & (0 \leqslant s \leqslant s_j) \\ 0 & (s > s_j) \end{cases}$

where the s_j are independent (of each other and of the t_j) with the common exponential density $\alpha e^{-\alpha s}$ ($s \geqslant 0$). Show that these have the same autocovariance function, but that version (ii) is reversible, while version (i) is not.

Discrimination between the two models is of interest in a biochemical context, in which the t_j are moments at which independent stimuli arrive. According to the model (i) the effect of a stimulus decays exponentially; according to model (ii) the stimulus has a constant effect for a random length of time, exponentially distributed.

5. Consider a discrete-time stochastic process $x(t)$ which has period n for all realizations, so that $x(t + n) = x(t)$ for all t. One can then effectively reduce all values of t modulo n to the set of standard values $t = 1, 2, \ldots, n$. These are to be regarded as n equidistant values on a circle rather than on an interval, however, in that $t = n + 1$ (for example) is meaningful, but is to be identified with $t = 1$.

Let us denote this cyclic set of time points by P_n. The translation operator T on the original time set $\mathbb{Z}$ induces a cyclic permutation on P_n. Such processes are termed *circulant* processes. We shall consider them later (section 18.6) because they allow us to discuss stationarity in a finite context (since $T^n = I$). Show that one can introduce a reversibility concept for such processes, in that there are values of the pivot a which satisfy (2).

6. Consider a continuous time model

$$x(t) = \begin{cases} 0 & t \neq \tau \\ 1 & t = \tau \end{cases}$$

where the time instant τ is random, with a prescribed probability density relative to Lebesgue measure. This is not even stationary, let alone reversible. However, the finite-dimensional distributions are consistent with reversibility in that they indicate $x(t)$ to be identically zero.

7. *Reversibility of Poisson streams.* A Poisson process of constant rate λ (see Exercise 2.7.14) could be characterized by the statement that the probability of an event in a given small time interval of length s conditional on the realization outside the interval is $\lambda s + o(s)$, and that the probability of more than one event is $o(s)$. This prescription is independent of both the origin and the direction of time, so the process is plainly both stationary and reversible. However, the choice of a process variable which exhibits this fact is not completely straightforward. The variable $x(t)$ of Exercise 2.7.14 was essentially the number of events occurring up to time t; this variable is not stationary. One could describe a realization by $y(t)$, where $y(t)$ is 1 or 0 according as an event occurs at t or not. However, y is then nearly always zero.

Typically one will be interested in linear functionals of the process:

$$\int \theta(t)\mathrm{d}x(t) = \sum_j \theta(t_j)$$

where the t_j are the instants at which events occur. Demonstrate the evaluation

$$\Phi(\theta) := E(\mathrm{e}^{\int \theta(t)\mathrm{d}x(t)}) = \exp[\lambda \int (\mathrm{e}^{\theta(t)} - 1)\mathrm{d}t]$$

of the *characteristic functional* of the Poisson process. The stationarity and reversibility of the process are manifested by the invariance of Φ under time translation and reversal, i.e. under the substitution of $\theta(t+s)$ or $\theta(-t)$ for $\theta(t)$.

3. MARKOV PROCESSES AND TIME REVERSAL

Theorem 3.1 *The Markov property is conserved under time reversal. For a process with discrete state space the reversed process has transition probability*

$$f(x(t-1)|x(t)) = \frac{f(x(t-1))f(x(t)|x(t-1))}{f(x(t))} \tag{1}$$

in discrete time and transition intensity

$$\lambda^*(x, x'; t) = \frac{\pi(x', t)\lambda(x', x; t)}{\pi(x, t)} \tag{2}$$

in continuous time.

Note that we are not asserting reversibility of the process, but merely conservation of the Markov property under time reversal. Rather than indicate an explicit time reversal about some pivot we have simply reversed the direction of conditioning, so that we evaluate $f(x(t-1)|x(t))$ rather than $f(x(t+1)|x(t))$. The quantity $\pi(x,t)$ gives the distribution of $x(t)$, and λ, λ^* are forward and backward transition intensities:

$$\lambda(x, x', t) = \lim_{s \downarrow 0} \frac{1}{s} P(x(t+s) = x'|x(t) = x)$$

$$\lambda^*(x, x', t) = \lim_{s \downarrow 0} \frac{1}{s} P(x(t-s) = x'|x(t) = x) \tag{3}$$

The assertions of the theorem have an obvious formal analogue for a general state space, see Exercise 4.

Proof The first assertion follows from the symmetric characterization of the Markov property of Theorem 2.1.5. If the condition of that theorem is satisfied then it is also satisfied for a time-translated or time-reversed process.

Relation (1) is valid as a statement about conditional densities whether the process is Markov or not. However, since the reversed process is Markov relation

(1) in fact expresses the transition probability of the reversed (or 'backward') process in terms of occupation and transition probabilities for the forward process. For the quotient (1) to be clearly defined we require that $f(x(t))$ be positive for the particular value of $x(t)$ considered, i.e. that the conditioning variable have a value which has positive probability. If the denominator were zero then the quotient in (1) would be of the form 0/0, to which one might or might not be able to assign a value.

Relation (2) is the continuous time analogue of relation (1), written rather in functional form. With the definitions (3) it follows formally from (1). ■

In the discrete-time discrete-state case one might have proved the theorem simply by noting that the Markov property is equivalent to the relation

$$f(x(1), x(2), \ldots, x(n)) = f(x(1)) \prod_{t=1}^{n-1} f(x(t+1)|x(t))$$

for all n and for all sequences with positive probability. In virtue of (1) this can be rewritten

$$f(x(1), x(2), \ldots, x(n)) = f(x(n)) \prod_{t=1}^{n-1} f(x(t)|x(t+1))$$

implying the Markov character of the reversed process.

For the stationary (and so time-homogeneous) case we shall revert to the functional forms: $\pi(x)$ for state distribution, $f(x, x')$ for one-step transition probability and $\lambda(x, x')$ for transition intensity.

Corollary 3.1 *Suppose $x(t)$ Markov on a discrete state space and stationary with strictly positive distribution $\pi(x)$. Then the time-reversed process (about any pivot) is also Markov and stationary with distribution $\pi(x)$. The transition probability is*

$$f^*(x, x') = \frac{\pi(x')f(x', x)}{\pi(x)} \tag{4}$$

in discrete time, the transition intensity is

$$\lambda^*(x, x') = \frac{\pi(x')\lambda(x', x)}{\pi(x)} \tag{5}$$

in continuous time.

For the process to be reversible it is plainly necessary that $f^*(x, x') = f(x, x')$ or $\lambda^*(x, x') = \lambda(x, x')$, respectively. In fact, with minor provisos these conditions are also sufficient.

For simplicity of exposition we shall now concentrate on the case of discrete time and discrete state space and return to continuous time and more general state spaces in section 7.

Theorem 3.2 *Suppose $\{x(t)\}$ a Markov process in discrete time on a discrete state space. For the process to be reversible it is necessary that it be time-homogeneous, stationary and that the stationary distribution π satisfy the detailed balance condition*

$$\pi(x)f(x, x') = \pi(x')f(x', x) \tag{6}$$

for all x, x'.

The conditions are also sufficient for reversibility in that, if the transition probabilities $f(\cdot, \cdot)$ satisfy (6) for some distribution π, then π is a stationary distribution for the process, and the process in this stationary regime, restricted to the set of x for which $\pi(x) > 0$, is reversible.

Proof Necessity is immediate. Reversibility implies stationarity implies time-homogeneity, and reversibility about the pivot $t + \frac{1}{2}$ requires that

$$P(x(t) = x, x(t+1) = x') = P(x(t) = x', x(t+1) = x)$$

which implies (6).

Turn now to sufficiency. Summing (6) over x' we see that it implies the total balance equation

$$\sum_{x'} \pi(x')f(x', x) = \pi(x) \tag{7}$$

so that π is indeed a stationary distribution. Consider the process in this stationary regime, restricted to the set of states x for which $\pi(x) > 0$. The corresponding reversed process has the same statistics as the forward process: it is stationary and Markov with distribution $\pi^*(x) = \pi(x)$ and transition probability $f^*(x, x') = f(x, x')$, by (4), (6). The process is thus reversible. ∎

Relation (6) is the celebrated *detailed balance equation*, stating that the fluxes $x \rightleftharpoons x'$ balance for any pair of states x, x'. In the notation of section (2.3) we would write it

$$[x', x](\pi f) = 0 \tag{8}$$

We pursue implications of the condition in the next section.

It is interesting that the existence of a relation (6) among the transition probabilities characterizes π as a possible equilibrium distribution, subject to regularity conditions. This is a welcome property of reversible Markov processes: that an equilibrium distribution is immediately deducible from the transition probabilities by inspection. Of course, all that is required is that the $\pi(x)$ satisfying (6) be non-negative and summable; they can then be normalized to a distribution.

If the process is irreducible then π is unique and constitutes the equilibrium distribution (see Theorem 2.4.2). However, we have not required uniqueness (see Exercise 2).

A stronger definition of reversibility is what one might term *transition-reversibility*: that

$$f(x, x') = f(x', x) \tag{9}$$

It may be argued that this corresponds more closely to what one means by time-reversible dynamics; see the next section. The condition is a special case of (6) and so implies reversibility with $\pi(x) = $ constant as a possible equilibrium distribution, and as the actual equilibrium distribution if the process is irreducible. This uniform distribution can make sense only if $\mathscr{X}$ is finite, unless one is willing to weaken the demand of positive recurrence (cf. Exercise 3).

Exercises and comments

1. Note that detailed balance is equivalent to partial balance on every pair of states.

2. Suppose a Markov process contains several irreducible classes, each individually positive recurrent and reversible. Show that the process is then always reversible on the set of states with positive equilibrium probability.

3. One can develop the notion of reversibility even for a null-recurrent process, although the $\pi(x)$ occurring in the detailed balance equation will not then be normalizable. The traditional example is the simple random walk on the integers, with $f(x, x+1) = f(x, x-1) = 1/2$. This is known to be recurrent, although with infinite expected recurrence time (see, for example, Feller (1971) p. 360). The detailed balance equation (6) is satisfied with constant $\pi(x)$. This reflects the fact that, in equilibrium, x is uniformly distributed over $\mathbb{Z}$. So, ratios are defined,

$\pi(x)/\pi(x') = 1$, although the only choice consistent with $\sum_x \pi(x) = 1$ is $\pi(x) = 0$.

4. Theorem 3.1 is formally valid for a general state space with relations (1), (2) replaced by

$$P^*(t) = F(t-1)^{-1} P(t-1)^{\mathsf{T}} F(t)$$

$$\Lambda^*(t) = F(t)^{-1} \Lambda^{\mathsf{T}}(t) F(t)$$

(10)

Here $F(t)$ is the operator on functions $\phi(x)$ with the action

$$(F(t)\phi)(x) = \pi(x, t)\phi(x)$$

where $\pi(x, t)$ is the density of $x(t)$ relative to the fixed measure μ, and the star uniformly denotes quantities relevant to a *backward* conditioning. The first relation for example follows from the required identity

$$\int \pi(x, t-1)\psi(x)(P(t-1)\phi)(x)\mu(dx) = \int \pi(x, t)\phi(x)(P^*(t)\psi)(x)\mu(dx)$$

4. REVERSIBLE MARKOV PROCESSES: DISCRETE TIME EXAMPLES

Assume that state values are discrete and labelled by $j = 1, 2, \ldots$. We shall change for a while to a subscript notation, writing π_j and f_{jk} rather than $\pi(j)$ and

$f(j, k)$. This is partly for typographical convenience and partly because, as in sections 3.4 and 3.5, this can be useful alternative notation when one is compounding Markov processes and wishes to distinguish between corresponding quantities for the component and compound processes.

A two-state Markov process is always reversible, because the total balance equation

$$\pi_1 f_{11} + \pi_2 f_{21} = \pi_1 \tag{1}$$

together with the normalization of π implies both the other balance equation and the only non-trivial detailed balance relation.

$$\pi_1 f_{12} = \pi_2 f_{21} \tag{2}$$

A collection of N simultaneous and statistically independent processes of this type will then also be reversible (cf. Exercise 2.2). These need not have the same transition matrix. However, if we suppose that they do then we have virtually constructed the 'Ehrenfest urn model' or 'dog-flea' model, used to resolve a classic reversibility paradox. This is a model of N particles which move independently but with the same transition rules between two boxes labelled 1 and 2. The two boxes correspond to the two states of the process considered above, and we suppose indeed that the transitions of each particle are governed by the transition matrix

$$\begin{pmatrix} f_{11} & f_{12} \\ f_{21} & f_{22} \end{pmatrix}.$$

The original Ehrenfest model was couched in terms of urns and balls rather than boxes and particles, doubtless to avoid confusing a conceptual experiment and physical reality. In a cruder physical analogue one speaks of dogs and fleas.

Our construction implies that we regard the particles as distinguishable, the state variable is then $x = (x_1, x_2, \ldots, x_N)$ where the component x_i takes values 1 or 2 according as to whether the ith particle is in box 1 or 2. Because of independence the equilibrium distribution is

$$P(x) = \pi_1^{n[x]} \pi_2^{N-n[x]} \tag{3}$$

where $n[x]$ is the number of particles specified as being in box 1 by x and (π_1, π_2) is the two-state distribution determined by (1).

The state variable x describes the configuration of the system completely. However, suppose that we are not interested in distinguishing between configurations which differ only by a permutation of the particles. The configuration at this reduced level of description is then specified by $n[x]$. The process $\{n(t)\} = \{n[x(t)]\}$, being derived from a reversible process, is itself reversible. It is also Markov, because n is the maximal invariant under statistics-preserving permutations of particles (see Theorem 3.7.2). Its equilibrium distribution is

$$\pi(n) = \binom{N}{n} \pi_1^n \pi_2^{N-n} \tag{4}$$

obtained by summing expression (3) over all x consistent with $n[x] = n$. This is of course the binomial distribution, with the distribution of the proportion n/N peaking ever more strongly about the value π_1 as N increases.

This model was proposed by the Ehrenfests (1907) in order to resolve an apparent paradox pointed out by Zermelo (1896). This was that, although the models of classical mechanics are time-reversible, the conclusions of statistical mechanics, supposedly derived from classical mechanics, demonstrate an irreversibility in that, for example, entropy always increases (see section 6.2). The Ehrenfests claimed that their model resolved the paradox in the following sense. The process $\{n(t)\}$ is time-reversible, and this shows itself in the fact that, in equilibrium

$$P(n(t) = n, n(t+1) = n') = P(n(t) = n', n(t+1) = n) \qquad (5)$$

for any n, n'. On the other hand, quite consistently with this reversibility, the model does demonstrate asymmetry in time in that

$$P(n(t+1) = n' | n(t) = n) \neq P(n(t+1) = n | n(t) = n') \qquad (6)$$

Indeed, $n(t)$ will tend to move from extreme values (near 0 and N) to values near $N\pi_1$, and will seldom move far against this trend.

This is then seen as the resolution of the paradox: that the fundamental reversibility of the process demonstrates itself in the time-symmetry of *joint* distributions, as in (5), and that the apparent time-asymmetry of sample-paths manifests itself in the asymmetry of *conditional* distributions, as in (6).

The argument is an appealing one, but one can query whether it really meets Zermelo's point. The reversibility of classical mechanics consists in invariance of the dynamical equations under a time-reversal (with qualifications, see section 6). There is no mention of equilibrium, and one can see this notion of reversibility of dynamics as corresponding to *transition-reversibility* (see (3.9)) rather than simple statistical reversibility.

The Ehrenfest model still resolves the dilemma, even with this changed characterization, and perhaps even more appealingly. Consider the case of symmetry between the two boxes, so that $f_{12} = f_{21} = p$, say, with p equal to neither zero nor unity. The equilibrium distribution (3) then becomes uniform:

$$P(x) = 2^{-N} \qquad (7)$$

Furthermore, the model is indeed then transition-reversible, for

$$P(x(t+1) = x' | x(t) = x) = P(x(t+1) = x | x(t) = x') = p^D (1-p)^{N-D} \qquad (8)$$

where D is the number of components in which x and x' differ. However, it will still be true that one has the asymmetry (6).

One may see the resolution of the Zermelo paradox in the fact that the model demonstrates its physical reversibility by the symmetry (8) of transition probabilities for the complete description x, and demonstrates its bulk irreversibility by the asymmetry (6) of the corresponding transition probabilities for the reduced description n.

The variable x could be said to describe a 'micro-state', the variable n a 'macro-state' (the term 'state' still being appropriate, since $\{n(t)\}$ is Markov). Micro-states are equiprobable in equilibrium; see (7). Macro-states are not; (4) will now reduce to

$$\pi(n) = \binom{N}{n} 2^{-N} \tag{9}$$

This is again a consequence of the reduction; a macro-state corresponds to an n-dependent number of micro-states.

The birth and death processes described in Exercise 2.3.2 are reversible, because the neighbour–balance relation (2.3.7) expresses just detailed balance. Detailed balance holds between non-neighbours in the trivial sense $0 = 0$. We see a more general version of this example in Theorem 5.3.

Exercises and comments

1. The transition rates for the reduced description n of the Ehrenfest model have a much simpler form in the continuous-time version of the process (see Exercise 7.2). However, note that the transition probability is indeed

$$f(n, n') = \sum_{j} \binom{n}{j} \binom{N-n}{n'-n+j} f_{11}^{n-j} f_{12}^{j} f_{21}^{n'-n+j} f_{22}^{N-n'+j}$$

5. IMPLICATIONS OF DETAILED BALANCE (DISCRETE TIME)

The detailed balance equation (3.6) asserts the symmetry in x and x' of $\pi(x) f(x, x')$, or equivalently, of $f(x, x') \pi(x')^{-1}$ or of $\pi(x)^{1/2} f(x, x') \pi(x')^{-1/2}$. The last characterization is perhaps the significant one, for it exhibits the transition matrix P as derived from a symmetric matrix by a similarity transformation.

Theorem 5.1

(i) *An irreducible Markov process is positive recurrent and reversible if and only if its transition matrix P can be represented*

$$P = W^{-1} A W \tag{1}$$

where A is symmetric and W is a diagonal matrix for which $\operatorname{tr} W^2 < \infty$.

(ii) *If $W = \operatorname{diag}(w(x))$ then $w(x)^2$ is proportional to the equilibrium distribution $\pi(x)$ of the process.*

(iii) *If state space is finite then the eigenvalues of P are confined to the real interval $[-1, 1]$.*

Proof Assertions (i), (ii) follow from Theorem 3.2, in that (1) is equivalent to assertion of a relation

$$w(x)^2 f(x, x') = w(x')^2 f(x', x).$$

Under the assumption of irreducibility positive recurrence will be equivalent to the conditions that $w(x)$ be nowhere zero and that $\Sigma w(x)^2 < \infty$, i.e. to the statements that W^{-1} exists and $\text{tr}\, W^2 < \infty$.

The matrices P and A have the same set of eigenvalues, since they are related by similarity transformations. Since P is a transition matrix its eigenvalues have modulus one at most (see, for example, Grimmet and Stirzaker (1982) p. 134, or Bellman (1960) p. 259). Since A is symmetric, its eigenvalues are real. Assertion (iii) thus follows. It is asserted only for the case of a finite state space simply to avoid the complications of the concept of eigenvalue for an infinite matrix. ∎

The property of transition-reversibility corresponds to the case $W = I$.

A periodicity of d in the process implies that P possesses an eigenvalue $\exp(2\pi i/d)$ (Grimmett and Stirzaker (1982), p. 134). Assertion (iii) of the theorem thus implies that reversibility excludes all periodicities except perhaps one of period 2, which would correspond to a simple alternation between groups of states (see Exercise 2). This statement is reinforced by the following result, first stated by Kolmogorov (1936).

Theorem 5.2 (*Kolmogorov's characterization of reversibility*) *Consider an irreducible positive recurrent Markov process with transition probability $f(x, x')$. This is reversible if and only if*

$$\prod_{t=0}^{n-1} f(c(t), c(t+1)) = \prod_{t=0}^{n-1} f(c(n-t), c(n-t-1)) \tag{2}$$

for all $c(0), c(1), \ldots, c(n-1), c(n) = c(0)$ and n.

Proof The condition is necessary because condition (1), which we may write elementwise as

$$f(x, x') = \frac{w(x')}{w(x)} a(x, x') \tag{3}$$

with $a(\cdot, \cdot)$ symmetric, implies (2). To prove sufficiency, set $c(0) = x$, $c(n-1) = x'$ and sum relation (2) over $c(1), c(2), \ldots, c(n-2)$. This yields

$$f(x, x'; n) f(x', x) = f(x', x; n) f(x, x') \tag{4}$$

As we let n tend to infinity then $f(x, x', n) \to \pi(x')$ (Theorem 2.4.2) and relation (4) becomes the detailed balance equation (3.6), sufficient for reversibility. ∎

Relation (2) states that, if one considers any path in state space returning to its initial point $c(0)$, then the probability of that path (conditional on the starting point) equals the probability of the reversed path. This could be expressed by saying that *a reversible process shows no net circulation in state space*. This property certainly excludes periods greater than two.

Finally, we have a characterization for a class of reversible processes which generalizes the birth and death example of the previous section.

Theorem 5.3 *Consider the undirected graph* Γ *whose nodes correspond to the states* x *of a Markov process and for which an* xx' *arc exists only if at least one of* $f(x, x')$ *and* $f(x', x)$ *is positive. If* Γ *is a tree then the process is reversible.*

Proof The detailed balance equation (3.6) is satisfied trivially if there is no xx' arc. Suppose there is such an arc, and let A be the set of states with which x would communicate if the arc were cut. We have then the balance relation

$$[A, \bar{A}] (\pi f) = 0 \tag{5}$$

by Theorem 2.3.1. However, because Γ is a tree, A and $\bar{A}$ are connected only by the xx' arc. Hence the cut-balance relation (5) reduces to the detailed balance relation (3.6). ∎

That is, balance of probability flux $A \rightleftharpoons \bar{A}$ implies balance of probability flux $x \rightleftharpoons x'$ because all $A/\bar{A}$ flux is along the xx' arc.

In the case of a birth and death process Γ is a linear graph and so certainly a tree.

Exercises and comments

1. Note that relation (2) is trivially valid for $n = 1, 2$.

2. If expression (1) is to be a transition matrix then W is essentially determined once A has been chosen. Indeed, the relation $P1 = 1$ implies that $Aw = w$, where w is the column vector with elements $w(x)$. If we choose all elements of w positive, so that all elements of A must be non-negative, then we can say that the non-negative matrix A has its maximal eigenvalue equal to unity with corresponding eigenvector w.

3. Reversibility implies that P has only real eigenvalues, the converse does not hold. Consider the matrix

$$P = \begin{bmatrix} \cdot & 1 & \cdot \\ 3/4 & \cdot & 1/4 \\ 1 & \cdot & \cdot \end{bmatrix}$$

This has eigenvalues $1, -1/2, -1/2$. The process with this transition matrix is not reversible; however, the sequence of states 1231 has positive probability and the reverse sequence 1321 has zero probability.

4. Consider the more general example

$$P = \begin{bmatrix} a_1 & b_1 & \cdot & \cdot & \cdots & \cdot \\ a_2 & \cdot & b_2 & \cdot & \cdots & \cdot \\ a_3 & \cdot & \cdot & b_3 & \cdots & \cdot \\ & & & \cdots & & \\ a_{n-1} & \cdot & \cdot & & \cdots & b_{n-1} \\ 1 & \cdot & \cdot & & \cdots & \cdot \end{bmatrix}$$

with $a_j + b_j = 1$ and the convention $a_n = 1$. Show that this matrix has eigenvalues $\theta = 1$ and $\theta_1, \theta_2, \ldots, \theta_{n-1}$ these latter being the roots of

$$\theta^{n-1} + b_1 \theta^{n-2} + b_1 b_2 \theta^{n-3} + \ldots + b_1 b_2 \ldots b_{n-1} = 0$$

Show that these roots can be any set of non-positive real values consistent with $\sum_1^{n-1} \theta_j > -1$. In particular, the choice $\theta_j = -1/n-1$ ($j = 1, 2, \ldots, n-1$) corresponds to $b_j = (n-j)/(n-1)j$.

The process (proposed and analysed in Whittle, 1975) is irreversible in that it shows a net circulation in state space; the state progresses through the values $1, 2, \ldots$ but drops out at some random point to restart from 1. However, the recurrence time to state 1 is so diffusely distributed that there is nothing like a periodicity, and P has only real eigenvalues.

5. Note that the Kolmogorov conditions (2) are not independent; quite a small subset of them may imply detailed balance and so imply all the others. Suppose, for example, that there is a state c which communicates directly with all others, in that $f(x, c) > 0$ and $f(c, x) > 0$ for all x. Relation (2) for $n = 3$ and $c(0) = c$ implies then that

$$f(c, x) f(x, x') f(x', c) = f(c, x') f(x', x) f(x, c)$$

which implies the detailed balance condition with

$$\pi(x) \propto \frac{f(c, x)}{f(x, c)}$$

6. DYNAMIC REVERSIBILITY

The reversibility property we have postulated cannot be identical with the reversibility associated with dynamic physical systems, for it precludes periodicity, whereas periodic behaviour is one of the commonest features of such systems. Let us make the classical assertion precise. The state variable of a Newtonian dynamical system has as components a vector of coordinates q and a corresponding vector of momenta p. (The notation infringes on that used for probabilities, but both uses are hallowed by tradition and the occasions of ambiguity few.) Let $(p(t), q(t))$ be a particular solution of the Newtonian equations. The assertion is that so also is $(-p(2a - t), q(2a - t))$, for arbitrary a. That is, that time reversal about an arbitrary pivot *plus momentum reversal* maps the set of solutions into itself.

In effect, $x = (p, q)$ is the state variable, and we are led to consider a *conjugated* state variable: $\bar{x} = (-p, q)$. The assertion is that a time reversal plus a conjugation leaves the set of solutions to the Newtonian equations invariant.

Let us generalize this particular case by defining a *conjugation operator C*, transforming realizations to realizations, with the properties:

(C1) $C^2 = I$
(C2) C commutes with T and with R

The momentum reversal considered certainly has these properties. In fact, we think of C as the operator induced by a permutation σ of state space. This structure would automatically imply property (C2). Property (C1) would imply that σ amounts to a number of simultaneous pairwise permutations of elements of $\mathscr{X}$, in that $\sigma^2 = 1$. We shall usually write σx as $\bar{x}$, the state value conjugate to x. The property $\sigma^2 = 1$ is then equivalent to the statement that x is also the value conjugate to $\bar{x}$. There may be some elements of $\mathscr{X}$ which are unchanged by the permutation, these are then self-conjugate, $\bar{x} = x$.

We now modify the definition (2.9) of reversibility by saying that a stochastic process $x(t)$ is *dynamically reversible* if

$$X \dashrightarrow CR_a X \tag{1}$$

for any admissible a.

The theorems proved for simple reversibility now have analogues so immediate that we give statement and proof only in sketch; differences in conclusions are, however, substantial enough to be significant. In analogue to Theorem 2.1:

Theorem 6.1 *Suppose $\{x(t)\}$ dynamically reversible. Then*

 (i) *Definition* (1) *implies stochastic equivalence*

$$X \sim CR_a X \tag{2}$$

   for admissible a.
(ii) *$\{x(t)\}$ is stationary.*

Proof Assertion (i) follows from (1) and the relation $(CR_a)^2 = I$. Assertion (ii) follows from (1) and the relation $(CR_b)(CR_a) = T^{2(a-b)}$; see the proof of Theorem 2.1. ∎

Theorems 2.2 and 2.3 have obvious analogues. For example, relation (2.11) must now become

$$P(x(t) = c(t); t = 0, 1, \ldots, n) = P(x(t) = \overline{c(n-t)}; t = 0, 1, \ldots, n) \tag{2}$$

The analogue of Theorem 3.2 is important enough to deserve full statement.

Theorem 6.2 *Suppose $\{x(t)\}$ a Markov process in discrete time on a discrete state space. For the process to be dynamically reversible it is necessary that it be time-homogeneous, stationary and that the stationary distribution π satisfy the conjugated detailed balance condition*

$$\pi(x) f(x, x') = \pi(\bar{x}') f(\bar{x}', x) \tag{3}$$

for all x, x', with consequence

$$\pi(\bar{x}) = \pi(x) \tag{4}$$

These conditions are also sufficient, in the sense explained in Theorem 3.2.

Proof Relation (3) follows from (2) and stationarity. Summing it over x' we deduce that

$$\pi(x) = \sum_{\bar{x}'} \pi(\bar{x}') f(\bar{x}', \bar{x}) = \pi(\bar{x})$$

establishing (4). The proof of sufficiency is an obvious version of that for Theorem 3.2. ∎

The Kolmogorov characterization has an immediate analogue; condition (5.2) is now to be replaced by

$$\prod_{t=0}^{n-1} f(c(t), c(t+1)) = \prod_{t=0}^{n-1} \overline{f(c(n-t), c(n-t-1))} \tag{5}$$

for all $c(0), c(1), \ldots, c(n-1), c(n) = c(0)$ and n. That is, the probability of any circuit in state space, conditional on starting point, is equal to the probability of the *reversed and conjugated* circuit. This criterion now no longer excludes periodic behaviour, as we see from the representation of $f(x, x')$, which now becomes

$$f(x, x') = w(x)^{-1} a(x, x') w(x') \tag{6}$$

with

$$a(x, x') = a(\bar{x}', \bar{x}) \tag{7}$$
$$w(x) = w(\bar{x}) \tag{8}$$

Again, $w(x)^2$ is identifiable as $\pi(x)$, a possible equilibrium distribution for the dynamically reversible regime (see Exercises 2, 3).

Dynamic transition-reversibility would correspond to the stronger condition

$$f(x, x') = f(\bar{x}', \bar{x}) \tag{9}$$

so that (6)–(8) hold with $w(x)$ constant (at least on an irreducible set).

Suppose that we assume that no state is self-conjugate (i.e. that 'momentum is never zero') and that state space is finite. We can then order and partition the states into mutually conjugate sets, $(1, 2, \ldots, m)$ and $(\bar{1}, \bar{2}, \ldots, \bar{m})$, say. It then follows from (6)–(8) that we can write the transition matrix in the form

$$P = \begin{pmatrix} W^{-1}AW & W^{-1}BW \\ W^{-1}B^{\mathsf{T}}W & W^{-1}A^{\mathsf{T}}W \end{pmatrix} \tag{10}$$

where A, B are $m \times m$ matrices and $W = \mathrm{diag}\,(w(1), w(2), \ldots, w(m))$. Such matrices can certainly have complex eigenvalues; see Exercise 1. There are constraints between the matrices A, B and W; see Exercise 4.

The material of this section appeared first in Whittle (1955) and was published in Whittle (1975).

Exercises and comments

1. Suppose that $x = (p, q)$, where p takes the values ± 1 and q takes values $1, 2, \ldots, m$ on the circle, in that $m + 1$ is to be identified with 1 and 0 with m, etc. Suppose that from (p, q) the possible transitions are to $(p, q + p)$ or $(-p, q)$ with respective probabilities α and $\beta = 1 - \alpha$. That is, as long as there is no transition in p the variable q cycles around the circle with velocity p.

Show that the eigenvalues θ of P are the roots of

$$\theta^2 - 2\theta\alpha \cos\left(\frac{2\pi r}{m}\right) + \alpha^2 - \beta^2 = 0 \qquad (r = 1, 2, \ldots, m)$$

and that there are always complex eigenvalues if $\beta < 1/2$ and m is large enough.
2. If $\beta = 0$ in the above example then p will not change ('momentum is conserved') and there is no communication between a state and its conjugate. The equilibrium distribution is now non-unique. However, the only π consistent with dynamic reversibility must satisfy (4), the only such is $\pi(x) = (2m)^{-1}$. That is, for dynamic reversibility we must assume the two ergodic classes ($p = \pm 1$) equally probable.
3. A more general example of the behaviour envisaged in Exercise 2 would be provided if B were zero in (10). Then, again, no state could communicate with its conjugate.
4. Note (cf. Exercise 5.3) that the elements of matrix (10) must satisfy

$$(A + B)w = (A + B)^{\mathsf{T}}w = w$$

where w is the column vector with elements $w(1), w(2), \ldots, w(m)$.
5. Suppose that $x(t)$ is a vector Gaussian process and that the effect of the conjugation operator C is to transform $x(t)$ to $Jx(t)$, where $J^2 = I$. Show that the condition for dynamic reversibility is then that

$$V(s) = JV(s)^{\mathsf{T}}J^{\mathsf{T}}$$

for appropriate s, where $V(s)$ is the autocovariance matrix defined in Exercise 2.3.

7. MARKOV REVERSIBILITY FOR A GENERAL STATE SPACE AND IN CONTINUOUS TIME

The formal analogue of the material of sections 4–6 for a general state space is clear. Stationarity, and so time-homogeneity, continues to be necessary for reversibility. The detailed balance equation (3.6) will continue to hold with $\pi(x)$ and $f(x, x')$ interpreted respectively as density and transition density relative to an understood measure μ on $\mathscr{X}$. One can write this relation as

$$P = \Pi^{-1}P^{\mathsf{T}}\Pi \tag{1}$$

where P is the one-step transition operator of section 2.2, P^T its adjoint and Π the operator with effect

$$(\Pi \phi)(x) = \pi(x)\phi(x) \tag{2}$$

(cf. Exercise 3.4). The representation (5.1) holds:

$$P = W^{-1}AW \tag{3}$$

with A a self-adjoint operator and $W \propto \Pi^{1/2}$. That is, $A^T = A$ and W has the action

$$(W\phi)(x) = w(x)\phi(x) \tag{4}$$

with $w(x)^2 \propto \pi(x)$.

The process is transition-reversible if $W = I$ in (3). It is dynamically reversible if (3) holds with the condition $A^T = A$ replaced by

$$JA^TJ = A \tag{5}$$

where J is the operator on functions of x induced by conjugation. That is, $(J\phi)(x) = \phi(\bar{x})$. It is then a consequence of (3), (5) that $w(\bar{x}) = w(x)$, i.e.

$$WJ = W \tag{6}$$

Formally again, all these relations remain valid in continuous time with the simple substitution of Λ for P, where Λ is the infinitesimal transition operator introduced in section 2.6. That is, continuous-time detailed balance and its consequences are necessary for reversibility in some almost-everywhere sense, and are sufficient under separability assumptions.

So, the detailed balance condition in the discrete-state case becomes

$$\pi(x)\lambda(x, x') = \pi(x')\lambda(x', x) \tag{7}$$

where λ is the transition intensity. The representation

$$\Lambda = W^{-1}AW$$

with $A^T = A$ implies that all the eigenvalues of the matrix Λ are real, and so confined to the negative real axis. The analogue of (7) for dynamic reversibility would be

$$\pi(x)\lambda(x, x') = \pi(\bar{x}')\lambda(\bar{x}', \bar{x}) \tag{8}$$

Exercises and comments

1. The birth and death process defined in Exercise 2.7.7 is again seen to be reversible, with relation (2.7.17) implying detailed balance.

2. Consider the continuous-time version of the Ehrenfest model of section 4, with the transition matrix for a single particle taken as

$$\begin{pmatrix} -\lambda_{12} & \lambda_{12} \\ \lambda_{21} & -\lambda_{21} \end{pmatrix}$$

Then the process $\{n(t)\}$ has possible transitions $n \to n+1$ and $n \to n-1$ with respective intensities $\lambda_{12}n$ and $\lambda_{21}(N-n)$. It is thus reversible simply in virtue of being a birth and death process.

3. *The Erlang distribution.* Suppose that a telephone exchange has c channels, of which x are in use at a given time. Suppose that transitions $x \to x+1$ occur with intensity v ($0 \leqslant x < c$) and that transitions $x \to x-1$ occur with intensity μx ($0 < x \leqslant c$). That is, new calls arrive in a Poisson stream of rate v, and are satisfied if there is a free channel, otherwise they are lost. Established calls have a termination intensity μ, independently of the states of other channels.

The process is a birth and death process, and so reversible, with equilibrium distribution

$$\pi(x) \propto \frac{(v/\mu)^x}{x!} \qquad (0 \leqslant x \leqslant c) \tag{9}$$

cf. (2.7.18). This is the *Erlang distribution*, a truncated Poisson distribution, first derived by Erlang in this context. The model may seem a rather simple and special case of a reversible process. However, it has both historical and technical importance. More than that, it can be seen as the simplest version of a whole class of interesting models (see the next exercise) and was perhaps the first model to be seen to demonstrate the phenomenon of *insensitivity* (see Chapter 12).

4. *Circuit-switched networks.* Consider a communication network consisting of a number of nodes between which calls are to be established. Rather than labelling the nodes (as will be natural for the queueing and processing networks of Part III) let us label possible point-to-point routes in the network by r. Suppose that demands for routes constitute independent Poisson streams, that for route r having rate v_r. (We are thus supposing that the demand is for a *route* rather than for a *connection*, which will be the case if operation is non-adaptive, and a given connection is always established along the same route.) A call for a route which cannot be satisfied is lost, and a call established along route r terminates with probability intensity μ_r, independently of the state of other connections. Let n_r be the number of calls established along route r, denote the vector of the n_r by n, and let F be the set of n which is physically feasible, so that calls whose acceptance would take n out of F are refused and lost.

Show that the process $\{n(t)\}$ is reversible with equilibrium distribution

$$\pi(n) \propto I(F, n) \prod_r \frac{(v_r/\mu_r)^{n_r}}{n_r!} \tag{10}$$

where $I(F, n)$ is the characteristic function of the set F. That is, the n_r are independent Poisson variables except in that they are subject to the constraint $n \in F$. Distribution (10) generalizes the Erlang distribution (9), and describes what we shall later refer to as *constrained Poisson statistics*.

The constraints which describe F may well be of the form

$$\sum_\alpha a_{\alpha r} n_r \leqslant M_\alpha \tag{11}$$

where α labels arcs on the network (i.e. one-sector routes), M_α is the number of lines the network provides on arc α and $a_{\alpha r}$ is the number of such lines called for by a connection via route r.

Distribution (10) was proved explicitly by Burman, Lehoczky and Lim (1984), although probably known more generally for some time. The term 'circuit-switching' refers to the fact that a whole route is called for rather than an arc at a time. In Exercises 5.8.2, 5.8.3 we shall consider the asymptotic version of (10), (11) for large networks, and in Chapter 12 shall discuss the insensitivity properties of the network.

5. The last exercise could be interpreted in terms of a system of processors, n_r being the number of jobs accepted by the rth processor. Restriction to the set F represents interference between processors as well as capacity limitations.

6. By first choosing an equilibrium distribution $\pi(x)$ and then choosing the transition intensities $\lambda(x, x')$ to be consistent with this and detailed balance one can easily construct a reversible process with known distribution. This inverts the usual idea of 'solution' but is not without art, as one must choose π and λ to be realistic for a given context.

As an example, consider a model of interspecific competition with state variable $\{n_j\}$, where n_j is the size of the jth population ($j = 1, 2, \ldots, d$). We suppose total birth and death rates for this population

$$\alpha_j(n) = v_j(n_j + 1)$$

$$\beta_j(n) = v_j n_j \exp\left(c_j + \sum_k h_{jk} n_k\right)$$

(i.e. these are the intensities for the transitions $n_j \to n_j + 1$ and $n_j \to n_j - 1$). The term $\sum_k h_{jk} n_k$ in the exponent represents the effect of competition. In particular, the term $\exp(h_{jk} n_k)$ is the factor by which the death rate of population j is increased by competition from population k. One readily finds the process reversible with equilibrium distribution

$$\pi(n) \propto \exp\left[-\frac{1}{2}\sum_j \sum_k h_{jk} n_j (n_k - \delta_{jk}) - \sum_j c_j n_j\right] \tag{12}$$

The rates α, β indeed constructed to be consistent with this π and detailed balance. They are not unrealistic, however, even though the competition effects are perhaps excessively strong. The distribution (12) is of course confined to the positive orthant $\mathbb{Z}_+^d$ so the quadratic form $\Sigma\Sigma h_{jk} n_j n_k$ need not and in general will not be positive definite. As one varies the nature of this quadratic form one can demonstrate a whole range of coexistence and exclusion effects. Roughly speaking, $\pi(n)$ can have local maxima in $\mathbb{Z}_+^d$ (or $\mathbb{R}_+^d$, if one neglects the integer character of n) in the interior or on the boundaries of this region. A local maximum at which n_j is zero for j in a set $\mathscr{G}$ corresponds to a regime (i.e. a locally stable point near which the random variable n will linger) in which the species of $\mathcal{G}$ have been effectively extinguished. See the next exercise.

In Chapter 8 we treat a competition model, also reversible, which is more realistic in that competition is given a mechanism: the depletion of critical resources.

7. Consider the competition example of Exercise 6 in the case $d = 2$. The contours of equiprobability of $\pi(n)$ are ellipses or hyperbolae according as h_{12}^2 is less than or greater than $h_{11} h_{22}$. That is, according as to whether interspecific competition is weaker or stronger than intraspecific competition. By examining the contours in the positive quadrant show that these cases correspond respectively to the cases when the two populations can coexist or exclude each other, statistically speaking. At least, this is true if the c_j are sufficiently negative, in other cases one or both of the populations may become inviable, and appear only at the token level maintained by immigration.

8. Quite a simple variation of the birth and death process can make all the difference between reversibility and irreversibility. Suppose that one has a birth and death process on the states $x = 0, 1, 2, \ldots, n$ except that states 0 and n are identified. The graph of possible transitions is then cyclic rather than linear. The detailed balance relation (7) then no longer necessarily holds, because there are now two paths between x and $x + 1$. Suppose, as an extreme case, that $\lambda(x, x - 1) = 0$ for all x, so that motion is unidirectional. The linear process is reversible, but with a degenerate equilibrium distribution concentrated entirely on n. The cyclic process is irreversible, with the distribution $\pi(x) \propto \lambda(x, x + 1)^{-1}$.

9. Consider the d-dimensional diffusion process defined in section 2.8, with the infinitesimal operator Λ specified by (2.8.7) and the measure μ on state space taken as Lebesgue measure. The detailed balance condition is then that $F\Lambda$ be self-adjoint. Show that, boundary conditions apart, this reduces to the condition

$$\frac{1}{2} \sum_k \frac{\partial}{\partial x_k} (c_{jk} \pi) = a_j \pi \qquad (j = 1, 2, \ldots, d) \tag{13}$$

However, the boundary conditions are part of the specification of Λ and do indeed matter, as Exercise 8 made plain. (F defined in Ex. 3.4).

10. One would expect the one-dimensional diffusion to be reversible, because it is the continuous equivalent of a birth and death process, at least if the boundary conditions have an appropriately local character. Show that this is so, in that (13) then has a solution for π which solves $\Lambda^\mathsf{T} \pi = 0$.

11. Consider the diffusion model with

$$\Lambda = -\sum_j \frac{\partial U(x)}{\partial x_j} \frac{\partial}{\partial x_j} + \frac{c}{2} \sum_j \left(\frac{\partial}{\partial x_j} \right)^2$$

This corresponds to dynamic equations

$$\dot{x}_j = -\frac{\partial U(x)}{\partial x_j} + \varepsilon_j \tag{14}$$

where the ε_j are independent white noise variables with zero mean and power density c. Show that this model is reversible with stationary density

$$\pi(x) \propto \exp\left[-\frac{2}{c} U(x) \right] \tag{15}$$

relative to Lebesgue measure.

One would interpret equation (14) as an attempt on the part of x to minimize a potential function $U(x)$, its descent of the potential gradient being disturbed, however, by white noise driving terms.

The quantity $U(x)$ is not quite the potential of Newtonian mechanics, in that we have incorporated no inertial effects, so that kinetic energy makes no appearance. We include these effects in the next exercise.

12. The condition of dynamic reversibility is that $F\Lambda = J(F\Lambda)^{\mathsf{T}} J$, with consequence $\pi(\bar{x}) = \pi(x)$, where J is the operator defined after (5).

Consider the model with $2d$-dimensional state variable $x = (p, q)$ $= (p_1, p_2, \ldots, p_d; q_1, q_2, \ldots, q_d)$, satisfying dynamic equations

$$\dot{q}_j = p_j$$

$$\dot{p}_j = -\frac{\partial U(q)}{\partial q_j} - \alpha p_j + \varepsilon_j \tag{16}$$

where the ε_j are again independent white noise variables with zero mean and power density c. That is

$$\Lambda = \sum_j \left(p_j \frac{\partial}{\partial q_j} - \left(\alpha + \frac{\partial U}{\partial q_j} \right) \frac{\partial}{\partial p_j} \right) + \frac{c}{2} \sum_j \left(\frac{\partial}{\partial p_j} \right)^2$$

Show that, if we define the state conjugate to $x = (p, q)$ as $\bar{x} = (-p, q)$, then the model is dynamically reversible with a stationary density

$$\pi(p, q) \propto \exp\left[-\frac{2\alpha}{c} H(p, q) \right] \tag{17}$$

relative to Lebesgue measure on $\mathbb{R}^{2d}$. Here

$$H(p, q) := U(q) + \frac{1}{2} \sum_j p_j^2$$

The interpretation is, of course, that the q_j are coordinates and the p_j corresponding momenta in a Newtonian mechanical model. The quantity $U(q)$ is the Newtonian potential at position q and $H(p, q)$ is the total energy. If the terms $-\alpha p_j + \varepsilon_j$ were absent in the equation for $\dot{p}_j$ then this would be a Hamiltonian system (see Appendix 2) and total energy H would be conserved. In fact, we have specified a Hamiltonian system with both random driving forces, ε_j, and velocity damping, $-\alpha p_j$. The effect of the random forcing is to increase H on average, and the effect of damping is to decrease it. The effects balance out to yield the equilibrium distribution (17).

13. *Onsager's relations.* Suppose that a statistical physical system is described macroscopically by a vector y which obeys the 'transport equation'

$$\dot{y} + My = 0$$

in a deterministic treatment. Suppose that in the stochastic treatment the relation becomes effectively

$$\dot{y} + My = \varepsilon \tag{18}$$

where ε is white noise of zero mean and power density matrix C. 'Effectively' is understood to mean that model (18) yields approximations to the autocovariance matrix $V(s) = \text{cov}(y(t), y(t-s))$ which are valid if $|s|$ is not too small. This autocovariance is then

$$V(s) = V(-s)^{\mathsf{T}} = e^{-sM} V(0) \qquad (s \geqslant 0) \tag{19}$$

Now, the fundamentals of statistical mechanics require that the model be reversible. (Actually, dynamically reversible, but we suppose that y contains only configurational rather than velocity variables.) Reversibility requires that the $V(s)$ be symmetric for all s (see Exercise 2.3) and we see from (19) that this will be so if and only if

$$MV = VM^{\mathsf{T}} \tag{20}$$

Here we have written $V(0)$ as V; this is a covariance matrix because $V^{\mathsf{T}} = V$. If one works with a normalized vector $V^{-1/2} y$ then M transforms to

$$L := V^{-1/2} M V^{1/2}$$

and (20) yields the *Onsager relations*

$$L = L^{\mathsf{T}} \tag{21}$$

These celebrated relations express symmetry of the matrix L of normalized transport coefficients, a necessary consequence of reversibility.

For an alternative expression of the Onsager relations, suppose that $C = RR^{\mathsf{T}}$ for invertible R. If we set $\varepsilon = R\tilde{\varepsilon}$, $y = R\tilde{y}$ then model (18) becomes

$$\dot{\tilde{y}} + \tilde{M}\tilde{y} = \tilde{\varepsilon}$$

where $\tilde{M} = R^{-1} MR$, $\tilde{C} = I$. Reversibility is then equivalent to symmetry of $\tilde{M} = R^{-1} MR$. As a particular case, it is equivalent to symmetry of $C^{-1/2} MC^{1/2}$.

8. REVERSIBLE MIGRATION PROCESSES

In Part III we shall be interested in what might be termed *migration processes*. These can be used to represent not merely the free wanderings of people or particles but also the movements of customers in a system of queues or the movements of jobs in a network of processors.

In its simplest version a migration process represents the movements of individuals of a single type between discrete sites labelled $j = 1, 2, \ldots, m$. Let $n_j(t)$ be the number of individuals at site j at time t and $n(t)$ the total *occupation number vector* $(n_1(t), n_2(t), \ldots, n_m(t))$. If one regards the individuals as indistinguishable then one will wish to work solely in terms of $n(t)$. If the migration model also regards individuals as indistinguishable, in that it is stochastically invariant under permutation of individuals, then it will allow one to do this, in that a Markov model implies Markov behaviour also of $\{n(t)\}$ (see section 3.7 and also the comments of Exercise 1).

If immigration into or emigration out of the system is forbidden then the system is *closed*, and

$$N := \sum_j n_j$$

conserves its value. Let e_j be the m-vector with a unit in the jth place and zeros elsewhere. The transition

$$n \to n - e_j + e_k \tag{1}$$

corresponds to migration of a single individual from site j to site k. In a *simple migration process* these are the only transitions permitted. We shall normally formulate the model in continuous time, when the assumption of simple migration is less artificial.

The model thus represents a flow of discrete units between discrete sites—we shall tend to use the neutral term 'units' rather than 'individuals'. One can introduce many variations: units may be identifiable, the system may be open, multiple transitions may be permitted, the space of sites may be continuous, there may be several types of units and units may change their type. Some of these will be considered in Chapters 9 and 10. We have already considered the simplest type of migration process in Exercise 2.7.5, that in which the units move independently.

We shall for consistency adopt the notation of Part III, in which the intensity of the transition $n \to n'$ is denoted by $\theta(n, n')$ rather than $\lambda(n, n')$. Lower case lambda is reserved to denote a factor of θ, the so-called routing intensity.

There is now some interest in determining conditions under which the model, a Markov process with possible transitions (1), is reversible. For one thing, the equilibrium behaviour of a reversible process is immediately determinable. For another, the Jackson networks of Chapter 9 do indeed turn out to be partially reversible (In fact, quasi-reversible in the sense used in section 1.)

Theorem 8.1 *Consider an irreducible Markov simple migration process in which the transition* (1) *has intensity* $\theta(n, n - e_j + e_k)$. *The process is reversible if and only if the intensity can be represented*

$$\theta(n, n - e_j + e_k) = \frac{\pi(n - e_j)}{\pi(n)} \chi_{jk}(n - e_j) \tag{2}$$

for some π *(identifiable with the equilibrium distribution) and* χ *satisfying*

$$\chi_{jk}(n) = \chi_{kj}(n) \tag{3}$$

identically.

Proof The detailed balance condition can be written in the symmetric form

$$\pi(n + e_j)\,\theta(n + e_j, n + e_k) = \pi(n + e_k)\,\theta(n + e_k, n + e_j)$$

That is, both sides of this equation are equal to a function of $n + e_j$ and $n + e_k$ or, equivalently, of n, j and k, which is unchanged when j and k are interchanged. Taking this function as $\pi(n)\,\chi_{jk}(n)$ we deduce the assertion of the theorem. ∎

The term $\pi(n - e_j)$ in representation (2) could be incorporated in the factor $\chi_{jk}(n - e_j)$, but the representation given is in many respects the natural one. Suppose we allow the process to be open in that we allow immigration $(n \to n + e_j)$ and emigration $(n \to n - e_j)$ transitions. The simplest way to do this formally is just to add a dummy site $j = 0$ to the system; this site represents the 'outside' or the 'environment'. For reversibility we shall still demand the same structure (2) for $j = 0, 1, \dots, m$. However, the site $j = 0$ will have a particular character: we shall assume all intensities functionally independent of n_0. This implies that the environment is effectively constant, whatever the situation at sites $1, 2, \dots, m$, and we shall indeed not include n_0 at all in the description n. It follows then from (2) that the immigration rate to site k is just

$$\theta(n, n + e_k) = \chi_{0k}(n) \tag{4}$$

That this should be independent of $\pi(n)$ seems right, in view of the fact that this rate will often be chosen independent of n, and equal to ν_k, say. The emigration rate would then, by (2), be of the form

$$\theta(n, n - e_j) = \frac{\nu_j \pi(n - e_j)}{\pi(n)} \tag{5}$$

but this indeed could not possibly be independent of n.

Kingman (1969) proposed a class of migration processes of the form

$$\theta(n, n - e_j + e_k) = \lambda_{jk}\,\phi_j(n_j)\,\psi_k(n_k) \tag{6}$$

The factor λ_{jk} represents what is often termed the *routing intensity* (for the route $j \to k$), while the other two factors give a dependence of rate on numbers in the sites being left and entered. It follows from Kolmogorov's criterion that this process will be reversible if and only if λ_{jk} is the transition rate for a reversible process; i.e. if there exist w_j such that

$$w_j \lambda_{jk} = w_k \lambda_{kj}. \tag{7}$$

One verifies then that

$$\pi(n) \propto \prod_j \left[\prod_{r=1}^{n_j} \frac{\psi_j(r-1)}{\phi_j(r)} \right] w_j^{n_j}$$

and that

$$\chi_{jk}(n) = w_j \lambda_{jk} \psi_j(n_j) \psi_k(n_k)$$

which is indeed symmetric in j, k.

Exercises and comments

1. That a model is invariant under permutation of individuals does not mean that it is not sensitive to ordering. For example, the site may be a queue at which there is a prescribed discipline (e.g. first come, first served, or last come, first served). The reduced description $\{n(t)\}$ will still be Markov, but will not in general be sufficient if one wishes to determine the statistics of the experience of an individual (e.g. the distribution of the time he spends in some part of the system).

More generally, the individuals at a site may form a *hierarchy*, which must rearrange itself when individuals join or leave that site (see section 9.6). Indistinguishability of individuals does not then mean that such hierarchies cannot form, but rather that an individual's place in them is determined by his timing rather than by his merits.

PART II

Abundance and transfer models

CHAPTER 5

Markov models and statistical mechanics: basic

1. INTRODUCTION

Any study of statistical equilibrium is bound to take the basic model of statistical mechanics as one of the most important special cases. It is in fact a model distinguished by its simplicity: with a few hypotheses on equipartition, communication and conservation one has enough to set up the theory. For this reason the model is of interest in contexts other than the familiar physical one. It is true that there is an extensive and subtle physical theory required to suggest and support the hypotheses, that the applications considered in specialist works are manifold and technical, and that thermodynamic considerations lead one on to a whole rich new trail of concepts. However, if one is prepared to accept the hypotheses and do no more than peer along the trails then one can develop the essential ideas with economy.

We construct then a Markov model which embodies the essential ideas in the barest fashion, but which permits simple and exact derivation of the core results of statistical mechanics. One begins by considering 'components' or 'sub-systems', which in isolation are described by Markov models in which energy is an invariant and whose ergodic classes correspond just to sets of states of the same energy. One then puts these component systems into communication by allowing 'energy quanta' to flow between them according to particular rules. These exchanges provide a special case of what we shall later refer to as 'weak coupling'. What one derives in particular is the celebrated *Gibbs distribution*: the probability distribution of a component system over its states when it is open to energy exchanges from without in this way. One can see the result as either an exact one when the components are immersed in a common sea of energy quanta (the 'open process', 'infinite heat bath' or 'canonical ensemble', see sections 4, 9 and 8.1) or as an asymptotic one when a number of sub-systems are restricted to exchanging energy with each other (the 'closed process' or 'micro-canonical ensemble'). The second approach is valuable in that it leads naturally to various extremal ideas and to the formulation of the concept of entropy. This interplay between open and closed versions of the process, one being in fact a conditioning of the other, will be a constant theme.

The programme may seem very much the stock one of any elementary introduction to statistical mechanics. However, the use of an appropriate Markov

107

model allows one to give a simple and exact treatment of matters which are often argued rather vaguely. The model gives the simplest example of effects which we shall encounter in applications far from statistical mechanics; see Part III. Finally, use of the modern theory of convex programming allows one to give a tidier account than usual of the various extremal characterizations (see sections 5, 6 and 8).

2. TRANSITION-REVERSIBLE COMPONENTS; ENERGY

Let us consider some kind of isolated physical system, which may be any assortment of particles, fields, springs or strings, provided only that it obeys the rules we shall now set down. In fact, we shall not call it a 'system', but rather a sub-system or *component*, as we shall soon be considering systems consisting of several such components in some kind of interaction.

Suppose the component is described by a Markov process with state variable x. We shall formulate the process in continuous time and shall, for simplicity, suppose the state-space discrete. Let us suppose the component *transition-reversible*, so that the transition intensity $\lambda(x, x')$ of the process describing it satisfies

$$\lambda(x, x') = \lambda(x', x) \tag{1}$$

(see sections 4.3, 4.6 and 4.7). This implies then that $\pi(x) =$ constant is a possible stationary distribution. However, all that condition (1) necessarily implies is that $\pi(x)$ should be constant within any given ergodic class of the process.

Suppose then that the ergodic classes of the process are labelled by a, and let $a(x)$ be the label of the class within which state x lies. The quantity $a(x)$ is then the maximal invariant of the motion, and we have

Theorem 2.1 *Suppose all states of the process positive recurrent. Then the transition-reversibility condition* (1) *implies that* $\pi(x)$ *is a stationary distribution if and only if it is of the form*

$$\pi(x) = h(a(x)) \tag{2}$$

for some h consistent only with the requirement that (2) *should constitute a distribution.*

Proof Expression (2) is certainly a stationary distribution, in that, in view of (1), it satisfies the detailed balance condition

$$\pi(x)\lambda(x, x') = \pi(x')\lambda(x', x) \tag{3}$$

and is normalizable in virtue of positive recurrence. It is also the most general form, since it assigns arbitrary probabilities to the ergodic classes (see Exercise 1), and the distribution within an ergodic class (i.e. conditional upon a) is unique. ∎

Suppose we identify $a(x)$ as proportional to the *energy* of the component in state x. Then energy is an invariant of the isolated component, and the theorem states that the equilibrium distribution is a function of energy alone. In particular, the distribution of x conditional on energy is uniform on the 'energy surface' (i.e. the ergodic class).

This identification is, of course, meaningless as yet. We have merely given a name 'energy' to the invariant. However, we shall add assumptions which strengthen the concept. We shall suppose that a takes the values $a = 0, 1, 2, \ldots$, to be identified with the number of 'energy quanta' in the component. When we put the component (hitherto isolated) into communication with other components we shall assume that the passage of quanta between components is possible, in a transition-reversible fashion. The total number of quanta is then conserved. However, for an individual component the emission and absorption transitions, $a \to a - 1$ and $a \to a + 1$, formerly impossible, are now permitted. With these additional hypotheses a structure emerges.

The motivation for the assumptions above are, of course, the properties of a classical mechanical system with its *Hamiltonian* structure (see Appendix 2). Such a system is dynamically transition-reversible, in that the set of solution-paths maps into itself under combined time- and momentum-reversal. This can be construed as simple transition-reversibility as far as the position variables are concerned, in a sense to be explained in Exercises 4.1 and 4.2. The system indeed has energy as an invariant (see (A2.4) and Exercise 2). State-space is continuous, and so the notion of a uniform distribution (and indeed of transition-reversibility in a statistical sense) is undefined until one has fixed an appropriate scaling in state-space. Such a scaling is provided by the choice of variables implied in a Hamiltonian formulation. 'Phase-space' is then state-space, so scaled that a uniform distribution is indeed a distribution with uniform density relative to Lebesgue measure; this is the import of the Liouville theorem, expressed by equation (A2.8).

Exercises and comments

1. The probability of a particular value of a on the basis of (2) is, of course, not $h(a)$ but $\omega(a)h(a)$, where $\omega(a)$ is the number of state values in the ergodic class a. (Alternatively, the number of states of the component corresponding to an energy level of a quanta).

2. The Hamiltonian equations allow invariants other than energy, e.g. total linear and angular momenta. One may ask why the values of these should not be included in a, the label of the ergodic classes. The conventional reply is that there are additional mechanisms (such as reflection of molecules of the system at the walls of the containing vessel) which conserve the energy of the contained system but change momenta. These mechanisms are then considered to leave energy as the only invariant; see the discussion at the end of Appendix 2.

3. Analyses of the logical progression from the purely deterministic model of Appendix 2 to statistical models of the type presented here has generated an enormous literature, of which ergodic theory is only a part. An awkward point is that the deterministic process cannot in general reach an equilibrium. This is because the values of $x(t)$ for different $x(0)$ remain distinct for all t; the detail and the information of the initial configuration is conserved, if in an increasingly convoluted and fine-grained form. The answer is to neglect fine detail in phase space: the average of $\phi(x(t))$ over a time-interval of length T will converge with increasing T to a limit showing a degenerate or zero dependence upon initial conditions if the function ϕ is sufficiently smooth. The justification is that a viewpoint taken for mathematical order generally turns out to have a physical basis.

3. WEAKLY COUPLED COMPONENTS

Let us formalize the assumptions of the previous section.

(S1) Components are statistically identical, in that the behaviour of each in isolation is governed by the same Markov process.

(S2) This Markov process is transition-reversible, all its states are positive recurrent and its ergodic states are considered to correspond to the number $a = 0, 1, 2, \ldots$ of energy quanta held in the component.

(S3) When N such components are conjoined into a system then, in addition to the *internal* transitions possible for an isolated component, *external* transitions are permitted which correspond to the passage of a quantum between two of the components. This transfer is also transition-reversible.

(S4) The total number of quanta in the system thus remains invariant. It is considered to be the only invariant of the system.

Formalizing somewhat more, then, one supposes that the system is described by a transition-reversible Markov process with state variable $\mathbf{x} = (x_1, x_2, \ldots, x_N)$, where x_j is the 'state' of the jth component. The additive functional $\sum_j a(x_j)$ is conserved, where $a(x_j)$ is the number of energy quanta in component j, and communication between states is rich enough that this is the only invariant of the system. More specifically, in an *internal* transition only one component changes state, $x_j \to x'_j$, say, but the transition must then be such that $a(x'_j) = a(x_j)$. In an external transition just two components change, $(x_j, x_k) \to (x'_j, x'_k)$, say, and this change is considered to imply passage of a quantum in that

$$a(x'_j) = a(x_j) - 1$$
$$a(x'_k) = a(x_k) + 1$$

The total number of quanta

$$M = \sum_j a(x_j) \tag{1}$$

is thus conserved.

We shall use x to denote the state of the system and x to denote the 'state' of a component picked out for attention. We shall find it useful to define the generating function

$$\phi(w) = \sum_x w^{a(x)} = \sum_a \omega(a)w^a \tag{2}$$

where $\omega(a)$ is the number of states x of a component consistent with $a(x) = a$. Positive recurrence requires that $\omega(a) < \infty$ for given a. We shall strengthen this to the requirement that the series (2) be convergent for some positive w, i.e. that it have a non-zero radius of convergence.

Let the number of states of the system of N components consistent with a total of M energy quanta be denoted $d(M, N)$.

Lemma 3.1 *The number $d(M, N)$ is equal to the coefficient of w^M in the power series expansion of $\phi(w)^N$.*

Proof Let a_j denote the number of quanta in component j. Then $d(M, N)$ equals the sum of $\omega(a_1)\omega(a_2) \ldots \omega(a_N)$ over all $a_1, a_2, \ldots, a_N$ consistent with $\sum_j a_j = M$. This is exactly the characterization of the lemma. ∎

Theorem 3.1 *Consider a system of N components obeying conditions (S1)–(S4). Then*

(i) *The equilibrium distribution of x for a prescribed value of M is uniform over the $d(M, N)$ values of x consistent with (1).*

(ii) *This distribution is identical with that for $(x_1, x_2, \ldots, x_N)$ conditional on the value M of $\sum_j a(x_j)$ if the unconditioned distribution prescribes the x_j as independent, each with distribution*

$$\pi(x) = \frac{w^x}{\phi(w)} \tag{3}$$

Here w may have any positive value consistent with $\phi(w) < \infty$.

(iii) *The actual state distribution for an individual component is*

$$\pi(x) = \frac{d(M - a(x), N - 1)}{d(M, N)} \tag{4}$$

Proof Assertion (i) follows simply from the fact that the Markov process governing the combined system is positive recurrent, transition-reversible and has

M as its only invariant. The unconditioned joint distribution suggested in (ii) implies that

$$\pi(\mathbf{x}) \propto \prod_j w^{a(x_j)} = w^M \tag{5}$$

whence uniformity of the conditioned distribution follows. Expression (4) gives the proportion of states x consistent with a given M which are also consistent with a prescribed x for a prescribed component. ■

A component in isolation would of course have an equilibrium distribution uniform on the set of x consistent with the prescribed a and zero elsewhere. The interest of formula (4) is that it shows how this distribution is modified when the component is put into communication with an environment consisting of $N - 1$ of its fellows. What we shall soon see is that distribution (4) converges to the simpler form (3), the celebrated *Gibbs distribution*, as the system becomes large in that M and N increase indefinitely in fixed ratio.

The term 'weak coupling' describes features of this model which will manifest themselves again in the more general models of Part III. The only interaction between the components is the transfer of quanta postulated in (S3). This form of interaction leads to the quasi-independence in equilibrium expressed in assertion (ii) of the theorem.

Exercises and comments

1. One has analogues of all the assertions of Theorem 3.1 if one simply postulates that the process $\{\mathbf{x}(t)\}$ is positive recurrent, transition-reversible and has as only invariant the additive functional $\sum_j a_j(x_j)$. That is, it is not necessary to postulate that components are statistically identical, or that communication is by transfer of single quanta only. However, the single quantum transfer makes for a simple model, and we assume identity of components in order to ease the statement of some of the asymptotic arguments which are to come.

2. For a given value of M it is immaterial what the value of w may be in (5). However, we postulate $\phi(w) < \infty$ in order that the distributions (2) may be well defined.

4. THE HEAT BATH

Suppose the model of section 3 modified in that single quanta are also allowed to enter and leave the system. We shall assume that there is a parameter w such that

$$\lambda(\mathbf{x}, \mathbf{x}') = w\lambda(\mathbf{x}', \mathbf{x}) \tag{1}$$

if $\mathbf{x} \to \mathbf{x}'$ is a reaction in which the number of quanta in the system is increased by one. Relation (1) seems to represent a departure from transition-reversibility, but in section 7 we shall in fact see it as a consequence of transition-reversibility at a higher level of specification.

In modifying the assumptions (S1)–(S4) of section 3 we may as well also remove the constraint that the components should be identical, which is unnecessary for many purposes. Let us therefore consider the revised set of assumptions:

(H1) Separate components are defined, in that the system state space $\mathscr{X}$ is the direct product of individual component state spaces $\mathscr{X}_j$ ($j = 1, 2, \ldots, N$).

(H2) $\{\underline{x}(t)\}$ is a Markov process; a variable defined on the process is $M(\mathbf{x}) = \sum_j a_j(x_j)$, where $a_j(x_j)$ is the number of energy quanta held in component j.

(H3) Transitions in which $M(\mathbf{x})$ is conserved are transition-reversible; transitions $\mathbf{x} \to \mathbf{x}'$ for which $M(\mathbf{x}') - M(\mathbf{x}) = 1$ obey the balance condition (1).

(H4) There are no other types of transition, but enough transitions are possible that all states of $\mathscr{X}$ intercommunicate.

Assumption (H3) leaves the mechanism of quantum-transfer open. It may be that quanta are transferred directly between components as before, with the additional option that a component can also acquire a quantum from outside the system, and similarly lose it. Alternatively, one can imagine that the components are immersed in a common sea of free quanta and that, rather than exchanging quanta directly, they simply emit quanta into or absorb quanta from the sea. This quantum sea constitutes the 'heat bath', an idea we shall develop in sections 7 and 6.1.

Theorem 4.1 *Consider the open system specified by conditions (H1)–(H4). It has the unique equilibrium distribution*

$$\pi(\mathbf{x}) \propto \prod_j w^{a_j(x_j)} \tag{2}$$

provided this expression is summable over $\mathscr{X}$. That is, if $\phi_j(w) < \infty$ for all j then components are independent in equilibrium with

$$\pi(x_j) = \frac{w^{a_j(x_j)}}{\phi_j(w)} \qquad (j = 1, 2, \ldots, N) \tag{3}$$

Proof By assumption (H3) expression (2) satisfies the detailed balance relations for the process. It thus provides a stationary distribution if summable and is then, in virtue of the irreducibility assumption (H4), the unique stationary and equilibrium distribution. ■

Here ϕ_j is the possibly j-dependent version of the generating function ϕ:

$$\phi_j(w) = \sum_{x \in \mathscr{X}_j} w^{a_j(x)}$$

The theorem tells us that the components are now independent, each following the Gibbs distribution (3). In the closed case they also followed this joint distribution, but subject to the conditioning $M(x) = M$, see Theorem 3.1 (ii). With the removal of the conditioning the components become independent, although with distributions depending upon a common parameter w, to be interpreted as something like the density of the quantum sea.

If the energy of a single quantum is taken as ε then the Gibbs distributions of system and component, (2) and (3), would normally be written

$$\pi(\mathbf{x}) \propto e^{-\beta U(\mathbf{x})} \tag{2'}$$

$$\pi(x_j) \propto e^{-\beta U_j(x_j)} \tag{3'}$$

where

$$U_j(x_j) = \varepsilon a_j(x_j) \tag{4}$$

is the energy of component j and

$$U(\mathbf{x}) = \sum_j U_j(x_j) \tag{5}$$

is the energy of the whole system. The parameter β is related to w by

$$w = e^{-\beta \varepsilon} \tag{6}$$

Thermodynamic considerations develop the idea of temperature and show that β is in fact inversely proportional to the temperature of the heat bath. The higher the temperature the more likely system and components are to be in high-energy states.

If energy transfers are considered to take place via the heat bath then one can understand why the components should be independent in equilibrium. They are coupled in so far as their distributions depend upon the common parameter w, which we shall indeed see (section 7) as proportional to the spatial density of free quanta. However, the components are statistically decoupled by the heat bath and so, given w, are independent. It is also true, however, that even direct exchanges between components produce much the same effect if the system is large enough, a point we shall pursue in the next two sections.

Suppose for simplicity that components are identical. The average number of quanta held by a component is then

$$E(a) = \frac{\sum_x a(x) w^{a(x)}}{\sum_x w^{a(x)}} = \frac{w\phi'(w)}{\phi(w)} \tag{7}$$

One might think that an effective value of w for the closed system would be that determined by

$$\frac{w\phi'(w)}{\phi(w)} = \frac{M}{N} \tag{8}$$

and this we shall find to be true.

Exercises and comments

1. To establish a complete correspondence with classical mechanical models we should postulate dynamic transition-reversibility rather than simple transition-reversibility. Replace then $\lambda(\mathbf{x}', \mathbf{x})$ by $\lambda(\bar{\mathbf{x}}', \bar{\mathbf{x}})$ in the right-hand members of equations (2.1) and (1). Note then that the Gibbs distribution (2), (2′) still holds as the solution of the modified detailed balance equation (4.7.8). We know (Th. 4.6.2) that $\pi(\bar{\mathbf{x}}) = \pi(\mathbf{x})$, which implies that $U(\bar{\mathbf{x}}) = U(\mathbf{x})$, even though it is not clear that x and $\bar{x}$ must be in the same ergodic class for the isolated component.

2. For the classic Hamiltonian system $x = (p, q)$ and

$$U(p, q) = \tfrac{1}{2}|p|^2 + V(q)$$

where p and q are the momentum and position vectors respectively and the two terms in U are recognizable as the kinetic and potential energies. The Gibbs distribution is

$$\pi(p, q) \propto e^{-\beta U(p,q)}$$

this being a density relative to Lebesgue measure in $\mathbb{R}^{2d}$, if p and q have dimension d. Since both density and measure factorize into p-dependent and q-dependent terms we can say that p and q are statistically independent in equilibrium and that q has equilibrium density

$$\pi(q) \propto e^{-\beta V(q)}$$

relative to Lebesgue measure in $\mathbb{R}^d$. Since q is unaffected by the conjugation operation $(p, q) \to (-p, q)$ then the position variable q has the same equilibrium statistics as it would have had if it had followed a simply (rather than dynamically) reversible process. This is an example of the phenomenon of *insensitivity* which we shall encounter repeatedly: that certain random variables have the same equilibrium statistics for models of differing levels of refinement (see section 7 and Chapter 12).

3. It is a feature of the kind of models we shall be examining that physically significant variables such as w or β will both adopt a numerical value in given cases and also be regarded as the marker variable of a generating function. The *partition function*

$$Z(\beta) := \sum_{\mathbf{x}} e^{-\beta U(\mathbf{x})},$$

regarded as a function of β, is important as summarizing many of the bulk-statistical and thermodynamic properties of the system. It can be regarded as an unnormalized m.g.f. of total energy U, in that the actual m.g.f. is

$$E(e^{\theta U(\mathbf{x})}) = Z(\beta - \theta)/Z(\beta)$$

if β is the actual numerical value occurring in the Gibbs distribution. For the model of this section

$$Z(\beta) = \phi(w)^N = \phi(e^{-\beta\varepsilon})^N$$

if the components are supposed statistically identical. Equivalently, the p.g.f. of M, the total number of quanta, is

$$E(z^M) = [\phi(wz)/\phi(w)]^N$$

5. PASSAGE TO THE THERMODYNAMIC LIMIT; ENTROPY AND EXTREMAL CONCEPTS

Let us return again to the model of section 3: a closed system with statistically identical components. It seems intuitively clear that for a large system the effect of the conditioning $M(\mathbf{x}) = M$ on any fixed set of components must be slight, and that the statistics of these components must be very much what they were for the open system. That is, that they independently follow the Gibbs distribution (3.3) with w chosen to make the total system energy correct in expectation, by (4.8). We shall now demonstrate these facts in a manner which may seem laborious, but which has the merit of revealing two important extremal aspects of the statistics.

Let $n(x)$ be the number of components in state x. Here x takes values in $\mathscr{X}$, the common state space of all N components, and $n(x)/N$ constitutes a random distribution over $\mathscr{X}$. Let us use n to denote the array of occupation numbers $\{n(x); x \in \mathscr{X}\}$. Then we know the statistics of n from Theorem 3.1.

Theorem 5.1 *The equilibrium distribution of n for the closed system of section 3 is*

$$\pi(n) \propto Q(n) := \prod_x \frac{1}{n(x)!} \tag{1}$$

for n in $\mathbb{Z}_+^{\mathscr{X}}$ satisfying

$$\sum_x n(x) = N \tag{2}$$

$$\sum_x a(x)n(x) = M \tag{3}$$

otherwise it is zero.

In words, n follows a multinomial distribution with uniform cell probabilities constrained by (3). The assertion follows simply from the fact that $\pi(\mathbf{x})$ is uniform on the set of $\mathbf{x}$ consistent with prescribed M, and that there are $N! \Big/ \Big[\prod_x n(x)! \Big]$ such states consistent with a given n.

One can write relation (3) equivalently as

$$\sum_x a(x)n(x) = N\bar{a} \tag{4}$$

where

$$\bar{a} = M/N \tag{5}$$

is the average number of quanta per component. Note that

$$E(n(x)) = N\pi(x) \tag{6}$$

where $\pi(x)$ is the probability that a given component is in state x, given by (3.4).

Let us suppose now that the system becomes indefinitely large, in that M and N both go to infinity in such a way that $\bar{a}$ maintains a constant prescribed value (or at least as nearly as is achievable with integral M, N). This is termed passage to the *thermodynamic limit*. The effect of this passage is to remove the constraining effects of a finite system (for example, that the energy of a given component could never exceed $M\varepsilon$).

One imagines that in the thermodynamic limit $\pi(x)$ will converge to some limit value, $\pi_\infty(x)$. Indeed it does, and $\pi_\infty(x)$ is just the Gibbs distribution, as we shall see in the next section by direct evaluation of expression (3.4). However, there is another method for evaluating $\pi_\infty(x)$, which relies upon the presumption that the random proportion $p(x) = n(x)/N$ will itself converge to $\pi_\infty(x)$ in some sufficiently strong stochastic sense. This is to calculate the *most probable* value of the empirical distribution $p = \{p(x)\}$. We shall simply carry out this evaluation now and confirm the details of stochastic convergence, etc. in section 10. For the moment, the interest is that there are two ways of calculating $\pi_\infty(x)$ and that one of these is associated with a natural extremal characterization (the characterization as the most probable value). This is a feature of a fully stochastic formulation which we shall exploit a number of times: that if there is a 'dominant' configuration then it is also very often the most probable one, and so endowed immediately with this extremal characterization.

In terms of the empirical distribution $p(x)$ the constraints (2), (4) become

$$\sum_x p(x) = 1 \tag{7}$$

$$\sum_x a(x)p(x) = \bar{a} \tag{8}$$

If we assume that the most probable values of the $n(x)$ are all of order N then one finds from expression (1) and an appeal to Stirling's approximation to the factorial that

$$-\frac{1}{N}\log Q(n) = S + \text{constant} + O(N^{-1}) \tag{9}$$

where

$$S := -\sum_x p(x)\log p(x) \tag{10}$$

is the *entropy* of the distribution p on $\mathscr{X}$. It is plausible, then, that the problem of finding n in $\mathbb{Z}_+^{\mathscr{X}}$ to maximize expression (1) subject to (2), (4) can be replaced in the thermodynamic limit by the problem of finding a distribution p on $\mathscr{X}$ with maximal entropy, subject to the average energy constraint (8). That is, that discreteness can legitimately be neglected in the limit.

Theorem 5.2 *Suppose that, for the given value of $\bar{a}$, the expression in θ*

$$L(\theta) = \sum_x e^{-\theta_1 - \theta_2 a(x)} + \theta_1 - 1 + \theta_2 \bar{a} \qquad (11)$$

possesses a stationary minimum. Then

(i) *The distribution over states π for an individual component which is most probable in the thermodynamic limit can be characterized as the distribution p maximizing entropy (10) subject to the specification (8) of average quantum abundance.*

(ii) *This distribution is the Gibbs distribution*

$$\pi(x) = e^{-\theta_1 - \theta_2 a(x)} \qquad (12)$$

where the parameters θ are to be chosen so that π satisfies the constraints imposed upon p in (7), (8).

(iii) *The parameter values θ can alternatively be characterized as those values minimizing $L(\theta)$, and the minimised value of $L(\theta)$ equals the maximal entropy*

$$S = -\sum_x \pi(x) \log \pi(x).$$

Proof Suppose initially that $\mathscr{X}$ is finite. Let us assume as plausible that any $n(x)$ solving the constrained maximization problem will indeed be of order N; the easy way to see this is indicated in section 10. Assertion (i) then follows from the discussion before the theorem.

Suppose the problem posed is *feasible* in that there exist distributions p satisfying (7), (8). (Essentially, that $\bar{a}$ lies between the minimal and maximal values of $a(x)$ which are possible.) The constrained maximization problem can then be solved by free maximization of the Lagrangian function

$$L(p, \theta) = -\sum_x p(x) \log p(x) + (\theta_1 - 1)\left(1 - \sum_x p(x)\right) + \theta_2\left(\bar{a} - \sum_x a(x)p(x)\right) \qquad (13)$$

The maximizing value is just the Gibbs distribution (12). There is the further constraint $p(x) \geqslant 0$, but solution (12) observes this without our requiring it. Treatment of the constrained maximization problem by Lagrangian methods is legitimate, since constraints (7), (8) are linear in p and the entropy (10) is concave (see Appendix 2).

The parameters θ can be determined by requiring that π as given by (12) satisfies the constraints (7), (8). Alternatively, they can be determined by solution of the 'dual problem': as the values of θ minimizing $L(\theta) := \max_p L(p, \theta)$, and the minimized value of $L(\theta)$ equals the constrained maximal value of S (see Appendix

2). $L(\theta)$ is easily found to have the evaluation (11), so that assertion (iii) is a statement of this dual characterization.

In the case of finite $\mathscr{X}$ the opening condition of the theorem is equivalent to the assumption that the conditions (7), (8) are feasible, as is again proved in Appendix 2. In the case of infinite $\mathscr{X}$ it implies also that the problem may be approximated arbitrarily closely by one with finite $\mathscr{X}$. The basis of the argument for this is fairly direct; we shall not give the details. ∎

Some of these calculations gain a more familiar form if we change dual variables to $v = e^{-\theta_1}$, $w = e^{-\theta_2}$. Assertion (iii) of the theorem then implies that v, w should adopt the minimizing values defined by

$$\hat{S} = \min_{v,w \geqslant 0} [v\phi(w) - 1 - \log v - \bar{a} \log w] \qquad (14)$$

Either by carrying out the minimization with respect to v or else by applying the normalization condition $\Sigma \pi(x) = 1$ we find that

$$v = \phi(w)^{-1}.$$

(cf. (3.3)). One is then left with the relation

$$\hat{S} = \min_{w \geqslant 0} [\log \phi(w) - \bar{a} \log w] \qquad (15)$$

implying an evaluation of both S and w. The minimizing value of w is determined by

$$\frac{w\phi'(w)}{\phi(w)} = \bar{a} \qquad (16)$$

(cf. (4.8)).

One may ask whether the dual problem of assertion (iii) has a physical interpretation; we shall see in the next two sections that it has two very clear interpretations.

Exercises and comments

1. The condition of Theorem 5.2 is equivalent to the condition that a minimizing w should exist for expression (15), and that $\phi(w)$ should converge in a neighbourhood of this value. Show that for finite $\mathscr{X}$ this is equivalent to the requirement that constraint (8) should be feasible, i.e. that $\bar{a}$ should lie between the minimal and maximal possible values of $a(x)$.

2. Note an implication of the condition for infinite $\mathscr{X}$: that $\phi(w)$ should have a non-zero radius of convergence. If this is not the case then, as the system becomes larger, some few components take up more and more of the total available energy.

3. Even if $\phi(w)$ has non-zero radius of convergence, it is possible that the condition of Theorem 5.2 is broken as $\bar{a}$ increases through a critical value. This transition corresponds to a change of state, from a regime in which all the components share the energy to one in which the energy becomes increasingly concentrated in a few components. We examine such transitions in a preliminary kind of way in section 6.3 and extensively in Part IV.

4. We have taken the Gibbs distribution in the form (12) for consistency with later generalizations, but in this context it would more usually be written

$$\pi(x) = e^{-\alpha - \beta U(x)}$$

where $U(x) = a(x)\varepsilon$ is the energy of the component in state x, and so $\alpha = \theta_1$, $\beta = \theta_2/\varepsilon$. The characterization of assertion (iii) then becomes

$$\hat{S} = \min_{\alpha, \beta} \left[\sum_x e^{-\alpha - \beta U(x)} + \alpha - 1 + \beta \bar{U} \right]$$

$$= \min_{\beta} \left[\log\left(\sum_x e^{-\beta U(x)} \right) + \beta \bar{U} \right] \tag{17}$$

where $\bar{U} = \bar{a}\varepsilon$ is the average energy of the component.

5. In thermodynamic studies one would regard the values of α and β as characterizing the system, in that they parametrize its equilibrium statistics completely. One can regard thermodynamics itself as a 'sensitivity analysis'; a study of how the parameters of equilibrium statistics (α and β in this case) vary as the conditions of the component are varied, e.g. by change in the ambient energy as measured by $\bar{U}$, by physical compression of the system or by application of a magnetic field. Otherwise expressed, one might describe equilibrium thermo-dynamics as a study of the sensitivity of the parameters of the statistics to the parameters of the model.

6. Such thermodynamic considerations lead to the identifications $\alpha = -F/T$, $\beta = 1/T$, where T is temperature (on an appropriate scale) and F is free energy. Note then that relations (17) imply the identity

$$S = \min_T \left[\frac{U - F}{T} \right]$$

where we have written $\bar{U}$ and $\hat{S}$ simply as U and S, and average energy U and free energy F are both regarded as functions of temperature T.

7. In the same notation, note from (17) the interpretation $\beta = \partial S/\partial U$, if S is regarded as a function of U.

6. THE DIRECT EVALUATION OF $\pi_\infty(X)$, THE DUAL PROBLEM AS A SADDLE-POINT EVALUATION

In expression (3.4) we have an evaluation of $\pi(x)$, the probability that a prescribed component of the closed process is in state x. The question is now whether this can be given a more transparent form in the thermodynamic limit.

We can write evaluation (3.4) as

$$\pi(x) = \frac{\mathscr{I}(w^{a(x)}/\phi(w))}{\mathscr{I}(1)} \tag{1}$$

where

$$\mathscr{I}(\psi(w)) := \frac{1}{2\pi i}\oint \psi(w)\phi(w)^N w^{-M-1}dw \tag{2}$$

The integration contour in (2) is a tight contour of the origin in the complex w-plane. More specifically, the contour can be taken as $|w| = \delta$, where δ is small enough that the only singularity of the integrand enclosed by the contour is that at the origin, due to the term w^{-M-1}. The integral (2) is then indeed equal to the coefficient of w^M in the power series expansion of $\psi(w)\phi(w)^N$.

Theorem 6.1 *Expression (1) for the distribution of component state has the evaluation*

$$\pi_\infty(w) = \frac{\bar{w}^{a(x)}}{\phi(\bar{w})} \tag{3}$$

in the thermodynamic limit, where $\bar{w}$ is the value of real positive w minimizing $\phi(w)w^{-a}$. Expression (3) is just the Gibbs distribution (5.12) with the substitutions $\theta_1 = \log\phi(\bar{w})$, $\theta_2 = -\log\bar{w}$.

Proof One can write (2) rather as

$$\mathscr{I}(\psi(w)) = \frac{1}{2\pi i}\oint \psi(w)[\phi(w)w^{-\bar{a}}]^N w^{-1}dw \tag{4}$$

and we wish to evaluate such integrals for large N. For N large the dominant contribution to the integral will come from the dominant saddle-point w of the expression $\phi(w)w^{-\bar{a}}$ raised to power N in the integrand, and ratio (1) will then indeed have the asymptotic evaluation (3). The saddle-point $\bar{w}$ will be real and positive with the maximizing and minimizing axes of the saddle normal to and coincident with the positive real axis respectively (see Exercise 1), and so has the character asserted. ∎

We thus indeed find the Gibbs form (3) for π_∞, with the saddle-point evaluation corresponding exactly to the reduced form (5.15) of the dual extremal problem stated in Theorem 5.2 (iii). We obtain the direct parallel with the unreduced form of the problem by considering the alternative expression for $d(M, N)$:

$$\frac{d(M,N)}{N!} = C_{MN}(e^{v\phi(w)}) \tag{5}$$

Here $C_{MN}(\cdot)$ is short for 'the coefficient of $w^M v^N$ in the power series expansion of $(\cdot)$'.

The alternative expression (5) is not just a piece of formalism, but corresponds to a new relaxation of concepts: the idea that N as well as M might be a random variable. That is, that one might consider a process open, not merely to the passage of quanta in and out, but also to the passage of components. This is the 'grand canonical ensemble' of classical statistical mechanics. It constitutes the natural background process, the closed process to be viewed as a constrained or conditioned version of it.

Specifically, suppose that

$$\Pi_{MN}(z) = E\left[\prod_x z(x)^{n(x)}\right] \tag{6}$$

is the joint p.g.f. of the totals $n(x)$ for the closed process with prescribed M, N. It follows then from (5.1)–(5.3) that we may set

$$\Pi_{MN}(z) \propto C_{MN}\left\{\exp\left[V\prod_x z(x)vw^{a(x)}\right]\right\} \tag{7}$$

the proportionality factor being a function of M, N alone. We have included a parameter V whose value will be irrelevant, since its only effect is to make a contribution V^N to the proportionality factor. However, when we come to consider the fully open process in the next section it will have an interpretation: the 'volume' of the spatial region within which the N components and N quanta are located. In fact, we shall regard the thermodynamic limit as the simultaneous passage of M, N and V to infinity in such a way that the component and quantum densities

$$\rho_C = N/V$$
$$\rho_Q = M/V \tag{8}$$

are held fixed.

Note the interpretation of (7): that the $n(x)$ are regarded as independent Poisson variables, constrained by (5.2), (5.3). This is a view again to be developed strongly in the next and later sections. One may say that the occurrence of the term $vw^{a(x)}$ in the exponent of (7) constitutes what one *means* by the Gibbs distribution, a pure consequence of the imposition of additive invariants on to a uniform distribution.

We deduce from expression (7) that

$$E(n(x)) = V\frac{C_{MN}(vw^{a(x)}e^{V v\phi(w)})}{C_{MN}(e^{V v\phi(w)})} = \frac{\mathscr{I}^*(vw^{a(x)})}{\mathscr{I}^*(1)} \tag{9}$$

Here $\mathscr{I}^*$ is the extended version of $\mathscr{I}$:

$$\mathscr{I}^*[\psi(v,w)] = \left(\frac{1}{2\pi i}\right)^2\oint\oint\psi(v,w)e^{V v\phi(w)}w^{-M-1}v^{-N-1}dwdv \tag{10}$$

in which w, v tightly circle the origin in their respective complex planes.

Theorem 6.2 *In the thermodynamic limit $E(n(x)/V)$ has the evaluation*

$$E(n(x)/V) = \bar{v}\bar{w}^{a(x)} = e^{-\theta_1 - \theta_2 a(x)} \tag{11}$$

where $(\bar{v}, \bar{w})$ is the saddle-point of the expression

$$v\phi(w) - \rho_C \log v - \rho_Q \log w \tag{12}$$

and minimizes this expression for real positive v, w. Alternatively, one can say that θ minimizes the expression

$$\sum_x e^{-\theta_1 - \theta_2 a(x)} + \rho_C \theta_1 + \rho_Q \theta_2 \tag{13}$$

The proof is just a two-variable version of that of Theorem 6.1. In (11) we recognize the Gibbs distribution again, now representing expected concentration rather than expected proportion. The two become the same if we set $\rho_C = 1$, in which case expression (13) reduces to the θ-dependent part of (5.12). That is, the dual problem of Theorem 5.2 (iii) characterizes exactly the location of the saddle-point $(\bar{v}, \bar{w})$. ■

Exercises and comments

1. The proof of Theorem 6.1 makes a number of assumptions about the location of the saddle-point of $H(w) := \phi(w)w^{-\bar{a}}$, which we now confirm. Since the coefficients in the expansion of $\phi(w)$ are non-negative, $|H(w)|$ will be maximal on any circle $|w| = \delta$ at real positive w. It will be maximal only at such a value if we have chosen the maximal quantum, i.e. if the values of the $a(x)$ are not restricted to multiples of some integer greater than unity. The dominant saddle-point thus lies on the real positive axis, and the reality of H for real w implies that the axes of the saddle will lie along and normal to this axis.

7. CONSTRAINED POISSON STATISTICS (BOLTZMANN STATISTICS)

In Chapter 7 we shall consider the kinetics of chemical reactions, and so construct a model for the combination and dissociation of molecules. We shall then have to incorporate at least two new features. Firstly, we must recognize the existence of several conserved quantities in a closed system: the total abundances of the various types of atom from which the molecules are formed (the 'elements') as well as of energy.

Secondly, the number of components itself may no longer be one of the conserved quantities. If components are to be identified with molecules, which combine and split up under possible transitions, then the total number will certainly vary. Indeed, in this context we are considerably relaxing the notion of what we mean by components. Before we considered them as fixed and permanently identifiable objects which interacted by the transfer of some kind of unit—energy quanta in sections 1–6. This is a picture to which we shall return for the processing networks of Part III. However, for the moment we are thinking of a component as a molecule, identifiable when it exists, but which may just have been formed, and may soon cease to exist, in that it either divides or combines with others. The description 'component' must then now include also free quanta and atoms, which are simply degenerate molecules.

We shall adopt the language of molecules and atoms in Chapter 7, but shall for the moment speak rather of components and units. A component is to be regarded as a recognizable 'package', identifiable even if impermanent. It has a state x which, in a full description, may specify not only its internal constitution but also variables such as its position and orientation in space. For brevity we shall refer to a component in state x as an x-*component*. The number of x-components will, as ever, be denoted $n(x)$. We shall assume that there are several types of unit which enter into a component, labelled by $\alpha = 1, 2, \ldots, p$. We shall then speak of α-*units*, and assume that an x-component contains $a_\alpha(x)$ α-units ($x \in \mathcal{X}$; $\alpha = 1, 2, \ldots, p$). The total number of α-units in the system is then

$$\sum_x a_\alpha(x)n(x) = :M_\alpha \tag{1}$$

where the M_α, $a_\alpha(x)$ and $n(x)$ are all non-negative integers. We shall usually write (1) in the vector form

$$\sum_x a(x)n(x) = M \tag{2}$$

M and $a(x)$ being the p-vectors with α^{th} element M_α, $a_\alpha(x)$.

Consider then a system with description $\mathbf{x}$, a listing of the complete descriptions x of all existing components, regarded as identifiable even if impermanent. We shall suppose that it has the following properties.

(B1) The process $\{\mathbf{x}(t)\}$ is Markov and transition-reversible.
(B2) It is stochastically invariant under permutations of components in the same state.
(B3) The total numbers of every type of unit are conserved, so that the vector M is invariant.
(B4) M is the only invariant, so that the process is irreducible on a set of states $\mathbf{x}$ consistent with prescribed M.

The set of conservation laws (1) (or (2)) is assumed to include the conservation law (5.2) for components if they are indeed conserved. However, they may very well not be, as in the case of molecular division and combination. Hypothesis (B2) plus the hypothesis that $\mathbf{x}$ is a state-description in the Markov sense imply more than that the identity of components is irrelevant. They imply that all components in the same state have 'equal opportunity', in that the same transitions at the same rates are open to them. We discuss this point in Exercise 3.

Suppose that the possible transitions $x \to x'$ of the system are labelled and enumerated by $i = 1, 2, \ldots$, so that what we shall term 'reaction i' has intensity λ_i. We shall denote the reverse reaction to i by i^*; the assumption of transition-reversibility then states that

$$\lambda_i = \lambda_{i^*} \tag{3}$$

Reaction i corresponds to a particular transition $x \to x'$, this will induce a corresponding transition $n \to n'$. We shall understand all these transitions to indeed be corresponding in what follows.

Because of assumption (B2) the process $\{n(t)\}$ is also Markov, and in fact reversible. If reaction i involves $m_i(x)$ x-components ($x \in \mathscr{X}$) in its initiation then the transition $n \to n'$ induced by it has intensity

$$\lambda(n, n') = \lambda_i \prod_x n(x)^{(m_i(x))} \tag{4}$$

Here the factor multiplying λ_i is just the number of ways of selecting $m_i(x)$ x-components from the $n(x)$ available ($x \in \mathscr{X}$).

Theorem 7.1 *Assume conditions (B1)–(B4). Then the equilibrium distribution of the Markov process $\{n(t)\}$ is*

$$\pi(n) \propto \prod_x \frac{1}{n(x)!} \tag{5}$$

on the set of n satisfying (2) for the prescribed M.

Proof Because of transition-reversibility (3) and the form (4) of $\lambda(n, n')$ we see that distribution (5) satisfies the detailed balance relation

$$\pi(n)\lambda(n, n') = \pi(n')\lambda(n', n) \tag{6}$$

It is then the unique equilibrium distribution on any given irreducible set, and the set of n consistent with a given M constitutes such a set, by hypothesis (B4). ■

Distribution (4) has of course the already familiar form (5.1). However, the direct combinatorial argument used earlier is not available if the number of components is variable; we appeal rather to the detailed balance relation (6). The distribution defined by (2), (5) constitutes what we shall refer to as *constrained Poisson* or *Boltzmann* statistics. The classic term 'Boltzmann' refers to the occurrence of the factorial factors $(n(x)!)^{-1}$, a consequence of the statistical identity and transition-reversibility of components. The distribution is indeed that which would hold if the $n(x)$ were independently Poisson distributed and then conditioned by the constraint (2); see Exercise 1.

Recall that x is a complete description of the component. Suppose that this description can be partitioned $x = (x_1, x_2)$, where x_1 is a 'relevant' part of the description (reflecting perhaps the actual internal constitution of the component) and x_2 is an 'irrelevant' part (reflecting perhaps its spatial position or orientation). Specifically, we shall add the hypotheses:

(B5) $a(x)$ is a function of x_1 alone, which we shall write $a(x_1)$.
(B6) $m_i(x)$ is a function of x_1 alone, which we shall write $m_i(x_1)$, for all i.

That is, the numbers of units required to make up an x-component and the number of x-components needed for a given reaction depend only upon the relevant component x_1. It still remains possible that the transition intensity $\lambda(\mathbf{x}, \mathbf{x}')$ depends upon both parts of component descriptions.

Let us also denote the number of x values corresponding to a given value of x_1 as $V\gamma(x_1)$. Here V is a parameter measuring the 'size' of the system; we shall regard it as the volume of the region within which the whole closed system is located. For example, a simple indication of position might be obtained by dividing the volume V up into V cells of unit volume, and taking x_2 as the label of the cell within which an x-component lies. In this case the number of x-values for a given x_1 would be exactly V for all x.

However, the size adopted for a cell is not arbitrary. The requirement (B1) of transition-reversibility fixes the 'volume quantum'; the size of the cell within which events may be considered transition-reversible. This could be different for different components; one could regard $\gamma(x)^{-1}$ as the effective size of the volume quantum for x-components.

Consider now the statistics of the occupation numbers

$$n_1(x_1) := \sum_{x_2} n(x)$$

for the reduced component description x_1. The process with description $n_1 := \{n_1(x_1)\}$ is reversible but *not* in general Markov. However, its equilibrium distribution is easily derived from that of n.

Theorem 7.2
(i) *Assume hypotheses (B1)–(B5). Then the non-Markov variable n_1 has equilibrium distribution*

$$\pi(n_1) \propto \prod_{x_1} \frac{(V\gamma(x_1))^{n_1(x_1)}}{n_1(x_1)!} \tag{7}$$

on the set of n_1 satisfying

$$\sum_{x_1} a(x_1) n_1(x_1) = M \tag{8}$$

for the prescribed value of M.
(ii) *Insensitivity. Assume hypotheses (B1)–(B6). Then the equilibrium distribution (7), (8) is just that which would hold if n_1 had been a Markov variable with the transition $n_1 \to n_1'$ induced by reaction i having intensity*

$$\lambda(n_1, n_1') = \lambda_i V \prod_{x_1} \frac{n_1(x_1)^{(m_i(x_1))}}{[V\gamma(x_1)]^{m_i(x_1)}} \tag{9}$$

Proof Constraint (8) is just a rewriting of constraints (2), and expression (7) follows from summation of expression (5) over values of n consistent with given n_1. For assertion (ii), expression (7) satisfies the detailed balance equation associated with transition intensity (9), and this intensity continues to be consistent with the conservation rules (8). ▪

One may say that in (7), (8) we have deduced an extended form of constrained Poisson (or Boltzmann) statistics for the case when x_1 provides a description of the state of the component which is adequate but with 'degeneracy' $V\gamma(x_1)$.

The point of assertion (ii) is that the occupation numbers n_1 for the reduced component description x_1 have the same equilibrium statistics whether one considers the full model or a reduced version of the model specified by (9). In other words, if x_1 is regarded as a coarse or aggregated (over x_2) description then the model is such that the coarse description has the same equilibrium statistics in what one might term coarse and fine (aggregated and non-aggregated) versions of the model. One would not generally expect equivalence of coarse and fine models in this regard, and, when it occurs, the model is said to demonstrate *insensitivity*. The theme of insensitivity has already appeared in Exercise 4.2. It will be a recurrent one, to be taken up systematically in Chapter 12.

In aggregating, first to a system description n and then to a description n_1, one of course loses transition-reversibility, although not reversibility. The interpretation of the rate (9) is clear. The term in the numerator of the product is the number of ways in which $m_i(x_1)$ x_1-components can be selected from the $n_1(x_1)$ available ($x_1 \in \mathscr{X}_1$). (Here we use the notations 'x_1-component' and $\mathscr{X}_1$ in obvious analogue of the full-description quantities.) The term $V^{-\Sigma m_i(x_1)}$ represents the probability that a given such set of components will achieve encounter in a prescribed cell, and the initial factor V takes account of the fact that encounter may take place in any one of V cells. The $\gamma(x)$ terms reflect the fact that effective cell size varies with component description.

In a more explicitly Newtonian model one would substitute dynamic transition-reversibility for transition-reversibility in assumption (B1). The fact that configurational variables then turn out to have the same equilibrium statistics as before (cf. Exercise 4.2) is again an example of insensitivity.

Exercises and comments

1. Note that, in virtue of the conservation relations (8), distribution (7) could as well be written

$$\pi(n_1) \propto \prod_{x_1} \frac{[V\gamma(x_1)e^{-\theta^T a(x_1)}]^{n_1(x_1)}}{n_1(x_1)!} \tag{10}$$

where $\theta = (\theta_\alpha)$ is an arbitrary p-vector. This is just the distribution that would be followed if the $n_1(x_1)$ were independently distributed with respective expectations

$$E(n_1(x_1)) = V\gamma(x_1)e^{-\theta^{\mathsf{T}}a(x_1)} \tag{11}$$

this joint distribution being conditioned by the constraints (8). Expression (11) is essentially a statement of the Gibbs distribution and the independent Poisson distribution a characterization of the open process. Under the conditioning (8) of the closed process the distribution of n_1 becomes independent of θ.

2. *Circuit-switched networks.* The two prime examples of constrained Poisson statistics that we shall consider are provided by the kinetics of molecular reactions (Chapter 7) and the behaviour of circuit-switched networks, described in Exercise 4.7.4. In both cases we shall label the reduced component description (molecular constitution or route respectively) by r rather than x_1 and the number of such entities will be denoted by n_r. Equation (10) exhibits the equilibrium distribution of the numbers of busy routes as constrained Poisson. The constraint will generally take the form of bounds on the numbers of lines available on various sectors of the route, expressed by linear inequalities (4.7.11). We explore the implications for the equivalent of the thermodynamic limit (high demands and capacities) in the exercises of the next section.

3. Suppose the rules changed, so that components in a given state x must queue up to enter into any reaction, only the component at the head of the queue being eligible for the next reaction involving an x-component. The transition rate (4) would then become

$$\lambda(n, n') = \lambda_i d_i(n)$$

where $d_i(n)$ is unity if $n(x) \geq m_i(x)$ for all x (i.e. if all ingredients required for the reaction are present) and is otherwise zero. This modification is consistent with hypotheses (B1)–(B4) except that one can no longer maintain that x is a complete description—the place of any component in its queue may also be relevant. Components may be identical in that components in the same internal state and same place in the queue follow the same rules, but components in the same internal state and different places follow different rules.

However, the n-process remains Markov, and has equilibrium distribution

$$\pi(n) = \text{constant}$$

on an irreducible set, instead of (5). That is, the system follows Bose–Einstein (geometric) statistics instead of Boltzmann (Poisson) statistics. We return to this topic in section 6.4.

4. Suppose that configurations in which there is more than one component in any given state are forbidden, in that transitions to such configurations have zero intensity. This is not inconsistent with hypotheses (B1)–(B3), and the equilibrium distribution (5) remains valid on any irreducible set. However, any such set is now also subject to the constraint $n(x) \leq 1$ for all x; we have Fermi–Dirac statistics. It is of course hypothesis (B4) which is transgressed.

8. ASYMPTOTICS FOR CONSTRAINED POISSON STATISTICS

Let us for simplicity write the x_1 and $n_1(x_1)$ of the previous section just as x and $n(x)$. That is, we now regard x as the reduced component state (considered as expressing the relevant aspects of state) and use $n(x)$ to denote the number of components of this reduced description. The equilibrium distribution is then

$$\pi(n) \propto Q(n) := \prod_x \frac{[V\gamma(x)]^{n(x)}}{n(x)!} \tag{1}$$

on the set of n (in $\mathbb{Z}_+^{\mathscr{X}}$) satisfying

$$\sum_x a(x)n(x) = M \tag{2}$$

We regard

$$\rho_\alpha = \frac{M_\alpha}{V} \tag{3}$$

as the density of α-units and consider the passage to the thermodynamic limit as corresponding to the passage of M and V to infinity in such a way as to maintain the density vector ρ fixed.

The entries $a_\alpha(x)$, $n(x)$ and M_α in formula (2) are of course all non-negative integers. It will be necessary to assume that equations (2) have solutions for n. This will be true if, for instance, α-units may exist in free form for any α. However, we shall make the hypothesis in another way.

Rather than consider empirical proportions $n(x)/N$ we now consider empirical concentrations $c(x) = n(x)/V$. Assuming that these are with probability one of order unity in the thermodynamic limit (so that the $n(x)$ are of order V) we find then that $V^{-1}\log Q(n)$ has the asymptotic evaluation

$$S = \sum_x c(x)[1 + \log\gamma(x) - \log c(x)] \tag{4}$$

This may be regarded as a generalized entropy. If we are looking for the most probable configuration then S is to be maximized with respect to the $c(x)$ subject to the constraints

$$\sum_x a(x)c(x) = \rho \tag{5}$$

and $c(x) \geqslant 0$ ($x \in \mathscr{X}$).

Theorem 8.1 *Suppose that, for the given value of ρ, the expression in θ*

$$L(\theta) = \sum_x \gamma(x)e^{-\theta^T a(x)} + \theta^T \rho \tag{6}$$

possesses a stationary minimum. Then

(i) *The most probable value of the empirical concentrations $c(x)$ in the thermo-dynamic limit is that obtained by maximizing the entropy expression S given by (4) subject to the constraints on unit abundances (5).*

(ii) *This maximizing value is given by the Gibbs formula*

$$c(x) = \gamma(x)e^{-\theta^{\mathsf{T}}a(x)} \tag{7}$$

where θ is a vector parameter chosen so that $c(x)$ as given by (7) satisfies constraints (5).

(iii) *The required value of θ can alternatively be characterized as that minimizing $L(\theta)$, and the minimized value of $L(\theta)$ equals the maximal constrained value of S.*

This is the analogue of Theorem 5.2 and the proof follows the same course: discussion of the Lagrangian form $S + \theta^{\mathsf{T}}(\rho - \sum_x a(x)c(x))$. The problem enunciated in (iii) is the dual of that of maximizing S subject to (5), and has again the interpretation of the location of a saddle-point in the evaluation of $Ec(x)$.

We note from (1), (2) that

$$\Pi(z) := E\left[\prod_x z(x)^{n(x)}\right] \propto C_M \exp\left[V \sum_x z(x)\gamma(x)\prod_\alpha w_\alpha{}^{a_\alpha(x)}\right] \tag{8}$$

where $C_M(\cdot)$ denotes 'coefficient of $\Pi_\alpha w_\alpha^M$ in the power series expansion of $(\cdot)$'; (cf. (6.7)). This direct consequence of the axioms in fact sums up everything: that the distribution of the $n(x)$ is constrained independent Poisson, with the coefficient of $z(x)$ amounting to a statement of the Gibbs distribution. From this expression we deduce, as in (6.9), that

$$E(c(x)) = \frac{\mathscr{I}\left(\gamma(x)\prod_\alpha w_\alpha{}^{a_\alpha(x)}\right)}{\mathscr{I}(1)} \tag{9}$$

where

$$\mathscr{I}[\psi(w)] := \frac{1}{(2\pi i)^p}\oint \cdots \oint \psi(w)e^{VJ(w)}\prod_\alpha \frac{dw_\alpha}{w_\alpha} \tag{10}$$

Here

$$J(w) := \sum_x \gamma(x)\prod_\alpha w_\alpha{}^{a_\alpha(x)} - \sum_\alpha \rho_\alpha \log w_\alpha \tag{11}$$

and the w_α integration path is a tight circuit of the origin in the complex plane $(\alpha = 1, 2, \ldots, p)$. In $J(w)$ we recognize expression (6) with the change of variables $w_\alpha = e^{-\theta_\alpha}$.

Theorem 8.2 *Suppose again the condition of Theorem 8.1. Then in the thermo-dynamic limit the expected value of the concentration of x-components has the evaluation*

$$E(c(x)) = \gamma(x)\prod_\alpha w_\alpha{}^{a_\alpha(x)} = \gamma(x)e^{-\theta^{\mathsf{T}}a(x)} \tag{12}$$

where w is the saddle-point of J, the real positive value of w minimizing J (w). Equivalently, θ has the real value minimizing expression (6).

This is the analogue of Theorem 6.2, and the proof follows by a p-dimensional version of the argument for Theorem 6.1.

The condition of Theorem 8.1 is essential. Its failure in the case of finite $\mathcal{X}$ is equivalent to *infeasibility*. That is, to the fact that the ρ_α have been prescribed inconsistently, so that (5) has no solution for c, and the equations

$$\sum_x a(x)\gamma(x)e^{-\theta^T a(x)} = \rho \tag{13}$$

no solution for θ. In the case of infinite $\mathcal{X}$ there is the additional possibility of *criticality*: that equations (13) may possess no θ-solution in virtue of the fact that values of θ which might solve (13) if the sum on the left were truncated cause the infinite sum to diverge. This corresponds to the phenomenon that the available units, instead of being spread in a comparable fashion over all existing components, are being drawn (at least for some α) largely into a single component. The x-value for this component will tend to 'infinity' in $\mathcal{X}$ as one goes to the thermodynamic limit. That is, in the molecular case a 'giant molecule' forms which takes up the bulk of the existing matter or energy. In the circuit-switching case the lines, instead of being allocated comparably over a large number of routes requested, are largely drawn into satisfying the needs of one exotic and infinitely-demanding route. We return to this matter in section 6.3 and Part IV.

Exercises and comments

1. In the molecular context the w_α would be termed 'activities' or 'fugacities', while θ_α is proportional to the chemical potential of an α-atom divided by absolute temperature.

2. As pointed out, equations (2) will always have solutions for the $n(x)$ if free units are permitted and such units are included in the listing x of possible components. If free units are possible but not included in the listing then constraint (2) takes rather the form of linear inequalities

$$\sum_x a(x)n(x) \leqslant M \tag{14}$$

The difference in the two sides of (14) is just the vector of numbers of free units (the 'slack variables' of linear programming).

3. *Circuit-switched networks.* Consider the model with restricted numbers of lines which leads to the distribution (4.7.10), (4.7.11) for the numbers n_r of routes in operation. That is,

$$\pi(n) \propto \prod_r \frac{(V\gamma_r)^{n_r}}{n_r!} \tag{15}$$

on the set of n specified by

$$\sum_\alpha a_{\alpha r} n_r \leqslant M_\alpha = V\rho_\alpha \tag{16}$$

Here we have set $v_r/\mu_r = V\gamma_r$, where V is a measure of the scale of the system, in that demand (measured by v_r) and numbers of lines installed (M_α) are both proportional to V. The problem is then exactly that treated in this section, with r replacing x, and the difference that the equality constraint (2) is replaced by the inequality constraint (16). As in the text, one finds that the most probable value of $c_r = n_r/V$ is given in the thermodynamic limit by

$$c_r = \gamma_r e^{-\theta^\mathsf{T} a_r} = \gamma_r \prod_\alpha w_\alpha{}^{a_{\alpha r}} \tag{17}$$

where the θ_α have the *non-negative* values minimizing

$$L(\theta) = \sum_r \gamma_r e^{-\theta^\mathsf{T} a_r} + \theta^\mathsf{T} \rho \tag{18}$$

The parameters w_α consequently lie between zero and unity, and formula (17) gives them an interesting interpretation. One can interpret w_α as the probability that a line is available on sector α. The product form of (17) implies that these availabilities are effectively independent for different sectors (or even repeated use of the same sector). That is, 'blocking' is independent for each line call.

4. The characterization of θ in Exercise 3 as the solution of the dual problem is again associated with location of a saddle-point in the integral representation of $E(c_r)$. The generating function $\Pi(z)$ of the n_r is proportional to the sum of the coefficients of $\prod_\alpha w_\alpha{}^{s_\alpha}$ for $0 \leqslant s_\alpha \leqslant M_\alpha$ $(\alpha = 1, 2, \ldots)$ in the expansion of $\exp[V\sum_r \gamma_r \prod_\alpha w_\alpha{}^{a_{\alpha r}}]$ in powers of the w_α. We thus deduce that $E(c_r)$ has the evaluation (17) in the thermodynamic limit, with w the saddle-point of the expression

$$\exp\left[V\sum_r \gamma_r \prod_\alpha w_\alpha{}^{a_{\alpha r}}\right] \prod_\alpha \left[\frac{1 - w_\alpha{}^{-V\rho_\alpha}}{1 - w_\alpha{}^{-1}}\right]$$

A further approximation is to take the Vth root of this expression as

$$\exp\left[\sum_r \gamma_r \prod_\alpha w_\alpha{}^{a_{\alpha r}}\right] \prod_\alpha \max(1, w^{-\rho_\alpha})$$

Assuming that the saddle-point minimizes this expression, we have exactly the dual problem: that of minimizing expression (18) over non-negative θ.

9. THE OPEN PROCESS

As pointed out in Exercise 7.1 the distribution (8.1) for the occupation numbers $n(x)$ for the closed process (x being the reduced component description) could just as well be written

$$\pi(n) \propto \prod_x \frac{[V\gamma(x)e^{-\theta^T a(x)}]^{n(x)}}{n(x)!} \tag{1}$$

for arbitrary θ. This distribution is subject to the constraint of the conservation relations (8.2).

We regard the open process as being that for which distribution (1) holds without constraint, for some θ, so that the $n(x)$ are indeed independent Poisson variables with respective expectations

$$E(n(x)) = V\gamma(x)e^{-\theta^T a(x)} \tag{2}$$

The open process certainly determines the closed process, in that the closed process has the statistics of the open process conditioned by the constraints (8.2). The question is whether the reverse holds: whether the class of processes given by (1) for varying θ are the only processes from which the closed process could thus have been derived. In fact they are not, unless a further assumption is added (see Exercise 1). The further assumption is that implicit in the 'cellular model', that for a closed model in a volume V the reduced description has a multiplicity of order V. That is, that the number of full component descriptions corresponding to a given reduced description is of order V. With this additional assumption the class of distributions (1) emerges as the only possible statistics for an open process.

Consider a cellular process on two disjoint regions of volumes V and V'. Components are identified by a reduced description x plus the statement of which of the two regions they occupy. If the numbers of x-components in the two regions are denoted $n(x)$ and $n'(x)$ then it follows, as for the derivation of (8.1), that the joint equilibrium distribution is

$$\pi(n, n') \propto \prod_x \frac{(V\gamma(x))^{n(x)}(V'\gamma(x))^{n'(x)}}{n(x)!\, n'(x)!} \tag{3}$$

on the set of (n, n') specified by

$$\sum_x a(x)[n(x) + n'(x)] = M \tag{4}$$

This is the only constraint, since by (B4) M is the only invariant. An implication of this assumption is that the two regions must be communicating, in that components can pass between them.

Summing distribution (3) over n' consistent with (4) and normalizing we deduce that

$$\pi(n) = \left[\prod_x \frac{(V\gamma(x))^{n(x)}}{n(x)!}\right] \frac{C_{M-\Sigma a(x)n(x)}\exp\left[V'\sum_x\gamma(x)\prod_\alpha w_\alpha{}^{a_\alpha(x)}\right]}{C_M\exp\left[(V+V')\sum_x\gamma(x)\prod_\alpha w_\alpha{}^{a_\alpha(x)}\right]} \tag{5}$$

(cf. (8.8)). We can write this as

$$\pi(n) = \left[\prod_x \frac{(V\gamma(x))^{n(x)}}{n(x)!}\right] \frac{\mathscr{I}\left[\exp\left\{-V\sum_x\gamma(x)\prod_\alpha w_\alpha{}^{a_\alpha(x)}\right\}\prod_x\prod_\alpha w_\alpha{}^{n(x)a_\alpha(x)}\right]}{\mathscr{I}[1]} \tag{6}$$

where $\mathscr{I}[\psi(w)]$ has the same definition as in (8.10), (8.11) except that V is replaced by $(V+V')$. The overall density of α-units is now given by

$$\rho_\alpha = \frac{M_\alpha}{V+V'} \tag{7}$$

Consider now the thermodynamic limit in which V remains fixed, but V' and M become infinite in such a way as to maintain constant unit densities ρ.

Theorem 9.1   *Suppose the condition of Theorem 8.1: Then:*

(i) *In the thermodynamic limit the occupation numbers $n(x)$ follow independent Poisson distributions with expectations (2), where the parameter θ can be chosen to give the correct expected unit densities ρ_α. Alternatively, θ has the value minimizing expression (8.6).*

(ii) *This joint Poisson distribution is consistent with a specification in which n follows a Markov process for which unit-conserving transitions have intensity (7.9) and x-components can also enter V in independent Poisson streams of rates $v(x)$ and leave V independently at rates $\mu(x)$ with*

$$\frac{v(x)}{\mu(x)} = V\gamma(x)e^{-\theta^{\mathsf{T}}a(x)} \tag{8}$$

Proof   By the now familiar saddle-point argument a ratio $\mathscr{I}[\psi(w)]/\mathscr{I}[1]$ becomes equal to $\psi(\bar{w})$ in the thermodynamic limit, where $\bar{w}$ is the real positive w-value minimizing $J(w)$. Applying this evaluation to expression (6) we deduce assertion (i). As for assertion (ii), the multivariate Poisson distribution asserted shows detailed balance under all the transitions indicated. ∎

The open process constituted by immersion of V in an infinitely larger volume with prescribed unit densities thus shows exactly the equilibrium statistics expressed by (1).

Assertion (ii) is again a statement of insensitivity. It states that there is a plausible but cruder model, with no developed spatial structure at all, which would yield the same equilibrium distribution as the open model. The unit-conserving reactions have the same intensities (7.9) as proposed before, with a form of dependence on V which we saw as plausible. We require 'immigration' and 'emigration' rates to show detailed balance for each x; see (8). The ratio of total emigration flux to immigration flux is $\mu(x)n(x)/v(x)$, which is of order $n(x)/V$ in $n(x)$ and V. That this should be so is plain (see Exercise 2). For the cellular model, closer to the actual physical mechanism, assumptions (B1)–(B4) of section 7 are sufficient.

Finally, the relationship between open and closed models has interesting inferential implications. Suppose one knows that the statistics of n are those of the open model asserted in (1). Suppose one then observes actual numbers $n(x)$ in a specimen volume V, and wishes to infer from them the value of θ. Let us write distribution (1) as $\pi(n|\theta)$ to emphasize its dependence upon the parameter vector θ. The normalized form of this distribution is

$$\pi(n|\theta) = \left[\prod_x \frac{(V\gamma(x))^{n(x)}}{n(x)!} \right] \exp\left[- V\hat{L}(\theta) \right] \tag{9}$$

where

$$\hat{L}(\theta) := \sum_x \gamma(x)e^{-\theta^{\mathsf{T}}a(x)} + \theta^{\mathsf{T}}\hat{\rho} \tag{10}$$

Expression $\hat{L}(\theta)$ differs only from the $L(\theta)$ of (8.6) in that limiting densities ρ_α are replaced by the empirically observed densities of α-units

$$\hat{\rho}_\alpha := M_\alpha/V = V^{-1}\sum_x a_\alpha(x)n(x)$$

Theorem 9.2

(i) *In equilibrium the quantity M (or $\hat{\rho}$) is sufficient for estimation of θ, in that the distribution of n conditional on M is proportional to the first factor of expression (9), and is functionally independent of θ.*

(ii) *The maximum likelihood estimate of θ is the value minimizing $\hat{L}(\theta)$.*

Both assertions follow immediately from the form of expression (9). Assertion (i) states what we know: that the closed model can be regarded as the conditional version of the open model for any θ. In assertion (ii) we recognize again the 'dual problem' of Theorem 8.1 (iii). The fact that the value of θ matching prescribed densities ρ minimizes $L(\theta)$ is paralleled by the fact that the maximum likelihood estimate of θ from observations minimizes $\hat{L}(\theta)$.

Exercises and comments

1. The distribution (8.1), (8.2) for the closed model is the conditioned form of an unconstrained 'open' distribution

$$\pi(n) = C(M)\prod_x \frac{(V\gamma(x)))^{n(x)}}{n(x)!}$$

for any normalizing $C(M)$. However, the $n(x)$ can be independent under this distribution only if

$$C(M) \propto e^{-\theta^T M} = e^{-\theta^T \Sigma a(x) n(x)}$$

for some θ, where the proportionality factor is independent of n. Of course, in the text independence is a conclusion rather than a demand. The demand is rather that a reduced description should have a multiplicity (or a degeneracy, one might say) of order V.

2. Emigration and immigration rates for x-components are assumed to be in the ratio $n(x)/V$. This is because if emigration and immigration take place out of and into a 'surface skin' of the region V of volume ΔV in unit time, then the immigration rate must be proportional to ΔV, and the emigration rate to $(\Delta V/V)n(x)$. However, this is again an 'aggregated' treatment. Derive the conclusion by considering statistics for a cellular model in which all events are transition-reversible with intensities independent of V.

10. STOCHASTIC CONVERGENCE IN THE THERMODYNAMIC LIMIT

Consider again the thermodynamic limit of a closed process in the general version of section 7. Behind the entropy-maximizing calculations of Theorem 7.1(i) (and, earlier, of Theorem 5.2) lay an assumption that the concentration $c(x) = n(x)/V$ did indeed converge stochastically in the thermodynamic limit. The fact that, under the condition of Theorem 8.1, the expected and most probable values of $c(x)$ seem to be identical in the thermodynamic limit supports a conjecture that there is a stochastic convergence in some quite strong sense, but the point has not yet been proved. Nor does it need to be: Theorem 9.1(i) makes as strong an assertion as one would wish about the statistics in a *fixed* volume V as one goes to the thermodynamic limit in a volume $V + V'$, without appeal anywhere to stochastic convergence.

However, if one wishes to prove stochastic convergence of the $c(x)$ to $\gamma(x)e^{-\theta^T a(x)}$ with θ correct for the prescribed ρ then the lines of a natural proof are now clear. Consider the *open* statistics with θ chosen to be consistent with the prescribed ρ. Then it is clear that the $c(x)$ converge to the values indicated above in almost any sense (e.g. strong convergence in probability for a prescribed finite set of x; convergence in r^{th} mean) because the $n(x)$ are independent and $c(x) = n(x)/V$ can be regarded as the average of V independent and identically distributed random variables. By this same convergence the constraints of the closed system

$$V^{-1} \sum_x a(x) n(x) = \rho$$

will be satisfied arbitrarily closely in the thermodynamic limit.

However, the condition of Theorem 8.1 (that $L(\theta)$ should possess a sufficiently regular minimum) is indeed necessary for this conclusion. Unless it is satisfied then the value of θ consistent with the prescribed ρ either fails to exist or is on the point of failing to exist. That is, the value of ρ is either internally inconsistent or would lead to critical behaviour.

CHAPTER 6

Markov models and statistical mechanics: variations

1. COUPLING VIA THE HEAT BATH

In this and the next section we shall return to the situation considered in sections 5.1–5.6. That is, of N permanent and identifiable components exchanging a single type of unit, which we shall again take as being an energy quantum, for definiteness.

An attractive and physically plausible method of achieving this exchange is to assume that free quanta of energy can occur in the system and that components can raise or lower their energy by absorbing or emitting a quantum. These transfers between components and the common quantum sea or *heat bath* replace direct exchanges between components, and constitute their only coupling.

The two models are not so different, in that statistics of a given set of components are the same in either case in the thermodynamic limit—again a manifestation of insensitivity. However, the heat bath is attractive as an explicit coupling mechanism, and, if the model does not contain it, the mathematics still suggest it.

We shall reserve the term 'components' in this section for the N permanent components, although the individual free quanta must also be viewed as components in the sense of section 5.7. We shall assume a cellular model in order to make assumptions explicit. Specifically, we suppose that the complete description of system state consists of a listing of the complete states $x_1, x_2, \ldots, x_N$ of the N components together with a listing of the positions (i.e. the cells occupied) by the free quanta. Transitions are supposed transition-reversible, so that all configurations in a given irreducible set have the same probability. If it is taken for granted that the components are permanent in identity then the only invariant of the process is supposed to be M, the total number of quanta in the system, free or bound. This conservation rule could be expressed

$$\sum_j a_j(x_j) + m = M \tag{1}$$

where $a_j(x_j)$ is the number of quanta held by component j if it is in state x_j (components are not necessarily supposed identical) and m is the number of free quanta. All system states consistent with (1) for given M are then intercommunicating and equiprobable in equilibrium.

Suppose the system description reduced to $(\mathbf{x}; \mathbf{m}) = (x_1, x_2, \ldots, x_N; m_1, m_2, \ldots, m_\nu)$. Here m_h is the number of free quanta in cell h, so that

$$\sum_h m_h = m \tag{2}$$

Summing the uniform distribution over permutations of the M quanta we deduce then that

$$\pi(\mathbf{x}, \mathbf{m}) \propto \prod_h \frac{1}{m_h!} \tag{3}$$

The reason why there is no combinatorial term in (3) for the bound quanta is that locations of the quanta within a component are effectively specified. If a component undergoes a transition $x \to x'$ in which it acquires a quantum then comparison of x and x' effectively gives the location of the acquired quantum. Alternatively, one can argue from detailed balance rather than combinatorially. If the transition $x \to x'$ is one in which a given ambient quantum (i.e. a free quantum in the same cell as the component) is absorbed then the reverse transition has the same intensity, by transition-reversibility. If there are m_h free quanta in the cell and we do not distinguish which is absorbed then the intensities of the transitions $x \rightleftharpoons x'$ are in the ratio m_h. Distribution (3) shows detailed balance under these transitions.

Summing distribution (3) further over all allocations of the m free quanta to cells we deduce that

$$\pi(\mathbf{x}, m) \propto \frac{V^m}{m!} \tag{4}$$

subject again to (1). This implies the distribution of x for given M as

$$\pi(\mathbf{x}) \propto \frac{V^{M - \Sigma a_j(x_j)}}{\left(M - \sum_j a_j(x_j)\right)!} \tag{5}$$

In (5) we have deduced the joint distribution of the states of N components immersed in a *finite heat bath*. We go to the thermodynamic limit or to the case of an *infinite heat bath* by allowing M, V to go to infinity in such a way that the density of quanta

$$\rho_Q = M/V \tag{6}$$

remains fixed. Distribution (5) can be written

$$\pi(\mathbf{x}) \propto \frac{M^{(r)}}{V^r} \tag{7}$$

where $r = \sum_j a_j(x_j)$ is the total number of bound quanta. We thus deduce that in the limit of an infinite heat bath the components have the joint distribution

$$\pi(\mathbf{x}) \propto \rho_Q^r = \prod_j \rho_Q^{a_j(x_j)}$$

$$= \exp\left[-\beta \sum_j U_j(x_j)\right] \qquad (8)$$

Here $U_j(x_j)$ is the energy of component j in state x_j, so that

$$\rho_Q = e^{-\beta\varepsilon}$$

where ε is the energy of a quantum. These calculations and their immediate implications can be summarized.

Theorem 1.1   *Assume the conditions listed before formula* (1). *Then*

(i) *The joint equilibrium distribution of component states is given by* (4), *or* (5). *This is the same distribution as for a Markov process in which the intensity of the energy-conserving transitions* $(x, m) \rightleftharpoons (x', m - 1)$ *for a given component are in ratio* m/V.

(ii) *In the limit of an infinite heat bath the joint equilibrium distribution of component states is given by* (8): *the components follow independent Gibbs distributions with the same parameter. This is the same distribution as for a Markov process in which the intensities of the transitions* $x \rightleftharpoons x'$ *for component* j *are in ratio* $\exp \beta[U_j(x') - U_j(x)]$.

So we end up, as ever, with the Gibbs distribution, and even with independence of components. However, the point of the theorem is very much in the assertions of insensitivity. As we have already stated, if $x \rightarrow x'$ is a transition of a component in which a given ambient quantum is absorbed then the reverse transition has the same intensity. If we do not specify the quantum then the transitions $x \rightleftharpoons x'$ have intensities in the ratio m_h, where m_h is the number of ambient quanta (i.e. the number of free quanta in the same unit cell as the component). One could regard m_h as the local (and statistically variable) density of free quanta. It is now asserted in (i) that component state distribution would be the same if this ratio of intensities had rather equalled m/V, the average density of free quanta over the whole region V. Finally, as asserted in (ii), for the infinite heat bath the component state distribution (8) would have been the same if the intensities of the transitions $x \rightleftharpoons x'$ had been in the ratio ρ_Q, the statistical average of free quantum density.

That is, distributional results are insensitive to quite drastic aggregation; to the replacement of local densities of free quanta by regional averages, and then indeed by statistical averages. These aggregations destroy transition-reversibility, but not reversibility.

2. COUPLING BY COLLISIONS

An alternative mechanism for weak coupling is the notion of *collision*, i.e. of an encounter between two components in which an energy transfer takes place. This concept goes back at least to Boltzmann, and is the obvious one when the components are the molecules of a gas. One can say that the mechanism is only a more graphic way of visualizing the direct transfers postulated in section 5.3. However, the picture is one that helped Boltzmann to a quasi-deterministic formulation of great interest.

In order to make the collision mechanism explicit, let us again adopt a cellular model. The region of volume V is divided into V cells of unit volume, labelled $h = 1, 2, \ldots, V$. The scaling of volume is such that events within a cell are transition-reversible. Suppose that a component has reduced description x and full description (x, h), where h is its position, i.e. the cell that it occupies. Let $\lambda(x, y; x', y')$ be the intensity with which a given x-component and y-component in a given cell collide and transform into an x'-component and a y'-component in the same cell. In fact, this intensity could be allowed to depend also upon h and upon the states of other components, just so long as the specification implies transition-reversibility. However, for simplicity we assume dependence upon (x, y, x', y') alone. Transition-reversibility then implies that

$$\lambda(x, y; x', y') = \lambda(x', y'; x, y) \tag{1}$$

We shall also suppose that

$$\lambda(x, y; x', y') = \lambda(y, x; y', x') \tag{2}$$

i.e. that the order of the two components in collision is immaterial. Finally, since energy is conserved then

$$a(x) + a(y) = a(x') + a(y') \tag{3}$$

For simplicity we suppose components statistically identical. There will be other transitions, also transition-reversible, corresponding to the motion of components between cells.

On the supposition of complete communication between all system-states consistent with a given total energy

$$\sum_j a(x_j) = M \tag{4}$$

one deduces then, as in section 5.3, that in equilibrium all values of $(\mathbf{x}, \mathbf{h}) = (x_1, x_2, \ldots, x_N, h_1, h_2, \ldots, h_N)$ consistent with (4) are equally likely, and hence that all values of $\mathbf{x}$ consistent with (4) are equally likely. This produces the standard Boltzmann statistics, constrained by (4), for $n := \{n(x, h)\}$ and for $n := \{n(x)\}$, where $n(x, h)$ is the number of x-components in cell h and

$$n(x) = \sum_h n(x, h)$$

is the number of x-components overall. However, we wish to take a rather more dynamic view of the passage to these equilibrium statistics.

The collision transition in which an x-component and a y-component in cell h produce an x'-component and a y'-component in the same cell induces transitions on $\mathbf{n}$ and n which could be written

$$\mathbf{n} \to \mathbf{n}' = \mathbf{n} - e(x, h) - e(y, h) + e(x', h) + e(y', h) \tag{5}$$

$$n \to n' = n - e(x) - e(y) + e(x') + e(y') \tag{6}$$

where $e(x, h)$ is the array with zero elements except for a unit element at position (x, h), correspondingly for $e(x)$.

The total rate for transition (5) is

$$\lambda(\mathbf{n}, \mathbf{n}') = \lambda(x, y; x', y')n(x, h)(n(y, h) - \delta(x, y)) \tag{7}$$

because of the assumed statistical identity of components with the same (x, h). One cannot speak of the intensity of the transition (6) because the process with n as description is not Markov. However, we know from Theorem 5.7.2 that the equilibrium distribution of n is the same as that for the process in which n is Markov with intensity

$$\lambda(n, n') = V^{-1}\lambda(x, y, x', y')n(x)(n(y) - \delta(x, y)) \tag{8}$$

for the transition (6); a stochastic version of the Boltzmann *Stosszahlansatz* model.

Boltzmann's model was a deterministic one for the concentrations

$$c(x) = n(x)/V \tag{9}$$

In analogue to the Markov model with intensity (8) he assumed the rate of decrease of $c(x)$ due to the $(x, y) \to (x', y')$ collision to be $\lambda(x, y; x', y')c(x)c(y)$. This leads to the deterministic kinetic equation

$$\dot{c}(x) = \sum_y \sum_{x'} \sum_{y'} \lambda(x, y, x', y')[c(x')c(y') - c(x)c(y)] \tag{10}$$

The total densities of components and energy

$$\rho_C = \sum_x c(x)$$

$$\rho_Q = \sum_x a(x)c(x)$$

are conserved in time under the flow (10), conservation of energy being a consequence of the fact that $\lambda(x, y, x', y')$ is zero unless (3) is satisfied. The following assertion is classical.

Theorem 2.1 Boltzmann's H-theorem. *If the concentrations $c(x)$ obey equations* (10) *then the entropy measure*

$$H = -\sum_x c(x) \log c(x) \tag{11}$$

increases with time.

Proof We have

$$\dot{H} = \sum_x \dot{c}(x)[-1 - \log c(x)] = -\sum_x \dot{c}(x) \log c(x)$$

$$= \sum_x \sum_y \sum_{x'} \sum_{y'} \lambda(x, y; x', y')[-c(x')c(y') + c(x)c(y)] \log c(x)$$

$$= \frac{1}{4} \sum_x \sum_y \sum_{x'} \sum_{y'} \lambda(x, y; x', y')[c(x)c(y) - c(x')c(y')] \log \left[\frac{c(x)c(y)}{c(x')c(y')} \right]$$

$$\geq 0 \tag{12}$$

where the last equality follows by applying the permutations $(x, y) \rightleftharpoons (x', y')$ and $(x, x') \rightleftharpoons (y, y')$ under the sum before it and appealing to the symmetries (1) and (2). ■

This result has so many implications that one is almost at a loss to list them. First, we note that the entropy measure (11) is a Liapunov function for the equation system (10), since it changes monotonically in time under the system. One can use this result to demonstrate convergence indeed to a Gibbs distribution over x. Furthermore, there is a complete analogue of all these results for the full stochastic model: that with transition intensities (7) or (8). We shall demonstrate all these assertions in the next chapter for the general case of constrained Poisson statistics.

This celebrated theorem obviously constitutes a Liapunov argument analogous to that employed in section 3.2. However, the Kolmogorov equation, linear in the variables $\pi(x, t)$, allowed a large choice of Liapunov functions. The choice is much more restricted for the *nonlinear* equations (10). The particular Liapunov function H thus has a particular role. However, one has a guide to this choice. In a stochastic model the concentration vector $c = \{c(x)\}$ is random, and one looks for the most probable value. As in section 5.4, H is an asymptotic measure of probability, close in definition to the entropy S defined there.

Exercises and comments

1. Concentrations $c(x)$ certainly constitute an equilibrium solution for the Boltzmann equation (10) if they satisfy

$$c(x)c(y) = c(x')c(y') \tag{13}$$

for all possible collision reactions. This is a deterministic form of detailed balance (for a transition-reversible model).

2. Relations (13) certainly have solutions

$$c(x) = vw^{a(x)} = e^{-\theta_1 - \theta_2 a(x)} \tag{14}$$

since total component and quantum numbers are conserved in a collision. Under irreducibility conditions (that M and N are the only invariants) these will be the only solutions.

3. The 'cellular' version of equations (10) would be

$$\dot{c}(x, h) = \sum_h \sum_y \sum_{x'} \sum_{y'} \lambda(x, y; x', y')[c(x', h)c(y', h) - c(x, h)c(y, h)]$$

$$+ \sum_i \lambda(x, h, i)[c(x, i) - c(x, h)]$$

Here $c(x, h)$ is the concentration of x-components in cell h and $\lambda(x, h, i)$ a transport rate of x-components from cell h to cell i. This rate also satisfies a condition of transition-reversibility:

$$\lambda(x, h, i) = \lambda(x, i, h).$$

Deduce a version of the Boltzmann theorem.

3. INFEASIBILITY AND CRITICALITY

Our most general formulation of Boltzmann (constrained Poisson) statistics up to now has been that of sections 5.7–5.9. There we saw that the equilibrium statistics of both the open and the closed system could be expressed in terms of the generating function

$$G(w) = \sum_x \gamma(x) \prod_\alpha w_\alpha{}^{a_\alpha(x)} \tag{1}$$

in fact the partition function for the component. As we have seen, it is also advantageous for some purposes to write this as a Laplace–Fourier series rather than a power series:

$$F(\theta) = \sum_x \gamma(x) e^{-\theta^T a(x)} \tag{2}$$

The parameter θ of the Gibbs distribution

$$c(x) = \gamma(x) e^{-\theta^T a(x)}$$

is matched to the vector ρ of prescribed unit densities by the equation

$$\sum_x a(x)\gamma(x) e^{-\theta^T a(x)} = \rho. \tag{3}$$

This matching condition can also be regarded as a minimality condition for the form

$$L(\theta) = F(\theta) + \theta^{\mathsf{T}}\rho \tag{4}$$

(cf. Theorem 8.1) The matching conditions (3) will fail to have a solution for θ for given ρ exactly in those cases when $L(\theta)$ does not have a minimum at a stationary point.

If $\mathscr{X}$ is finite then such failure occurs exactly in those cases when ρ has been inconsistently prescribed (i.e. inconsistently with *any* distribution over $\mathscr{X}$, not merely with the Gibbs distributions). For example, suppose that there are just two conserved quantities, components and energy quanta, and that a component carries exactly one energy quantum in all states. Then $a_1(x) = a_2(x) = 1$ and

$$L(\theta) = \mathrm{e}^{-\theta_1 - \theta_2} \sum_x \gamma(x) + \theta_1\rho_1 + \theta_2\rho_2.$$

Unless $\rho_1 = \rho_2$ (so that abundances of components and quanta match) then the stationarity conditions for $L(\theta)$ are mutually inconsistent and $L(\theta)$ has in fact the infimal value $-\infty$.

If $\mathscr{X}$ is infinite then one can have failures of a more subtle type. Let us for definiteness consider the case when there is only a single conserved quantity, so that $a(x)$, θ and ρ are all scalars. Let us indeed suppose that $a(x) = x$ ($x = 0, 1, 2, \ldots$). We might, for example, be thinking of telephone calls through a network, and supposing that the number of calls in effect at a given time which demand x circuits would be Poisson distributed with expectation proportional to $\gamma(x)$ (independently for different x) were it not for the existence of a constraint on the total number of circuits available.

One possibility is that $\gamma(x)$ increases faster than exponentially, so that $F(\theta)$ converges for no θ. In this case all our discussion of the thermodynamic limit fails, in that all resources become concentrated in one or a few components. Thus, as one increases the scale of operations in the telephone network then operations themselves do not scale correspondingly: one or a few calls hog the bulk of the increased number of available circuits.

That $F(\theta)$ should diverge for some θ need not in itself indicate the possibility of any kind of singular behaviour. Suppose, for example, that $\gamma(x) = \mathrm{e}^{\phi x}$, so that

$$F(\theta) = (1 - \mathrm{e}^{\phi - \theta})^{-1}$$

diverges for $\theta \leqslant \phi$. The matching condition (3) becomes

$$\frac{\mathrm{e}^{\phi - \theta}}{(1 - \mathrm{e}^{\phi - \theta})^2} = \rho$$

This has a solution for θ whatever the value of ρ. As ρ increases then θ approaches the value ϕ marking the boundary of the region of convergence, but never reaches it.

Consider, however, the case $\gamma(x) = e^{\phi x}/x^3$ $(x > 0)$. The matching condition (3) becomes

$$\sum_{x > 0} \frac{e^{(\phi - \theta)x}}{x^2} = \rho \tag{5}$$

The infinite sum diverges for $\theta < \phi$, but actually converges at the transitional value $\theta = \phi$. So, suppose we define a critical value of ρ:

$$\rho' = \sum_{x > 0} x^{-2}$$

For $\rho \leqslant \rho'$ equation (5) has a solution for θ, decreasing with increasing ρ. As ρ reaches the critical value ρ' then θ actually reaches the boundary value ϕ of the region of convergence. For $\rho > \rho'$ equation (5) has no solution for θ, and the treatment we have given of the thermodynamic limit again breaks down.

For the hypothetical example of the telephone network the parameter ρ measures the availability of circuits. For $\rho < \rho'$ the available circuits are spread comparably over different types of call. As ρ increases and availability improves then more of the more demanding calls are accepted. However, when ρ increases through the critical value ρ' then a call is accepted so exotically demanding that it swamps the network.

This effect is truly a change in phase, a transition to a qualitatively different regime. We shall analyse such phase changes in Part IV and support the vague characterization we have given of conditions beyond criticality.

4. NON-POISSON STATISTICS

We saw that hypotheses (B1)–(B3) of section 5.7 implied Poisson equilibrium statistics on any irreducible set. One essential supposition was that of transition-reversibility in the complete description. However, once one gets away from a stochasticization of classical mechanics then transition-reversibility is no longer so natural, and there are models for which it is natural to accept the hypotheses (B1)–(B4) of section 5.7, except that the hypothesis of transition-reversibility is weakened to that of reversibility. In particular, forward and backward rates for a given reaction between specified components may depend upon n in different ways. For example, for the ecological competition model of Chapter 8 one has statistical identity of individual species, conservation of biochemical resources, and reversibility (formally plausible at a superficial level) but not transition-reversibility. Otherwise expressed, the rates at which a given individual dies and is born (in the sense of being 'reincarnated') depend upon population numbers in quite different ways.

Suppose, then, that the process with variable $n = \{n(x)\}$ is Markov and reversible, so that the detailed balance equation

$$\pi(n)\lambda(n, n') = \pi(n')\lambda(n', n) \tag{1}$$

admits a solution

$$\pi(n) = \Phi(n) \tag{2}$$

Suppose we again assume that the numbers of p types of units are conserved in any transition, so that the p-vector $\sum_x a(x)n(x)$ is invariant under possible transitions. Then relations (1) also admit a solution

$$\pi(n) \propto \Phi(n)\exp\left(-\theta^{\mathrm{T}}\sum_x a(x)n(x)\right) \tag{3}$$

for any θ. If $M = \sum_x a(x)n(x)$ is the only invariant then the most general form of $\pi(n)$ consists of linear combinations of solution (3) for varying θ.

Suppose that $\Phi(n)$ factorizes

$$\Phi(n) = \prod_x \phi_x(n(x)) \tag{4}$$

so that (3) becomes

$$\pi(n) \propto \prod_x [\phi_x(n(x))e^{-\theta^{\mathrm{T}}a(x)n(x)}] \tag{5}$$

It is then natural to consider expression (5) as an equilibrium distribution for all n in $\mathbb{Z}_+^{\mathscr{X}}$, this describing the statistics of an open version of the process in which the $n(x)$ are independent random variables.

The case $\phi_x(n(x)) = (n(x)!)^{-1}$ is the only one we have yet encountered; for this the $n(x)$ are Poisson distributed in the open model. In the ecological example of Chapter 8 it turns out to be natural to take $\phi_x(n(x))$ as having something like exponential growth in $n(x)$. In the case of exact exponential growth the $n(x)$ are then geometrically distributed in the open process.

Exercises and comments

1. Ideas of volume-dependence and the thermodynamic limit are less simple for the non-Poisson case. The Poisson distribution has the property that sums of independent Poisson variables are also Poisson; one may say that the Poisson property is invariant under aggregation. This explains why we found in section 5.7 that bulk variables had the same equilibrium statistics for both a cellular model and an aggregated model. In the non-Poisson case this is no longer in general true. For example, sums of independent identically distributed geometric variables have a gamma distribution. We return to these matters in Chapter 8; see also Exercise 3.

2. Suppose that free units are not included in the listing of components, so that the conservation relation must be rewritten

$$\sum_x a(x)n(x) + m = M \tag{6}$$

where $m = (m_\alpha)$ is the vector of numbers of free units. This relation could be written in the full matrix form

$$An + m = M \tag{7}$$

Then the full Markov variable is (m, n) and solution (3) becomes

$$\pi(m, n) \propto \Phi(m, n)e^{-\theta^T M} = \Phi(m, n)e^{-\theta^T(An+m)} \tag{8}$$

It is now natural to assume the limited factorization

$$\Phi(m, n) = \psi(m) \prod_x \phi_x(n(x)) \tag{9}$$

This implies that the components interact only via the sea (m) of free units. It amounts to a more general expression of the concept of interaction via the heat bath and is useful in Chapter 8, when we wish to express an assumption that organisms interact only via resource limitations.

3. Suppose we look for a family $\mathcal{D}$ of distributions on the natural numbers $\mathbb{Z}_+$ which is closed under aggregation. That is, a sum of independent random variables with distributions drawn from $\mathcal{D}$ has a distribution also in $\mathcal{D}$. Suppose that such a distribution has p.g.f. $\chi(z|v)$, where v is a parameter which labels the members of $\mathcal{D}$ as it varies in a set $\mathcal{N}$. We require then that for $v^1, v^2, \ldots$ in $\mathcal{N}$ there exist a $\bar{v}$ in $\mathcal{N}$ such that

$$\prod_j \chi(z|v^j) = \chi(z|\bar{v}) \tag{10}$$

Condition (10) implies that, under some rescaling of v, one can take v as a vector with functionally independent elements v_k, and that $\log \chi(z|v)$ is linear in v. That is

$$\chi(z|v) = \exp\left[\sum_k v_k(f_k(z) - f_k(1)) \right]$$

where $f_k(z)$ is itself a p.g.f. of a distribution on the natural numbers. The Poisson distribution is the simplest example of such a case, the compound Poisson the next.

CHAPTER 7

Chemical kinetics and equilibrium

In this chapter we assume the components of the system considered in Chapter 5 to be molecules. The full state x of the molecule perhaps specifies its position, orientation, etc., but also expresses of what compound r it is a molecule. The label r then describes the internal constitution or *type*; we shall regard it as equivalent to the reduced state x_1 of section 5.7. We suppose that there is a bounded number of types (so that critical effects are excluded for the moment) each of known structure, binding energy, etc. The molecules are made up from basic constituents (atoms of various elements, energy quanta) which are immutable and in restricted abundance. The problem is then to determine the distribution of the numbers of various types of molecules under prescription of the abundances of the elements.

If we make the assumptions of section 5.7 then this is just the problem addressed and solved (in equilibrium) in sections 5.7–5.9. We shall make those assumptions, and the mathematical analysis is little more than a recapitulation of that of those sections. However, we shall rephrase the results in the language of compound formation, consider models which are more highly aggregated, discuss kinetics and investigate the extremal role of entropy in more detail.

1. CHEMICAL KINETICS; DETERMINISTIC MODELS

There is a literature on stochastic formulations of chemical kinetics, some of it by chemists and some by probabilists (see the reference list at the end of the chapter). Not all the probabilistic authors, at least, are aware of the implications of statistical mechanics for the necessary form of the equilibrium solution, and so some of them become much embroiled with solutions to unnatural models in terms of special functions. Of course, it is true that transient (time-dependent) solution will scarcely ever be tractable, but one will expect the form of equilibrium solution to be fixed by the Gibbs distribution.

We shall assume the full description x of the molecule to be reduced to the minimal description r on the basis of which one can construct a Markov model, at least with some averaging ('aggregation') over position, orientation, etc. So, in reducing from x to r we are aggregating over aspects of the molecule which are assumed not to affect transition rates (e.g. over *isomers*—differences in molecular structure which produce no relevant observable effect). We are also aggregating over aspects that may indeed affect reaction rates, such as just position and orientation. The justification must be the 'insensitivity' argument of section 5.7: that the equilibrium statistics of the aggregated description are the same for the aggregated and non-aggregated models. Suppose that r takes values in a finite set $\mathcal{R}$.

In speaking of models we shall often use 'high-level' and 'low-level' as synonyms for 'unaggregated' and 'aggregated' respectively, just as in speaking of descriptions we shall often use 'fine' and 'coarse'. In reducing a high-level model to a low-level model we lose transition-reversibility, but retain reversibility. To postulate just reversibility is quite convenient, because it means that our model then covers both adiabatic and isothermal equilibrium: the cases when energy is fixed or temperature is fixed. Nevertheless, the fact that there is a high-level model which is transition-reversible has implications for the low-level model: that it is reversible and has reaction rates of the particular form (5.7.9).

We shall denote a single molecule of type r by A_r, and use n_r to denote the number of A_r in the closed region of volume V within which reactions take place. Relation (5.7.1) then becomes

$$\sum_r a_{\alpha r} n_r = M_\alpha \qquad (\alpha = 1, 2, \ldots, p) \tag{1}$$

if M_α is the total number of α-atoms and $a_{\alpha r}$ the number of α-atoms in an A_r. We can write (1) in matrix notation as

$$An = M \tag{2}$$

Equivalently, we can write

$$Ac = \rho \tag{3}$$

where

$$c_r = n_r/V, \qquad \rho_\alpha = M_\alpha/V \tag{4}$$

are the corresponding spatial concentrations. The molecular concentrations c_r are the variables of interest, and in this section we shall construct a deterministic model for their evolution.

Suppose that a finite number of types of reaction are possible, labelled by $i = 1, 2, \ldots$, and that the ith reaction takes the form

$$\sum_r m_{ir} A_r \rightleftharpoons \sum_r m'_{ir} A_r \tag{5}$$

That is, that m_{ir} r-molecules ($r \in \mathcal{R}$) become converted into m'_{ir} r-molecules ($r \in \mathcal{R}$) in the forward reaction, the reverse taking place in the reverse reaction. A reaction is sometimes spoken of as 'reversible' in the literature if it can proceed in either direction. In accordance with the principles of statistical mechanics we shall assume reversibility in the much stronger technical sense of Chapter 4. Indeed, as in section 5.7 we shall assume that the reaction rates are deduced from those for a more refined model which is transition-reversible.

The simplest reactions are the single-molecule reactions

$$A_r \rightleftharpoons A_{r'} \tag{6}$$

or two-molecule reactions

$$A_r + A_s \rightleftharpoons A_{r'} + A_{s'} \tag{7}$$

$$A_r + A_s \rightleftharpoons A_u \tag{8}$$

These correspond respectively to an internal change in the molecule, to encounter and interaction, and to encounter/combination and splitting. Indeed, one imagines that these are the 'elementary' reactions and that the multi-molecule reaction (5) in fact consists of a rapid succession of reactions of types (6)–(8). Nevertheless, it is simpler to give a treatment which allows the more general reactions.

Let us define the *forward rate* Λ_i of reaction i as the expected number of occurrences of the forward reaction i per unit time and volume. We assume the model aggregated over space and so, effectively, spatially homogeneous. In a Markov model with multiple events excluded the probability intensity of the event 'an occurrence of the forward reaction i' is $V\Lambda_i$. We shall denote the rate of the reverse reaction by Λ_i'. These rates will of course be functions of the concentrations c_r.

Under these assumptions the equations giving the kinetics of the reaction are, in the deterministic case,

$$\dot{c}_r = \sum_i (m_{ir}' - m_{ir})\,(\Lambda_i - \Lambda_i') \tag{9}$$

and these rates of change must be zero at equilibrium. In fact, we expect detailed balance, and so validity of the stronger equilibrium conditions

$$\Lambda_i = \Lambda_i' \tag{10}$$

If the model is deduced by aggregation from a high-level model obeying assumptions (B1)–(B6) of section 5.7 then the total rate $V\Lambda_i$ must be of the form (5.7.9). In a deterministic version, in which $n_r^{(m)}/V^m$ is replaced by c_r^m, we have then

$$\Lambda_i = \lambda_i \prod_r c_r^{m_{ir}}$$
$$\Lambda_i' = \lambda_i' \prod_r c_r^{m_{ir}'} \tag{11}$$

Here the constants λ_i, λ_i' have values compatible with reversibility, i.e. with the existence of a solution c to all the detailed balance relations (10). (Note that λ_i is not the λ_i of (5.7.9), partly because the γ-factors of (5.7.9) have been absorbed into it, and partly because we may be assuming isothermal conditions, so that the process is open energetically.) The term $\prod_r c_r^{m_{ir}}$ represents something like an encounter rate per unit volume for the m_{ir} r-molecules $(r \in \mathcal{R})$ required for the forward reaction to take place.

Under assumption (11) the detailed balance reactions (10) take the form

$$\lambda_i \prod_r c_r^{m_{ir}} = \lambda_i' \prod_r c_r^{m_{ir}'} \tag{12}$$

The hypothesis of reversibility implies the possibility of detailed balance, and so the assurance that equations (12) have a solution

$$c_r = \gamma_r \tag{13}$$

say.

Now, the conservation condition (1) implies that

$$\sum_r m_{ir} a_{\alpha r} = \sum_r m_{ir}' a_{\alpha r} \tag{14}$$

for all α for any possible reaction i. It follows that if (13) is a solution of (12) then so is

$$c_r = \gamma_r \prod_\alpha w_\alpha^{a_{\alpha r}} = \gamma_r e^{-\theta^T a_r} \tag{15}$$

for arbitrary $w_\alpha = \exp(-\theta_\alpha)$. Here a_r is the column vector with elements $a_{\alpha r}$.

Solution (15) is of course the familiar Gibbs distribution, in a non-statistical context. To prove its true equilibrium character and its extremal characterization we follow the lines suggested by the stochastic treatment of Chapter 5.

Exercises and comments

1. Suppose the process closed to energy quanta as well as to atoms, so that conditions are adiabatic. Note by comparison of (5.7.9) and (11) that the γ_r of (13) may, indeed, be identified with the analogue of $\gamma(x_1)$: the number of forms that an r-molecule can take in a complete description. Under isothermal conditions it would be this quantity multiplied by $\exp(-\beta U_r)$ where U_r is the energy of an r-molecule and β is inversely proportional to temperature.

2. EQUILIBRIUM AND ENTROPY FOR THE DETERMINISTIC MODEL

We consider then a deterministic model which is

(D1) *Derived* from a transition-reversible model in that the reactions (1.5) have rates of the form (1.11) for which the detailed balance relations (1.12) possess a solution,

(D2) *Conserving*, in that Ac is invariant under all reactions,

(D3) *Irreducible*, in that there is no other conserved quantity, and

(D4) *Non-degenerate*, in that the equation $Ac = \rho$ has solutions $c \geqslant 0$ for all $\rho \geqslant 0$.

The irreducibility condition excludes the possibility that there are two groups of compounds not connected by any reaction path. It implies, we shall see, that the equilibrium c for a given value of ρ is unique.

The non-degeneracy condition is designed to obviate the need for annoying conditions on ρ, which were required, for instance, in the first example of section 6.3 if the specification was to be feasible. One certainly has non-degeneracy if all elements can exist in their atomic forms (see Exercise 5.8.2). The condition implies that A is of full row-rank, in that there is no non-zero θ such that $\theta^{\mathsf{T}} A = 0$. That is, no θ such that $\theta^{\mathsf{T}} a_r$ is zero for all r.

Let us define d_i as the $\mathscr{R}$-vector with r^{th} element $m'_{ir} - m_{ir}$ and D as the matrix $(d_1\, d_2\, d_3\, \ldots)$. Then relation (1.14), expressing conservation of elements in a reaction, can be written

$$AD = 0 \tag{1}$$

We shall require the converse of this assertion.

Lemma 2.1 *Suppose conditions (D1)–(D4) satisfied. Then*

(i) *If an $\mathscr{R}$-vector Δ satisfies $A\Delta = 0$ then the equation*

$$D\xi = \Delta \tag{2}$$

for ξ possesses solutions.
(ii) *If an $\mathscr{R}$-vector η satisfies*

$$\eta^{\mathsf{T}} D = 0 \tag{3}$$

then it is of the form

$$\eta^{\mathsf{T}} = \theta^{\mathsf{T}} A \tag{4}$$

for some η.

Proof Let n and n' be two values of the occupation number vector consistent with the same value of M, so that $A(n - n') = 0$. We know then from (D3) that there is a sequence of reactions taking the system state from n to n'. That is, there are integers ξ_i such that

$$n' = n + \sum_i \xi_i d_i \tag{5}$$

Here $|\xi_i|$ is the number of times reaction (1.5) occurs on the reaction path, in the forward direction if ξ_i is positive, in the reverse direction if it is negative. But if we set $n' - n = \Delta$ then relation (5) just amounts to (2). We have established the existence of solutions ξ to (2) only for Δ with integer elements, but, seeing the elements can be arbitrarily large and we can divide ξ and Δ in (2) by a common large integer, we see that (2) has solutions for any Δ in the null space of A. Assertion (i) is thus established.

Suppose now that there is a vector η satisfying (3) which is not of the form (4). Then we can find a Δ satisfying $A\Delta = 0$ such that $\eta^{\mathsf{T}}\Delta \neq 0$. Premultiplying (2) by η^{T} we deduce that $\eta^{\mathsf{T}}\Delta = 0$, which is a contradiction. Assertion (ii) is thus established. ■

From this we deduce

Theorem 2.1 *Suppose conditions (D1)–(D4) satisfied, and let $c = \gamma$ be a solution of the detailed balance equations* (1.12). *Then any other solution is of the form* (1.15).

Proof If c and γ are both solutions of (1.12) then

$$\prod_r (c_r/\gamma_r)^{m_{ir} - m'_{ir}} = 1 \tag{6}$$

If we define $\eta_r = \log(c_r/\gamma_r)$ then (6) amounts just to (3) with the implication (4) for some θ. That is, c must have the form (1.15). ■

We need one more preparatory result.

Lemma 2.2 *Suppose condition D4 satisfied. Then the function*

$$F(\theta) = \sum_r \gamma_r e^{-\theta^{\mathsf{T}} a_r} \tag{7}$$

is strictly convex.

Proof The function will be strictly convex if the form in real θ

$$\sum_\alpha \sum_\beta \theta_\alpha \theta_\beta \frac{\partial^2 F}{\partial\theta_\alpha \partial\theta_\beta} = \sum_r (\theta^{\mathsf{T}} a_r)^2 \gamma_r e^{-\theta^{\mathsf{T}} a_r}$$

is strictly positive. It is certainly non-negative, and can be zero only if $\theta^{\mathsf{T}} a_r = 0$ for all r. But this is excluded by hypothesis (D4). ■

Now we come to a deterministic analogue of Theorem 5.8.1.

Theorem 2.2 *Consider the model specified by hypotheses (D1)–(D4). Let $c = \gamma$ be any solution of the detailed balance equations. Then*

(i) *The vector c which maximizes the 'entropy'*

$$S(c) = \sum_r c_r(1 + \log\gamma_r - \log c_r) \tag{8}$$

subject to
$$Ac = \rho \tag{9}$$

has the form
$$c_r = \gamma_r e^{-\theta^{\mathsf{T}} a_r}, \tag{10}$$

known itself to be a solution of the detailed balance equations.

(ii) *The parameter vector θ is determined by enforcement of conditions (9) or, equivalently, by minimization of the form $L(\theta) = F(\theta) + \theta^{\mathsf{T}}\rho$. This determination has a solution, which is unique.*

(iii) *The minimized value of $L(\theta)$ equals the constrained maximum of $S(c)$.*

Proof All assertions follow from the now familiar discussion of the Lagrangian form $L(c, \theta) = S(c) + \theta^{\mathsf{T}}(\rho - Ac)$ except for the statement in (ii): that the determination of θ is unique. The truth of this assertion follows from Lemma 2.2, which implies that $L(\theta)$ also is strictly convex, and so has a unique minimum. ∎

It is interesting that, not only the Gibbs distribution (10), but also the entropy measure (8) should appear in a non-stochastic context. The notion that there should be a function $S(c)$ of concentrations which is extremal in equilibrium is, of course, suggested by the idea of a most probable configuration for the stochastic model. It is also from the stochastic model that one obtains a guide to the form this functional should take. Since the deterministic model is the 'thermodynamic limit' of the stochastic model in some senses, it is not surprising that ideas useful for one should be useful for the other.

They are, indeed, useful to the extent that the entropy functional (3) is a Liapunov function for the nonlinear dynamic system specified by (1.9), (1.11). We have demonstrated in Theorem 2.1 that all possible solutions of the detailed balance equations are of the form (10) and in Theorem 2.2 (ii) that there is only one such solution compatible with given ρ. There is thus only one possible solution of the detailed balance equations, but we have not yet demonstrated this to be truly the equilibrium solution.

Theorem 2.3 The H-theorem. *Under conditions (D1)–(D4) the entropy measure $S(c)$ defined in (8) is non-decreasing in time. The concentration vector c converges to the unique solution (10) compatible with the prescribed ρ, which is consequently the equilibrium solution.*

Proof Let us write

$$f_i(c) = \prod_r c_r^{m_{ir}}$$

$$f'_i(c) = \prod_r c_r^{m'_{ir}}$$

so that $c = \gamma$ is a solution of the detailed balance equations

$$\lambda_i f_i(c) = \lambda'_i f'_i(c) \tag{11}$$

We have then from (1.9), (1.11)

$$\dot{S} = \sum_r \frac{\partial S}{\partial c_r} \dot{c}_r = \sum_r \sum_i (\lambda_i f_i - \lambda'_i f'_i)(m_{ir} - m'_{ir}) \log(c_r/\gamma_r)$$

$$= \sum_r \sum_i (\lambda_i f_i - \lambda'_i f'_i) \log[f_i f'_i(\gamma)/f'_i f_i(\gamma)]$$

$$= \sum_r \sum_i (\lambda_i f_i - \lambda'_i f'_i) \log(\lambda_i f_i/\lambda'_i f'_i) \qquad (12)$$

where f_i and f'_i have argument c if no argument is indicated. Expression (6) is non-negative, so that S is non-decreasing in time. Being bounded above it reaches a finite limit, at which $\dot{S} = 0$. We see from (12) that this can hold only when c satisfies (11) for all i. That is, c must satisfy the detailed balance equations. We know that there is only one such solution compatible with prescribed ρ, and this is of the Gibbs form (10) with appropriately determined θ. ∎

Exercises and comments

1. There can be complications with the Lagrangian theory if ρ is at the boundary of its permitted region, which means in this case that $\rho_\alpha = 0$ for some α. That is, that certain elements are missing. This will then mean that the corresponding θ_α may take the value $+\infty$, so that compounds containing the missing elements are forbidden by the Gibbs distribution. However, the simplest course in such a case is to start from a model which envisages only those elements which are actually present.

3. CHEMICAL KINETICS; STOCHASTIC MODELS

This section will amount very much to a restatement of the material of sections (5.7)–(5.9), differing only in that we employ the molecular nomenclature and admit the possibility of isothermal conditions (i.e. that the system is not closed to the passage of energy quanta). We also discuss the kinetic equation (8) which embodies the model in full, and is the stochastic analogue of the corresponding deterministic equations (1.9), (1.11).

As in section 1 we shall suppose that the i th type of reaction has form (1.5), which we shall also write as $n \rightleftharpoons n' = n - m_i + m'_i$. We shall suppose that the total transition intensity for the forward and reverse reactions has the forms

$$\lambda(n, n') = V \lambda_i \prod_r [n_r^{(m_{ir})}/V^{m_{ir}}]$$

$$\lambda(n', n) = V \lambda'_i \prod_r [(n'_r)^{(m'_{ir})}/V^{m'_{ir}}] \qquad (1)$$

These correspond to the forms (1.11) assumed for the deterministic reaction rates except that there is an extra factor V, since the intensities (1), (2) refer to events in the whole volume rather than per unit volume.

Theorem 3.1 *Suppose that $n = \{n_r\}$ represents the aggregated description of a high-level model obeying conditions (B1)–(B6) of section 5.7, the model being closed to the passage of atoms and either closed or open to the passage of energy. Then the rates (1), (2) for the aggregated model, with λ_i and λ_i' having values permitting reversibility, are consistent with the high-level model, in that for appropriate λ_i, λ_i' they lead to the same equilibrium statistics for n, parametrized in the same way by element abundances.*

Proof Consider first the adiabatic case, when energy is just another 'element', with a fixed number of quanta in the volume V. Then the theorem is simply a restatement of Theorem 5.7.2(ii): that the equilibrium statistics of a model satisfying (B1)–(B6) are the same as those of a model with transition intensities (5.7.9). In translating from (5.7.9) to (1) one makes the substitutions $x_1 \to r$, $\gamma(x_1) \to \gamma_r$, $m_i(x_1) \to m_{ir}$, $\lambda_i \prod_{x_1} \gamma(x_1)^{-m_i(x_1)} \to \lambda_i$. The transition-reversibility statement $\lambda_i = \lambda_i'$ becomes a statement that the deterministic balance equations (1.12) possess a solution $c = \gamma$.

In the isothermal case the only difference is that what was $\gamma(x_1)$ is now to be replaced by $\gamma(x_1) \exp[-\beta U(x_1)]$, where $U(x_1)$ is the energy of a component with reduced description x_1 and the parameter β is inversely proportional to temperature. ∎

We now have a result which may seem natural or remarkable according to viewpoint.

Lemma 3.1 *The process with reaction rates (1) is reversible if and only if the deterministic detailed balance equations*

$$\lambda_i \prod_r c_r^{m_{ir}} = \lambda_i' \prod_r c_r^{m_{ir}'} \qquad (i = 1, 2, \ldots) \qquad (2)$$

possess a solution for c.

That is, deterministic reversibility (in the sense of deterministic detailed balance) is both necessary and sufficient for stochastic reversibility. Sufficiency is quickly verified, as we shall see. However, the proof of necessity is less direct, and we shall defer the proof of the Lemma until the next section.

We now postulate the stochastic analogue of the conditions (D1)–(D4) of section 2. That is, we assume a stochastic kinetic model which is

(M1) *Derived* from a transition-reversible model, in that reaction intensities have the form (1) and the detailed balance equations (2) possess a solution $c = \gamma$.

(M2) *Conserving*, in that *An* is invariant under all reactions.
(M3) *Irreducible*, in that there is no other conserved quantity.
(M4) *Non-degenerate*, in that *A* has full row-rank.

The now familiar assertions can be listed almost without further proof.

Theorem 3.2 *Assume that the process obeys conditions* (M1)–(M4). *Then*

(i) *It is reversible, with equilibrium distribution*

$$\pi(n) \propto \prod_r \frac{(V \gamma_r)^n}{n_r!} \tag{3}$$

on the set

$$An = M \tag{4}$$

for the prescribed value M of element abundances.
(ii) *Natural open processes are those for which the n_r are independently Poisson distributed with expectations $E(n_r) = Vc_r$, where*

$$c_r = \gamma_r e^{-\theta^{T} a_r} \tag{5}$$

for some θ, in that this distribution constrained by (4) *yields* (3), *whatever θ.*
(iii) *Consider the closed process in the thermodynamic limit with limiting element concentrations $M/V \to \rho$. Then the random molecular concentrations n_r/V converge to the limit values c_r determined in Theorem 2.2.*

Proof Assertion (i) follows from the fact that, with transition intensities given by (1), distribution (3) gives stochastic detailed balance. Assertion (ii) is then evident and familiar. Assertion (iii) refers to the now familiar complex of ideas: that the n_r/V converge to the values c_r maximizing $S(c)$ subject to

$$Ac = \rho \tag{6}$$

where $S(c)$ is the entropy measure defined in (2.8); that this c_r is of the form (5) with θ determined by (6); that θ can alternatively be determined by the dual characterization of minimizing $L(\theta) = F(\theta) + \theta^{T}\rho$ with $F(\theta)$ defined by (2.7), and that this latter characterization can be interpreted either as location of a saddle-point in the evaluation of $E(n)$ or as a maximum likelihood estimation of θ from observations of n. The 'convergence of n/V' refers to convergence of the most probable value of n/V, convergence of the expected value $E(n/V)$, or, as indicated in section 5.10, convergence of n/V itself in almost any stochastic sense. ∎

The 'natural open processes' referred to in (ii) are actually the only possible open processes, in that the process obtained if V is regarded as part of an infinitely large volume with free passage of matter throughout is necessarily of this form (see section 5.9). However, this point cannot be made unless one has an expression for the joint distribution of n in two disjoint regions and this cannot be obtained from assumptions (M1)–(M4), but only by reference back to the assumptions of the high-level (cellular) model; see Exercise 2.

Finally, we should set down the stochastic kinetic equation. This would be just the Kolmogorov forward equation, but is best stated in terms of the p.g.f.

$$\Pi(z) = E\left(\prod_r z_r^{n_r} \right) \qquad (7)$$

We shall use the simplifying notation

$$D_r = \frac{\partial}{\partial z_r}$$

Theorem 3.3 *The p.g.f. (z) of molecular totals n_r obeys the equation*

$$\dot{\pi} = V \sum_i \left(\prod_r z_r^{m_{ir}'} - \prod_r z_r^{m_{ir}} \right) \left(\lambda_i \prod_r (D_r/V)^{m_{ir}} - \lambda_i' \prod_r (D_r/V)^{m_{ir}'} \right) \Pi \qquad (8)$$

This has an equilibrium solution

$$\Pi(z) \propto \exp\left[V \sum_r z_r \gamma_r e^{-\theta^T a_r} \right] \qquad (9)$$

for arbitrary θ.

Proof Relation (8) follows from the Kolmogorov forward equation by the formalism of section 2.9; see (2.9.9). Solution (9) is the familiar open process 'independent Poisson' solution. It satisfies detailed balance in that expression (9) makes the last bracket in (8) zero for every i. ∎

If we set $\exp(-\theta_\alpha) = w_\alpha$ in (9) then w_α can be regarded as the marker variable for M_α, the number of α-atoms (bound or free), just as z_r is the marker variable for n_r, the number of r-molecules. If M is the only invariant of the closed system then the most general equilibrium solution for π is obtained by linear combination of

solutions (9). By extracting the coefficient of $\prod_\alpha w_\alpha^M$ from (9) one derives the p.g.f.

of n for given M. By averaging this over M one obtains the p.g.f. of n for a given general distribution of M.

Exercises and comments

1. Suppose that free α-atoms are allowed to migrate in and out the system at respective rates $V v_\alpha$ and $\mu_\alpha m_\alpha$, where m_α is the number of free α-atoms in the volume V ($\alpha = 1, 2, \ldots, p$). Show that then the open distribution (9) holds with

$$w_\alpha = e^{-\theta_\alpha} = v_\alpha/\mu_\alpha$$

2. Consider a cellular model with n_{hr} r-molecules in cell h. Show that solution (9) is then replaced by

$$\pi(z) \propto \exp\left[\sum_h \sum_r z_{hr} \gamma_r e^{-\theta^T a_r}\right]$$

Deduce from this that the open distribution (9) will indeed hold in a volume V in communication with an infinite volume, the same cellular statistics holding throughout.

3. Suppose that all elements α can indeed occur in the free atomic form. Note that under the distribution (9) the amounts m_α in free form are independent Poisson variables with expectations Vw_α. However, the distribution of the total amounts M_α of α-atoms follow independent *compound* Poisson distributions, $P(M)$ being proportional to the term in $\prod_\alpha w_\alpha^{M_\alpha}$ in the expansion of $\exp\left[V \sum_r \gamma_r \prod_\alpha w_\alpha^{a_{\alpha r}}\right]$.

4. Consider a reaction $e_1 + e_2 \rightleftharpoons A$, where a molecule A thus consists of one atom each of elements 1 and 2. Suppose 1-atoms and molecules may not enter or leave the reaction vessel, but consider the two following situations.

(1) The reaction is buffered by maintenance of the concentration of free 2-atoms at a constant value w.
(2) 2-atoms may migrate in and out freely (following the rules of Exercise 1) at rates which give an *expected* concentration w of free 2-atoms in the vessel in the absence of any reaction.

Show that the distribution of the number of molecules is the same in both cases, and that the expected concentration of free 2-atoms equals w in both cases.

5. *Adsorption.* Consider a model in which there are M_0 sites and M_1 particles, of which n have become adsorbed (i.e. attached to a site; each site can accommodate only one particle). The 'particles' are molecules, but we avoid using the second term for reasons which will soon be clear. Assume the adsorption $(n \rightarrow n+1)$ and desorption $(n \rightarrow n-1)$ rates $\lambda(M_0 - n)(M_1 - n)$ and $\lambda'n$ respectively. Then the deterministic balance condition gives the *Langmuir adsorption isotherm*

$$\lambda(M_0 - n)(M_0 - n) = \lambda'n$$

and one derives easily the equilibrium distribution

$$\pi(n) \propto \frac{1}{(M_0 - n)! \, (M_1 - n)! \, n!} \left(\frac{\lambda}{\lambda'}\right)^n$$

The spatial element of 'sites' seems to bring in something not envisaged in the models of this chapter. However, one can regard 'sites' and 'particles' as being the atoms of two kinds of elements, and an occupied site as constituting a 'molecule' composed of a site and a particle. The rates proposed are then consistent with (1) and $\pi(n)$ consistent with (3).

6. *Competitive adsorption.* Suppose there are p types of particle with the αth

having abundance M_α and adsorption and desorption rates $\lambda_\alpha \left(M_0 - \sum_\beta n_\beta \right)(M_\alpha - n_\alpha)$ and $\lambda'_\alpha n_\alpha$, where n_α is the number of adsorbed α-particles $(\alpha = 1, 2, \ldots, p)$. Then the process is reversible with equilibrium distribution

$$\pi(n) \propto \frac{1}{\left(M_0 - \sum_\beta n_\beta \right)!} \prod_\alpha \frac{(\lambda_\alpha / \lambda'_\alpha)^{n_\alpha}}{(M_\alpha - n_\alpha)! \, n_\alpha!}$$

This is again a special case of (3), and gives the joint distribution of the $2p + 1$ types of 'compound': free and bound particles and vacant sites.

7. *Layer adsorption.* Suppose that there are M_α particles of type α, $(\alpha = 0, i, 2, \ldots, p)$, particles of type 0 being interpreted as sites. Suppose that a particle of type 1 can become adsorbed to a vacant site, a particle of type 2 to a particle of type 1 which is itself adsorbed, and in general a particle of type α to particle of type $\alpha - 1$ which is itself adsorbed $(\alpha = 1, 2, \ldots, p)$. A site on which adsorptions have taken place up to level α constitutes a 'chain' of length $\alpha + 1$. There are $(n_{\alpha-1} - n_\alpha)$ of these if n_α particles of type α are adsorbed (with $n_0 = M_0$). Suppose only the last particle in a chain can desorb, and adsorption and desorption rates are $\lambda_\alpha (n_{\alpha-2} - n_{\alpha-1})(M_\alpha - n_\alpha)$ and $\lambda'_\alpha (n_{\alpha-1} - n_\alpha)$ for an α-particle. The process is reversible with

$$\pi(n) \propto \frac{\prod_1^p (\lambda_\alpha / \lambda'_\alpha)^{n_\alpha}}{\left(M_0 - \sum_1^p n_\alpha \right)! \prod_0^p (n_\alpha - n_{\alpha+1})!}$$

$(n_0 = M_0, n_{p+1} = 0)$.

One can again regard the chains of various lengths as possible types of molecule. One has in effect a very special model of linear molecules in which the atoms must always occur in the α-sequence $0, 1, 2, \ldots$. See Part IV for much more general polymerization models.

4. DETERMINISTIC AND STOCHASTIC REVERSIBILITY

We come now to the proof of Lemma 3.1: the statement of the equivalence of deterministic and stochastic reversibility for the model of section 3. We must demonstrate that the existence of a solution c to the equations (3.2) implies the existence of a solution $\pi(n)$ to the stochastic detailed balance equations

$$\lambda_i \pi(n + m_i) \prod_r \frac{(n_r + m_{ir})^{(m_{ir})}}{V^{m_{ir}}} = \lambda'_i \pi(n + m'_i) \prod_r \frac{(n_r + m'_{ir})^{(m'_{ir})}}{V^{m'_{ir}}} \tag{1}$$

and conversely.

The forward implication is clear: if (3.2) has a solution $c = \gamma$ then (1) has a solution (3.3). One might think that the reverse implication would also be fairly immediate, because if we sum relation (1) over all n we obtain

$$\lambda_i E\left[\prod_r (n_r/V)^{m_{ir}} \right] = \lambda_i' E\left[\prod_r (n_r/V)^{m_{ir}'} \right] + O(V^{-1}) \tag{2}$$

If in the thermodynamic limit n/V converges sufficiently strongly in distribution to a constant c then (2) would imply the existence of a solution c to (3.2). However, one can scarcely assert this convergence before the form of $\pi(n)$ has been determined.

If we set

$$\pi(n) = \frac{V^{\sum n_r} Q(n)}{\prod_r n_r!} \tag{3}$$

then (1) becomes

$$\lambda_i Q(n + m_i) = \lambda_i' Q(n + m_i') \tag{4}$$

Let $\bar{n}$ be a fixed value of n and set $\phi_i = \log(\lambda_i/\lambda_i')$. Then we deduce from (4) that

$$Q(\bar{n} + D\xi) = Q(\bar{n}) e^{\phi' \xi} \tag{5}$$

Here D is the matrix with $(ri)^{\text{th}}$ element $d_{ri} = m_{ir}' - m_{ir}$.

The elements ξ_i of ξ indicate the number of times the ith reaction occurs in passage from $\bar{n}$ to

$$n = \bar{n} + D\xi \tag{6}$$

the sign being positive or negative according as the reaction is in the forward or reverse direction (cf. (2.5)). Of course, the reaction path must be such that the vector n remains non-negative at all times. Now, for prescribed n and $\bar{n}$ the equation (6) may have a solution ξ or it may not. (That is, there may or may not be a reaction path which links the two system-states.) If it has a solution then this will be of the form

$$\xi = D^{-1}(n - \bar{n}) + \zeta \tag{7}$$

Here ζ is a vector which satisfies

$$D\zeta = 0 \tag{8}$$

and so lies in the null space of D. It may depend upon both n and $\bar{n}$, and essentially describes a reaction path which brings the state back to where it started. The operator D^{-1} is a quasi-inverse to D, made determinate by requiring that, for possible passages $\bar{n} \to n$, the vector $D^{-1}(n - \bar{n})$ should lie in the orthogonal complement of the null space of D. It gives a more definite prescription of the required reaction path. The set of n for which (7) has a solution constitutes the set $\mathcal{N}(\bar{n})$ of values of n reachable from $\bar{n}$ by some reaction-path, and constitutes the ergodic class of the process which contains $\bar{n}$. If the process is conserving, irreducible and non-degenerate then it is just the set of n for which $A(n - \bar{n}) = 0$, but we are not necessarily making these assumptions.

Inserting (7) into (5) we deduce that

$$\phi' \xi = 0 \tag{9}$$

and

$$Q(n) = Q(\bar{n}) \exp[\phi' D^{-1}(n - \bar{n})] \quad (n \in \mathcal{N}(\bar{n})) \tag{10}$$

With appropriate definition of c we can write (10) as

$$Q(n) = Q(\bar{n}) \prod_r c_r^{n_r - \bar{n}_r} \quad (n \in \mathcal{N}(\bar{n})) \tag{11}$$

Relation (9) expresses a constraint on the rates λ_i, λ_i'. However, by substituting (11) into (4) we obtain the constraints in the form we want: that there exists a c satisfying the deterministic balance equation (3.2).

With this Lemma 3.1 is proved, and under wider conditions than envisaged in section 3. Equation (11) establishes the necessity of the form (3.3) for a reversible equilibrium solution of the process, not for all n, but only for all n in a given ergodic class of the process. We have not assumed these ergodic classes specified in any particular way (e.g. by the conserving, irreducibility and non-degeneracy conditions assumed in section 3).

5. REFERENCES

The literature is of course enormous. Specimen papers on stochastic models for chemical reactions are those by Bartholomay (1958), Darvey, Ninham and Staff (1966), McQuarrie (1967), Orriss (1969), Tallis and Leslie (1969) and Hall (1983a, b). The material of Exercises 3.5–3.7 is taken from Orriss. The situation of Exercise 3.4 was encountered in experimental work by Drs Lamb and McBurney of the Department of Physiology, University of Cambridge. Whittle (1965a) proposed models of the type considered here (Markov, with reversibility ensured by a deterministic reversibility condition) for a polymerization model. There is a very extensive literature on stochastic models of polymerization, to which we shall refer in Part IV. The treatment of this chapter is to some extent original, in its use of an explicit Markov model, but of course is in other ways very standard. Morgan (1976) proved the equivalence of deterministic and stochastic reversibility for these models under the assumption that passage between all molecules and the simple atomic state is possible. The rather more general treatment of section 4 is believed to be new.

CHAPTER 8

Resource-induced ecological competition

There have been many models of competition between species, deterministic and stochastic. The most satisfactory of these are those in which the mechanism of competition is made explicit, as competition for limited resources. Deterministic versions of such models lead to the conclusion termed the *principle of competitive exclusion* (cf. Phillips, 1973, 1974). The principle is that, if several species are competing for p critical resources, then in equilibrium at most p species will survive. One must exclude circumstances which are so special as to be structurally unstable.

In this chapter we give a stochastic version of the model which indeed demonstrates the principle of competitive exclusion. It also gives an extremal criterion which indicates just which species will survive in the equilibrium state.

Remarkably enough, this model is close to the model for chemical equilibrium of Chapter 7, but with Bose–Einstein statistics replacing Boltzmann statistics (i.e. distributions of geometric type replacing Poisson distributions). This semi-analogy does not follow from some facile and probably fallacious transfer of ideas, but from a consideration of reasonable ecological kinetics.

It must be emphasized, however, that the model we work with is time-reversible, and so necessarily has its limitations. Growth, reproduction and death are intrinsically time-directed processes, as are all processes concerned with the development and conservation of structure. However, our model would be a fair representation of the case of primitive organisms feeding in a chemical soup, and interacting only via their effect on that soup. Indeed, the chemical soup is the ecological equivalent of the heat bath. Higher interactions such as predation and sexual reproduction require an irreversible model.

The deterministic version of the principle of competitive exclusion is well established in the literature (see the references to Phillips). The work of remaining sections is either taken from or developed from Whittle (1983a).

1. ECOLOGICAL KINETICS; DETERMINISTIC MODELS

We begin rather similarly to the formulation of the chemical model of section 1, with an individual of species r corresponding to a molecule of type r, and a quantum of nutritional resource α corresponding an atom of element α. So, we denote an individual of species r by A_r and assume that there are p basic nutritional resources, labelled $\alpha = 1, 2, \ldots, p$ and assumed immutable and non-substitutable. We assume that an A_r (or an r-individual) contains an amount $a_{\alpha r}$ of resource r, and that such an individual cannot be constituted unless at least these amounts of resources are free in the system. We consider a closed system in which resources are recycled rather than consumed so that, upon death, an individual returns all his constituent chemical resources to general availability. There must then be a conservation equation

$$\sum_r a_{\alpha r} n_r + m_\alpha = M_\alpha \tag{1}$$

where n_r is the total number of r-individuals (i.e. the population size for species r), M_α is the prescribed total abundance of resource α, and m_α is the amount of *free* resource α. In the chemical context of section 7.1 we did not usually distinguish free element atoms, since these could be regarded as special cases of molecules, to be included in the listing by r. However, we can now hardly regard a quantum of free resource as a special case of an organism, and so must distinguish it.

We can write the conservation relations (1) in matrix form

$$An + m = M$$

One might refer to M and m as the *abundance* and *availability* vectors respectively, and to A as the *composition* matrix. For simplicity we shall assume all resources quantized, so that A, m and M all have non-negative integer elements, as of course does n. This is no constraint if the quanta are taken small enough.

Of course, there are resources which are indeed consumed rather than recycled; energy is the obvious example. In such cases the resource limitation is one of rate of supply (e.g. of sunlight) rather than of total amount.

One may again regard the system as occupying a spatial region of content V, which might still be referred to as volume if the environment is aquatic (i.e. in the soup rather than on the land). We may then speak of concentrations $c_r = n_r/V$ of species, $\rho_\alpha = M_\alpha/V$ of total resources, and

$$\zeta_\alpha = m_\alpha/V = \rho_\alpha - \sum_r a_{\alpha r} c_r \tag{2}$$

of free resources (available resources, ambient resources).

In a deterministic model one does not represent the events of birth and death for an individual, but simply regards population density c_r as a continuous variable obeying a dynamic equation. The simplest interesting model would be

$$\dot{c}_r = g_r(\zeta) c_r \tag{3}$$

where $g_r(\zeta)$ is a net reproduction rate for species r, dependent only on r itself and on ζ, the vector of free resource concentrations. This model formulates the idea that individuals would develop independently (although in a species-dependent fashion) if ambient resource concentrations ζ could be held fixed by some external buffering mechanism. However, for the closed system one will have the dependence of ζ upon c indicated by (2), and individuals will interact by this mechanism alone. In this way do resource limitations induce competition both between and within species.

Theorem 1.1 The principle of competitive exclusion. *For the model specified by* (2), (3) *not more than p species can survive in equilibrium, if one disregards structurally unstable cases.*

Proof This is simplicity itself. Let $\mathcal{R}_s$ and $\mathcal{R}_e$ be the sets of r for species which respectively survive or are extinct in equilibrium, so that

$$c_r = 0 \qquad (r \in \mathcal{R}_e) \tag{4}$$

$$g_r(\zeta) = 0 \qquad (r \in \mathcal{R}_s) \tag{5}$$

If q species survive then relations (5) constitute q equations in p variables. There will be no solution if $q > p$, unless one has a chance compatibility, which would be destroyed by a perturbation of the model. ∎

The argument is so simple that it has a certain robustness; it is not dependent upon very special conditions (see Exercise 1). There would be circumstances which would invalidate it, however. One would be if there were some other coupling mechanism between organisms, such as the existence of a common predator. Another would be the possibility that a species might evolve in time and so change its reproduction rate $g_r(\zeta)$ in such a way that equations (5) indeed become compatible for more than p values of r. That is, niches are created for more than p species, and the 'degeneracy' which we regarded as structurally unstable would be made structurally stable by the system's actively compensating for perturbations.

However, in the absence of other interactions and on a time scale which is less than evolutionary the theorem must be accepted. One would then like a criterion indicating *which* species may survive. The obvious criterion is that of stability: that the linearized model in the neighbourhood of a given equilibrium point $\bar{c}$ should have $\bar{c}$ as a stable equilibrium (cf. Exercise 1.4.2). The matrix of this linearized model has $(rs)^{\text{th}}$ element

$$\frac{\partial}{\partial c_s}(c_r g_r(\zeta(c))) = \delta_{rs} g_r - c_r \sum_\alpha a_{\alpha s} \frac{\partial g_r}{\partial \zeta_\alpha} \tag{6}$$

where all arguments are given their equilibrium values. We require that all eigenvalues of this matrix have negative real part. This condition immediately yields a very appealing criterion in the case of a single resource, when $p = 1$.

Theorem 1.2 *Consider the case of a single resource and assume all the reproduction rates $g_r(\zeta)$ increasing in ζ. Then the surviving species is that which can survive at the lowest concentration of free nutrient.*

Proof Suppose the surviving species is that for $r = 1$, so that

$$g_1(\zeta) = 0 \tag{7}$$

The matrix with elements (6) is diagonal below the first row; a necessary condition for stability is then that all diagonal elements below the first should be negative, so that

$$g_r(\zeta) < 0 \qquad (r > 1) \tag{8}$$

We see from equations (7), (8) that the surviving species is just viable at the ambient concentration ζ, and that the others are non-viable. No species is viable at a lower concentration, by the monotonicity assumption. ∎

The conclusion is exactly what one would expect, and is illustrated in Fig. 8.1.1. Populations grow, the free resource falls, and falls until the least-demanding population is just viable.

One would like to extend this extremal characterization of the equilibrium configuration to general p; to do so is one of our aims.

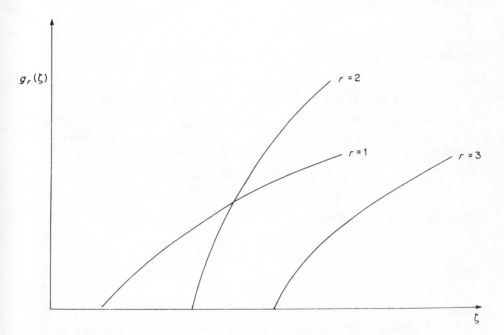

Fig. 8.1.1 The criterion for survival in the case of a single resource. Species 1 is the viable species, because it can survive at the lowest ambient concentration of resources.

Exercises and comments

1. A generalization of the model which retains the feature of resource-induced competition *between* species is

$$\dot{c}_r = g_r(c_r, \zeta(c))$$

where $g_r(0, \zeta(c)) = 0$ for all r and $\zeta(c)$ is a more general specification of the dependence of ζ upon c than (2). The principle of competitive exclusion will still hold.

2. One might allow model (3) to have the coupled form

$$\dot{c}_r = \sum_s g_{rs}(\zeta)c_s$$

if an r-individual could have mixed offspring. That is, if there were mutation or if different values of r represented the same organism at different ages or stages. For the principle of competitive exclusion to continue to hold one would have to regard a species as a cluster of such interrelated variants.

3. One could also assume that resources are consumed rather than recycled; ρ_α would then be a rate of supply per unit time. The conservation equations (2) would then be replaced by

$$\dot{\zeta}_\alpha = k_\alpha(\rho_\alpha - \sum_r a_{\alpha r} c_r)$$

where k_α is a response rate to an imbalance between supply and demand.

2. ECOLOGICAL KINETICS, STOCHASTIC MODELS

Let $\lambda_r(n)$ and $\mu_r(n)$ be the total birth and death intensities for species r as a function of the vector $n = \{n_r\}$ of population sizes for all species. We know from section 6.4 (see Exercise 6.4.2) that if functions $\Phi_r(n_r)$, $\Psi(m)$ exist such that

$$\frac{\lambda_r(n)}{\mu_r(n + e_r)} = \frac{\Phi_r(n_r + 1)}{\Phi_r(n_r)} \frac{\Psi(m - a_r)}{\Psi(m)} \tag{1}$$

then the process is reversible and has an equilibrium distribution

$$\pi(n) \propto \Psi(m) \prod_r \Phi_r(n_r) \tag{2}$$

Here

$$m = M - An \tag{3}$$

is the vector of free resources and a_r is the r th column of A, the vector of amounts of resources incorporated in an r-individual. One can either regard (2) as giving the distribution of n with m given in terms of n by (3), or as giving the joint distribution of m and n constrained by (3).

One may say that the ratio of Φ-functions in (1) reflects the ratio of intrinsic birth and death rates for species r, while the ratio of Ψ-functions represents the modifying effect of resource availability.

Consider the case of a single species and resource, for simplicity. One will expect total birth and death rates to be of order n_r (for fixed m) for anything but small n_r. This implies that $\Phi_r(n_r + 1)/\Phi_r(n_r)$ is roughly constant for large n_r, or that $\Phi_r(n_r)$ is roughly exponential.

$$\Phi_r(n_r) \sim \text{const.} \, \gamma_r^{n_r} \tag{4}$$

for large n_r.

This contrasts with the 'molecular' case of Chapter 7, where the 'birth' and 'death' rates would be constant and of order n_r respectively. That is, although r-molecules are lost at a rate proportional to n_r, other things being equal, they are created at a rate determined only by availability of constituents. This leads to a relation

$$\Phi_r(n_r) \propto \frac{\gamma_r^{n_r}}{n_r!} \tag{5}$$

In other words, in the molecular case we have the Poisson statistics (5) and in the population case the geometric-type (i.e. geometric for large enough n_r) statistics (4), at least if resource availability m is held fixed by external buffering. That is, the two modes of growth, by accretion and by reproduction, induce Boltzmann and Bose–Einstein statistics respectively.

The ratio $\Psi(m - a_r)/\Psi(m)$ in (1) will presumably be an increasing function of m, reflecting the fact that increased resources are helpful to life. However, there would presumably be a saturation effect, in that for large enough m a further increase would have little beneficial effect. One expects, then, that $\Psi(m - a_r)/\Psi(m)$ will become roughly constant for large m, so that $\Psi(m)$ is roughly exponential

$$\Psi(m) \sim \text{const.} \, \delta^m \tag{6}$$

for large m. This assertion will need qualification for the case of vector m, but the idea is plain.

Again, this contrasts with the molecular case of Chapter 7, or the case when m represented the number of free energy quanta. In these cases the birth rate $\lambda_r(n)$ and death rate $\mu_r(n)$ are respectively proportional to and independent of m. This is because the rate of uptake of free constituents or energy quanta is proportional to their numbers, the rate of release independent of their numbers. The ratio (1) would thus be proportional to m for fixed n, leading to a conclusion

$$\Psi(m) \propto \frac{\delta^m}{m!} \tag{7}$$

rather than (6).

Expressions (6) and (7) again represent a contrast between geometric-type and Poisson distributions. The interpretation is that, in an appropriate open version of the process, the number of free resource quanta is geometrically distributed, or at least that the tail of its distribution decays exponentially. This is again a shift from Boltzmann to Bose–Einstein statistics, induced now by a saturation effect. That is, by the fact that uptake rate of nutrient by a given organism will reach a limit as nutrient concentration in the medium is increased.

To recapitulate, the difference in kinetics that make geometric-type distributions appropriate in the present context are that (a) species increase by reproduction, not by association as in the molecular case, and (b) resources can only be absorbed by individuals of the species at a limited rate, even if these resources are present in gross excess.

That distributions should be *exactly* geometric is not crucial. One requires only geometric-type behaviour, i.e. that the ratios in (1) should become independent of n_r and of $|m|$ respectively as n_r becomes large and as the elements of m become large in fixed ratio. Otherwise expressed, the generating functions

$$\tilde{\Phi}_r(z_r) := \sum_{n_r} \Phi_r(n_r) z_r^{n_r} \tag{8}$$

$$\tilde{\Psi}(w) := \sum_m \Psi(m) \prod_\alpha w_\alpha^{m_\alpha} \tag{9}$$

should have finite regions of convergence. In the Poisson cases (5), (7) this was not so; the generating functions $\tilde{\Phi}_r$ and $\tilde{\Psi}$ were exponential in their arguments and converged everywhere. It is the finiteness of the domain of convergence of these functions that marks the qualitative difference of the ecological case.

In (3) we have an explicit expression for the equilibrium distribution of n. Note that the corresponding p.g.f. $\Pi(z)$ is proportional to the coefficient of $\prod_\alpha w_\alpha^{M_\alpha}$ in the expansion of

$$\Pi(z, w) = \tilde{\Psi}(w) \prod_r \tilde{\Phi}_r \left(z_r \prod_\alpha w_\alpha^{a_{\alpha r}} \right) \tag{10}$$

in powers of the w_α. As ever, w_α is a marker for M_α and expression (10) a joint p.g.f. (at least if normalized) for the n_r and M_α in the open process, i.e. in an appropriate 'grand canonical ensemble'.

We have spoken of 'resource concentrations' but have not made the spatial element explicit. In fact, the notion of a cellular model, for example, is not as straightforward as in the molecular case. In the molecular case the counts for different cells could be independent Poisson and this was consistent with Poisson statistics for the whole region. However, it is not true that a sum of independent geometric variables is geometric, although it does remain of geometric type (see Exercise 1). Indeed, one would expect that passage of organisms between cells would not be the blind migration of the molecular model, but a deliberate search for food, so that the spatial rules may be rather different.

We shall incorporate spatial effects simply by making the abundance M proportional to V; the 'thermodynamic limit' is again obtained by letting V become large for a fixed value of the abundance concentration vector

$$\rho = M/V$$

We shall not, as a rule, let dynamics otherwise depend upon V. The assumption is then that organisms hunt nutrient actively enough that they respond to the free resource total m rather than to the free resource density m/V.

Exercises and comments

Consider the geometric distribution with p.g.f. proportional to $(1 - \gamma z)^{-1}$. Then the sum of V independent such variables has p.g.f. proportional to $(1 - \gamma z)^{-V}$. The distribution is changed but, since the region of convergence is unaffected by a powering of the generating function, the distribution remains of geometric type.
2. The reversible stochastic specification (1) corresponds to a deterministic model

$$\dot{n}_r = \lambda_r(n) - \mu_r(n)$$

$$= k_r(m, n) \frac{\partial}{\partial n_r} U(n) \qquad (11)$$

where

$$U(n) = \log\left[\Psi(M - An) \prod_r \Phi_r(n_r)\right]$$

and $k_r(m, n)$ is a rate factor, required only to stay positive. The kinetics (11) seek a maximum of $U(n)$; i.e. of the distribution (2) constrained by (3).
3. If one does not postulate purposeful and unfatiguing hunting of resources by the organism, but rather local browsing plus random migration, then one must go to the thermodynamic limit by study of a cellular model. If we postulate spatial homogeneity then distribution (2) will generalize to

$$\pi(\mathbf{m}, \mathbf{n}) \propto \prod_{h=1}^{V} \Psi(m_h) \prod_r \Phi_r(n_{hr}) \qquad (12)$$

constrained by

$$\sum_h \sum_r a_{\alpha r} n_{hr} + m_{h\alpha} = M_\alpha \qquad (13)$$

The additional subscript h indicates the cell. That (13) should be the only constraint indicates mobility, either of organisms or of resources. If distribution (12) with geometric-type Φ and Ψ is to satisfy detailed balance relations on migration then total migration rates for a given species or resource on a given route must be intrinsically bounded above. One could imagine mechanisms which would make this plausible.

It follows from (12), (13) that the p.g.f. of total population numbers

$$\Pi(z) = E\left(\prod_r z_r^{n_r}\right), \quad n_r = \sum_h n_{hr}$$

is proportional to the coefficient of $\prod_\alpha w_\alpha^{M_\alpha}$ in the expansion of $\Pi(z, w)^V$, with $\Pi(z, w)$ still given by expression (10).

3. THE SIMPLEST CASE

Let us consider the simplest case, when the geometric behaviour (2.4), (2.6) is followed exactly. The generating functions (2.8), (2.9) then have the forms

$$\tilde{\Phi}_r(z_r) = (1 - e^{\phi_r} z_r)^{-1} \tag{1}$$

$$\tilde{\Psi}(w) = \prod_\alpha (1 - e^{\psi_\alpha} w_\alpha)^{-1} \tag{2}$$

equivalent to an assumption that the n_r and m_α follow independent geometric distributions in the open process. As ever, one determines the p.g.f. for the closed process from

$$\Pi(z) \propto C_M \Pi(z, w) \tag{3}$$

where $\Pi(z, w)$ is given by expression (2.10) and C_M indicates the operation of extracting the coefficient of $\prod_\alpha w_\alpha^{M_\alpha}$.

The parameters ϕ_r and ψ_α fix levels statistically, they also fix the regions of convergence of these generating functions as $|z_r| < e^{-\phi_r}$, $|w_\alpha| < e^{-\psi_\alpha}$. The point of choosing the constants in exponential form will soon become apparent.

We have

$$\Phi_r(n_r) = \exp(\phi_r n_r)$$

$$\Psi(m) = \exp\left(\sum_\alpha \psi_\alpha m_\alpha\right) \tag{4}$$

for non-negative integral arguments, zero otherwise. Suppose that we choose the simple form

$$\mu_r(n) = \mu_r n_r \tag{5}$$

for the death rate, where μ_r is a constant. Then (2.1) and expressions (4) imply the form

$$\lambda_r(n) = \begin{cases} \gamma_r(n_r + 1) & (m \geqslant a_r) \\ 0 & \text{otherwise} \end{cases} \tag{6}$$

for the birth-rate, with

$$\gamma_r = \mu_r \exp\left(\phi_r - \sum_\alpha \psi_\alpha a_{\alpha r}\right)$$

The form of birth-rate (6) indeed expresses in starkest form the type of behaviour we should wish. The total birth-rate is of order n_r, but is actually proportional to $n_r + 1$. The extra unit can be regarded as a trickle of immigration that saves the population from chance extinction. The total birth-rate is zero if the constituents for a new individual are not available from free resources; it is otherwise independent of m.

Let us again denote concentrations by $c_r = n_r/V$, $\rho_\alpha = M_\alpha/V$, and use c, ρ, ϕ and ψ to denote column vectors of the relevant scalars.

Theorem 3.1

(i) *The model with total birth- and death-rates* (6), (5) *has equilibrium distribution*

$$\pi(m, n) \propto \exp(\phi^\mathsf{T} n + \psi^\mathsf{T} m) \tag{7}$$

in its open version. The equilibrium distribution for the closed model is then

$$\pi(n) \propto \exp[\phi^\mathsf{T} n + \psi^\mathsf{T}(M - An)] \tag{8}$$

for non-negative integral n_r satisfying

$$An \leqslant M \tag{9}$$

otherwise it is zero.

(ii) *The most probable equilibrium value of $c = n/V$ for the closed model is determined by solution of the linear programme: Maximize $(\phi^\mathsf{T} - \psi^\mathsf{T} A)c$ with respect to c subject to $c \geqslant 0$, $Ac \leqslant \rho$.*

(iii) (The principle of competitive exclusion) *For the most probable equilibrium configuration the number of species surviving does not exceed the number of critical (i.e. fully utilized) resources, save in structurally unstable special cases.*

Proof Expression (7) is indeed the general solution (2.2) for this particular case, and (8) is its constrained version. In passing to (ii) we have simply taken the scaled variable $c = n/V$ and neglected the constraints of integrality; justifiable in the thermodynamic limit.

The extremal problem of maximizing $\pi(n)$ is then indeed a linear programme, in the classic form of maximizing a linear form in the positive orthant subject to linear inequalities. Assertion (iii) reflects a well-known property of such problems: that the optimal solution is *basic* save in fortuitously degenerate cases. That is, that the total number of variables c_r and slack variables $\rho_\alpha - \sum_r c_r a_{\alpha r}$ which are positive in the optimal solution does not exceed the number of constraints p. This is equivalent to assertion (iii). ∎

The theorem thus fulfils our two aims, at least for this particular model. In (iii) we have a statement of the principle of competitive exclusion, and in (ii) an extremal characterization of the equilibrium configuration.

Of course, the term $(n_r + 1)$ in the birth-rate (6) ensures that no species will actually become extinct. In assertion (iii) we are implying the understanding that species r survives or is extinct according as to whether c_r is positive or zero, i.e. according as to whether n_r is $O(V)$ or $o(V)$.

The term $\Psi(m - a_r)/\Psi(m)$ occurring in (2.1) indicates, at least for a death-rate (5), the rate at which an individual of species r takes up resources in order to reproduce. For the simple model of this section this rate is constant if $m \geqslant a_r$, zero otherwise, and so has the rather stark form of curve (i) in Fig. 8.3.1. A rather more realistic response would be (ii): for the reproduction rate to increase with increasing resource availability but then to level out as factors other than resource limitation become critical. For the molecular case one presumes that there are no other such limitations, and that the function follows the linear course (iii).

We shall consider the effect of relaxing assumptions in the direction of (ii) in section 5, but there is another interesting point to be covered first.

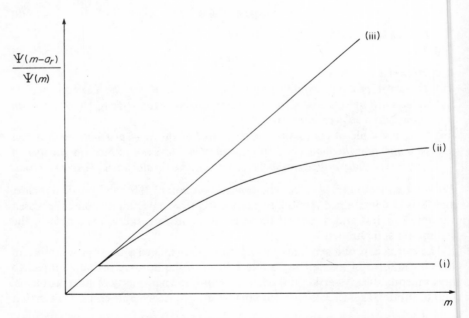

Fig. 8.3.1 The rate of uptake of resources by individuals as a function of free resource availability m for (i) the simple case of this section, (ii) the more realistic case of an increasing but ultimately bounded uptake rate, and (iii) the 'molecular' (chemical) case, in which rate remains proportional to availability.

Exercises and comments

1. Assertion (iii) of the theorem is elementary in the single-resource case $p = 1$. In this case c_r will be positive in the most probable configuration only for the r which maximize $(\phi_r - \psi a_r)/a_r$, i.e. which maximize ϕ_r/a_r. Only in special cases will there be more than one such value.

2. Note that the 'entropy' measure for this problem is the linear form $S(c, \zeta) = \phi^{\mathsf{T}}c + \psi^{\mathsf{T}}\zeta$.

3. To prove convergence (in probability, say) of n/V to the maximizing c in the thermodynamic limit requires methods other than those used (see section 5.10) in the case of Poisson statistics. The simplest course is to note that the constrained distribution (8) has a unique maximum (save in degenerate cases) and that it falls away exponentially fast as n departs from the maximizing value.

4. THE INTERPRETATION OF THE DUAL PROBLEM

The extremal problem dual to the linear programme of Theorem 3.1 (ii) is: choose θ to minimize $\theta^{\mathsf{T}}\rho$ subject to

$$\theta^{\mathsf{T}}A \geqslant \phi^{\mathsf{T}}$$
$$\theta \geqslant \psi \tag{1}$$

The two problems have related solutions in that for c_r to be positive (i.e. for species r to survive) in the first problem requires that $\sum_\alpha \theta_\alpha a_{\alpha r} = \phi_r$ in the second, and for $\sum_r a_{\alpha r} c_r$ to be less than ρ_α (i.e. for resource α to be present in excess) requires that $\theta_\alpha = \psi_\alpha$.

The experience of section 5.6 leads us to expect that the dual problem can be interpreted in terms of an evaluation of $E(n/V)$ in the thermodynamic limit. This is true, but there are new features.

Note the general formula

$$E(n_r) = \frac{\oint \ldots \oint \left[\dfrac{\partial}{\partial z_r} \Pi(z, w)\right]_{z = 1} \prod_\alpha w_\alpha^{-M_\alpha - 1} dw_\alpha}{\oint \ldots \oint \Pi(1, w) \prod_\alpha w_\alpha^{-M_\alpha - 1} dw_\alpha} \tag{2}$$

where $\Pi(z, w)$ is given by expression (2.10). The two integrals generally share singular points, saddle-points, etc. so that examination of the lower integral is sufficient to reveal the state of affairs.

For the Poisson statistics of sections 5.6 and 5.8 the whole integrand $\Pi(1, w)\prod_\alpha w_\alpha^{-M_\alpha}$ was of the form $\exp(VJ(w))$ for V-independent $J(w)$, and asymptotic evaluation became a matter of determining the relevant saddle-points of $J(w)$. This is no longer the case. For the special model of the last section we have

$$\Pi(1, w) = \prod_\alpha (1 - e^{\psi_\alpha} w_\alpha)^{-1} \prod_r (1 - e^{\phi_r} \prod_\alpha w_\alpha^{a_{\alpha r}})^{-1} \tag{3}$$

There is thus no V-dependence, but there is a failure of convergence outside a finite w-region.

Let us make the familiar change of variable $w_\alpha = \exp(-\theta_\alpha)$. In the general case the integrand then becomes $K(\theta) \exp(V\theta^\mathsf{T}\rho)$, where

$$K(\theta) := \Pi(1, w) = \left(\sum_m \Psi(m) e^{-\theta^\mathsf{T} m} \right) \left(\prod_r \sum_{n_r} \Phi_r(n_r) e^{-n_r \theta^\mathsf{T} a_r} \right) \tag{4}$$

Let D be the set of real θ for which $K(\theta)$ can be expanded in a convergent power series in the $w_\alpha = \exp(-\theta_\alpha)$. Then the generalization of the saddle-point evaluation of section 5.8 to this case is: choose real θ to minimize $\theta^\mathsf{T}\rho$ subject to $\theta \in D$. For the particular case (3)

$$K(\theta) = \prod_\alpha (1 - e^{\psi_\alpha - \theta_\alpha})^{-1} \prod_r (1 - e^{\phi_r - \theta^\mathsf{T} a_r})^{-1} \tag{5}$$

so that the set D is exactly that prescribed by constraints (1). So, the dual problem corresponds, indeed, to location of a restricted saddle-point in integrals (2).

The boundary of D marks the limit of the region of convergence of $K(\theta)$ for real θ. Since $\theta^\mathsf{T}\rho$ has no free minimum in D the saddle-point θ will lie against the boundary of D. Just where it does so (i.e. which inequalities in (1) become equalities) is a matter of prime interest. If the rth inequality of the first set is an equality, so that

$$\theta^\mathsf{T} a_r = \phi_r \tag{7}$$

then the rth species survives in that $E(n_r)$ is of order V (see Exercise 1). If the αth inequality of the second set becomes an equality, so that $\theta_\alpha = \psi_\alpha$ then there is a surplus of resource α in equilibrium.

We shall see in another context (section 9.4) that divergence of a generating function can characterize occurrence of a bottleneck, i.e. of a tendency of units to accumulate overwhelmingly at some point in the system. The point of divergence in transform space has some relation to the point of accumulation in the system. In this case, the location of the saddle-point on the boundary of D is described by the constraints which are active in (1), and these correspond respectively to the species which survive and the resources which are in surplus. These represent the points at which resources accumulate in the equilibrium configuration, and so represent what one would regard as 'bottlenecks' for resources if circulation of resources were the aim.

For Poisson systems the saddle-point is a free one. Correspondingly there are no bottlenecks; there is free flow in that an accumulation clears itself at a rate proportional to its size.

The dual problem also has the interpretation of a maximum likelihood estimation of parameters from observed m_α and n_r, as in section 5.9. The joint p.g.f. of m and n for the open system is

$$E\left(\prod_\alpha \prod_r y_\alpha^{m_\alpha} z_r^{n_r}\right) \propto \prod_\alpha \prod_r (1 - y_\alpha e^{\psi_\alpha - \theta_\alpha})^{-1} (1 - z_r e^{\phi_r - \theta^\mathsf{T} a_r})^{-1} \qquad (8)$$

Let us regard the quantities $w_\alpha = \exp(-\theta_\alpha)$ as parameters, determining levels, rather than as marker variables. Now, a p.g.f. proportional to $(1 - yz)^{-1}$ for n corresponds to a distribution $P(n) = (1 - \gamma)\gamma^n$ and the parameter γ necessarily satisfies $0 \leqslant \gamma < 1$. The p.g.f. (8) hence corresponds to a distribution

$$\pi(m, n) = \left[\prod_\alpha \prod_r (1 - e^{\psi_\alpha - \theta_\alpha})(1 - e^{\phi_r - \theta^\mathsf{T} a_r})\right] e^S \qquad (9)$$

where

$$\begin{aligned} S &= \sum_\alpha (\psi_\alpha - \theta_\alpha)m_\alpha + \sum_r (\phi_r - \theta^\mathsf{T} a_r)M_r \\ &= \psi^\mathsf{T} m + \phi^\mathsf{T} n - \theta^\mathsf{T} M \\ &= \psi^\mathsf{T} m + \phi^\mathsf{T} n - V \theta^\mathsf{T} \rho \end{aligned} \qquad (10)$$

The location of the maximum likelihood estimate of θ is then just the dual minimization problem stated at the beginning of the section, with observed density $\rho = M/V$ replacing the theoretical limiting density.

This then gives an interpretation of the dual variables. The change in parameters of the problem

$$\psi \to \psi - \theta$$
$$\phi \to \phi - A^\mathsf{T}\theta$$

does not affect the distribution of m and n conditional on M, but does affect the unconditional distribution. We choose the value of θ which maximizes the likelihood calculated on the basis of the open distribution.

Exercises and comments

1. Prove from expression (2) for $E(n_r)$ and the corresponding one for $E(m_\alpha)$ that $E(n_r)$ is of order V or smaller according as equality (7) holds or not at the saddle-point θ, and that $E(m_\alpha)$ is of order V or smaller according as θ_α is equal to or greater than ψ_α.

2. For the 'local' version of the model formulated in Exercise 2.3 formula (2) will be modified by the substitution of $\Pi(z, w)^V$ for $\Pi(z, w)$. The saddle-point of the integrand is then the value of θ minimizing $K(\theta) \exp(-\theta^\mathsf{T}\rho)$, where $K(\theta)$ is still defined by (4). If the matrix A is of full rank and the numbers of species and resources are finite then this minimum always exists, and is attained at a value θ in the strict interior of the region of convergence D. Thus all species are represented and all resources present in free form to some extent. In other words, the assumption that food-gathering is random rather than purposeful invalidates the principle of competitive exclusion.

3. For the local version of the simplest model the distribution of cell counts $m_{h\alpha}$ and n_{hr} in the open process will be independent for different cells, the distribution in a given cell h still being given by expression (9) (with m_α, n_r and M_α all made h-dependent). Show that the saddle-point evaluation stated in (2) corresponds again to the maximum likelihood evaluation of θ, and also to the calculation of the most probable value of (m, n) subject to $An + m = M$, where these quantities refer to totals over cells (i.e. $n_r = \sum\limits_h n_{hr}$, etc.).

5. THE MORE GENERAL CASE

Formulations are more natural if the factors Φ_r, Ψ are put into exponential form

$$\Phi_r(n_r) = \exp[\phi_r(n_r)]$$

$$\Psi(m) = \exp[\psi(m)]$$

so that the equilibrium distribution becomes

$$\pi(n) \propto \exp\left[\sum_r \phi_r(n_r) + \psi(m)\right] \qquad (1)$$

(cf. (2.2)) where, as ever, $m = M - An$.

Exponential behaviour of $\Phi_r(n_r)$ then corresponds to linear behaviour of $\phi_r(n_r)$. What one would expect of $\phi_r(n_r)$ in the most regular cases would be that it is concave and with a linear asymptote of finite slope as n_r becomes large. We formalize this assumption:

(C) The functions $\phi_r(n_r)$ and $\psi(m)$ are concave and have a linear asymptote of finite slope as the argument increases in any given direction.

The last assumption implies that, for a fixed unit vector u, the ratio $\psi(su)/s$ has a finite limit as the positive scalar s becomes infinite: the asymptotic slope of ψ in direction u. Since this slope can depend upon u in any manner consistent with concavity, one is not assuming that $\psi(m)$ is asymptotically linear, but that it is asymptotically linear in scale. In other words, the tendency for an organism to take up a given resource may depend upon the proportions in which free resources are present, but not on their scale if this is large enough.

Minor deviations from concavity and asymptotic linearity would be unimportant, as they would not affect the essential character of the most probable configuration in the thermodynamic limit. Indeed, in determining behaviour in the thermodynamic limit one might as well replace the functions ϕ_r and ψ by their least concave majorants. One should regard Assumption C as a convenient regularizing condition, which can be considerably relaxed without vitiating the essential asymptotic validity of the analysis to come.

A quantity that arises naturally is the *maximum transform* of a function, that of ψ being

$$\psi^-(\theta) = \max_{m \geqslant 0} [\psi(m) - \theta^\mathsf{T} m] \tag{2}$$

The maximum in (2) can be taken over all m if ψ is continued by being given the 'forbidding' value of $-\infty$ outside the positive orthant. The *minimum transform* of this expression

$$(\psi^-)^+(m) = \min_\theta [\psi^-(\theta) + \theta^\mathsf{T} m] \tag{3}$$

is just the least concave majorant of ψ (see Appendix 3), so that relation (3) supplies the inversion of (2) if ψ is itself concave (i.e. coincides with its own least concave majorant on $\mathbb{Z}_+^p$).

Assumption C will imply that the maximum transform $\psi^-(\theta)$ of ψ and the *Laplace transform*

$$\hat{\Psi}(\theta) = \sum_m \Psi(m) e^{-\theta^\mathsf{T} m} \tag{4}$$

of Ψ have the same convex region F of convergence for real θ. This implies the evaluation

$$\lim_{s \to \infty} \frac{\psi(su)}{s} = \inf_{\theta \in F} (\theta^\mathsf{T} u)$$

for the asymptotic slope of ψ in direction u. The Laplace transform $\hat{\Psi}(\theta)$ is of course derived from the power transform $\tilde{\Psi}(w)$ by the substitutions $w_\alpha = \exp(-\theta_\alpha)$.

Theorem 5.1   (i) *The most probable configuration (m, n) is obtained by maximizing* $\sum_r \phi_r(n_r) + \psi(m)$ *subject to*

$$An + m = V \tag{5}$$

(ii) *If one neglects integrality of (m, n) then the dual of the extremal problem (i) is: minimize*

$$\sum_r \phi_r^-(\theta^\mathsf{T} a_r) + \psi^-(\theta) + V\theta^\mathsf{T}\rho \tag{6}$$

with respect to θ.

(iii) *(The principle of competitive exclusion.) Under Assumption C the most probable population size n_r is of order V for large V for at most p values of r, save for structurally unstable special cases.*

Proof Assertion (i) is just a statement of the maximization problem and assertion (ii) a statement of its dual, with θ the Lagrange multiplier vector for constraints (5).

Assertion (iii) appeals to the fact that, if n_r is large, then $\phi_r(n_r)$ is on its linear asymptote and we may set

$$\phi_r(n_r) \sim \phi_r n_r$$

where the constant $\exp(\phi_r)$ is, as in section 3, the reciprocal of the radius of convergence of the generating function $\tilde{\Phi}_r(z_r)$. The problem then becomes one of maximizing

$$\sum_r' \phi_r n_r + \psi\left(V\rho - \sum_r' a_r n_r\right) \qquad (7)$$

where the sum $\sum'$ covers the values $r \in \mathscr{R}_s$ corresponding to species which survive. Suppose we continue $\psi(m)$ differentiably outside the positive orthant in such a way that it quickly acquires a large negative value, essentially prohibiting values of m outside the positive orthant. The maximality condition for expression (7) then becomes a stationarity condition

$$\phi_r = \delta a_r \qquad (r \in \mathscr{R}_s) \qquad (8)$$

where δ is the row p-vector of differentials of ψ at the most probable m-value. One cannot expect to satisfy (8) for more than p values of r, the dimensionality of the parameter vector δ occurring in the solution. ∎

The statement (iii) of the principle of competitive exclusion is weaker than the corresponding statement (iii) of Theorem 3.1. This is because we are not now asserting that $\psi(m)$ is linear for large m, merely that it is linear in scale.

Evaluation of $E(n_r)$ gives us an asymptotic form of the dual rather than the dual itself. Under Assumption C the set of real θ for which $\Phi_r(\theta)$ is finite is just

$$\theta_r > \phi_r \qquad (r \in \mathscr{R})$$

and the set of real θ for which $\hat{\Psi}(\theta)$ converges is just F.

Theorem 5.2 (i) *The modified saddle-point θ in the integrals of expression* (4.2) *is obtained by minimizing* $\theta^\mathsf{T}\rho$ *subject to*

$$\left.\begin{array}{l} \theta^\mathsf{T}a_r > \phi_r \qquad (r \in \mathscr{R}) \\ \theta \in F \end{array}\right\} \qquad (9)$$

which is an asymptotic form of the dual of Theorem 5.1 (ii).
(ii) *The dual of this dual is: choose n to maximize*

$$\sum_r \phi_r n_r + \Delta(V\rho - An) \qquad (10)$$

in $n \geqslant 0$, where

$$\Delta(v) := \inf_{\theta \in F} (\theta^T v)$$

Proof   Assertion (i) follows from the general considerations of section 4; the set (9) is just the set D of real θ for which the generating function $K(\theta)$ converges. We recognize the problem formulated in (i) as an asymptotic (for large V) form of the problem of minimizing expression (6), in which, for example, one makes the effective approximation

$$\psi^-(\theta) = \begin{cases} 0 & \theta \in F \\ +\infty & \theta \notin F \end{cases}$$

The problem characterized in (ii) is indeed the formal dual of (i), with its variable identified with n. Explicitly, if we associate a Lagrangian multiplier n_r with the rth inequality constraint in (9) we then have the problem of minimizing the expression

$$V\theta^T \rho + \sum_r n_r (u_r + \phi_r - \theta^T a_r)$$

with respect to u, θ in $u \geqslant 0$, $\theta \in F$. The minimized value is just expression (10) if $n \geqslant 0$ and $-\infty$ otherwise. The vector n must have the value maximizing this expression, whence we deduce assertion (ii). ∎

Note that the effect of taking the dual of the asymptotic dual is to recover a 'stripped-down' form of the original problem of maximizing $\pi(n)$, in which $\phi_r(n_r)$ and $\psi(m)$ are replaced by large-argument forms of their least concave majorants: $\phi_r n_r$ and $\Delta(m)$.

Exercises and comments

1. Suppose that $\tilde{\Phi}_r(z_r)$ is given by the v_r th power of the form assumed in (3.1) and that the death-rates continue to be given by (3.5). Show that assumption (2.1) then yields the expression

$$\lambda_r(n) = \mu_r(n_r + v_r)\Psi(m - a_r)/\Psi(m)$$

for the birth-rate. The effect of powering up the form of $\tilde{\Phi}_r$ (so making $\phi_r(n_r)$ rise initially at a faster rate than its linear asymptote) is to increase the effective immigration rate v_r. This will not affect asymptotic conclusions, if v_r is independent of V.

Deduce the corresponding result if $\tilde{\Psi}$ factors into similar terms.

2. Calculate ϕ_r^-, ψ^- and Δ for the particular choices of $\tilde{\Phi}_r$ and $\tilde{\Psi}$ suggested in Exercise 1.

3. It is interesting to note what happens if any of the functions Φ_r or Ψ does *not* show geometric-type behaviour, in that its power (or Laplace) transform converges everywhere. Suppose, for example, that the Φ_r have the simple geometric forms (3.4) but that we choose

$$\Psi(m) = \prod_\alpha \frac{\exp(m_\alpha \psi_\alpha)}{m_\alpha!} \tag{11}$$

The effect of this will be that organisms respond proportionately to the concentration of ambient resources, and that their birth-rates increase indefinitely as levels of free resources rise. In fact, under Assumptions (2.1), (3.1), (3.5) and (11) we have

$$\lambda_r(n) = \mu_r(n_r + 1) \prod_\alpha m_\alpha^{(a_{\alpha r})} e^{\psi_\alpha a_{\alpha r}}$$

The primal problem is then the maximization of

$$\sum_r \phi_r n_r + \sum_\alpha [\psi_\alpha m_\alpha - \log(m_\alpha!)]$$

subject to $An + m = M$. The dual is the minimization of $\theta^\mathsf{T} \rho$ subject only to $\theta^\mathsf{T} A \geqslant \phi^\mathsf{T}$, and the 'dual dual' the maximization of $\sum_r \phi_r n_r$ subject to $n \geqslant 0$ and

$$An = M \tag{12}$$

The dual has minimum $-\infty$ and the 'dual dual' is infeasible unless (12) has a non-negative solution for n. That is, unless there are values of population levels which exactly use all available resources (at least in that the proportion unused tends to zero as V increases). The reason for this is that the choice (11) implies an overwhelming incentive to reduce free resources to zero. Determine the actual character of the solution of the primal.

4. Again, the modification of (2.2) to

$$\pi(m, n) \propto [\Psi(m) \exp(-\theta^\mathsf{T} m)] \left[\prod_r \Phi_r(n_r) \exp(-n_r \theta^\mathsf{T} a_r) \right] \tag{13}$$

does not affect the distribution of (m, n) constrained by $An + m = M$, but does affect the open (unconstrained) distribution. If we normalize the unconstrained distribution (13) then we see that the maximum likelihood estimate of θ is that maximizing $[\exp(-V\theta^\mathsf{T}\rho)]/[\hat{\Psi}(\theta) \prod_r \hat{\Phi}_r(\theta^\mathsf{T} a_r)]$. For large V this will again amount to the solution of the dual problem stated in Theorem 5.2(i).

PART III

Network models

A system could be considered to constitute a *network* if it consists of a fixed set of components, these components interacting by transmission of units of some kinds through a fixed pattern of interconnections. Such a system can be seen as a graph, the components being identified with the nodes and the interconnections with the arcs.

Electrical networks provide the obvious example. These are not usually seen stochastically, but they certainly can be formulated as stochastic networks, with particular symmetry and linearity properties (at least in the case of resistance-capacity networks; see Exercise 1.2).

However, the kind of example we have in mind is something like a network of queues. The queues themselves constitute the components (or nodes) and the prescribed routing pattern between them constitutes the arcs. The customers constitute the units, and these may indeed be of several and changing kinds, according to their needs and their histories in the system.

For another example, consider physical production in a factory. The various fixed working stations with their machines, workers or stores constitute the components. They are effectively the nodes of a network the arcs of which are the routing rules which prescribe the sequence of operations. The units themselves are the jobs, of different and changing types according to what is being made and its degree of completion.

However, as soon as one speaks of networks of processors one thinks of computers, and it is the analysis of computer operation which has provided enormous and renewed impetus to the study of networks. In this case the components are of course the various processors and the units are the jobs passed between them. It is a computer that perhaps provides the most vivid picture of a large complex system simultaneously handling many jobs. The jobs are independent in that they constitute separate 'packets', interacting in that they compete for processing capacity and with a course through the system unpredictable enough to be best viewed statistically. A statistical view of the system is natural, as is the consciousness of the overall efficiency of the system and the factors which limit it.

Other examples are provided by communication, traffic and distribution networks. However, the example that beats the computer for vividness, complexity, subtlety and economy is the animal brain. This is a network of neurons, exchanging impulses. Such neural nets constitute the standing challenge: of operation to be understood, of large-scale high-level phenomena to be conceived and of performance to be matched.

It must be admitted that this enumeration has become a list of examples one would *like* to be able to treat rather than of those which *can* adequately be treated at the moment. We shall be giving the theory of the so-called *Jackson networks*, which have been found useful in representing and analysing the behaviour of production, queueing and processing networks. We shall also, through the 'weak coupling' concept, construct strongly generalized forms of such networks. However, it remains true, as we shall see, that even these generalized versions cannot represent true 'blocking', or competition for processors (although they can represent congestion). Neither can they represent fundamental operations of the nature of assembly or disassembly (although they can represent a progressive processing of individual units). Nor is any model available at the moment which gives substantial insight into the neural network.

Nevertheless, the Jackson networks do represent a clear and definite stage of attainment, with a natural relation to much of the material of Part II, and with quite penetrating general implications (see Chapters 11 and 12). It is moreover possible that they represent the first level of a hierarchy of progressively more 'intelligent' networks which are natural and tractable; see section 9.9.

The components of the statistical mechanical models of Part II could be regarded as constituting a fixed network so long as they are not allowed to vary in number and position, and so in mutual interaction pattern. Now that components are fixed in mutual relation we regard them as identifiable and, in general, different. One might see the processors of a computer, exchanging jobs, as a development of the picture of statistically identical molecules exchanging energy. However, the processors are individually identifiable and, in general, quite different in form and function. Moreover, their exchanges are in some sense more purposeful, and we shall find it necessary and possible to greatly relax the reversibility hypotheses of Part II.

This introduction to Part III began by defining a network. One cannot then resist repeating Dr Samuel Johnson's definition of the same object, one of the celebrated entries in his Dictionary. 'Network: Any thing reticulated or decussated, at equal distances, with interstices between the intersections.' This is a very spatial conception, with a regular geometry imposed. In our days the concept has been extended to any abstract pattern of connections, with the addition of the dynamic element of some kind of communication or transport. Could Dr Johnson now rewrite his definition in these terms, just imagine its majesty!

CHAPTER 9

Jackson networks

1. THE BASIC MODEL

One can for concreteness think of the model as describing the migration of customers between queues or of jobs between processors. However, to keep nomenclature standard we shall simply speak of units moving between nodes. This movement constitutes a stochastic flow through the network.

Suppose that there is only one type of unit and that there are n_j units at node j ($j = 1, 2, \ldots, m$). For the moment we shall assume the situation at node j adequately described by specification of n_j, in that the m-vector of occupation numbers $n = \{n_j\}$ is the state variable of a continuous-time Markov process. The intensity of the transition $n \to n'$ will be denoted $\theta(n, n')$; the more usual symbols λ or Λ will be required for other service, as we shall see.

It will already be apparent that there is something of a shift in notation from Part II. The total number of components is denoted by m (rather by N) and the total number of units denoted N (rather than M); we shall later denote the number of units of type α by N_α. The symbol n_j denotes the number of units at node j rather than the number of components of a particular type. Notation will certainly be consistent within a Part. Complete consistency between Parts is difficult to achieve, partly because of history in different fields and partly because of the finiteness of the alphabet.

Let e_j be an m-vector with a unit in the jth place, zeros elsewhere. It thus represents a system with a single unit, this being sited at node j. Indeed, we shall often use e_j simply to denote such a unit. The transition $n \to n - e_j + e_k$ represents migration of a single unit from node j to node k. Let this be the only type of transition permitted, for varying j, k. The system is then *closed*, in that the total number of units

$$N = \sum_j n_j \tag{1}$$

is invariant. Let us suppose, more specifically, that the migration transition has intensity

$$\theta(n, n - e_j + e_k) = \lambda_{jk} \phi_j(n_j) \tag{2}$$

So, the exit rate of units from node j is a function of n_j alone, with a dependence represented by $\phi_j(n_j)$. We assume negative occupation numbers forbidden (so that $n \in \mathbb{Z}_+^m$) and must then have

$$\phi_j(0) = 0 \tag{3}$$

The matrix (λ_{jk}) is referred to as the *routing matrix*; it specifies the way a unit is routed around the network once it has been emitted from a node. Indeed, one may say that the λ_{jk} specify the arc-characteristics of the network and the $\phi_j(n_j)$ the node-characteristics.

Theorem 1.1 *Consider the simple migration process with transition intensity* (2). *This has equilibrium distribution*

$$\pi(n) \propto \Phi(n) \prod_j w_j^{n_j} \tag{4}$$

on a given irreducible set of states, where

$$\Phi(n) = \prod_j \Phi_j(n_j) := \prod_j \prod_{r=1}^{n_j} \phi_j(r)^{-1} \tag{5}$$

and the w_j solve the linear equations

$$\sum_k (w_k \lambda_{kj} - w_j \lambda_{jk}) = 0 \qquad (j = 1, 2, \ldots, m) \tag{6}$$

Distribution (4) *satisfies the partial balance relations*

$$\sum_k \left[\pi(n - e_j + e_k)\theta(n - e_j + e_k, n) - \pi(n)\theta(n, n - e_j + e_k) \right] = 0$$

$$(j = 1, 2, \ldots, m) \tag{7}$$

Proof The distribution $\pi(n)$ specified by (4)–(6) is readily found to satisfy the partial balance equations (7). It then also satisfies the full balance equation, obtained by summing relation (7) over j. Being a stationary distribution, it constitutes the equilibrium distribution on any prescribed irreducible closed set of states on which it is not identically zero. ∎

If the process is indeed irreducible in that $\sum_j n_j$ is the only invariant, then distribution (4) is the equilibrium distribution on the set specified by (1), for prescribed N. Were it not for this constraint the factored form of (4) (see (5)) would indicate that the n_j were independent $(j = 1, 2, \ldots, m)$. One is then inclined to regard the unconstrained distribution (4) as the equilibrium distribution for an as yet hypothetical *open* form of the process, a distribution conditioned in the closed case by constraint (1). We shall confirm this interpretation in section 3.

By choosing $\phi_j(n_j)$ appropriately one can represent various types of behaviour. The choice

$$\phi_j(n_j) = n_j \tag{8}$$

would be the consequence of an assumption that the units at node j behaved independently, the exit intensity having the same constant value $\lambda_j := \sum_k \lambda_{jk}$ for any unit. One might call such a node a *depôt*, as there is no aspect of queueing. For the same reason it is often termed an 'infinite-server queue'. The 'compartmental models' of metabolism, radioactivity, etc. envisage just such behaviour (see Exercise 1). For such a model one has

$$\Phi_j(n_j) = (n_j!)^{-1} \tag{9}$$

so that n_j is Poisson distributed in the open model.
The choice

$$\phi_j(n_j) = \begin{cases} 0 & n_j = 0 \\ 1 & n_j > 0 \end{cases} \tag{10}$$

would correspond to the assumption that node j behaves as a *simple queue*, i.e. a single-server queue with exponential service times. This behaviour implies that

$$\Phi_j(n_j) = 1 \tag{11}$$

so that n_j is geometrically distributed in the open process.
The choice

$$\phi_j(n_j) = \min(n_j, s), \tag{12}$$

intermediate between (8) and (10), would be the consequence of assuming that the node j behaved as an s-server queue (for positive integral s). This family spans cases (10) to (8) as s varies between 1 and $+\infty$. See Exercise 6 for another useful spanning family.

The parameters w_j are not determined in scale by equations (6). If we normalize their scale so that $\sum_j w_j = 1$ then w could be regarded as the equilibrium distribution of what we shall term the *singleton process*: the process in which a single unit moves around the sites of the network, the transition $j \to k$ having intensity λ_{jk}. This characterization is a useful one, but what is probably the essential characterization will be given in Theorem 3.3.

We shall increasingly employ the useful notation of sections 2.3 and 2.7, according to which we would write the balance equation (6) as

$$[\mathcal{J}, j](\lambda w) = 0 \tag{13}$$

Here $\mathcal{J}$ is the set of sites $\{1, 2, \ldots, m\}$, the state space of the singleton process. The partial balance equations (7) would correspondingly be written

$$[C_j(n), n](\theta \pi) = 0 \qquad (j = 1, 2, \ldots, m) \tag{14}$$

where $C_j(n)$ is the set of states $\{n - e_j + e_k; k = 1, 2, \ldots, m\}$. In the nomenclature of section 3.1, equations (14) assert partial balance of n against $C_j(n)$, but do not assert partial balance over a set (i.e. of members of that set against the set). They do, however, imply partial balance in the second sense in a fuller description.

Consider a description x in which the units are identified, so that x describes the position (i.e. the node) at which each of the N units is to be found. This description will imply the value of n, as $n(x)$, say. Let us suppose that a transition $x \to x'$ which implies a $j \to k$ migration, i.e. for which

$$n(x') = n(x) - e_j + e_k \tag{15}$$

has intensity

$$\Lambda(x, x') = n_j^{-1}\theta(n, n') \tag{16}$$

where θ has the previous prescription (2). Here we have written $n(x)$ and $n(x')$ simply as n and n', a convention we shall follow fairly consistently. Assumption (16) implies a symmetry in the treatment of all individuals at node j: if there is to be a migration out of node j then the migrant is equally likely to be any one of the n_j units there. Otherwise expressed, the total *service rate* $\sum_k \theta(n, n - e_j + e_k)$ at node j is equally divided between all units there.

Theorem 1.2 *Under assumptions* (2), (16) *the full description x has equilibrium distribution*

$$\pi(x) = \frac{N!}{\prod_j n_j!} \pi(n) \tag{17}$$

where $\pi(n)$ is given by (4). *If A_{ij} is the set of x for which individual i occupies node j, then $\pi(x)$ shows joint partial balance over all the sets A_{ij} ($i = 1, 2, \ldots, N$; $j = 1, 2, \ldots, m$).*

Proof Equations (7) (or (14)) for $\pi(n)$ are readily seen to amount to

$$[\bar{A}_{ij}, x](\Lambda\pi) = 0 \qquad (x \in A_{ij}) \tag{18}$$

for $\pi(x)$, with Λ given by (16) and $\pi(x)$ given by (17). A given x will belong to several A_{ij}; by adding relation (18) over all such (i, j) for a given x we deduce that the $\pi(x)$ given by (17) indeed satisfies the full balance equation. Relation (18) then implies that this π also shows partial balance over each of the A_{ij}. Because only one migration can take place at a time $\pi(x)$ then shows joint partial balance over all the A_{ij}, in virtue of Theorem 3.1.2. ∎

The 'time-sharing' rule of (16) may of course not be appropriate; one may wish to have a queueing discipline at a node which recognizes factors such as order of arrival. This is a matter to which we shall return in the next chapter.

The Jackson model represents a number of practical situations well or passably; Exercises 1–5 give some examples. In Chapter 4 of his 1979 text Kelly gives an illuminating discussion of a number of examples, including those of Exercises 3 and 4. An adequate formulation of some of these models requires the incorporation of features (e.g. queueing discipline, stage of processing) which can only be achieved by an enrichment of the description of the state of affairs at a node. We consider these matters in the next chapter. However, it must also be recognized that there are effects which cannot be represented by a Jackson network, however much it is elaborated. These include blocking and buffering (Theorem 3.4) and assembly and disassembly (Theorem 10.9.2). Thus, for example, the Erlang and circuit-switching models of Exercises 4.7.3 and 4.7.4 cannot be regarded as Jackson networks. They are nevertheless amenable, as we have seen, and even more can be said about them (see Chapter 12).

Exercises and comments

1. *Compartmental models.* These are the models in which units move independently, so that the nodes all behave as depôts, i.e. as infinite-server queues. They can represent the flow of metabolites through an animal's body, the flow of employees through a hierarchy of grades or the transmutation of radioactive molecules through several states. Since (8) holds for all j then the equilibrium distribution (4) of n becomes

$$\pi(n) \propto \prod_j \frac{w_j^{n_j}}{n_j!}.$$

That is, the n_j follow a multinomial distribution in the closed case, or are independent Poisson in the open case (yet to be clearly formulated; see section 3).

Since the units are following independent Markov processes (their 'state' being the node they occupy) then the equilibrium distribution of occupation numbers n will be unchanged if the Markov process is replaced by a semi-Markov process with the same expected sojourn times at a node and the same routing probabilities between nodes (Corollary 3.5.1). This semi-Markov process can even be specific to the unit. That is, the distribution of 'service time' for each unit at each node may be arbitrary and node/unit-specific without affecting the equilibrium distribution of n, provided that these service times are independent and have the prescribed expectations: $\left(\sum_k \lambda_{jk}\right)^{-1}$ at node j. Jackson networks show such insensitivity more generally, a point we return to in Chapters 10 and 12.

The closed compartmental model can be regarded as a stochastic version of the deterministic linear system

$$\dot{n}_j = \sum_k (n_k \lambda_{kj} - n_j \lambda_{jk}) \tag{19}$$

2. *Electrical networks.* Consider the simple network in which nodes j and k are connected by resistor of conductance a_{jk} and node j is connected to earth by a capacitor of capacity C_j ($j, k = 1, 2, \ldots, m$). Let node j have a potential V_j relative to earth, suppose C_j carries charge Q_j, and let i_{jk} be the net current on the direct link from j to k (so that $i_{kj} = -i_{jk}$ and of course $a_{kj} = a_{jk}$). Then

$$\dot{Q}_j = \sum_k i_{kj}$$

$Q_j = C_j V_j$ and $i_{jk} = a_{jk}(V_j - V_k)$. Eliminating the V_j and i_{jk} from these relations we deduce that

$$\dot{Q}_j = \sum_k (Q_k \lambda_{kj} - Q_j \lambda_{jk}) \tag{20}$$

where $\lambda_{jk} := a_{jk}/C_j$. Comparing (19) and (20) we see the n_j of the compartmental model as a 'stochastic charge' at node j. However, for this interpretation to be possible the model must be *route-reversible* in that a solution w exists to

$$w_j \lambda_{jk} = w_k \lambda_{kj} \tag{21}$$

The w_j solving (21) also solves (6) and can be taken as proportional to C_j, the effective capacity of node j.

3. *Time-sharing processors.* Suppose that N terminals of a computer compete for time on a central processing unit. One can regard this situation as one of N units each migrating between the two nodes corresponding to the states of being 'busy' ($j = 1$) or 'idle' ($j = 2$). The idle node is a depôt, in that each terminal may become active with an intensity λ independent of the states of other terminals. The busy node is a queue with a single server (the processor) dispensing service at rate μ. As things stand at the moment, we can only deal with a time-sharing processor, in which service is shared in that each busy terminal may have its service completed with intensity μ/n_1.

4. *Machine interference.* This is the situation in which each of N machines can be either running or faulty; if faulty it must queue for repair. The model is then the same as that of Exercise 3, with running and faulty machines being the analogue of idle and busy terminals respectively. However, the assumption of time-sharing at the repair shop is no longer realistic; machines will probably be repaired in order of arrival, with pre-emptions or priority in some cases. We return to this model in section 10.5.

5. *Exchange congestion.* We consider the general question of congestion in section 4, but the following closed model has its own particular features. Consider a model in which n_1, n_2 and n_3 denote respectively the numbers of telephone subscribers not seeking connection, seeking connection and connected, through a given telephone exchange. Let the rates of migration be $\lambda_{12} n_1$, $\lambda_{23} n_2 \phi(n_2)$ and $\lambda_{31} n_3$, where the subscripts on λ indicate the migration route. The $\phi(n_2)$ term in the rate of connection indicates an interference effect: that the more subscribers there are seeking connection, the more likely they are to block the lines for each other.

The equilibrium distribution is then

$$\pi(n) \propto \Phi(n_2) \prod_1^3 \frac{w_j^n}{n_j!} \quad \left(\sum_1^3 n_j = N \right)$$

for appropriate w, where $\Phi(n_2) = [\phi(1)\phi(2) \dots \phi(n_2)]^{-1}$. The distribution of the number of callers seeking connection is then

$$\pi(n_2) \propto \Phi(n_2) \binom{N}{n} p^{n_2} (1-p)^{N-n_2}$$

for appropriate p. That is, a binomial distribution modified by the factor $\Phi(n_2)$. Since $\phi(n_2)$ becomes small as n_2 increases then $\Phi(n_2)$ becomes large, and one might expect a distribution of the form illustrated in Fig. 9.1.1. The presence of two peaks in the distribution indicates the existence of two distinguishable regimes, the one at the lower value of n_2 corresponding to relatively uncongested operating conditions, the one at the higher value corresponding indeed to a congested state.

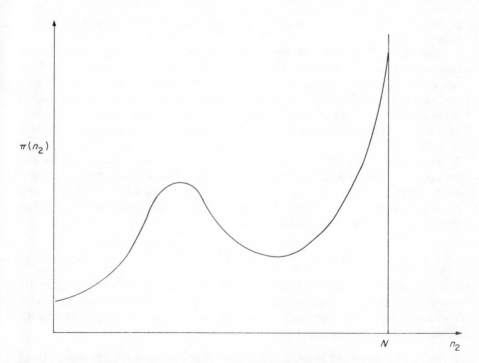

Fig. 9.1.1 The distribution of the numbers of subscribers seeking connection. The bimodal form corresponds to the two distinguishable regimes: of free and of congested working.

The fact that the two regimes communicate indicate that congestion is an occasional rather than a permanent condition, but it is typical for such situations that the system can be stuck in one of the regimes for relatively long periods.

Clearly the $\phi(n_2)$ term should also include a dependence on n_3, because connected subscribers also reduce available channels. However, such a modification destroys the Jackson property.

6. The s-server assumption (12) is a natural intermediate between the cases (8) and (10), but not in fact the most tractable one. Consider the prescription

$$\phi_j(n_j) = \frac{n_j s}{n_j + s - 1} \tag{21}$$

for which

$$\Phi_j(n_j) = \frac{(n_j + s - 1)!}{s^{n_j} n_j!}$$

with corresponding generating function $(1 - z_j/s)^{-s}$. That is, n_j would follow a negative binomial distribution in the open process. This case also spans cases (10) to (8) as s varies between 1 and $+\infty$, and the distribution of n_j is in fact that which would come about if the input to the node were divided equally (i.e. with probabilities $1/s$) between s parallel single-server queues, n_j then being the total in these s queues.

The exit rate from such a system would not in fact be given by (21), but by the number of these s sub-queues which are occupied. Show, nevertheless, that the two prescriptions are equivalent, in that if in a Jackson network the node j with rule (21) is replaced by the system of s parallel queues envisaged (their output being amalgamated) then the joint distribution of n_j (the total number at the node, in either its simple or compound form) and of numbers at other nodes is the same in both cases.

7. The solution w of (6) is determined only up to a scale factor. As for the processes of Chapter 5, this scale factor can be regarded either as a marker variable for N, the number of units in the system, or as a parameter which determines unit abundance in the open process. That is, if $w_j = \gamma_j$ is a non-negative solution of (6) then so is $w\gamma_j$ for positive scalar w, and

$$\pi(n) \propto \Phi(n) \prod_j (w \gamma_j)^{n_j}$$

is also a possible equilibrium distribution. This expression, when normalized, constitutes a possible distribution for the open process for any positive scalar w. The unnormalized distribution for the closed process is obtained by extracting the coefficient of w^N in this expression.

8. The total routing intensity $\lambda_j = \sum_k \lambda_{jk}$ could be absorbed into the definition of $\phi_j(n_j)$, so that we write the migration intensity (2) rather as $\bar{\phi}_j(n_j)p_{jk}$, where $\bar{\phi}_j = \lambda_j \phi_j$ and $p_{jk} = \lambda_{jk}/\lambda_j$. The new routing matrix (p_{jk}) is then a probability transition matrix, p_{jk} being the probability that a migrant emerging from node j will seek node k. The new w is $\bar{w}_j = \lambda_j w_j$.

9. The prescriptions (8) and (10), leading to Poisson and geometric distributions, could be regarded as corresponding to Boltzmann and Bose–Einstein statistics respectively. For Fermi–Dirac statistics one would have to take the specification

$$\phi_j(n_j) = \left\{ \begin{array}{c} 0 \\ 1 \\ +\infty \end{array} \right\} \quad \text{for } n_j \left\{ \begin{array}{c} = 0 \\ = 1 \\ > 1 \end{array} \right\}$$

corresponding to $\Phi_j(n_j)$ equal to 1, 1 and 0, respectively. The rule that a node may not have more than one occupant cannot be enforced by refusing admission, under a rule of the form (2), but only by requiring immediate ejection of surplus occupants. The effect of the two mechanisms is different, because ejection does not imply return to the point of origin, but a re-routing under the rules λ.

2. FUNCTIONALLY DEPENDENT NODES

The equilibrium distribution

$$\pi(n) \propto \prod_j \left[\Phi_j(n_j) w_j^{n_j} \right] \tag{1}$$

of the Jackson network set up in section 1 is celebrated as the *product form distribution*. The product form is seen as particularly significant, expressing as it does the surprising independence of the n_j in an appropriate open model. However, the product form may not be significant, as the following observation (due to Kelly 1975a) makes plain. Suppose the formula (1.2) for migration intensity modifies to

$$\theta(n, n - e_j + e_k) = \lambda_{jk} \frac{\Phi(n - e_j)}{\Phi(n)} \tag{2}$$

for some $\Phi(\cdot)$. We assume $\Phi(n) = 0$ for n not in the positive orthant $\mathbb{Z}_+^m$, so that passage of n out of this orthant is forbidden.

Theorem 2.1 *Consider the process with transition intensity (2). This has equilibrium distribution*

$$\pi(n) \propto \Phi(n) \prod_j w_j^{n_j} \tag{3}$$

on a given irreducible set, where w is a solution of (1.6). Distribution (2) satisfies the partial balance equations (1.7).

Proof is again a matter of easy verification. Theorem 2.1 implies Theorem 1.1, and reduces to it in the case when $\Phi(n)$ factorizes

$$\Phi(n) = \prod_j \Phi_j(n_j) \tag{4}$$

We shall say in case (4) that the nodes are *functionally independent*, in that the total migration rate out of node j then depends upon n only through n_j. In the functionally dependent case there is an interaction between nodes, although of the form restricted by the form (2). One could characterize this form by saying that the intensity of expulsion of a unit from node j is a function of the 'difference in potential' of the two configurations n and $(n - e_j)$; once the unit has been expelled its further course is determined totally by the routing matrix λ, and is not at all sensitive to n. However, this rule preserves all the features of the simpler case of section 1 with the exception of product form.

If we write relation (3) as

$$\pi(n) \propto \exp\left[-U(n) - \sum_j \zeta_j n_j\right] \tag{5}$$

then $U(n)$ could indeed be regarded as a *configuration potential* (equal to $-\log \Phi(n)$) and ζ_j as a *site potential* (determined from the routing matrix by (1.6)). However, these are not necessarily potentials associated with the usual ideas of statistical mechanics, since the process is not in general reversible. If we ask for detailed balance

$$\pi(n)\theta(n, n') = \pi(n')\theta(n', n) \tag{6}$$

with θ, π having evaluations (2), (3) we find that this is indeed satisfied as far as the Φ factors are concerned, but not as far as the λ and w factors are concerned unless the *route-reversibility* relation

$$w_j \lambda_{jk} = w_k \lambda_{kj} \tag{7}$$

is satisfied. In general the w_j satisfy only the more general balance relation (1.6). So, the Jackson network shows only a partial reversibility, reflected in the fact that it satisfies only the partial balance relations (1.7). One might indeed identify $U(n)$ as a conventional potential in (5) (since (6) is satisfied as far as this component is concerned) but the site potentials ζ_j cannot be thus identified unless (7) is satisfied. The routing parameters λ will in general induce a net circulation of units in parts of the system which is inconsistent with reversibility. We return to these matters in section 5.

Recall the definition of the *singleton process* as one in which a single unit migrates between the nodes, the $j \to k$ transition having intensity λ_{jk}.

Theorem 2.2 *Suppose $\Phi(n)$ strictly positive in $\mathbb{Z}_+^m$. Then the migration process specified by (2) is irreducible for a given value of N if and only if the singleton process is irreducible.*

Proof The process is irreducible in the sense claimed if and only if the same is true for the process with

$$\Phi(n) = \prod_j (n_j!)^{-1} \tag{8}$$

since the two processes have the same pattern of communication between states. But the case (8) describes a process in which all units independently follow the rules of the singleton process, and this is irreducible if and only if the singleton process is irreducible. ■

If $\Phi(n)$ were indeed zero for some value $\bar{n}$ then (3) would seem to indicate that the value $\bar{n}$ is forbidden. In fact, relation (2) does not define transition rates between states thus forbidden.

However, the interest is now to determine what kinds of behaviour can be represented by transition rules of the form (2). This is best done for open networks, which provide the most natural examples. We shall consider in sections 3, 7 and 10.9 what effects can be represented by a Jackson network. However, one may unfortunately well say that Jackson networks are the least intelligent of all networks, in that they show the least anticipation of the effects of a migration transition. As we have already noted, intensity (2) implies that system state n affects only the rate of ejection of a unit from a node, and not at all the subsequent routing of that unit. Once the unit has been ejected from a node it is caught up by the network and distributed to other nodes in a manner blindly determined by the routing rule λ.

Nevertheless, the model has proved practically useful just because its equilibrium properties are so explicit and simple, and also because there are indeed contexts in which it has some plausibility as a model. One wonders whether the model might not one day be seen as the natural 'precursor': the lowest in a natural series of progressively more intelligent networks. One would certainly upgrade the intelligence of the network if one could make the routing rules adaptive, in that they were responsive to the state of the network. This is a point we examine in section 10.

Theorem 1.2 obviously remains valid for the more general case of this section, with the definitions of $\theta(n, n')$ in (1.16) and $\pi(n)$ in (1.17) now replaced by expressions (2) and (3) respectively.

A view of the model which we shall later find useful is the following, due in this context to Kelly (1979). The probability intensity that a unit is emitted from node j is

$$c_j(n) = \lambda_j \frac{\Phi(n - e_j)}{\Phi(n)}$$

where $\lambda_j = \sum_k \lambda_{jk}$; units once emitted are then distributed through the network with probabilities $p_{jk} = \lambda_{jk}/\lambda_j$. One can view $c_j(n)$ as an n-dependent service rate for units at node j. We can now state

Theorem 2.3 *The statement that $c_j(n)$ is the intensity of emission of a unit from node j is equivalent to the supposition that units enter node j with independent exponentially distributed service requirements of unit mean, and that the total service rate of $c_j(n)$ is distributed over the n_j units at node j.*

Proof Suppose the n_j units at node j are labelled $i = 1, 2, \ldots, n_j$. Then the statement that the total service rate $c_j(n)$ is distributed over these is understood as saying that the intensity of the event 'emission of unit i' is $f_i c_j(n)$, where $f_i \geqslant 0$, $\sum_i f_i = 1$. The quantity f_i is the fraction of service effort devoted to unit i, which will in general be a function of j, n and indeed of other factors, such as order of arrival of the units. The total rate of emission is then $\sum_i f_i c_j(n) = c_j(n)$, as postulated. By Theorem 3.4.3 the statement that emission of unit i has intensity $f_i c_j(n)$ can indeed be construed as the working off of a standard exponential service requirement at rate $f_i c_j(n)$. ∎

Exercises and comments

1. *Several types of unit.* One can generalize to the case of several types of unit simply by replacing the site label j by the compound type/site label (α, j) with $\alpha = 1, 2, \ldots, p; j = 1, 2, \ldots, m$. That is, n is now the array $\{n_{\alpha j}\}$ of the numbers of α-units at node j. One assumes the intensity

$$\theta(n, n - e_{\alpha j} + e_{\beta k}) = \lambda_{\alpha j, \beta k} \frac{\Phi(n - e_{\alpha j})}{\Phi(n)}$$

for the transition in which an α-unit simultaneously migrates from node j to node k and mutates to a β-unit. The equilibrium distribution is then

$$\pi(n) \propto \Phi(n) \prod_\alpha \prod_j w_{\alpha j}^{n_{\alpha j}}$$

where w solves

$$\sum_\beta \sum_k (w_{\beta k} \lambda_{\beta k, \alpha j} - w_{\alpha j} \lambda_{\alpha j, \beta k}) = 0$$

Note that mutation is allowed only on an arc, not at a node.

2. *Node sequencing.* By choosing the routing parameters $\lambda_{\alpha j, \beta k}$ appropriately one can dictate, either wholly or partially, the route which a unit will take through the network. This was the device which Kelly (1979) employed to ensure that a unit went through the stages of processing in the desired sequence. To take an extreme example, suppose that $\lambda_{\alpha j, \beta k}$ is zero unless β and k have the (α, j) dependent values $\beta = T(\alpha, j)$, $k = R(\alpha, j)$. Type is then a label which can record all relevant current history of the individual in the system and determine where he should go next.

3. As an example of sequencing by history, consider the situation in which arrivals at an airport are required to go through the entry formalities of passport, currency, health and customs examinations before release. In practice the examinations are taken in a definite sequence, with several parallel streams for each, as in Fig. 9.2.1. However, in addition to the sequencing the immigrant will require a record (on documents, baggage or person) of the formalities he has

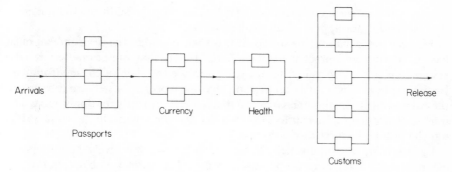

Fig. 9.2.1 The 'series–parallel' network of conventional immigration formalities.

satisfied as he passes through a station of each type. He will certainly not be released until his record shows that he has completed every formality.

One could envisage a network less rigidly sequenced than that of the figure, but with the immigrant's 'type' (record of formalities completed) guiding him to which stations he might yet attend and when he may leave the system. If it is possible that individuals can fail some of the formalities in some sense then 'type' must include a description of the problematic aspects of the individual in question, and these may lead to his being differently routed upon examination.

4. Note that the condition (7) of route-reversibility is just the condition that the singleton process be reversible.

3. OPEN JACKSON NETWORKS

Suppose we regard the closed Jackson network as being specified by the pattern of transitions and intensities embodied in formula (2.2). Then one may ask whether there is a natural way of opening the network (i.e. of allowing immigrations and emigrations) which preserves this pattern. The natural way to do so is to add a dummy node, labelled $j = 0$, say, which represents the 'outside world' or the 'environment'. Migrations between the environment $(j = 0)$ and the system $(j = 1, 2, \ldots, m)$ constitute immigrations into and emigrations out of the system.

As a node the environment must have a particular character. We shall assume in fact that $\Phi(n)$ is independent of n_0, and indeed that n_0 does not appear at all in the description n of Markov state. The consequence of this is that migration transitions within the system will still have intensity (2.2), while immigration and emigration transitions will have intensities

$$\theta(n, n + e_k) = \lambda_{0k} = v_k$$

$$\theta(n, n - e_j) = \lambda_{j0} \frac{\Phi(n - e_j)}{\Phi(n)} = \mu_j \frac{\Phi(n - e_j)}{\Phi(n)} \tag{1}$$

These are indeed the intensities derived from (2.2) with the additional supposition that n_0 should not appear as an explicit variable.

The reason for making the supposition is that, as a node, the environment must have the particular character that it is *infinite*. That is, that migrations between the system and the environment do not really change the state of the environment. For example, however many immigrants may have been released into the system, the environment remains unchanged in that it is equally ready to supply more. The infinite environment thus buffers the system of m components in the same way as did the infinite heat bath of section 6.1.

By assuming the intensities (1) and (2.2) we are effectively assuming the environment not only *infinite* but also *constant*. The distinction between the two ideas will be made explicit in the next chapter. Note also that the open process can in fact also be derived from the closed one: by allowing one of the components to become infinite (see Exercises 1 and 2).

Theorem 3.1 *Consider the open Jackson network, the process with transitions and intensities given by* (1) *and* (2.2). *This has a stationary distribution*

$$\pi(n) \propto \Phi(n) \prod_j w_j^{n_j} \tag{2}$$

where the parameters w are determined by

$$\sum_{k=0}^{m} (w_k \lambda_{kj} - w_j \lambda_{jk}) = 0 \qquad (j = 0, 1, \ldots, m) \tag{3}$$

with the scaling condition

$$w_0 = 1 \tag{4}$$

Distribution (3) *then constitutes the equilibrium distribution if it is summable and the process irreducible. It satisfies the partial balance relations*

$$\sum_{k=0}^{m} [\pi(n - e_j + e_k)\theta(n - e_j + e_k, n) - \pi(n)\theta(n, n - e_j + e_k)] = 0$$

$$(j = 0, 1, \ldots, m) \tag{5}$$

where the transitions $n \rightleftharpoons n \pm e_j$ have been formally written as $n \rightleftharpoons n \pm e_j \mp e_0$.

The proof is again a matter of immediate verification. One of relations (3) is redundant, since the relations sum to $0 = 0$. This corresponds to an indeterminacy in scale, which is settled by the additional condition (4). Note that (3) and (4) can alternatively be written

$$v_j - \mu_j w_j + \sum_{k=1}^{m} (w_k \lambda_{kj} - w_j \lambda_{jk}) = 0 \qquad (j = 1, 2, \ldots, m) \tag{6}$$

$$\sum_{j=1}^{m} (v_j - \mu_j w_j) = 0 \tag{7}$$

equation (7) again following from equations (6).

The results follow exactly the pattern of the closed case, the new condition (4) expressing the fact that n_0 does not occur as a component of Markov state. Note that in the functionally independent case (2.4) the n_j are indeed independent in the equilibrium of the open process.

Of course, the distribution is now over an infinite set of states. The possibilities that either the solution of (3), (4) is not finite or that expression (2) is not summable correspond to the possibility of *congestion* of the network, a matter we examine in the next section. The question of irreducibility is easily settled.

Theorem 3.2 *Suppose* $\Phi(n)$ *strictly positive on* $\mathbb{Z}_+^m$. *Then the process defined in Theorem* 3.1 *is irreducible if and only if equations* (3) *and* (4) *have a unique solution. This condition is equivalent to the requirement that the singleton process on the nodes* $j = 0, 1, \ldots, m$ *be irreducible and that the state* $j = 0$ *be recurrent.*

Proof Irreducibility of the singleton process is equivalent to requiring that equations (3) have a solution unique up to a scale factor. Recurrence of $j = 0$ is equivalent to requiring that w_0 be non-zero in this solution. Equivalence of the two sets of conditions is thus established.

Since expression (2) is a stationary distribution for any solution w of (3) and (4), uniqueness of w is certainly necessary for uniqueness of π. For the process to be irreducible it is thus necessary that w be uniquely determined by (3), (4). Delete all nodes for which $w_j = 0$. This excludes all nodes which inevitably become empty in the course of time and play no further role. The process is then irreducible if and only if all states n, n' communicate in this reduced version. This will be true if and only if all states communicate for some N if one has N units independently following the singleton process (on the $m + 1$ nodes) and this will be true if and only if the singleton process is irreducible on the reduced set of states. It is necessary that $j = 0$ should occur in this reduced set if states n, n' with different totals $\sum_{1}^{m} n_j, \sum_{1}^{m} n'_j$ are to communicate. ■

We can now state the natural interpretation of the parameters w_j.

Theorem 3.3 *Suppose that expression* (2) *gives the unique and proper equilibrium distribution of the process. Then under this distribution*

$$E\left[\frac{\Phi(n - e_j)}{\Phi(n)}\right] = w_j \tag{8}$$

The expected equilibrium rate of the migration stream j *to* k *is thus* $w_j \lambda_{jk}$.

Assertion (8) of course follows immediately from (2). It can be otherwise expressed: $\lambda_j w_j$ is the equilibrium average service rate at node j.

The Jackson network specified by the rates (1) and (2.2) does not imply product form but does imply partial balance, as demonstrated in Theorem 3.1. One is now curious as to what interactions can be represented by these rates. An interaction that unfortunately cannot be represented is that of *blocking*, in which a busy component refuses to accept input until it is free.

Theorem 3.4 *The partial balance property (5) is inconsistent with the phenomenon of blocking.*

Proof We demonstrate that consistency is impossible by the counter-example illustrated in Fig. 9.3.1. This constitutes the simplest possible case of blocking: an open network of two nodes in sequence. Node 1 is interpretable as a buffer store for node 2, which is a processor. Jobs enter the buffer in a Poisson stream of rate v and are despatched from it to the processor at rate $\phi_1(n_1, n_2)$. They are then despatched from the processor at rate $\phi_2(n_2)$. It is the dependence of ϕ_1 upon n_2 that admits the possibility of blocking: the fact that jobs are admitted from the buffer store at a rate dependent upon the processor's own state.

If this process showed partial balance then its equilibrium distribution $\pi(n_1, n_2)$ would satisfy the relations

$$v\pi(n_1 - 1, n_2) = \phi_1(n_1, n_2)\pi(n_1, n_2) = \phi_2(n_2 + 1)\pi(n_1 - 1, n_2 + 1) \qquad (9)$$

From the outer terms of (9) we deduce that $\pi(n_1, n_2)$ is of the product form $\Phi_1(n_1)\Phi_2(n_2)$. From the first equality of (9) we then deduce that $\phi_1(n_1, n_2)$ must be independent of n_2. That is, partial balance is inconsistent with a one-way flow through the processor with an input rate dependent upon the processor's own state. ■

However, congestion can certainly occur, we study the phenomenon in the next section. That is, in a Jackson network a component may not refuse input, but may accept it and then be unable to cope. A degree of blocking is also possible if the buffer store is infinite (see Exercise 2).

Exercises and comments

1. To see the open system as a limit case of the closed system, consider a closed system on nodes $j = 0, 1, \ldots, m$ with $\Phi(n_0, n) = \Phi_0(n_0)\Phi(n)$. Here n is, as before,

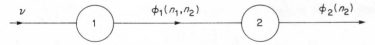

Fig. 9.3.1 The simplest example of blocking. Blocking turns out to be inconsistent with the properties of a Jackson network.

the vector $(n_1, n_2, \ldots, n_m)$. Suppose that component 0 is a simple queue, so that $\Phi_0(n_0) = 1$. Then we deduce from (2.3) the equilibrium distribution

$$\pi(n_0, n) \propto \Phi(n) \prod_{j=1}^{m} w_j^{n_j} \qquad \left(\sum_{j=0}^{m} n_j = N \right)$$

where w is the solution of (3), (4). If we let N tend to infinity then this solution converges to the unrestricted distribution (2) for n. If distribution (2) is summable (so that n is finite with probability one) then this is an indication that component 0 has indeed become infinite under the passage $N \to \infty$, in that this component contains infinitely many units in equilibrium.

2. Suppose one takes a closed system of components as in Exercise 1, but chooses $\Phi_0(n_0) = \Psi(N - n_0)$, where Ψ is such that the distribution

$$\pi(n) \propto \Psi\left(\sum_{1}^{m} n_j \right) \Phi(n) \prod_{1}^{m} w_j^{n_j} \tag{10}$$

is summable on $\mathbb{Z}_+^m$. Then (10) is indeed an equilibrium open distribution for the case when the rates specified by (1) and (2.2) are changed only in that the immigration rate is modified to

$$\theta(n, n + e_k) = \lambda_{0k} \frac{\Psi\left(\sum_{1}^{m} n_j + 1 \right)}{\Psi\left(\sum_{1}^{m} n_j \right)}$$

One may still regard the environment as infinite, but the system exerts a back-reaction on it in that the immigration rate is affected by the number already in the system. In particular, if $\Psi(N) = 0$ for $N > N_0$ then the system will not accept input at all once it is holding N_0 units. This example is due to Kelly (1979).

3. Generalize the above example by the choice $\Phi(n_0, n) = \Psi(N - n_0, n)$, with the form of Ψ independent of N.

4. Suppose that a relation (3) holds for given j and for k summed over a set A_j. Show that the same is then true for a partial balance relation (6).

5. One would regard the open and closed processes of this section and the last as corresponding only if the closed process could be derived from the open one by a conditioning on the value of $\sum_{1}^{m} n_j$. This implies that the values of the w_j must be proportional in the two cases, which implies that the w_j satisfying (1.6) must also satisfy

$$w_0 \lambda_{0j} - w_j \lambda_{j0} = 0$$

for $j = 1, 2, \ldots, m$ and some scalar w_0. These can be regarded as deterministic detailed balance conditions for immigration/emigration.

6. One wonders whether there might not exist a time-dependent n-distribution of the form (2), in that it can be written

$$\pi(n, t) = C(t)\Phi(n)\prod_j w_j(t)^{n_j} \tag{11}$$

Here the $w_j(t)$ obey the dynamic analogue (4.1) of (3), (4) and $C(t)$ is a time-dependent normalization factor. Insertion of expression (11) in the Kolmogorov forward equation yields

$$\frac{\dot{C}}{C} + \sum_{j \neq 0} \frac{\dot{w}_j}{w_j}\left[n_j - \frac{\Phi(n - e_j)}{\Phi(n)}\right] + \sum_k (\lambda_{0k} - w_k \lambda_{k0}) = 0 \tag{12}$$

It would seem that the only genuinely dynamic case in which (12) can hold identically in n is the Poisson case $\left(\Phi(n) = \left(\prod_1^m n_j!\right)^{-1}, C = \exp\left(-\sum_1^m w_j\right)\right)$. See section 13.7.

7. An open compartmental model might represent numbers in the different grades of an organization, movements corresponding to recruitment, promotion, demotion, sideways moves or losses.

4. CONGESTION AND BOTTLENECKS

Suppose, for simplicity, that the open model of section 3 is irreducible. Then expression (3.2) constitutes the only possible stationary distribution. If this expression is not summable then the indication is that at least some components of n are infinite with probability one. In other words, that the network is irretrievably *congested*. This is a feature specific to the open case, and its study is plainly a matter of practical importance.

The fact that a simple queue will become congested if the input traffic is too heavy is well-known. The same will be true of any network containing components of limited handling capacity, in some sense. It is obviously useful to know which components are critical in that congestion cannot possibly be relieved until their capacity is increased.

Congestion could occur because some of the w_j solving (3.3), (3.4) are infinite, or because they are all finite but expression (3.2) is not summable. The first situation is a relatively crude one: the routing rules are such that a unit can find its way into the system but cannot find its way out. The second expresses a subtler combined property of both routing and components: passage is possible but insufficient.

To dispose of the first possibility, consider the dynamic version

$$\dot{w}_j = \sum_k (w_k \lambda_{kj} - w_j \lambda_{jk}) \qquad (j > 0)$$
$$w_0 = 1 \tag{1}$$

of equations (3.3), (3.4). We shall write these rather as

$$\dot{w}_j = v_j + \sum_{k \neq 0} \lambda_{kj} w_j - \left(\sum_k \lambda_{jk} \right) w_j \qquad (j > 0) \tag{1'}$$

or

$$\dot{w} = v + (L - L_0) w \tag{2}$$

Here w is the column m-vector $(w_1, w_2, \ldots, w_m)$, v is the corresponding vector of immigration rates $v_j = \lambda_{0j}$, and L, L_0 are the diagonal-free and diagonal matrices implicitly defined by (1') and (2), both with non-negative elements.

Theorem 4.1

(i) *Equations (3.3), (3.4) have a finite solution if and only if the system (2) is dissipative, i.e. if the eigenvalues of $L_0 - L$ are strictly positive.*

(ii) *Under this condition the solution for w can be expressed*

$$w = (L_0 - L)^{-1} v \tag{3}$$

where the matrix $(L_0 - L)^{-1}$ has all its elements non-negative.

Proof The stochastic model of the last section can have an equilibrium only if the deterministic system (2) has a stable equilibrium. This is because w can be identified as $E(n)$ in the case of independent movement of the units. Stability of (2) is equivalent to requiring that all eigenvalues of $L_0 - L$ be strictly positive, which indeed implies that the equilibrium solution (3) of (2) is finite.

Now, the matrix LL_0^{-1} is non-negative. Its eigenvalue of largest modulus is consequently real and positive and not greater than the greatest of its column sums (see, for example, the section on positive matrices in Bellman, 1960), which cannot be greater than unity. But if LL_0^{-1} has eigenvalue unity then $L_0 - L$ has eigenvalue zero, which we have excluded. Thus LL_0^{-1} has all its eigenvalues less than unity in modulus and we can write

$$(L_0 - L)^{-1} = L_0^{-1} (I - LL_0^{-1})^{-1} = L_0^{-1} \sum_{j=0}^{\infty} (LL_0^{-1})^j$$

the sum being both finite and non-negative. ∎

Define now a generating function

$$\tilde{\Phi}(z) = \sum_n \Phi(n) \prod_j z_j^{n_j} \tag{4}$$

Let F denote the set of non-negative real z for which the series (4) converges. Then the occurrence of congestion is equivalent to the statement that w, determined by (3), lies outside F.

One can view the onset of congestion as a kind of criticality. Consider for simplicity the case of functionally independent nodes, when

$$\tilde{\Phi}(z) = \prod_j \tilde{\Phi}_j(z_j) \tag{5}$$

Denote the radius of convergence of $\tilde{\Phi}_j(z_j)$ by ρ_j. Suppose the immigration rates v_j small enough that $w < \rho$. The system is then free-flowing and n is finite with probability one. Suppose now that the input vector v is increased continuously along some path, slowly enough that the system stays in equilibrium, although with changing parameters w. The vector w will in fact also increase, by Theorem 4.1(ii). If loading is increased to the point that $w_j > \rho_j$ for all j in some non-empty set S of components then at this point the system can no longer cope and becomes congested. The values of n_j for j in S will be infinite, and it is just the components in S which constitute the congestion points or *bottlenecks* of the system (cf. section 8.4).

A rather stronger statement is to note that the p.g.f. of n is

$$\Pi(z) \propto \tilde{\Phi}(w_1 z_1, w_2 z_2, \ldots, w_m z_m) = \prod_j \tilde{\Phi}_j(w_j z_j)$$

and that $\tilde{\Phi}_j(w_j z_j)$ ceases to define a valid p.g.f. if w_j exceeds ρ_j. Whether congestion occurs at node j when w_j first attains the value ρ_j or when it first surpasses it depends upon circumstances (see Exercise 3). There is a whole more refined theory of degrees of congestion and character of bottlenecks (see Pollett, 1982). The only observation we would make is that *location* of a bottleneck is obvious in the case of functional independence (all the members of S); less so in the case of functional independence.

Theorem 4.2 *Let F denote the region of convergence of the generating function* (4) *for non-negative real z. As v increases so does $w = (L_0 - L)^{-1} v$; suppose that w (initially in F) then approaches the boundary of F. This heralds the onset of congestion, and*

 (i) *node j is involved in the bottleneck if a further increase in w_j takes w out of F*
(ii) *node j is critical to the bottleneck if a further increase in v_j takes w out of F.*

Proof   Consider the equilibrium distribution for a value of w just inside F, so that the network is on the brink of congestion. If a further increase in w_j would take w out of F then this is equivalent to saying that

$$E\left(\frac{w_j + \delta}{w_j}\right)^{n_j} = \infty$$

for arbitrarily small positive δ. In this sense, n_j is infinite, and so node j is involved in the bottleneck.

Assertion (ii) is virtually a definition of what one means by 'critical'. Node j is critical if a bottleneck which is incipient would become actual if node j were loaded further. ∎

Exercises and comments

1. A depôt, for which $\Phi_j(n_j) = \theta^{n_j}/n_j!$, say, will never congest, because $\tilde{\Phi}_j(z_j) = \exp(\theta z_j)$ has infinite radius of convergence. A simple queue, for which $\Phi_j(n_j) = \theta^{n_j}$, say, will congest if $w_j > \theta^{-1}$, because $\tilde{\Phi}_j(z_j) = (1 - \theta z_j)^{-1}$ has radius of convergence θ^{-1}.

2. Consider a sequence of m functionally independent nodes in series (i.e. such that internal migrations from node j are to node $j + 1$). Note that, against intuition, a congestion at one node does not necessarily spread to nodes feeding it. This is because, in a Jackson network (at least for the functionally independent case), a node will continue to emit units whatever the state of the system elsewhere, and even a congested node is bound to accept a unit routed to it. Since the point of congestion continues to accept input, there is no backing up at earlier points in the network.

3. Consider a functionally independent component, which we may then consider singly and so for convenience drop the subscript j. Suppose that the component has intrinsic departure rate

$$
\phi(n) = \begin{cases} 0 & n = 0 \\ \left(\dfrac{n+r}{n}\right)\rho & n = 1, 2, \ldots \end{cases}
$$

The distribution of n in equilibrium is then

$$\pi(n) \propto \Phi(n)w^n$$

where w is the parameter derived from the routing matrix and

$$\Phi(n) = \frac{\Gamma(n+1)\Gamma(r+1)}{\Gamma(n+r+1)}\rho^{-n}$$

Thus $\Phi(n)$ is of order $n^{-r}\rho^{-n}$ for large n, and $\tilde{\Phi}(z)$ has radius of convergence ρ and the component congests if $w > \rho$. For $r \leqslant 1$ it congests actually at $w = \rho$; for $r > 1$ the random variable n is finite even for $w = \rho$. One has $E(n^\alpha) < \infty$ for $\alpha < r - 1$. So, increasing r implies an increasing softening of (but not shifting of) the threshold.

4. Consider the simple case of functional dependence specified by

$$\tilde{\Phi}(z) = \left(1 - \sum_j \theta_j z_j\right)^{-1}$$

and so for which

$$\Phi(n) = N! \prod_j \frac{\theta_j^{n_j}}{n_j!}$$

and

$$\frac{\Phi(n - e_j)}{\Phi(n)} = \frac{\theta_j n_j}{N}$$

This is like a system of m queues with a single server for the whole system, who divides his time in proportion to the length of queues and works at rate θ_j in queue j. The distribution of n is in fact multinomial for the closed system, just as it would have been for Poisson statistics (functionally independent depôts). However, the limitations of server capacity are shown in the finiteness of F, specified by $z \geqslant 0$, $\theta^{\mathsf{T}} z < 1$.

5. TIME-REVERSED JACKSON NETWORKS

The model of section 3 is the most general network we have yet formulated, including models which are closed or have functionally independent nodes as special cases. Let us then take it for the moment as defining what we mean by a 'Jackson network'. A particular such network will be specified by (Φ, λ), the service function $\Phi(n)$ and the routings λ_{jk} $(j, k = 1, 2, \ldots, m)$.

Let us consider what the model becomes under time reversal. We shall indicate corresponding quantities for the time-reversed version by a star, so that, by the continuous-time version of Corollary 4.3.1

$$\pi^*(n) = \pi(n) \tag{1}$$

$$\theta^*(n, n') = \frac{\pi(n')\theta(n', n)}{\pi(n)} \tag{2}$$

Let us for simplicity suppose irreducibility, either unconditionally in the open case, or conditional upon the value of N in the closed case.

Theorem 5.1   *The time-reverse of a Jackson network* (Φ, λ) *is a Jackson network* (Φ^*, λ^*), *where*

$$\Phi^*(n) = \Phi(n) \tag{3}$$

$$\lambda_{jk}^* = \frac{w_k \lambda_{kj}}{w_j} \tag{4}$$

and so $w^* = w$. *The process is thus reversible if and only if it is route-reversible, i.e. if the* w_j *satisfy*

$$w_j \lambda_{jk} = w_k \lambda_{kj} \tag{5}$$

Verification of the first assertion follows immediately from formula (2) and the prescription (2.2) of intensities. The open case, when (2.2) is supplemented by (3.1), is formally no more general. The second assertion follows from the condition $\lambda^* = \lambda$.

We thus have an example of what is sometimes loosely called quasi-reversibility: that a class of processes transforms into itself under time-reversal, though the members of the class may not be individually reversible. The quasi-reversibility and the preservation $\Phi^* = \Phi$ stem in the present case from the fact that detailed balance holds for the intensities (2.2), (3.1) as far as the n-dependent factors are concerned. However, reversibility does not hold for the route-dependent factors in general, and one has full reversibility if and only if there is route-reversibility. Note the particular cases of relation (4) for the open process

$$v_j^* = \lambda_{0j}^* = \lambda_{j0} w_j = \mu_j w_j \tag{6}$$

$$\mu_j^* = \lambda_{j0}^* = \lambda_{0j}/w_j = v_j/w_j \tag{7}$$

An argument first used by Reich (1957) and applied in this context by Kelly (1979) now gives powerful conclusions with great economy.

Theorem 5.2 *Consider the Jackson network* (Φ, λ). *The output streams from the system are mutually independent Poisson streams of rates* $v_j^* = \mu_j w_j$, *and the streams before time t are independent of system state* $n(t)$ *at time t.*

Proof It follows from the specification of the process that the input streams are Poisson of rates v_j and are independent of $n(t)$ after time t. Since the time-reversed process is also a Jackson network, the same is true for it with the substitution of v_j^* for v_j. But the input streams of the reversed process can be identified as the time-reversed output streams of the original process, whence the assertion follows. Note the appeal to the fact that Poisson character is retained under time-reversal (cf. Exercise 4.2.7). ∎

It appears then that one characteristic of a Jackson network is its *Poisson-preserving property*: that independent Poisson input streams are converted into independent Poisson output streams. Several authors (notably Muntz (1972) and Kelly (1979)) have tried to take this property as a defining characteristic. More properly, *quasi-reversibility* is adopted as a definite technical term for a slightly stronger property (see section 10.8). In the next chapter we shall consider the twin properties of weak coupling and of quasi-reversibility, both of which seem to be important in this context, without either quite including the other.

The internal migration streams of the process cannot be jointly Poisson in general; one sees this by consideration of a closed process.

Exercises and comments

1. Suppose we ask of a migration process only that it should satisfy the partial balance relations (3.5), which we might write in the more symmetric form

$$\sum_k [\pi(n+e_k)\theta(n+e_k, n+e_j) - \pi(n+e_j)\theta(n+e_j, n+e_k)] = 0.$$

Show that the same is true of the time-reversed process.

6. NETWORKS AS COMPONENTS

This section somewhat foreshadows the material of the next chapter, in which a component is endowed with more structure than has hitherto been envisaged.

An open network of the kind specified in section 3 could itself be regarded as a component, as an entity whose output streams can provide the inputs of similar composite components to constitute even larger networks. The first feature of such a component is, of course, that it has a composite state: the distribution $n = (n_1, n_2, \ldots, n_m)$ of units over its interior nodes rather than simply the total number of units held in the component. The second feature follows from Theorem 3.1: in an open system of functionally independent such components the states of components will be independently distributed in equilibrium, the state distribution of the typical component being

$$\pi(n|\mu, v) \propto \Phi(n) \prod_{j=1}^{m} w_j^{n_j} \tag{1}$$

Here the w_j solve

$$v_j - \mu_j w_j + \sum_{k=1}^{m} (w_k \lambda_{kj} - w_j \lambda_{jk}) = 0 \qquad (j = 1, 2, \ldots, m) \tag{2}$$

where v_j is the expected rate of flow into node j of the component from other components of the network and μ_j measures the drain exerted by the network on node j of the component. We have indicated μ, v as parametrizing the distribution, because these quantities express the coupling of the composite component with the rest of the system. It is a feature of a Jackson network that the coupling can thus be summarized. The inputs are from other parts of the system and not in general jointly Poisson or independent, but an application of Theorem 3.1 to the whole network nevertheless indicates that the sub-network consisting of the component has the statistics indicated by (1), (2).

Note a third feature of a composite component: it has several input and output streams, which must be regarded as distinct, even if they handle the same type of unit. It may be, however, that one expects relatively few of the nodes of the component to act as ports, i.e. as input or output points. In such a case the coupling of the component with the rest of the system is summed up in relatively few parameters. Sometimes it seems reasonable to demand that it also be summed up in relatively few *statistics*, in that the equilibrium distribution (1) can be factorized

$$\pi(n|\mu, v) = P(\tilde{n}|\mu, v) P(n|\tilde{n}) \tag{3}$$

Here $\tilde{n}$ is low-dimensional function of n which is to be regarded as a *sufficient statistic* for the effect of outside coupling, in that the equilibrium distribution of n conditional on $\tilde{n}$ is independent of the coupling parameters μ, v. An alternative expression of this condition would be to demand that, for given μ, v, the conditional distribution $P(n|\tilde{n})$ be identical with that for the isolated component, when μ and v are both zero. A prime candidate for $\tilde{n}$ in such a case would seem to be $N = \sum_j n_j$, which is the invariant of n for the isolated component. We find easily

Theorem 6.1 *The conditional equilibrium distribution $P(n|N)$ is identical for the isolated and coupled component if and only if*

$$v_j = \mu_j w_j \qquad (j = 1, 2, \ldots, m) \tag{4}$$

where w is a solution of

$$\sum_k (w_k \lambda_{kj} - w_k \lambda_{jk}) = 0 \qquad (j = 1, 2, \ldots, m) \tag{5}$$

That is, input and output must balance at each node for the coupled component. This is, of course, the criterion we have already derived in Exercise 3.5 when demanding that the distribution for a closed system indeed be identifiable with the conditional distribution for the corresponding open system. However, we now prove the result (simple enough) formally.

Proof The demand is that the solution w of (2) be proportional to the solution of those equations holding in the special case (5) when μ and v are set equal to zero. ∎

Corollary 6.1 *The identity of distributions required in Theorem 6.1 will always hold for the single-port case: the case when all input and output streams are restricted to a single node. The identity will never hold if there is a single input node and a single output node, and these are distinct.*

Proof Suppose the single input/output node in the first case is that for $j = 1$, so that $\mu_j = v_j = 0$ for $j \neq 1$. Then (4) holds trivially for $j \neq 1$, and holds for $j = 1$ in virtue of the overall balance relation

$$\sum_j (v_j - \mu_j w_j) = 0 \tag{6}$$

obtained by summing relations (2) over j.

For the second case, we may suppose v_1 the only non-zero v_j and μ_m the only non-zero μ_j. Relation (4) cannot then hold for $j = 1$. ∎

The two cases represent two distinct characters. In the second case the component is 'progressive' in its operation in that units are accepted at node 1 and discharged at node m. We may consider that they have indeed been 'processed' between these two nodes. In the first case units leave by the same door that they came in, and have effectively just wandered around the internal states of the component in the interim. There is no notion of processing, in that a unit at node 1 faces the same future whether it has just arrived there from outside or just arrived there from other nodes of the component. It is natural to term such components 'non-progressive'.

The components for the statistical-mechanical models of Chapters 5 and 6 were non-progressive in that all their transitions, both internal and external, were reversible. The archetype of a progressive component is a linear system of depôts. That is, units arriving at node 1 pass independently and successively through nodes 1, 2, ..., m at the rates indicated in Fig. 9.6.1. The equilibrium distribution is easily found to be

$$\pi(n) \propto \frac{v^N \mu^{-n_m}}{n_m!} \prod_{j=1}^{m-1} \frac{\lambda_j^{-n_j}}{n_j!} \tag{7}$$

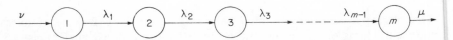

Fig. 9.6.1 A one-way linear system; the archetype of a progressive component.

The pair (N, n_m) is thus a sufficient statistic for the coupling parameters, but N itself is not.

Exercises and comments

1. Generalize the last example by considering a composite component with a single input node ($j = 1$, say) and a single and distinct output node ($j = m$, say) which is such that the output node cannot feed back into the component, but only to the outside. Show then that the coupling parameters appear in $\pi(n)$ only through a factor $v^N \mu^{-n_m}$, so that N and n_m are sufficient.

2. We shall see in the next chapter that a 'Jackson component' seems not to be able to represent a first-come first-served queue. The linear progressive system of Fig. 9.6.1 does represent an approximation to such a queue, however. While one unit can certainly overtake another in the system, the order of emission will be close to the order of arrival, the more so as the number of stages increase for a given expected transit time. If the transition $n \rightarrow n - e_j + e_{j+1}$ is given rate $\lambda_j n_j / N$ rather than $\lambda_j n_j$ (cf. Exercise (4.4)) then expression (7) is multiplied by $N!$ and the distribution of total number in the queue is geometric. However, the distribution of waiting time is not that of the single-server first-come first-served queue, and individuals are indeed sharing service at all stages.

7. THE OPTIMAL DESIGN OF A JACKSON NETWORK

Sections 7–9 are concerned with optimization and constitute something of a diversion from the main theme. However, it may be of some interest that the very features which give a Jackson network a particular character in equilibrium also give its optimization rules a special structure. This fact may strengthen the hope, earlier expressed, that the Jackson network is the first in a natural sequence of progressively more 'intelligent' networks.

We shall consider the choice of routing rules for an open Jackson network which will induce it to clear the given input traffic, specified by the v_j, as efficiently as possible. Specifically, we shall suppose that a network in state n incurs a cost $a(n)$ per unit time, and that a cost b_{jk} is also incurred whenever there is a direct $j \rightarrow k$ migration. One might, for example, assume $a(n)$ to have the form

$$a(n) = \sum_j a_j n_j \tag{1}$$

if waiting time at node j is costed at a_j per unit time. However, there is no need to specialize as yet. The b_{jk} component of cost reflects the cost of a $j \rightarrow k$ routing. For example, $b_{j0} = +\infty$ would imply that the system cannot be left directly from node j. We are assuming units of a single type, but one can choose the b_{jk} to constrain the path of a unit to ensure, if desired, that it follows some processing sequence in an acceptable order.

Appealing to the expressions for equilibrium distribution and flow rates derived in Theorems 3.1 and 3.3 we deduce that the expected rate of cost for a given network in equilibrium is

$$C = A(w) + \sum_j \sum_k b_{jk} \lambda_{jk} w_j \tag{2}$$

where

$$A(w) := \frac{\sum_n a(n)\Phi(n)\prod_j w_j^{n_j}}{\sum_n \Phi(n)\prod_j w_j^{n_j}}. \tag{3}$$

Certain aspects of the problem are prescribed; we assume these to be the set of nodes (but see Exercise 1), the input rates $v_j = \lambda_{0j}$ and the 'configuration' function $\Phi(n)$. We shall also assume that capacity constraints are placed on exit routing at nodes:

$$\lambda_j := \sum_k \lambda_{jk} \leqslant \bar{\lambda}_j \qquad (j = 1, 2, \ldots, m) \tag{4}$$

Here the bounds $\bar{\lambda}_j$ are prescribed. We regard the design problem as the task of minimizing expression (2) for C with respect to the routing matrix $\lambda = (\lambda_{jk})$ subject to the constraints listed and, of course, that of non-negativity: $\lambda_{jk} \geqslant 0$.

Let us rather regard it as the task of minimizing C with respect to the λ_{jk} and $w_j (j \neq 0)$ subject to the constraints (4) and the determining constraints

$$\sum_{k=0}^m (w_k \lambda_{kj} - w_j \lambda_{jk}) = 0 \qquad (j = 1, 2, \ldots, m) \tag{5}$$

plus non-negativity.

The values $w_0 = 1$ and $\lambda_{0k} = v_k$ are of course prescribed. We apply constraints (4), (5) only for $j \neq 0$; there is no constraint (4) for $j = 0$ and the constraint (5) for $j = 0$ is redundant. We have then the Lagrangian form

$$L = C + \sum_j \sum_k \xi_j (w_k \lambda_{kj} - w_j \lambda_{jk}) + \sum_j \eta_j \left(\sum_k \lambda_{jk} - \bar{\lambda}_j \right) \tag{6}$$

where ξ_j, η_j are Lagrangian multipliers associated with the constraints (5), (4). If we adopt the convention $\xi_0 = \eta_0 = 0$ then all summations in (6) can be taken over the range $(0, 1, \ldots, m)$. These multipliers have the interpretations

$$\xi_j = \frac{\partial \bar{C}}{\partial v_j}$$

$$\eta_j = -\frac{\partial \bar{C}}{\partial \bar{\lambda}_j}, \tag{7}$$

where $\bar{C}$ is the minimal constrained value of C (see Appendix 3). That is, ξ_j is the marginal cost of additional input at node j and η_j the marginal benefit of increased capacity at node j.

Now, the full formalism of convex programming described in Appendix 3 is not available, because the function $A(w)$ defined by (3) is not necessarily convex and the constraints (5) are not linear in (λ, w). One can nevertheless appeal to the Kuhn–Tucker optimality conditions (see Appendix 3) to deduce the necessary conditions asserted in the following theorem.

Theorem 7.1

(i) *The marginal costs of input* ξ_j *satisfy* $\xi_0 = 0$ *and*

$$\xi_j \leqslant c_j + \min_k (b_{jk} + \xi_k) \qquad (j \neq 0) \tag{8}$$

with equality for those nodes j *which are used in the optimal design. Here*

$$c_j := \frac{1}{\lambda_j} \frac{\partial A(w)}{\partial w_j} \tag{9}$$

(ii) *If the routing* $j \to k$ *is used in the optimal design then equality holds in (8) for the given* j *and the minimum in the right-hand member is attained at the given* k.

(iii) *If node* j *is not used in the optimal design then* $w_j = 0$. *If it is used but at less than full capacity then* $c_j = 0$.

Proof Applying the Kuhn–Tucker conditions in the region $(\lambda, w) \geqslant 0$ we deduce that

$$\lambda_j c_j + \sum_k \lambda_{jk} b_{jk} - \lambda_j \xi_j + \sum_k \lambda_{jk} \xi_k \geqslant 0 \tag{10}$$

with equality if $w_j > 0$, and

$$w_j(b_{jk} - \xi_j + \xi_k) + \eta_j \geq 0 \qquad (11)$$

with equality if $\lambda_{jk} > 0$. We deduce from (11) that $b_{jk} + \xi_k$ will have the same value for all k such that $\lambda_{jk} > 0$; the value minimizing this expression over k. The second part of assertion (ii) is thus demonstrated. The remainder of assertions (i) and (ii) then follow by appeal to this and the condition associated with (10). The evaluation $\xi_0 = 0$ is a consistent convention, associated with the fact that there is no term for $j = 0$ in the first summation of expression (6). Finally, certainly $w_j = 0$ if node j is not used, and $\eta_j = 0$ if node j is used at less than full capacity. If the node is used but at less than full capacity then we deduce from (11) that

$$\min_k (b_{jk} - \xi_j + \xi_k) = 0 \qquad (12)$$

equality following from the fact that $\lambda_{jk} > 0$ for some k. Comparing (11) and the equality version of (8) we deduce that $c_j = 0$.∎

If we reduce the set of nodes to those which are actually used in the optimal design then inequality (8) takes the equality form

$$\xi_j = c_j + \min_k (b_{jk} + \xi_k) \qquad (13)$$

with boundary condition $\xi_0 = 0$. We recognize in (13) a *dynamic programming equation* of the type that turns up in time- and path-optimization problems (see, e.g. Whittle, 1982c). The reason for this is clear, once one thinks about it. Recall the interpretation of ξ_j as the marginal cost of accepting new additional input at node j, for the optimal design. Equation (3) represents this cost as the sum of the marginal cost c_j of passing this input through node j, the transit cost b_{jk} of passing it to node k and the marginal cost ξ_k of accepting it at node k. The only following nodes k which are employed are those for which the sum of these cost components is minimal.

If the c_j were prescribed, as are the b_{jk}, then it would be exceptional for the minimum in (13) to be attained at more than one value of k, and so for the output from node j to be split. However, c_j is a function of w. The fact that some nodes can handle traffic only at a limited rate (if they are queue-like) will mean that traffic from a given node will often have to be split. That is, w, and so the c_j, will adjust so that the minimum is attained in (13) at several values of k. These multiple equalities give extra equations which determine the actual λ_{jk}. To fully determine the solution one must couple the w-determining relations (5) with the network-determining relations (4), (8) and (9).

The solution is plainly far from complete, but one can deduce a number of features already.

Corollary 7.1 *Suppose $c_j + b_{jk} > 0$ for $j \neq 0$ and all k. Then the optimal network has no cycles.*

Proof The assumption and relation (13) imply that ξ_j decreases strictly as j follows a possible path through the optimal network to the termination point $j = 0$. Cycles are thus excluded. ∎

Note also

Theorem 7.2   *The marginal cost c_j of passage through node j has the interpretation*

$$c_j = \lambda_j^{-1} \, \text{cov} \, (a \, (n), n_j) \tag{14}$$

where the covariance is calculated under the equilibrium statistics of the network, whether optimal or not. If $a(n)$ and $\Phi(n)$ are such that these covariances are necessarily strictly positive then all nodes which are used in the optimal design are used at full capacity.

Proof   Relation (14) follows immediately from (3), (9). The conclusion $c_j > 0$ would imply the second assertion, by Theorem 7.1(iii). ∎

Exercises and comments

1. The set of nodes is supposed given. However, nodes can be dropped from the optimal design, so if one begins with a dense set of nodes then there is virtually no constraint on the set of nodes actually to be employed. Suppose, for example, that any point in $\mathbb{R}^d$ could be a node, with the functions a, Φ and b all appropriately defined. Then, depending upon the convexity/concavity properties of these functions, the nodes of the optimal network will form either a continuous or a discrete set.

8. SOCIAL, INDIVIDUAL AND BUREAUCRATIC OPTIMA

We could well describe the optimization of the last section as a *social* optimization, since it is understood as the attempt of a planner to minimize collective costs. Let us assume in this section, for simplicity of argument, that all nodes are used and at full capacity in the optimal design.

Suppose now that

$$\Phi(n) = \prod_j (n_j!)^{-1} \tag{1}$$

so that all units move independently through the network. Suppose also that $a(n)$ takes the linear form (7.1). Then one finds that c_j, defined by (7.9), has the simple evaluation,

$$c_j = a_j/\bar{\lambda}_j \tag{2}$$

independent of w. The dynamic programming equation (7.13) will then yield an evaluation of the marginal input costs ξ_j and of the optimal routing which are also independent of w, i.e. independent of expected traffic conditions.

These costs are just those which would apply to an individual unit if it were understood that the individual himself bore a cost a_j per unit time while waiting at node j (when (2) would give the expected total waiting cost at node j) and himself bore the cost b_{jk} of passage between nodes. The route recommended by the minimizing option in (7.13) (with c given by (2)) is then exactly the route that an individual should take on leaving node j, in order to minimize his individual costs, and the w-independent solution $\xi_j = F_j$ of (7.13) is then exactly the minimal future cost faced by an individual entering node j.

Let us refer to this optimization as an *individual* optimization. Our conclusion may then be expressed.

Theorem 8.1 *Suppose that units move independently and that the cost function $a(n)$ has the linear form (7.1). Then social and individual optimizations agree, in that the optimal routing is the same in both cases, and the marginal costs ξ_j of the social case are exactly the minimal individual costs F_j.*

Suppose now one allows $\Phi(n)$ to be general and chooses

$$a(n) = \sum_j a_j \Phi(n - e_j)/\Phi(n) \tag{3}$$

Then we see by appeal to Theorem 3.3 that $A(w) = \sum_j a_j w_j$ and evaluation (2) still holds. The optimal social routing will then again be that recommended by the individual optimization. This may seem strange, because individuals now interact in general, and nodes may congest.

The optimization with choice (3) might be termed a *bureaucratic* optimization. The ratio $\Phi(n - e_j)/\Phi(n)$ is proportional to the rate of exit (i.e. of 'service') at node j. If cost (3) plus the transit costs b_{jk} are regarded as costs borne by the *operators* of the system then one can say that the operators are trying to choose a routing which gives them least work, consistent with full capacity ($\lambda_j = \bar{\lambda}_j$) working. That is, the routing is one that suits the bureaucracy best, a bureaucracy concerned by its own work-rate rather than by customer waiting times, etc.

With this understanding we have established

Theorem 8.2 *The bureaucratic optimal routing is that determined by the individual optimization rule, whatever $\Phi(n)$.*

9. ADAPTIVE ROUTING RULES

It was said in section 2 that a Jackson network is the least intelligent of all networks. To bring it to a higher level one must make the routing responsive to the general state n of the system. In other words, one must use a routing rule which is *adaptive*, in that it depends upon n.

The fact that the non-adaptive optimization of routing led to the dynamic programming equation (7.13) is now most suggestive. Suppose we can find functions $c_j(n)$ such that we can make the identification

$$c_j = E(c_j(n)) \tag{1}$$

where c_j is the function of w determined by (7.9). The expectation in (1) is that for equilibrium conditions under the optimal fixed routing. If we now replaced recursion (7.13) by

$$\xi_j(n) = c_j(n) + \min_k (b_{jk} + \xi_k(n)) \tag{2}$$

with terminal condition $\xi_0(n) = 0$, then this would suggest an n-dependent routing, in that the minimizing k in (2) would depend on n. In other words, one is making routing responsive to actual traffic conditions rather than expected traffic conditions.

The rule has been determined by analogy, and the only evidence for its merit is the naturalness of the analogy. If $\xi_j(n)$ is to be interpreted as a social cost created by the entry of an immigrant to node j when the system is in state n then, instead of $\xi_k(n)$ in (2), we should really have some kind of conditional expectation $E[\xi_k(n')|n]$, where n' is the value of system state when the immigrant leaves node j. However, this would be a secondary effect if movement through the network were faster than changes in n (which, surprisingly, is the case: see Reiman (1984).

A more cautious modification of the routing rule is to replace (2) by

$$\xi_j^{(r)}(n) = c_j(n) + \min_k (b_{jk} + \xi_k^{(r-1)}(n)) \tag{3}$$

with $\xi_j^{(0)}(n) = \xi_j$ and $\xi_0^{(r)}(n) = 0$. The minimizing value of k in (3) would give the route recommendation (on leaving node j) corresponding to an r-step look-ahead rule. That is, to a rule in which one takes account of system state n for r steps ahead and reverts to an optimal non-adaptive rule after that. Of course, one would operate (3) for a fixed value of r at all times, so regarding r as the time needed for predicted system state to settle effectively from its current value n to its asymptotic value $E(n)$.

Of course, there will now usually be only a single minimizing value of k in (2), or (3). That is, there will be a definite recommendation of the route to be followed by an individual emerging from a node. Presumably the relative frequencies of routes from a given node adjust themselves correctly as n fluctuates.

The quantity $c_j(n)$ could be regarded as the *social cost* of passage of an extra individual through node j. For the cost function (7.1) with node j functionally independent of the others and with service function (1.21) one finds that

$$c_j(n) = \frac{a_j}{\bar{\lambda}_j} \frac{(n_j + s)(n_j + s + 1)}{s(s + 1)}$$

The extreme case $s = +\infty$ corresponds to the case of independent movement and the social cost reduces to the individual cost $a_j/\bar{\lambda}_j$. The other extreme $s = 1$ corresponds to the case of a simple queue, when

$$c_j(n) = \frac{a_j}{\bar{\lambda}_j} \frac{(n_j+1)(n_j+2)}{2} \qquad (4)$$

The n_j^2 dependence indicates the social costs of queueing and potential congestion.

Evaluation of these adaptive rules remains a matter for the future, but there is one simple assertion that can be made. The true optimal adaptive routing rule would be determined by solution of the stochastic dynamic programming equation

$$\gamma = a(n) + \sum_{j \neq 0} [v_j(f(n+e_j) - f(n)) + \bar{\lambda}_j \frac{\Phi(n-e_j)}{\Phi(n)} \min_k (b_{jk} + f(n-e_j+e_k) - f(n))] \qquad (5)$$

Here γ is the minimal average cost incurred, $f(n)$ a transient cost associated with a current state n, and the minimizing value of k in the jth summand is the optimal immediate destination for a unit leaving node j when the system is currently in state n.

Theorem 9.1 *For the case of bureaucratic optimization the dynamic programming equation* (5) *has the solution*

$$f(n) = \sum_j n_j F_j$$

$$\gamma = \sum_j v_j F_j$$

and the optimal adaptive routing rule reduces again to use of the fixed routing determined by solution of

$$F_j = (a_j/\bar{\lambda}_j) + \min_k (b_{jk} + F_k)$$

with terminal condition $F_0 = 0$.

The assertion follows by verification.

Exercises and comments

1. For the case of a fixed-routing Jackson network the equation analogous to (5) is

$$C = a(n) + \sum_{j,k=0}^{m} \frac{\Phi(n-e_j)}{\Phi(n)} \lambda_{jk}[b_{jk} + f(n-e_j+e_k) - f(n)] \qquad (6)$$

where C is the average cost per unit time associated with operation of the network and $f(n)$ is again the transient cost associated with a current state n. Recall that Φ, f do not include n_0 as argument, so e_0 is disregarded where it appears.

It should be a consequence of (6) that C has exactly the evaluation (7.2). To see that this is indeed so, multiply (6) by $\Phi(n) \prod_j w_j^{n_j}$ and add over n. One then obtains

$$C = A(w) + \sum_j \sum_k b_{jk} \lambda_{jk} w_j + \sum_j (w_k \lambda_{kj} - w_j \lambda_{jk}) D_j(w) \tag{7}$$

where

$$D_j(w) = \frac{\sum \Phi(n) \prod_j w_j^{n_j} (f(n+e_j) - f(n))}{\sum \Phi(n) \prod_j w_j^{n_j}}$$

Giving w the value satisfying (7.5) we deduce (7.2) from (7).

2. Consider the case of a single node, containing n units, with immigration intensity v, emigration intensity $\phi(n) = \Phi(n-1)/\Phi(n)$ and incurring cost $a(n)$ per unit time. For this case equation (6) has the particular form

$$C = a(n) + v(f(n+1) - f(n)) + \phi(n)(f(n-1) - f(n))$$

Multiplying by $\Phi(n)w^n$ and adding over n one deduces the generating function relation

$$CF(w) = G(w) + (v-w)D(w) \tag{8}$$

where

$F(w) = \Sigma \Phi(n)w^n$
$G(w) = \Sigma a(n)\Phi(n)w^n$
$D(w) = \Sigma[f(n+1) - f(n)]\Phi(n)w^n$

From (8) we obtain

$$C = G(v)/F(v) = A(v)$$

$$\frac{D(w)}{F(w)} = \frac{A(w) - A(v)}{w - v}$$

3. Suppose the node in Exercise 2 is a single-server queue with unit service rate, so that $\Phi(n) = 1$. Calculate $D(w)$ and show that $f(n)$ is proportional to $n(n+1)$ cf. (4).

10. LITERATURE

The celebrated product-form solution (1.4) was obtained first by Jackson (1963) for the case of a system of single-server queues. Two papers appeared in 1967, independent of each other and (regrettably) of Jackson: Whittle (1967) and

Gordon and Newell (1967). The paper by Gordon and Newell gave the Jackson results for a system of more general queues; the paper by Whittle gave the full product-form generalization of section 1 and observed the phenomenon of partial balance. This work was generalized to the open case by Whittle (1968). Kelly (1975a, 1979) deduced the generalization of section 2, showing that the notion of a 'potential' is more significant than that of product-form. Kelly (1975a, b, 1976) also elaborated the theory in a number of ways: the material of section 5, the use of a type label to programme units, and the study of queues with more detailed structure (cf. Chapter 10). A rounded account of this work with many examples is given in Kelly (1979).

The paper by Kingman (1969) not only constructs the reversible migration model of (4.8.6), more general in significant respects than a Jackson network, but gave the most explicit treatment yet available of a model in which blocking is an essential effect. The material of sections 7–9 is a development of that appearing in Whittle (1984b, 1985b).

CHAPTER 10

Jackson networks with structured nodes

1. THE CONCEPT OF WEAK COUPLING

The components which actually occur in networks of processors or queues need much more in the way of description than the bald n_j, the number of units held in the component. If one is to have an adequate description of a processor, for example, one must probably say something about the types of job in progress, their current stage of processing and their order of arrival. We have then to construct network models with much more structure at the nodes. This will also be the point at which we shall begin to formulate the concept of weak coupling more generally.

To recapitulate our view of the network first stated in the introduction to Part III, at each node j of a network is sited a component ($j = 1, 2, \ldots, m$). The jobs or quanta which circulate between components are termed units. These may be of several types and may even be individually identifiable. The streams of units which enter (leave) a component in some well-defined fashion are referred to as input streams (output streams) for that component.

Let the state of component j be denoted x_j. We shall suppose that the system state $x = \{x_j; j = 1, 2, \ldots, m\}$ is the state variable of a continuous-time Markov process with intensity $\Lambda(x, x')$ for the transition $x \to x'$. The number n_j of units held in component j will be a function $n_j(x_j)$ of x_j. Correspondingly, the whole vector of component totals $n = \{n_j\}$ is a function $n(x)$ of n. The distinguishing feature of the random variable n is that it can change only in transitions in which there is a transfer of units between components. Indeed, we shall term a transition $x \to x'$ *external* or *internal* according as to whether it permits the transfer of a unit between components or not. The transitions $x \to x'$ and $x' \to x$ are then either both external or both internal. The value of n can then only change in an external transition. However, we shall not assume that it necessarily does so. We shall assume that there may be transitions in which the unit leaves the component only to immediately return to it. These are to be regarded as *shunting* transitions, in which the component's output has been directly routed to its own input.

It is useful to define $\mathcal{N}$ as the set of possible values of n, and $\mathscr{X}(n)$ as the set of values of x for which $n(x) = n$. The *classes* $\mathscr{X}(n)$ ($n \in \mathcal{N}$) then constitute a decomposition of $\mathscr{X}$, and a transition is internal or external according as to

220

whether it is within a class or between classes. (Again, 'between classes' can include exit from and immediate return to a given class.)

The addition of structure to the nodes amounts to the elaboration of a Markov process with state variable n to one with state variable x, the random variable $n = n(x)$ remaining defined and meaningful. We shall refer to the processes of unstructured and structured nodes as the *n-process* and the *x-process* respectively. One might say that the x-process *imbeds* the n-process; we shall give this term a definite technical meaning in section 11.5.

We wish the x-process to have the characteristic properties of a Jackson network, and we must now ask what these characteristic properties are. One is certainly that the equilibrium distribution should have the simple form (9.2.3), but this seems to be associated with ideas of partial balance, quasi-reversibility, conservation of Poisson flows and, perhaps much less inevitably, of product form.

We imagine that, as for the unstructured case, the external transition intensities must depend upon a *routing matrix* $\lambda = (\lambda_{jk})$. The routing λ can be chosen from some permissible class $\mathscr{L}$ of routings and parametrises the network. Let $\pi(x|\lambda)$ denote the equilibrium distribution of x for a given value of λ. For simplicity we shall assume $\mathscr{X}$ discrete, so that $\pi(x|\lambda)$ is the simple probability of the state-value x. Then a generalization of the equilibrium form (9.2.3) might be to assume that $\pi(x|\lambda)$ may be written

$$\pi(x|\lambda) = \Phi(x)h(n(x)|\lambda) \tag{1}$$

That is, λ occurs in $\pi(x|\lambda)$ only in an n-dependent factor. Otherwise expressed, the routing λ affects the equilibrium distribution of x only through its effect on the equilibrium distribution of n. This is the property introduced in Whittle (1984a) as defining weak coupling. It is a concept which turns out to be natural in a much more general context and which naturally unites the concepts of higher structurings of a system and of insensitivity (see section 11.5 and Chapter 12). A statistician would express the basic weak coupling property (1) by saying that, in equilibrium, the statistic n is *sufficient* for inference on the parameter λ.

Representation (1) is of course not unique, since a factor dependent on n alone could be transferred between the factors Φ and h. Various normalizations are possible. The natural one is that given in the following theorem.

Theorem 1.1 *Suppose $\mathscr{X}$ discrete. If one so normalizes Φ that*

$$\sum_{\mathscr{X}(n)} \Phi(x) = 1 \qquad (n \in \mathscr{N}) \tag{2}$$

then one has the identifications

$$h(n|\lambda) = P(n|\lambda) \tag{3}$$

$$\Phi(x) = P(x|n, \lambda) \tag{4}$$

where the probabilities refer to equilibrium conditions. The property of weak coupling is thus equivalent to the statement that $P(x|n, \lambda)$ is independent of λ.

Proof One has the universally valid relation

$$\pi(x|\lambda) = P(n|\lambda)P(x|n, \lambda) \tag{5}$$

If we assume $P(x|n, \lambda)$ independent of λ then we may define h, Φ by (3), (4) which certainly then imply (1), (2). Conversely (1) and (2) imply (3), and (1), (3), (5) then imply (4). ∎

We are supposing $P(x|n, \lambda)$ uniquely determined so that, with this normalization, $\Phi(x)$ is uniquely determined. That is, relation (1) is regarded as valid for all λ in $\mathscr{L}$, any indeterminacy of π (due to reducibility of the process) being attributable to the corresponding indeterminacy in h.

Normalization (2) is indeed the one we shall adopt in Chapter 12. However, before that, it is more natural to adopt the normalization that $h(n|\lambda)$ can be identified with the term $\prod_j w_j^{n_j}$ of (9.2.3) or whatever its equivalent may be in more general cases. This alternative normalization does not of course invalidate the final conclusion of Theorem 1.1.

The weak coupling assumption may seem far removed from the properties we have hitherto felt to be characteristic of a Jackson network. However, the assumption that relation (1) should hold identically in λ implies strong balance conditions on the x-process, which in turn imply some of the more familiar properties, as we shall see. However, one property which is not implied is that of product form. If both Φ and h factorize so that representation (1) becomes

$$\pi(x|\lambda) = \prod_j [\Phi_j(x_j)h_j(n_j|\lambda)] \tag{6}$$

then we shall say that the nodes are *functionally independent*. However, this property is quite a special one, and one that is distinct from the 'Jackson property', as we have already seen in section 9.2. Nevertheless, it is a natural case to consider first.

The models of section 5 will essentially be found in Kelly (1979). Much of section 8, on quasi-reversibility, is derived from Kelly's (1979, 1982a) pioneering development of the topic. The rest of the chapter is development of material set forth in Whittle (1983b, 1984a, 1985b).

Exercises and comments

1. Suppose that only external transitions of the x-process depend upon λ. Suppose further that there is a value of λ in $\mathscr{L}$ under which all external transitions are forbidden; it is natural to denote this by $\lambda = 0$. It must then be a consequence of (1) that

$$[\mathscr{X}, x](\Phi\Lambda^1) = 0 \tag{7}$$

where Λ^1 is the transition intensity for internal transitions. One can express (7) by saying that Φ *balances internal transitions*.

2. Relation (7) implies that states x which are positive recurrent for the communicating system are also positive recurrent for the 'severed' system. This may simply not be true. In other words, the possibility of complete prohibition of transfer may be inconsistent with weak coupling. We saw in section 9.6 that a component cannot show the same distribution of the internal state in open and closed cases, conditional upon the number of units it contains, if its operation has a 'progressive' character.

If component j is of the progressive type considered in section 9.6 then one may say that its internal statistics, when conditioned by n, still depend upon λ through the total component output rate $\lambda_j = \sum_k \lambda_{jk}$. So, either one must assume that λ_j is constant in $\mathscr{L}$ (so that permissible changes in λ can only redirect the total output from component j, but not affect its rate) or one must assume that the rates of some internal transitions are also affected by λ (so that the internal working of the component responds to a changed demand for component output). The idea that there are two types of internal transition, 'progressive' and 'non-progressive', will find its expression in the next section.

2. TRANSITION HYPOTHESES FOR FUNCTIONALLY INDEPENDENT COMPONENTS

A natural class of systems to begin with is that for which components are functionally independent, and for which properties such as weak coupling can be seen as properties of individual components rather than as properties of the system. We look then for transition rules, for both internal and external transitions, which ensure this.

To begin with, we assume components independent in their internal transitions, so that components do not have simultaneous internal transitions, and an internal transition in component j has intensity $\Lambda_j(x_j, x_j')$. To set the pattern we shall also begin with the simplest type of external transitions. We shall assume that there is only one kind of unit and that an external transition involves the transfer of a single unit. If the transition $x \to x'$ implies such a transfer from component j to component k then, in the notation already employed, one has

$$n \to n' = n - e_j + e_k \tag{1}$$

Here by n, n' we shall consistently understand the values $n(x), n(x')$ corresponding to x, x' respectively. We shall in fact then also assume that

$$x_i' = \begin{cases} x_j' & i = j \\ x_k' & i = k \\ x_i & \text{otherwise} \end{cases} \tag{2}$$

so that only the states of the two components concerned change. We shall suppose that this transition has intensity

$$\Lambda_{jk}(x, x') = \lambda_{jk}\phi_j(x_j, x'_j)\psi_k(x_k, x'_k) \qquad (3)$$

In a sense there is no need for the j, k subscripts, since the nature of the transition is implicit in the values of x and x'. The notation makes it explicit, however, and it is a convenience to make the convention that $\Lambda_{jk}(x, x')$ is zero if x and x' are not consistent with (1) and (2). It is also understood that $\phi_j(x_j, x'_j)$ is zero unless $n_j(x'_j) = n_j(x_j) + 1$, correspondingly for $\psi_k(x_k, x'_k)$.

The factored form (3) is somewhat special, but it also has special properties in that it decouples emission of a unit by component j and the absorption of the unit by component k. The factor $\phi_j(x_j, x'_j)$ is an intensity factor for emission in the particular mode $x_j \rightarrow x'_j$; the factor $\psi_k(x_k, x'_k)$ is an intensity factor for absorption in the particular mode $x_k \rightarrow x'_k$. The coefficient λ_{jk} is again a routing rate, measuring in some sense the load capacity of the jk route.

The factored form (3) permits us to consider the component in isolation, $\psi_j(x_j, x'_j)$ and $\phi_j(x_j, x'_j)$ being *input* and *output rates* for component j respectively, and λ_{jk} being a characteristic of the network. It is convenient to define

$$\psi_j(x_j) = \sum_{x'_j} \psi_j(x_j, x'_j)$$

$$\phi_j(x_j) = \sum_{x'_j} \phi_j(x_j, x'_j) \qquad (4)$$

where the summations are necessarily restricted to absorption and emission transitions respectively.

We can make certain statements and normalizations. We may assume, for a given component, that $\psi_j(x_j) > 0$ for some x_j. If not, the component would never accept units, and be of little interest. We can change $\psi_j(x_j)$ by a factor independent of x_j and absorb this scale change in redefined routing rates. Let us then assume that there is some x_j for which $\psi_j(x_j) = 1$. Under slightly stronger conditions we might indeed impose the normalization

$$\max_{x_j} \psi_j(x_j) = 1 \qquad (5)$$

for each j. That is, that $\psi_j(x_j) \leqslant 1$, with equality for some x_j.

We may assume that $\phi_j(x_j) > 0$ for some x_j, otherwise the component could never release units, and so never reach equilibrium. However, it must also be true that $\phi_j(x_j) = 0$ for x_j such that $n_j(x_j) = 0$. That is, no unit can leave an empty component, if we forbid negative n_j.

If we say that a component (usually that for $j = 0$) is *infinite* then we mean that its state can be changed only by its own internal transitions (see section 9.3). In particular, it is not affected by external transitions, so that

$$\psi_0(x_0, x'_0) = \psi_0(x_0)\delta(x_0, x'_0)$$
$$\phi_0(x_0, x'_0) = \phi_0(x_0)\delta(x_0, x'_0) \qquad (6)$$

If the component is both infinite and *constant* then its state changes neither under external nor internal transitions. That is, x_0 is fixed, and ψ_0, ϕ_0 can be regarded as constant. One can then achieve the normalization

$$\psi_0 = \phi_0 = 1$$

by a redefinition of the routing intensities.

We need not specify whether the system is open or closed, because a closed network can effectively be opened by inclusion of an infinite component.

If the network consists only of component j and an infinite constant component 0 then component j suffers absorption transitions $x_j \rightarrow x'_j$ at rate $\lambda_{0j}\psi_j(x_j, x'_j)$ and emission transitions at rate $\lambda_{j0}\phi_j(x_j, x'_j)$. In this sense, we can indeed regard ψ_j and ϕ_j as input and output rates for component j. One can view the input as a Poisson stream of units with state-dependent rate $\lambda_{0j}\psi_j(x_j)$, the component being taken to state x'_j with probability $\psi_j(x_j, x'_j)/\psi_j(x_j)$ if an absorption occurs in state x_j.

Formula (3) expresses, among other things, the dependence of external transition rates upon routing. One might imagine that internal transition rates should not depend explicitly on the routing scheme, but there is reason to allow the possibility; see the discussion of Exercise 1.2. We shall assume a dependence

$$\Lambda_j(x_j, x'_j) = \left(\sum_k \lambda_{jk}\right)\Lambda_j^p(x_j, x'_j) + \Lambda_j^0(x_j, x'_j) \tag{7}$$

where the terms Λ^p and Λ^0 are associated with *progressive* and *non-progressive* transitions respectively. We saw the need for this kind of distinction in section 9.6. Non-progressive transitions are altogether unaffected by circumstances outside the component. Progressive transitions are affected in the very limited sense that their rates are accelerated by a factor $\lambda_j = \sum_k \lambda_{jk}$. In other words, progressive transitions are sensitive to routings in the same way that emission transitions are; they respond to 'down-stream demand'.

Since an infinite component cannot be sensitive to routing rates we must have

$$\Lambda_0^p = 0 \tag{8}$$

That is, the environment, being both stationary and autonomous, can show no progressive transitions.

One might ask whether one might also include a term in $\sum_k \lambda_{kj}$ in (6), i.e. a term responsive to 'up-stream demand'. This is indeed natural to some extent, but there are reasons for not including it (see Exercise 4.1). Despite the form of (3), input and output rates are not entirely mutually symmetric in their properties, as we have already seen in the discussion of normalizations, etc. Of course, one can argue that a component responds to input by changes in its internal state. However, at this level there is no anticipative response.

The reader will feel that it is time for some examples. We shall take these in section 5, having first determined conditions that linked components shall indeed constitute a weakly coupled system, and so a generalized Jackson network.

Exercises and comments

1. Suppose that $\psi_j(x_j) = 1$ identically for all x_j and j, so that a component accepts all input offered to it. This is what we shall in the next section term the *reactionless* case. Then rule (3) indeed implies decoupling of emission and absorption reactions, in that emission rate depends only upon the state of the emitting component, and the unit emitted is then guided completely by the routing rules.
2. The infinite component is considered to constitute the 'environment'. To give the concept some definiteness let us consider this to be the atmosphere. The atmosphere is 'infinite' in so far as it is unaffected by our puny exchanges of energy and matter with it (although pollution demonstrates that this hypothesis is strained). Even if we could regard it as infinite, we could scarcely regard it as constant if weather matters. If it were not subject to solar or lunar excitation then it would presumably settle down and indeed be constant.

3. BALANCE CONDITIONS ON COMPONENTS

The assumptions of the last section allow us to consider the properties of a component in isolation. In doing so, let us for typographical convenience drop the j subscript. So, component j has a state variable x_j taking values in a state space $\mathscr{X}_j$, which can be decomposed into the sets $\mathscr{X}_j(n_j)$ for which $n_j(x_j) = n_j$, etc. However, we shall provisionally drop the subscript j from all these quantities, and restore it when we subsequently discuss systems of such components.

There are a few properties which turn out to be significant. Let us say that the component is *structureless* if $x = n$. That is, if the state of the component is fully described by the number (or numbers, in the multi-type case) of units it is holding. Then one can say that only in section 9.6 did Chapter 9 consider anything but structureless components.

Let us say that the component is *reactionless* if $\psi(x)$ is constant in x (and equal to unity, by the normalization of (2.5)). This is an assumption that the component exerts no back-reaction on the flow, in that its state does not affect flow to it from other components.

The point of these two definitions is fairly clear. The point of the next is less obvious, but, as we shall see in the next section, it describes the principal property of components of a Jackson network. Let us say that the component is *balanced* if there exists a function $\Phi(x)$ satisfying the relations

$$\sum_{x'} [\Phi(x')\Lambda^0(x', x) - \Phi(x)\Lambda^0(x, x')] = 0 \qquad (1)$$

$$\sum_{x'} [\Phi(x')\phi(x', x) - \Phi(x)\psi(x, x')] = 0 \qquad (2)$$

$$\sum_{x'} \left[\Phi(x')\Lambda^P(x', x) - \Phi(x)\Lambda^P(x, x') \right] + \sum_{x'} \left[\Phi(x')\psi(x', x) - \Phi(x)\phi(x, x') \right] = 0 \quad (3)$$

These can be written more compactly

$$[\mathcal{X}(n), x] (\Phi\Lambda^0) = 0 \quad (4)$$

$$[\mathcal{X}(n+1), x] (\Phi\Lambda^e) = 0 \quad (5)$$

$$[\mathcal{X}(n), x](\Phi\Lambda^P) + [\mathcal{X}(n-1), x] (\Phi\Lambda^e) = 0 \quad (6)$$

where it is understood that x lies in $\mathcal{X}(n)$ and Λ^e is an effective external transition intensity with the definition

$$\Lambda^e(x, x') = \begin{cases} \psi(x, x') & x' \in \mathcal{X}(n+1) \\ \phi(x, x') & x' \in \mathcal{X}(n-1) \end{cases}$$

In words, one can say that, if $\Phi(x)$ is regarded as an equilibrium distribution for component state, then (4) asserts that Φ balances the internal non-progressive transitions, (5) asserts that the transitions $x \rightleftharpoons \mathcal{X}(n+1)$ balance under the external transition intensity Λ^e, and (6) asserts that the transitions $x \rightleftharpoons \mathcal{X}(n) + \mathcal{X}(n-1)$ balance under internal and external transition intensities Λ^P and Λ^e. So, progressive internal transitions are again grouped with emission transitions.

Of course, if Φ were an equilibrium distribution under a combined transition intensity $\Lambda^e + \Lambda^P + \Lambda^0$ then the sum of relations (1)–(3) would hold. We require that the relations hold separately because some classes of transitions are accelerated relative to others when the component is made part of a network.

We shall wish to write the balance equations (1)–(3) yet more compactly. Let us write them as

$$\Delta^0 = 0 \quad (7)$$

$$S^+\Phi - \psi\Phi = 0 \quad (8)$$

$$\Delta^P + S^-\Phi - \phi\Phi = 0 \quad (9)$$

All these quantities are functions of x, this is to be understood. So Φ is $\Phi(x)$ and ψ, ϕ are the total input and output rates $\psi(x)$, $\phi(x)$ defined in (2.4). The definition of the quantities Δ^0, Δ^P is implicit, as is that of the linear operators S^+, S^-, having effect

$$S^+\Phi(x) = \sum_{x'} \Phi(x')\phi(x', x)$$

$$S^-\Phi(x) = \sum_{x'} \Phi(x')\psi(x', x)$$

We shall suppose that, not merely can relations (7)–(9) be satisfied by some Φ, but also that they determine Φ to within a scale factor. This is equivalent to demanding that the component be irreducible, in that any state can be reached from any other if the component is driven by a Poisson stream (i.e. connected to an infinite constant component). See Exercise 2.

For an infinite component the balance relations (8), (9) amount simply to

$$\phi_0(x_0) = \psi_0(x_0) \tag{10}$$

in view of (2.6), (2.8). That is, input and output rates for the environment should show the same dependence upon environmental state. We may then make the simple assertion:

Theorem 3.1 *An infinite component is balanced and reactionless if and only if it is effectively constant.*

Proof We see from (10) that reactionlessness and balance imply that ϕ_0 and ψ_0 are independent of x_0. The interactions of the component with other components are then independent of x_0 and, in this sense, the component is effectively constant. The reverse implications are clear. ■

Exercises and comments

1. For an example of a component which *is* reducible, suppose that $\psi(x) = 0$ if $n(x) = s$ and $\phi(x) = 0$ if $n(x) = s+1$. The sets $n \leqslant s$ and $n > s$ are then closed. The component may start in either, but cannot cross from one to the other.

2. Suppose the component has total transition rate $\Lambda(x, x') = v\psi(x, x') + \mu(\Lambda^P(x, x') + \phi(x, x')) + \xi\Lambda^0(x, x')$ for constants v, μ and ξ. That is, absorption reactions are accelerated by a factor v, progressive and emission reactions by a factor μ, and the non-progressive reactions by a factor ξ. Loosely speaking, the component is driven by a Poisson stream of rate v (of which it accepts a proportion $\psi(x)$) and has its throughput accelerated by a factor μ. Show that, if the component is balanced, then its equilibrium balance equations are satisfied by

$$\pi(x) \propto \Phi(x)(v/\mu)^n$$

The assumption of irreducibility stated in the text is just that this should be the unique solution of the balance equations.

4. NETWORKS OF STRUCTURED COMPONENTS

Consider now a system of components whose external and internal transitions have intensities respectively of the forms (2.3) and (2.7). This constitutes a network whose nodes are the components and whose arcs are the routes by which units flow. Components interact only via the flow of units. We assume the structure a fixed one (as distinct from the random graphs of Part IV); the study of evolving or self-reinforcing networks is fascinating but for the future.

The network formulation gives an embryonic spatial structure to the model, a structure in which the flows along arcs are as significant as the processing taking place within components. Of course, if the network really represents space then most of it will be 'empty'. That is, most of the nodes will be mere intermediate holding points, without internal structure, and so rather degenerate as components. Such holding points are indeed needed if one is not to assume that transfer of units between the real working processors is instantaneous. By incorporating them one rather returns to the concept of section 6.1, of regarding units in transit as constituting a cloud or sea within which the components are immersed. However, the cloud is not now a passively wandering collection of free units, but is purposefully directed by the rules implicit in the routing matrix.

However, for the moment we need not distinguish between 'functioning' and 'holding' components, but merely assume that a number of functionally independent components with internal transitions governed by the intensities (2.7) are linked into a network by prescription of the external transition intensities (2.3). The components will usually be labelled $j = 1, 2, \ldots, m$, but we shall leave open the possibility of an infinite component, labelled $j = 0$, whose inclusion would effectively open the system. The jth component is characterized by the quantities $\psi_j(x_j, x'_j)$, $\phi_j(x_j, x'_j)$, $\Lambda^p_j(x_j, x'_j)$ and $\Lambda^0_j(x_j, x'_j)$. The linking is characterized by the routing matrix $\lambda = (\lambda_{jk})$.

Theorem 4.1 *Suppose that all the components of the system with transition intensities* (2.3), (2.7) *are balanced and reactionless. Then*
(i) *The system has equilibrium distribution*

$$\pi(x) \propto \prod_j [\Phi_j(x_j) w_j^{n_j(x_j)}] \tag{1}$$

on any given irreducible set, where Φ_j is the solution of the balance equations (3.1)–(3.3) *and w the solution of the traffic equations*

$$\sum_k (w_k \lambda_{kj} - w_j \lambda_{jk}) = 0 \tag{2}$$

with the additional prescription $w_0 = 1$ if an infinite component ($j = 0$) is included.
(ii) *The system is weakly coupled, and shows partial balance over the non-progressive transitions of any given component, and over the progressive plus external transitions of any given component.*

Proof If we can prove that expression (1) satisfies the partial balance relations asserted in (ii) then it certainly satisfies the total balance relation and all assertions follow. If we assume an equilibrium distribution (1) then, in the notation of (3.7)–(3.9), these partial balance relations can be written

$$\Delta^0_j = 0 \tag{3}$$

$$\left(\sum_k \lambda_{jk}\right) \Delta^p_j \Phi_j^{-1} + \sum_k [\lambda_{kj} w_k w_j^{-1} (S_k^+ \Phi_k)(S_j^- \Phi_j)(\Phi_k \Phi_j)^{-1} - \lambda_{jk} \phi_j \psi_k] = 0 \tag{4}$$

The balance relations (3.7)–(3.9) imply the validity of (3) and that (4) reduces to

$$\sum_k \psi_k(x_k)[w_k \lambda_{kj} - w_j \lambda_{jk}] = 0 \tag{5}$$

The assumption of reactionlessness now implies that (5) reduces to (2). Distribution (1) thus satisfies the total and partial balance relations asserted, and its form implies weak coupling of the components. ∎

As we saw in the last section, if the system is open then balance and reactionlessness require that the environment (the infinite component) be effectively constant, with $\phi_0(x_0) = \psi_0(x_0) = 1$, whatever the environmental state x_0.

Many of the points made in Chapter 9 now have obvious analogues, we list these rather formally in the exercises.

Exercises and comments

1. *Uniqueness.* The distribution (1) will be unique if the components are irreducible and if the routing matrix satisfies the conditions of Theorem 9.2.2 (closed case) or Theorem 9.3.2 (open case).

2. *Congestion.* Expression (1) must be summable over the relevant irreducible set if it is to constitute a distribution. If we define

$$\Phi_j(n_j) = \sum_{\mathcal{X}_j(n_j)} \Phi_j(x_j) \tag{6}$$

then (in the open case) the sums $\sum_{n_j} \Phi_j(n_j) w_j^n$ must be convergent; failure to be so indicates some kind of congestion in the network. However, failure might now occur because the sum (6) is itself non-convergent. That is, because for a given value of n_j the component state x_j may become lost in 'remote' state space.

3. As in (9.2.5) one might write distribution (1) as

$$\pi(x) \propto \exp\left[-U(x) - \sum_j \zeta_j n_j\right]$$

where $U(x)$ is a 'configuration potential' and the ζ_j 'site potentials'. However, again, one must be cautious; reversibility is assumed in the model only to the very limited extent that one postulates the balance relations (3.7)–(3.9). See Exercise 5.

4. In view of the considerable degree of symmetry between input and output of a component one might imagine that it would be natural to generalize (2.7) to

$$\Lambda_j = \left(\sum_k \lambda_{kj}\right)\Lambda_j^+ + \left(\sum_k \lambda_{jk}\right)\Lambda_j^- + \Lambda_j^0$$

and the balance relations (3.7)–(3.9) to

$$\Delta_j^0 = 0$$
$$S_j^+ \Phi_j - \psi_j \Phi_j + \Delta_j^+ = 0$$
$$S_j^- \Phi_j - \phi_j \Phi_j + \Delta_j^- = 0$$

where Λ^+ and Λ^- are components of the intensity for progressive internal transitions which are associated with external transitions $x_j \rightleftharpoons \mathscr{X}_j(n_j + 1)$ and $x_j \rightleftharpoons \mathscr{X}_j(n_j - 1)$ respectively. Show that Theorem 4.1 does not in general hold for such a system of components, although it holds, for example, for a system consisting of one such component and a single infinite component.

5. There are circumstances under which the condition of reactionlessness imposed in Theorem 4.1 can be somewhat relaxed. Suppose that for a given component i the solution w of (2) satisfies

$$w_i \lambda_{ij} - w_j \lambda_{ji} = 0$$

for all j. One then sees that (5) can be satisfied however $\psi_i(x_i)$ behaves. In words, if component i shows detailed balance in routings against all other components then it need not be reactionless for the conclusions of Theorem 4.1 to apply. This was observed (for the structureless case) by Kingman (1969). In particular, the condition holds for a system of two components in which the total output of one is led to the other (i.e. there is no shunting).

This observation is a significant example of the fact that a strengthening of balance conditions may permit a weakening of other restrictions on the process. See Chapter 12 for a systematic treatment.

5. EXAMPLES

We shall use c to label a component in this section, since it will be convenient to use j and k to label internal nodes of a component. We know from Theorem 4.1 that if a balanced reactionless component c is incorporated in an open network of such components then its state x_c will in equilibrium be independent of that of other components of the network and have distribution

$$\pi(x_c) \propto \Phi_c(x_c) w_c^{n_c} \tag{1}$$

where w_c is determined by the equation system

$$\sum_{c'} (w_{c'} \lambda_{c'c} - w_c \lambda_{cc'}) = 0$$

(with $w_0 = 1$). However, w_c itself is best seen simply as a parameter that relates component c to the rest of the network, in that it is the ratio

$$w_c = \rho_c / \lambda_c \tag{2}$$

of the input rate

$$\rho_c = \sum_{c'} w_{c'} \lambda_{c'c} \tag{3}$$

and the 'drain rate'

$$\lambda_c = \sum_{c'} \lambda_{cc'} \tag{4}$$

The two rates can be varied arbitrarily as the component is incorporated in different networks, but it is characteristic of a Jackson network that the effect of this coupling is represented totally by the factor $w_c^{n_c}$ in (1). However, the factor $\Phi_c(x_c)$ represents a property of the component alone, and it is this that we shall now examine in special cases. Since for this purpose we can restrict attention to the single component we shall drop the subscript c and, as in section 3, write $\Phi_c(x_c)$ simply as $\Phi(x)$, etc.

(a) Simple symmetric components, societies and queues

A very degenerate case of a balanced component is one in which there are no internal transitions, so that $\Lambda^p = \Lambda^0 = 0$, and Λ^e satisfies the balance equations (3.5), (3.6) in the strong sense that it satisfies a detailed balance relation

$$\Phi(x)\phi(x, x') = \Phi(x')\psi(x', x) \tag{5}$$

One might term this a *symmetric* component in that it is a slight generalization of what is normally termed a 'symmetric queue' (see Exercise 1). We shall term it a 'simple symmetric component' in that example (c) demonstrates something of the same type but with more structure. If the component is reactionless then $\psi(x', x)$ is a distribution over x for given x': the probability that a component in state x' and suffering an absorption transition will be taken to state x.

As a particular case, suppose that x represents the mutual relationships between the $n = n(x)$ units present: a *society*. We do not suppose that units can be identified as individuals (which would pose difficulties for an open process) but rather that each can be identified by his current place in society (e.g. 'ith in the queue' or 'Assistant Secretary to the Minister of Transport'). He can be identified to the extent that he can be traced through transitions. For example, if someone leaves the queue then one knows from what position he left and how the remaining individuals reorder themselves. If someone is recruited into the hierarchy of the Ministry of Transport then one knows to what post he goes and how previous post-holders are reallocated over the now enlarged set of posts.

It will make distinctions clearer later if we use s (for 'society') rather than x to indicate the state of the component and write $\psi(x', x)$ as $a(s', s)$. Then $a(s', s)$ is a distribution over s, the societies that can be formed when a new member joins society s'. Relation (4) would then imply that

$$\phi(s, s') = \frac{\Phi(s')}{\Phi(s)} a(s', s) \tag{6}$$

So, suppose, for example, that $\Phi(s)$ is a function $\Omega(n)$ of $n = n(s)$, the number in society s. Relation (5) could then be seen as stating that there is an 'expulsion pressure' $\Omega(n-1)/\Omega(n)$ on a society of size n, and that this is exerted differentially on members of the society in that the individual i whose departure would change s to s' experiences pressure proportional to $a(s', s)$. It is the fact that this factor is the same as the joining probability which makes the society a 'symmetric' one.

A queue is a special case of a society in which members are ordered linearly. For these it turns out that $a(s', s)$ can be regarded as a distribution over s' (and so over i) for given s as well as a distribution over s for given s'. That is, it indicates the proportion of the service effort that is directed to the individual in position i as well as the probability that an arrival to the queue joins it in position i. We consider symmetric queues in Exercise 1.

(b) Jackson components

Suppose that an open Jackson network of the type discussed in section 9.3 is itself taken as a component, a possibility already raised in section 9.6. The state variable x of the component is then just the distribution $n = \{n_j\}$ of units over its internal nodes j. One would then make the identifications

$$\psi(x, x') = \theta(n, n + e_j) = v_j \tag{7}$$

$$\phi(x, x') = \theta(n, n - e_j) = \mu_j \frac{\chi(n - e_j)}{\chi(n)} \tag{8}$$

$$\Lambda^P(x, x') = \theta(n, n - e_j + e_k) = \lambda_{jk} \frac{\chi(n - e_j)}{\chi(n)} \tag{9}$$

$$\Lambda^0(x, x') = 0 \tag{10}$$

Recall here that j and k refer to *internal* nodes, and all the routing parameters λ, μ and v refer to *internal* routings, and would be distinguished by an additional subscript c if we had to make explicit the fact that we are discussing component c. We have replaced the function Φ of section 9.3 by χ to avoid a conflict of notation.

Theorem 5.1 *The 'Jackson component' defined by (7)–(10) is reactionless and balanced, with*

$$\Phi(x) = \chi(n) \prod_j u_j^{n_j} \tag{11}$$

where the u_j satisfy

$$v_j - \mu_j u_j + \sum_k (u_k \lambda_{kj} - u_j \lambda_{jk}) = 0 \tag{12}$$

Proof   The component is reactionless because $\psi(x) = \sum_j v_j$ is independent of x.

We know (Theorem 9.3.1) that expression (11) balances the transitions with intensity θ. In fact, it balances them partially against only those transitions to and from a given state n which affect a given node j. The partial balance relation for $j = 0$ implies (3.2), and (3.3) then follows as the complementary balance relation. ∎

One can in fact normalize v so that

$$\sum_j v_j = 1$$

(the normalization of $\psi(x)$ considered in section 2) and v_j is then interpretable as the probability that a unit entering the component is directed to the internal node j.

By regarding the internal nodes as stages in the passage of a unit through the component one can derive insensitivity conclusions of the type derived systematically in Chapter 12.

Define $|n| = \sum_j n_j$ as the total number of units in the component and suppose that $\chi(n)$ has the particular form

$$\chi(n) = \frac{|n|!}{\prod_j n_j!}\, \Omega(|n|) \tag{13}$$

The n-dependent factor occurring in intensities (8), (9) then takes the form

$$\frac{\chi(n - e_j)}{\chi(n)} = \frac{n_j}{|n|}\, \frac{\Omega(|n| - 1)}{\Omega(|n|)} \tag{14}$$

We see from (1), (11) and (13) that the equilibrium distribution of $|n|$ then becomes

$$\pi(|n|) \propto \Omega(|n|)\,(wu)^{|n|} \tag{15}$$

where

$$u = \sum_j u_j \tag{16}$$

This result can be expressed:

Theorem 5.2   *Consider a simple node with intrinsic service rate*

$$\phi(|n|) = \frac{\Omega(|n| - 1)}{\Omega(|n|)} \tag{17}$$

and so with equilibrium distribution

$$\pi(|n|) \propto \Omega(|n|)w^{|n|} \tag{18}$$

in a network. The distribution (18) *is unchanged if the service requirements of units entering the node are independent with an arbitrary common distribution with unit mean, and the service effort* (17) *is distributed uniformly over all units at the node.*

Proof Relations (8), (9) and (13) imply that the units are passing independently through the auxiliary process constituted by the internal nodes of the component at a working rate $\phi(|n|)/|n|$. This corresponds to a uniform distribution of the service rate (17) over all units currently present. If the auxiliary process is chosen so that expected service requirement is unity, i.e. so that $u = 1$, then distribution (15) indeed reduces to (18). We know, moreover, by Theorem 3.4.2, that we can achieve an arbitrary distribution of service time requirement by appropriate choice of the auxiliary process. The theorem is thus proved. ∎

The auxiliary process could be regarded either as a device for achieving a desired service requirement distribution or as representing the actual physical processing in the component. We already have the analysis needed to weaken the assumptions of Theorem 5.2.

Corollary 5.1 *The conclusions of Theorem* 5.2 *remain valid even if service requirement distribution is specific to the unit* (*but of unit mean and independent for different units*).

We give the supporting argument in Exercise 3.

However, this model of a compound node will be unsatisfactory for many applications in that all units at the node are treated equally, i.e. on a time-sharing basis. If one is to represent queues, for instance, one will require differential treatment of individuals according to some notion of precedence. Let us then hybridize models (a) and (b).

(c) *Symmetric components, societies and queues*

Let us suppose that the units in the component both have a mutual relationship (i.e. constitute a society) and can be distributed over internal nodes of the component (i.e. be at different stages of processing or development). The state variable x of the component should then indicate both the structure $s = s(x)$ of the society and the location of each identified member of this society on the internal nodes (i.e. the stage of his development).

When we speak of 'individual i' we shall mean more exactly 'the unit in place i of society'. If x is an arbitrary value of state for which $s(x)$ indeed possesses a place i then we shall use $T_j^i x$ to denote the state value x modified only in that individual i has been shifted to node (i.e. internal node) j. Any state for which individual i is at node j can be represented as $T_j^i x$ for a variety of x. We can consistently use $T_0^i x$ to denote the state to which x moves if individual i leaves the component. Note that $s(T_j^i x) = s(x)$, for $j \neq 0$.

Assume that the internal node-transitions have intensity

$$\Lambda^{\mathrm{P}}(T_j^i x, T_k^i x) = \lambda_{jk} \frac{\chi(T_0^i x)}{\chi(T_j^i x)} a(s_i, s) \tag{19}$$

Here s is the existing society $s(x) = s(T_j^i x)$ and s_i is the society $s(T_0^i x)$ that it would become if individual i were lost from it. As before, we suppose that $a(s', s)$ is a distribution over societies s that can be formed if society s' gains a member. We shall assume that $\Lambda^0 = 0$ and that society changes only in the external transitions, which are assumed to have the intensities

$$\phi(T_j^i x, T_0^i x) = \mu_j \frac{\chi(T_0^i x)}{\chi(T_j^i x)} a(s_i, s) \tag{20}$$

$$\psi(T_0^i x, T_k^i x) = v_k a(s_i, s) \tag{21}$$

As previously, these rates are consistent with the form (19) if we make the identifications $\mu_j = \lambda_{j0}$ and $v_k = \lambda_{0k}$.

The model with transitions and intensities (19)–(21) constitutes then a hybrid between the simple symmetric society models (a) and the Jackson models (b) in which the factor $a(s_i, s)$ not merely determines the probability that a new arrival shall take up place i but also dictates the pace of progression of individual i subsequently.

Theorem 5.3 *The component defined by (19)–(21) is reactionless and balanced, with*

$$\Phi(x) = \chi(x) \prod_j u_j^{n_j} \tag{22}$$

where the u_j satisfy (12).

Proof We see from (21) that $\psi(x) = \sum_k v_k$, so the component is certainly reactionless. The Φ defined by (22) is readily found to satisfy the balance conditions

$$\sum_{k=0}^{m} [\Phi(T_k^i x) \Lambda (T_k^i x, T_j^i x) - \Phi(T_j^i x) \Lambda (T_j^i x, T_k^i x)] = 0 \qquad (j = 0, 1, 2, \ldots, m) \tag{23}$$

where the transition intensity Λ is given by (19) for $j, k \neq 0$, by (20) for $k = 0$ and by (21) for $j = 0$. Summing relation (23) for a given $j \neq 0$ over all i in $s(x)$ we deduce the balance relations (3.3). Summing relation (23) for $j = 0$ over all x consistent with a prescribed value of $T_0^i x$ for some i (i.e. over all structures that would reduce to the given value of $T_0^i x$ if they lost a unit) we deduce the balance relation (3.2). The component is thus balanced, with Φ having evaluation (22). ∎

The fact that the component obeys balance conditions so much more detailed than are needed shows that it is very far from being the most general balanced, reactionless component. However, we can deduce the following theorem from Theorem 5.3 just as we deduced Theorem 5.2 and Corollary 5.1 from Theorem 5.1.

Theorem 5.4 *Consider a component which accepts all arriving units and which, if in state s', moves to state s with probability $a(s', s)$ on the arrival of a unit. Units arrive with independently distributed service requirements of unit mean, the distribution otherwise being arbitrary and specific to the unit. This service requirement is worked off at rate $a(s', s)\Phi(s')/\Phi(s)$, where s is the current state of the component and s' the state to which it would revert if the unit in question left. Units leave on completion of their service requirement. Then s has the equilibrium distribution*

$$\pi(s) \propto \Phi(s)w^{n(s)} \tag{24}$$

where w is the network coupling parameter (2).

This is just model (a), but with a general distribution of service requirement replacing the exponential distribution. The 'structural state' s is then not the full Markov state of the component, but what would be the Markov state if service requirements were exponential, i.e. if the work-rate were a termination intensity.

So, for example, the simple nodes in the examples of section 9.1, such as the time-sharing processor and the machine repair model, can be replaced by symmetric queues, and the service requirement of a job or machine entering one of these nodes can have a distribution which is arbitrary and specific to the unit itself (subject to independence and unit mean) without affecting the equilibrium distribution of the 'structural states' of the queues.

Exercises and comments

Symmetric queues

1. Consider a queue whose intrinsic service rate is a function $\phi(n) = \Omega(n-1)/\Omega(n)$ of its size n. The *queueing discipline* is the set of rules which govern order of waiting and order of service. This normally changes only on an arrival or a departure, and those not joining or leaving the queue normally retain their relative positions. The queueing discipline is then specified by the probabiliites $\gamma(i, n)$ that an individual joining a queue of length $n-1$ does so in i^{th} place, and the probabilities $\gamma^*(i, n)$ that an individual leaving a queue of length n does so from the ith place $(i = 1, 2, \ldots, n)$. So, for a first-come first-served queue, in which individuals are served in order of arrival, one has $\gamma(i, n) = \delta_{in}$ and $\gamma^*(i, n) = \delta_{i1}$.

A symmetric queue is one for which $\gamma(i, n) = \gamma^*(i, n)$ identically in i and n. This has an equilibrium distribution $\pi(s) = \pi(n) \propto \Omega(n)w^n$ when coupled into a network. In our terminology, it is a balanced, reactionless component with

$\Phi = \Omega$. The reason why we can identify $\pi(s)$ with $\pi(n)$ is that s does not state the order of *specified* individuals, but merely labels members of the queue by their places so that these can be appropriately transformed on an arrival or a departure.

In the terminology of example (a) one can also regard it as a symmetric society, with s denoting the linear order of the members of the queue, $\Phi(s)$ identified with $\Omega(n)$, and $a(s', s)$ identified with $\gamma(i, n)$ if s' is the ordering in a queue of $n - 1$ and s the ordering that then results if an individual joins in i^{th} place.

A time-sharing node could be regarded as a symmetric queue (with $\gamma(i, n) = 1/n$), as could a last-come first-served queue ($\gamma(i, n) = \delta_{in}$), but not, as we have seen, a first-come first-served queue.

2. Consider the symmetric society of example (a) and suppose that the transition $s' \to s$ is one in which the arrival to a society in state s' takes up place i. This has probability $a(s', s)$ conditional on s'. The probability that, conditional on s, a departure which occurs is from place i is then, by (6), $\Phi(s') a(s', s)/[\sum_{s'} \Phi(s') a(s', s)]$ where the summation in the denominator is over all s' consistent with s. Note that this departure probability is not in general equal to the arrival probability $a(s', s)$. For the case of a symmetric queue in Exercise 1 it was, because all possible $\Phi(s')$ had the same value $(\Omega(n-1))$ and the matrix $(a(s', s))$ was doubly stochastic in that $\sum_{s'} a(s', s) = 1$.

3. The sketch proof of Corollary 5.1 is as follows. First, by having several non-communicating sets of internal nodes one can represent several types of unit, and the service requirement distribution can be made type-specific. Second, by increasing the number of types indefinitely one can effectively make this distribution unit-specific.

To be explicit, suppose that the units can be divided into types α which are immutable and such that the service rate (17) and the routing between components are independent of type. Suppose that a fraction f_α of units is of type α. Replace the internal nodel label j for the component by the compound label (α, j), which represents both type and stage. Then the internal routing parameters v_j, λ_{jk} and μ_j will be replaced by α-dependent versions $f_\alpha v_{\alpha j}$, $\lambda_{\alpha jk}$ and $\mu_{\alpha j}$. With the analogues of assumptions (7)–(10) and (13) one then deduces, in analogue to (15), that

$$\pi(|n|) \propto \Omega(|n|) \left(w \sum_\alpha f_\alpha u_\alpha \right)^{|n|} \tag{25}$$

where

$$u_\alpha = \sum_j u_{\alpha j}$$

Expression (25) reduces to (18) if $u_\alpha = 1$ for all α. The assumptions of Theorem 5.2 can thus be weakened in that the distribution of service requirement of a unit can be made type-specific and, in the limit, unit-specific.

6. TIME REVERSAL AND FLOW PROPERTIES

The material of section 9.5 has a complete analogue for the network of structured components envisaged in Theorem 4.1. That is, that the system retains all its characteristic features under time reversal, with the implications that this has for the statistical properties of the flows. As usual, the time-reversed version of any quantity such as θ is written θ^*. It is convenient to introduce the notation

$$R_j \theta(x_j, x'_j) := \theta(x'_j, x_j) \Phi(x'_j)/\Phi(x_j) \tag{1}$$

for a transformation of a transition intensity $\theta(x_j, x'_j)$.

Theorem 6.1 *The system of Theorem 4.1 retains its character under time reversal in that the transition rates remain of the forms (2.3) and (2.7) and the components remain balanced and reactionless. The detailed analogues are*

$$\phi_j^* = R_j \psi_j \qquad \psi_j^* = R_j \phi_j$$
$$\Lambda_j^{p^*} = R_j \Lambda_j^p \qquad \Lambda_j^{0*} = R_j \Lambda_j^0 \tag{2}$$

with the implication $\Phi_j^* = \Phi_j$; also

$$\lambda_{jk}^* = \lambda_{kj} w_k / w_j \tag{3}$$

with the implications $w_j^* = w_j$ and $\lambda_j^* = \lambda_j$.

Proof This follows as usual by verification from the general formula (9.5.2) for transformation of a transition intensity under time reversal and the form (4.1) of the equilibrium distribution. The correspondences are interesting: relations (3.1) and (3.3) imply their own reverse forms, whereas (3.2) and the reactionlessness condition imply each other's reversed forms. The final assertion of the theorem is indeed necessary if representation (2.7) is to conserve its form under time-reversal. ∎

It follows from (4.6) and the transformations (2) that the time-reversed environment is again effectively constant:

$$\phi_0^*(x_0, x'_0) = \psi_0^*(x_0, x'_0) = \delta(x_0, x'_0) \tag{4}$$

The conclusion of Theorem 9.5.2 is again immediate.

Theorem 6.2 *Consider the system of Theorem 4.1 in the open case. Then the output streams from the nodes $j = 1, 2, \ldots, m$ are mutually independent Poisson streams of rates $\lambda_{0j}^* = \lambda_{j0} w_j$, and the streams before time t are independent of system state at time t.*

Proof This result is in part a consequence of the reactionlessness of the reversed system. We have

$$\Lambda_{0j}^*(x, x') = \lambda_{0j}^* \phi_0^*(x_0, x'_0) \psi_j^*(x_j, x'_j)$$
$$= \lambda_{j0} w_j \delta(x_0, x'_0) \phi_j(x'_j, x_j) \Phi_j(x'_j)/\Phi_j(x_j)$$

The input stream to node j in the reversed system thus has rate

$$\sum_{x'} \Lambda_{0j}^*(x, x') = \lambda_{j0} w_j \psi_j(x_j) = \lambda_{j0} w_j$$

and this constant rate can again be identified with the rate of the output stream from node j in the original system. ∎

So, if we regard an open network of balanced, reactionless components as a generalized Jackson network then this again shows the Poisson-preserving property.

7. MULTI-TYPE FLOWS

There are a number of reasons for wishing to consider several types of unit. In physical contexts there will indeed be several types of unit; e.g. energy quanta, or atoms of different elements. These will usually be regarded as immutable, with totals for each type conserved in a closed system. In commodity flows one will have several types of goods; in traffic or information flows units will be distinguishable at least by their destination.

However, type may well be mutable. In computing and industrial contexts 'type' can indicate not merely the nature of a job but also the stage that it has reached in its required processing. Kelly (1979) has indicated how type can be used in this way to programme the routing of a job through the system. If type indicates the stage which a job has reached and the routing intensity can be made type-dependent then one can direct a job which has reached a given stage to the next appropriate process (cf. Exercises 9.2.1–9.2.3).

As Kelly formulated the model, type-changes occurred in routing, i.e. on emergence from the component rather than within the component. If the component represents a processor then it would seem more natural that mutation of types or merging or division of type-streams should take place during processing, i.e. within the processor. However, the whole concept of balance developed in section 3 becomes trickier in this case, and can indeed be inconsistent with the phenomena one is trying to represent. We say something on these matters in the next two sections.

For the case when type is conserved within the component the analogues of the material of sections 2 to 6 is fairly immediate. The generalization of the external transition intensity (2.3) will be

$$\Lambda_{\alpha j, \beta k}(x, x') = \lambda_{\alpha j, \beta k}\, \phi_{\alpha j}(x_j, x_j')\, \psi_{\beta k}(x_k, x_k') \tag{1}$$

This represents a migration/mutation transition in which an α-unit leaves component j and arrives at component k as a β-unit. Only the states of components j and k are affected in the transition, and $\lambda_{\alpha j, \beta k}$ is the associated routing intensity.

The generalization of the internal transition intensity (2.7) will be

$$\Lambda_j(x_j, x_j') = \sum_\alpha \left(\sum_\beta \sum_k \lambda_{\alpha j, \, \beta k} \right) \Lambda_{\alpha j}^p(x_j, x_j') + \Lambda_j^0(x_j, x_j') \tag{2}$$

This transition is indeed internal; no units move outside the component or change their type. The term Λ_j^0 is an intensity for non-progressive transitions. The term $\Lambda_{\alpha j}^p$ is an intensity for progressive transitions associated with α-units; the term $\lambda_{\alpha j}$ multiplying it indicates that the rate of this transition is affected by the rate at which α-units are drained from the component.

In discussing the properties of a single component we may again drop the subscript j. The concepts of reactionlessness and balance generalize naturally. The component is reactionless if

$$\psi_\alpha(x) := \sum_{x'} \psi_\alpha(x, x') = 1 \tag{3}$$

for all α relevant to the component. We speak of 'all α relevant to the component' because it may not be that all components have to deal with all types of units; the routing intensities may be such that a given component will only ever be called on to admit a sub-set of types. Relation (3) states that it must then indeed admit these, without back-reaction.

The generalization of the balance conditions in the form (3.7)–(3.9) would then be that there should exist a function $\Phi(x)$ of component state such that

$$\Delta^0 = 0 \tag{4}$$

$$S_\alpha^+ \Phi - \psi_\alpha \Phi = 0 \tag{5}$$

$$\Delta_\alpha^p + S_\alpha^- \Phi - \phi_\alpha \Phi = 0 \tag{6}$$

for all α relevant to the component. Here ψ_α denotes $\psi_\alpha(x)$, correspondingly $\phi_\alpha = \phi_\alpha(x)$ and

$$S_\alpha^+ \Phi(x) = \sum_{x'} \Phi(x') \phi_\alpha(x', x) \tag{7}$$

$$S_\alpha^- \Phi(x) = \sum_{x'} \Phi(x') \psi_\alpha(x', x) \tag{8}$$

as in section 3. Conformability is assumed throughout: for example, that $\psi_\alpha(x, x')$ is zero unless x' is a state that could be reached from x by absorption of an α-unit.

With these definitions Theorem 4.1 has a complete analogue: a network of balanced, reactionless components has equilibrium distribution

$$\pi(x) \propto \prod_j \left[\Phi_j(x_j) \prod_\alpha w_{\alpha j}^{n_{\alpha j}} \right] \tag{9}$$

where the $w_{\alpha j}$ solve

$$\sum_\beta \sum_k (w_{\beta k} \lambda_{\beta k, \alpha j} - w_{\alpha j} \lambda_{\alpha j, \beta k}) = 0 \tag{10}$$

with $w_{\alpha 0} = 1$ for all relevant α if the system is open. This distribution implies weak coupling, and also satisfies partial balance conditions: probability fluxes balance for movement of α-units in and out of component j for arbitrary given α and j.

Note that, if there are type mutations which cannot occur even in the routing, then the equation system (10) will decompose into separate systems in separate sets of variables. Indeed, in the closed case it will decompose into as many sub-systems, with as many arbitrary parameters in the solution, as the network has invariants in unit numbers.

Again, the simplest structured case is the analogue of example (a) of section 5, in which the component shows no internal transitions and shows detailed balance in its external transitions in that

$$\phi_\alpha(x, x') = \frac{\Phi(x')}{\Phi(x)} \, \psi_\alpha(x', x) \qquad (11)$$

for all relevant α (cf. (5.5), (5.6)). This assures balance, and the component will be reactionless if $\psi_\alpha(x', x)$ is a distribution in x for given x'. By examining even simpler cases of this model one sees, for example, that a balanced reactionless component cannot represent a queue with priority scheduling, i.e. a queue that will favour some customer types over others (see Exercise 1).

Examples (b) and (c) of section 5 also have multi-type versions; all that is necessary is that one labels the internal nodes by (α, j) rather than simply by j, and that nodes of differing α do not communicate.

It must now be said that the condition of weak coupling, with its associated demands of balance and reactionlessness, is in some ways too strong. Suppose we regard an open Jackson network (either of simple nodes, as in Chapter 9, or of balanced, reactionless components, as in section 4) as itself a composite multi-input multi-output component. Then this is certainly something we can handle, in that we can calculate the equilibrium distribution of interconnected such components, and say something about the statistics of flows, etc. However, such a component does not in general show weak coupling with the others. As we observed already in section 9.6, it will not in general be true that the numbers of units (of different types) held in the component constitute sufficient statistics for its external routing parameters. Just because the component has several distinct input and output streams, even for units of a given type, and because there is no reason that there should be some fortunate relationship between its internal and external routing intensities, the equilibrium distribution of the component varies with changes in external routing in a much more complicated fashion than could be summed up in a few statistics. More specifically, the composite component is reactionless, but not in general balanced.

Nevertheless, one would gain welcome flexibility if one could regard such sub-systems indeed as components. One would then have components within which type-streams could split and mutate. Furthermore, the class of such components would be closed under *networking*, i.e. a network of such components would itself be a component of the same class. Kelly (1979) terms this the class of *quasi-reversible* components. Quasi-reversibility, in this definite technical sense, is a concept weaker in some ways than that of weak coupling, with an elegant theory which we shall now discuss. However, neither concept includes the other, and the weak coupling concept is closely related with the important concept of insensitivity, taken up in the next two chapters.

Exercises and comments

1. Consider a queue with two types of customers, and suppose its state adequately described by the numbers (n_1, n_2) of each of the two classes which it contains. That is, ordering of customers other than by type is not considered relevant. If the queue is reactionless (i.e. accepts all arrivals) then $\psi_\alpha(n_1, n_2) = 1$. If it is balanced then there exists a Φ such that the intrinsic service rates are given by

$$\phi_1(n_1, n_2) = \Phi(n_1 - 1, n_2)/\Phi(n_1, n_2) \tag{12}$$

$$\phi_2(n_1, n_2) = \Phi(n_1, n_2 - 1)/\Phi(n_1, n_2) \tag{13}$$

Suppose the queue a single-server queue with constant service rate, and suppose we wish to express an absolute priority for customers of type 1, so that such customers are always served first when present. This corresponds to service rates

$$\phi_1 = 1 \qquad (n_1 > 0) \tag{14}$$

$$\phi_2 = 0 \qquad (n_1 > 0) \tag{15}$$

$$\phi_1(0, n_2) = 0 \qquad \phi_2(0, n_2) = H(n_2) \tag{16}$$

where $H(n_2)$ has value 0 or 1 according as n_2 is zero or positive.

From (12), (14) and (16) we find that $\Phi(n_1, n_2) = H(n_1)H(n_2)$. But this expression for Φ is inconsistent with (13), (15). The conclusion is then that a model of this type cannot represent priority scheduling.

8. QUASI-REVERSIBILITY

We adapt Kelly's (1979) discussion to our notation and context. The property of quasi-reversibility is defined by requiring that a component show something of the same behaviour that Theorem 9.5.2 asserts for an open Jackson network.

We shall consider a component in isolation and so dispense with component label, denoting component state simply by x, etc. Units can be of types $\alpha = 1, 2, \ldots, p$.

The component is also considered to be in equilibrium, with stationary input streams. Nothing more for the moment is supposed of the input streams than stationarity. The component and its input streams together constitute the *process*, and it will turn out that the property of quasi-reversibility places constraints upon the whole process. Let the input rate for α-units be v_α. This is an *average* rate; the instantaneous rate (i.e. the intensity of an α-input conditional upon past history of the process) could conceivably depend upon history in an arbitrary fashion.

The component is said to be *quasi-reversible* for given $v_\alpha (\alpha = 1, 2, \ldots, p)$ if there exists an input such that the process satisfies the following conditions at an arbitrary time t.

(Q1) Input and output histories can be deduced from component state history $X(t) = \{x(s), s \leqslant t\}$.
(Q2) Component state $x(t)$ is independent of input after time t.
(Q3) $x(t)$ is independent of output before time t.
(Q4) $\{x(t)\}$ is Markov and stationary.

To these we shall add an assumption which is simplifying rather than essential.

(Q5) The input and output streams are streams of single units, in that the probability that more than one unit either enters or leaves the system in a small time interval of length dt is $o(dt)$.

A useful convention is the following: that if we are considering a pair of states x, x' then $\Sigma_{x'}^{\alpha+}$ represents a sum over x' such that $n(x') = n(x) + e_\alpha$, and $\Sigma_{x'}^{\alpha-}$ represents a sum over x' such that $n(x') = n(x) - e_\alpha$.

The following assertions hold for the component in equilibrium when its input streams are those giving it the properties (Q1)–(Q5).

Theorem 8.1
  (i) *A quasi-reversible component remains quasi-reversible under time-reversal.*
  (ii) *The instantaneous input and output rates of the component are independent of* x, *so that*

$$\sum_{x'}^{\alpha+} \Lambda(x, x') = v_\alpha \tag{1}$$

$$\pi(x)^{-1} \sum_{x'}^{\alpha-} \pi(x') \Lambda(x', x) = v_\alpha^* \tag{2}$$

    where v_α *and* v_α^* *are the average input and output rates for* α-*units.*
  (iii) *Property* (ii) *is characteristic of quasi-reversibility, in that* (Q1), (Q4) *and* (ii) *imply* (Q2) *and* (Q3).
  (iv) *The input streams of a quasi-reversible component are independent Poisson, as are its output streams.*

Proof Statement (i) follows from the fact that the definition of quasi-reversibility is invariant to time-reversal.

The intensities $\Lambda(x, x')$ mentioned in (ii) are those for the component. These exist, by (Q4) and are equivalent to those of the process, by (Q1). Let $v_\alpha(X(t))$ be the instantaneous α-input rate, so that the probability of the event 'entry of an α-

unit' in $(t, t + dt)$ conditional on $X(t)$ is $v_\alpha(X(t)) dt + o(dt)$. By (Q5) input will occur only in such single units. By (Q4) this rate is a function only of $x(t)$. By (Q2) it is independent even of $x(t)$. It is thus constant, and can be equated with v_α.

We have thus established (1), and (2) follows by application of the same argument to the time-reversed version, establishing (ii).

Statement (1) and properties (Q1) and (Q4) imply on their own that behaviour of the input stream in $(t, t + dt)$ is independent of process history $X(t)$, and so of past state at any time and of past input streams. This implies both property (Q2) and that the independent streams are independent Poisson. (Hypothesis (Q5) excludes the possibility of compound Poisson streams.) Coupling these assertions with their time-reversed versions we deduce assertions (iii) and (iv). ∎

Relation (1) amounts to a reactionlessness condition: that all input offered is accepted. Relations (1) and (2) between them imply some kind of partial balance in that, if there is some kind of deterministic balance between the average rates v and v^*, relations (1) and (2) will imply a corresponding stochastic balance.

To be specific, let us say that a class $\mathscr{A}$ of types is *closed* for the component if all input in $\mathscr{A}$ appears as output in $\mathscr{A}$, and output in $\mathscr{A}$ derives only from input in $\mathscr{A}$. There must then be a conservation relation

$$\sum_{\alpha \in \mathscr{A}} (v_\alpha - v_\alpha^*) = 0 \tag{3}$$

Let us denote by $x + e_{\mathscr{A}}$ the set of x' for which $n_\alpha(x') - n_\alpha(x)$ is zero for all α except some one value in $\mathscr{A}$, for which it equals unity. That is, there is a type in $\mathscr{A}$ for which x' possesses one more unit than x, and the two states possess the same numbers of units of other types.

Theorem 8.2 *Suppose the class of states $\mathscr{A}$ is closed for a given quasi-reversible component. Then the probability fluxes of the transitions $x \rightleftharpoons x + e_{\mathscr{A}}$ balance.*

Proof Relations (1)–(3) imply that

$$[x + e_{\mathscr{A}}, x](\pi\Lambda) = 0 \tag{4}$$

which is the partial balance asserted. ∎

The interest is then that, for a quasi-reversible component, every conservation law (3) implies a partial balance relation (4). If the closed class $\mathscr{A}$ consists of a single type α, which then neither loses nor gains by mutation, then relation (4) amounts exactly to the balance relation (7.5). That (7.6) is not demanded is a reflection of the fact that quasi-reversibility is in some senses a weaker property than weak coupling.

The equilibrium distribution of component state $\pi(x)$ will depend upon the input rates v; we shall sometimes emphasize this by writing it $\pi(x|v)$. The output rates v_α^* will show a similar dependence, which we shall write in vector form as

$$v^* = g(v) \tag{5}$$

We shall achieve consistency with earlier notation if we set

$$\Lambda(x, x') = v_\alpha \psi_\alpha(x, x') \qquad (6)$$

if $x \to x'$ is an α-input transition, and

$$\Lambda(x, x') = \lambda_\alpha \phi_\alpha(x, x') \qquad (7)$$

if $x \to x'$ is an α-output transition. We know then from (1) that $\psi_\alpha(x, x')$ is a distribution over x' for given x.

The parameter λ_α in (7) is to be regarded as a 'drain rate' for α-units upon the component. The point of introducing it is that, when we connect components into a network, then a transition $x \to x'$ of *system* state in which an $e_{\alpha j}$ becomes an $e_{\beta k}$ will be assumed to have intensity (7.1). Here we have added subscripts j, k to distinguish components, and the total α-drain rate on component j would indeed be just

$$\lambda_{\alpha j} = \sum_\beta \sum_k \lambda_{\alpha j, \beta k} \qquad (8)$$

The notation employed in relations (7.4)–(7.6) is again very useful. Reverting for the moment to the case of a single component, we define the S operators by (7.7) and (7.8) and use ϕ_α to denote $\phi_\alpha(x) = \sum_{x'} \phi_\alpha(x, x')$, analogously for ψ_α. We no longer distinguish between various types of internal transitions; all internal transitions are lumped and we define

$$\Delta\pi = [\mathcal{X}^{\mathrm{I}}, x](\pi\Lambda) \qquad (9)$$

where $\mathcal{X}^{\mathrm{I}}$ is the set of component states communicating with x, one way or the other, by internal transition.

In this notation relations (1) and (2) become

$$\psi_\alpha = 1 \qquad (10)$$

a reactionlessness condition, and

$$\lambda_\alpha S_\alpha^+ \pi = v_\alpha^* \pi \qquad (11)$$

a balance condition. The overall balance condition for the component (i.e. the equilibrium forward Kolmogorov equation for the process consisting of the component and its input streams) can be written

$$\Delta\pi + \sum_\alpha v_\alpha (S_\alpha^- - 1)\pi + \sum_\alpha \lambda_\alpha (S_\alpha^+ - \phi_\alpha)\pi = 0 \qquad (12)$$

Suppose units as a whole are conserved in the component, so that $\sum_\alpha (v_\alpha^* - v_\alpha) = 0$, or

$$\sum_\alpha (\lambda_\alpha S_\alpha^+ - v_\alpha)\pi = 0 \qquad (13)$$

Relation (12) then becomes

$$\Delta\pi + \sum_\alpha (v_\alpha S_\alpha^- - \lambda_\alpha \phi_\alpha)\pi = 0 \qquad (14)$$

We now come to the principal conclusion: that a network of quasi-reversible components has the significant properties of a Jackson network. We consider the network specified by the statements that the internal transitions of different components are independent and that the transition $x \to x'$ of system state in which an α-unit leaves component j and arrives at component k as a β-unit has intensity (7.1), the factor $\lambda_{\alpha j, \beta k}$ then being interpreted as the routing matrix of the network.

Theorem 8.3 *Consider a network of quasi-reversible components, each of which conserves units as a whole. Suppose that the equations*

$$v_j^* = g_j(v_j) \tag{15}$$

$$\sum_k \sum_\beta (\lambda_{\beta k, \alpha j} v_{\beta k}^* / \lambda_{\beta k}) = v_{\alpha j} \tag{16}$$

have solutions for v, v^*. Then

(i) *The balance equations for the system are solved by*

$$\pi(x) = \prod_j \pi_j(x_j | v_j) \tag{17}$$

where v is the solution of (15) and (16).

(ii) *Distribution (17) shows partial balance over the transitions of any given component.*

Here v_j denotes the p-vector with α^{th} element $v_{\alpha j}$, correspondingly for v_j^*.

Proof The equation stating balance for the transitions of component j would be

$$\Delta_j \pi + \sum_k \sum_\alpha \sum_\beta (\lambda_{\beta k, \alpha j} S_{\beta k}^+ S_{\alpha j}^- - \lambda_{\alpha j, \beta k} \phi_{\alpha j} \psi_{\beta k}) \pi = 0 \tag{18}$$

where the operators with subscript j operate only on the argument x_j of $\pi(x)$, in an obvious fashion.

The sum of these relations over j is the full balance equation, i.e the equilibrium form of the Kolmogorov forward equation for the system. If we can prove that expression (17) satisfies (18) we shall have proved both parts of the theorem. Test then expression (17) as a solution of (18). Substituting for $\Delta_j \pi$ in (18) from (14) and substituting $S_{\beta k}^+ \pi_k = \lambda_{\beta k}^{-1} v_{\beta k}^* \pi_k$ (cf. (11)) and $\psi_{\beta k} = 1$ we see then that (18) reduces to

$$\sum_\alpha (S_{\alpha j}^- \pi_j) \left[\sum_\beta \sum_k (\lambda_{\beta k, \alpha j} v_{\beta k}^* / \lambda_{\beta k}) - v_{\alpha j} \right] = 0 \tag{19}$$

Relation (19) is indeed satisfied if v and v^* satisfy (16). ∎

Theorem 8.3 expresses two properties of the simple Jackson network of section 9.1: product form of the solution (implying independence of the components in the open case) and a partial balance relation for individual components.

If all types of unit are conserved by all components then (15) reduces to $v_j^* = v_j$. Relation (16) then reduces to the familiar form

$$\sum_\beta \sum_k (w_{\beta k} \lambda_{\beta k, \alpha j} - w_{\alpha j} \lambda_{\alpha j, \beta k}) = 0 \tag{20}$$

if we set

$$w_{\alpha j} = v_{\alpha j} / \lambda_{\alpha j} \tag{21}$$

Under the stronger hypotheses of weak coupling it could be further asserted that $\pi_j(x_j|v_j)$ would take the form

$$\pi_j(x_j|v_j) \propto \Phi_j(x_j) \prod_\alpha w_{\alpha j}^{n_{\alpha j}} \tag{22}$$

where $\Phi_j(x_j)$ is independent of the routing parameters. If one postulates merely quasi-reversibility then there will be no such relation: the values of routing parameters may directly affect all the internal statistics of the component. Despite this, the assumption of quasi-reversibility does indeed lead to the strong and significant conclusions of Theorem 8.3.

The open case is formally subsumed in the discussion of Theorem 8.3, but is best stated explicitly. Equations (16) would then become

$$v_{\alpha j}^E + \sum_\beta \sum_{k=1}^m (\lambda_{\beta k, \alpha j} v_{\beta k}^* / \lambda_{\beta k}) = v_{\alpha j} \qquad (j = 1, 2, \ldots, m) \tag{23}$$

where $v_{\alpha j}^E$ is the rate of the Poisson stream of α-units into component j from outside the system. There is thus an effective identification

$$v_{\alpha j}^E = \sum_\beta (\lambda_{\beta 0, \alpha j} v_{\beta 0}^* / \lambda_{\beta 0})$$

Equation (16) also holds for $j = 0$, giving an effective definition of $v_{\alpha 0}$ as simply the rate of α-output from the system to the environment. There is no relation (15) for $j = 0$; there need be no relation between the input and output of an infinite component.

One may now ask what components demonstrate quasi-reversibility. There is as yet no complete characterization of the class of quasi-reversible components, but there are certainly ways of generating members of the class.

Theorem 8.4

(i) *The balanced, reactionless components of section 7 are quasi-reversible.*
(ii) *An open network of quasi-reversible components is itself quasi-reversible.*

(iii) *Suppose that all balance relations for a given quasi-reversible component are balance relations for the movement of some unit. Then the component remains quasi-reversible if its transition intensity $\Lambda(x, x')$ is replaced by $\Lambda^{\Phi}(x, x')$ = $\Lambda(x, x')\Phi(x_-)/\Phi(x)$, where Φ is arbitrary (regularity conditions apart) and x_- is the configuration left when the unit in question is removed from configuration x.*

Proof Assertion (i) follows from the fact that a balanced, reactionless component remains reactionless under time reversal. The same-argument serves for assertion (ii): the network is reactionless as a component, and transforms into an open network of quasi-reversible components (hence also reactionless as a composite component) under time reversal.

The operation described in (iii) is significant, because it indicates a way of transforming a network of functionally independent components into a network of functionally dependent components, consistent with retention of quasi-reversibility.

The balance condition for the component is

$$\sum_{x'} [\pi(x')\Lambda(x', x) - \pi(x)\Lambda(x, x')] = 0 \qquad (24)$$

Note that $x'_- = x_-$ for all x' occurring here, because, since x and x' differ only in that a unit has migrated, then both configurations will be the same if that unit is deleted. Note also that x' indeed equals x_- if $x' \to x$ is an immigration transition or $x \to x'$ an emigration transition.

Because of these facts the version of (24) in which Λ and π are replaced by Λ^{Φ} and π^{Φ} is solved by

$$\pi^{\Phi}(x) = \pi(x)\Phi(x) \qquad (25)$$

Now, if $x \to x'$ is an immigration transition then $x_- = x$, so that

$$\Lambda^{\Phi}(x, x') = \Lambda(x, x') \qquad (26)$$

and the reactionlessness condition (1) satisfied by the component is still satisfied after transformation. If $x' \to x$ is an emigration transition then $x'_- = x$, so that

$$\pi^{\Phi}(x')\Lambda^{\Phi}(x', x)/\pi^{\Phi}(x) = \pi(x')\Lambda(x', x)/\pi(x) \qquad (27)$$

and the time-reversed reactionlessness condition (2) satisfied by the component is still satisfied after transformation. The other properties required for quasi-reversibility plainly also remain satisfied. ∎

Exercises and comments

1. Suppose the class $\mathscr{A}$ of types closed for component j, so that

$$\sum_{\alpha \in \mathscr{A}} (v_{\alpha j} - v_{\alpha j}^*) = 0$$

for the given j. Show from this equation, (11) and (16), that

$$\sum_{\alpha \in \mathscr{A}} \left[\left(\sum_{\beta} \sum_{k} \lambda_{\beta k, \alpha j} S^+_{\beta k} \pi \right) - \lambda_{\alpha j} S^+_{\alpha j} \pi \right] = 0.$$

This can be construed as a partial balance equation whose verbal expression would be: 'the probability that the system is in state x and that there is a member of $\mathscr{A}$ in transit to component j equals the probability that the system is in state x and there is a member of $\mathscr{A}$ in transit from component j'.

9. CLUSTERING NODES

It is natural in many contexts to assume, not merely that units may be of many types, but also that units can associate to form clusters (which can also dissociate), and it is these clusters which are the packages shuttled between components. That is, if the units are 'atoms' then we are allowing the formation of 'molecules' and supposing that molecules can be transported as such around the network.

The literal example of molecular formation and transport is indeed a case in point. For another, we could consider a commodity distribution system under which goods are made up into packages (e.g. containers); these packages are then consigned elsewhere, presumably to be broken up when appropriate. A manufacturing network will at some point assemble items of different types into a composite: the manufactured or semi-manufactured article. A computer will certainly bring parallel computation streams together at some point, and the essence of a neural network is presumably that it allows interaction between impulse trains.

Expectation of being able to deal with all these cases must be dashed. The most general behaviour which the Jackson formalism seems to be able to represent could be viewed as a model in which molecular association and dissociation proceed locally and reversibly, but molecular transport follows rules which can show both density-dependence and some irreversibility.

We consider, then, the possibility that, not merely may units change type at nodes, but they may combine into clusters at nodes, perhaps to divide at others. Hence the term *clustering nodes*. Clustering must indeed occur at the node rather than in transmission, because we must postulate a way in which units (or clusters) can meet. Having postulated this, we have virtually defined a node.

We keep nomenclature more consistent with that of Part IV if we use the term *item* for the 'molecules', the composites which actually flow between nodes, and use the term *unit* for the 'atoms', the ultimate units from which the composites are formed. Units may well be a special type of item, of course, if they also occur in the inter-node flows.

As in Chapter 7 we shall label item-types by r, and denote an r-item by A_r ($r \in \mathscr{R}$). The list of possible A_r will in general include the simple units e_α. An α-unit at node j will be denoted $e_{\alpha j}$, an r-item at node j will be denoted A_{rj}. The number of A_{rj} will be denoted by n_{rj}, the vector of occupation numbers $\{n_{rj}, r \in \mathcal{R}\}$ will be denoted by n_j, and the full array of occupation numbers $\{n_{rj}\}$ will be denoted by n. We shall also use A_{rj} to denote the array with a unit in the $(rj)^{\text{th}}$ place and zeros elsewhere. The migration/mutation which converts an A_r at node j to an A_s at node k thus implies the transformation

$$n \to n' = n - A_{rj} + A_{sk} \tag{1}$$

For reasons which will transpire we shall allow some degree of mutation along routes as well as at nodes.

We shall assume that the network is one of simple nodes, in that n is itself a state variable for the system, and so n_j a state variable for node j. Nodes are not wholly simple though. There are external transitions, in which there is flow between nodes, and internal transitions, in which n_j may change because of association/dissociation reactions at node j. We shall follow the pattern of section 9.2 in that we assume the only external transitions are the simple migration/mutations (1), with intensity

$$\Lambda(n, n - A_{rj} + A_{sk}) = \lambda_{rj,\,sk} \frac{\Phi(n - A_{rj})}{\Phi(n)} \tag{2}$$

for some Φ. Here $\lambda_{rj,\,sk}$ is a routing intensity for the transition (1), and the formulation includes the open case if we allow the existence of an infinite node, conventionally at $j = 0$. We shall assume that Φ balances internal transitions, in that

$$\sum_{n'} [\Phi(n')\Lambda^0(n', n) - \Phi(n)\Lambda^0(n, n')] = 0 \tag{3}$$

where Λ^0 is the intensity for internal transitions.

Suppose that in such internal transitions the quantities

$$b_{\alpha j} = \sum_r a_{\alpha r} n_{rj} \tag{4}$$

are conserved. These are to be regarded as the abundances of some basic immutable constituents (not to be identified with the units, unless the units are themselves immutable and conserved in all transitions). Then, if $\Phi(n)$ solves (3), so too does

$$\pi(x) \propto \Phi(x) \prod_j \prod_\alpha w_{\alpha j}^{b_{\alpha j}} \tag{5}$$

$$= \Phi(x) \prod_j \prod_r u_{rj}^{n_{rj}}$$

where

$$u_{rj} = \prod_\alpha w_{\alpha j}^{a_{\alpha r}} \tag{6}$$

Theorem 9.1 *Suppose that the equations in the $w_{\alpha j}$*

$$\sum_s \sum_k (u_{sk} \lambda_{sk,rj} - u_{rj} \lambda_{rj,sk}) = 0 \qquad (7)$$

with the u_{rj} given by (6) possess solutions. Then the balance equations for the process have solution (5).

Proof Verification is immediate. If there are p invariants in an internal transition and q types of item are mobile (i.e. allowed to move in the network) then (7) constitutes mq equations in mp unknowns. If $p > q$ then there are invariants even under external transitions. If $p < q$ then the rates λ will have to obey certain consistency conditions if equations (6) and (7) are to have solutions.

The assumptions (2) and (3) imply a degree of reversibility in transitions in so far as these depend upon Φ, but not as far as they depend upon the routings λ. However, there will be restrictions on the routings which lead to restrictions on the behaviour of the system. Suppose that there is a class $\mathscr{C}$ of item-types which is *closed under routing* in that the class $\mathscr{C}$ can neither be left nor entered in routing mutations. That is $\lambda_{rj,sk} = 0$ for all j and k if one of r, s belongs to $\mathscr{C}$ and the other does not.

Theorem 9.2 *Consider an open network of clustering nodes for which the class $\mathscr{C}$ of item-types is closed under routing. Then the total input and output of class $\mathscr{C}$ items balances for the network.*

Proof The input rate of A_{rj}s is $\sum_s \lambda_{s0,rj}$; the output rate in equilibrium is $\sum_s \lambda_{rj,s0} u_{rj}$ (cf. Theorem 9.5.2). Adding relations (7) over $j \neq 0$ and over r in $\mathscr{C}$ we deduce, in virtue of the constraints on λ, that

$$\sum_{j \neq 0} \sum_{r,s \in \mathscr{C}} (\lambda_{s0,rj} - \lambda_{rj,s0} u_{rj}) = 0 \qquad (8)$$

This expresses the total input/output balance for class $\mathscr{C}$ asserted. ∎

The assertion has consequences. Suppose, for example, that the total number of units (irrespective of type) is unchanged in any transition, internal or external. Let the number of units in an item be termed its *weight*. Then Theorem 9.2 implies that total input and output must balance for items of a given weight. One cannot, for example, see the network as taking inputs of items of weight one and assembling them to produce an output of items of weight two. So, a Jackson network cannot achieve the net effect of 'assembly'; the effect is too far from being reversible.

10. FUNCTIONALLY DEPENDENT NODES

The property of 'product form' solution and of independence of the nodes in equilibrium is often seen as one of the significant features of a Jackson network. However, it is not one required if the system is to show weak coupling and, as for the structureless case of section 9.3, the material of sections 3 to 7 has a natural version even for functionally dependent components. The communication hypotheses of section 2 were so formulated that weak coupling could be seen as a consequence of the properties of the component rather than of the system. However, we can assume analogous system properties and still recover all the results of sections 3 to 7 except the product form of the equilibrium distribution (4.1). For simplicity, we shall discuss only the case of a single type of unit.

We shall generalize the external transition rule (2.3) by assuming that an external transition $x \rightarrow x'$ which implies a $j \rightarrow k$ transfer has intensity

$$\Lambda_{jk}(x, x') = \lambda_{jk}(n)\kappa_{jk}(x, x') \tag{1}$$

Here the jk subscripts indicate the particular type of external transition; one for which $n \rightarrow n - e_j + e_k$. However, we assume the state values x and x' otherwise unconstrained; even components not directly involved in the transfer may change state because of it.

The factor $\lambda_{jk}(n)$ in (1) again represents the routing rule, now permitted to be n-dependent. The factor $\kappa_{jk}(x, x')$ represents a characteristic of the system independent of routing. To within rate factors it can be regarded as the probability of a transition $x \rightarrow x'$ conditional on the fact that the transition has the character of a $j \rightarrow k$ transfer. We assume $\kappa_{jk}(x, x') = 0$ unless the transition has this character. Assumption (1) states that the effect of adopting a general routing scheme is to multiply this 'intrinsic' transition intensity by $\lambda_{jk}(n)$.

We also allow internal transitions, in which n does not change. The appropriate generalization of assumption (2.7) turns out to be that the intensity $\Lambda(x, x')$ of such an internal transition shows the following dependence upon routing.

$$\Lambda(x, x') = \sum_j \sum_k \lambda_{jk}(n) \Lambda_{jk}^{\mathrm{p}}(x, x') + \Lambda^0(x, x') \tag{2}$$

Here Λ^0 is the component of Λ corresponding to non-progressive transitions and $\Lambda_{jk}^{\mathrm{p}}$ is the component of Λ corresponding to progressive internal transitions associated with an external transfer $j \rightarrow k$. Transitions with intensity (1) may be allowed even in the case $k = j$. They are then still regarded as external; as a transition in which a unit has been emitted and then immediately reabsorbed by the same component.

The various concepts developed for individual components in section 3 now have system analogues which are natural, even if not immediately evident. The question of normalization is best deferred; we return to it in sections 11.2 and 12.3.

We shall say that the kj route is *balanced* if there exists a $\Phi(x)$ which balances non-progressive transitions

$$[\mathscr{X}, x]\,(\Phi\Lambda^0) = 0 \tag{3}$$

and which also satisfies the balance equation

$$\sum_{x'}\left[\Phi(x')\,(\kappa_{kj}(x', x) + \Lambda_{jk}^{\text{p}}(x', x)) - \Phi(x)\,(\kappa_{jk}(x, x') + \Lambda_{jk}^{\text{p}}(x, x'))\right] = 0 \tag{4}$$

That is, Φ balances the fluxes $x \rightleftharpoons \mathscr{X}$ associated with the intensities κ_{jk}, κ_{kj} and Λ_{jk}^{p}. It is convenient to write (4) rather in the form

$$S_{kj}\Phi(x) - \mu_{jk}(x)\Phi(x) + [\mathscr{X}, x]\,(\Phi\Lambda_{jk}^{\text{p}}) = 0 \tag{5}$$

where

$$\mu_{jk}(x) := \sum_{x'}\kappa_{jk}(x, x') \tag{6}$$

$$S_{kj}\Phi(x) := \sum_{x'}\Phi(x')\kappa_{kj}(x', x) \tag{7}$$

We shall say that component j is *indiscriminate* if μ_{jk} and Λ_{jk}^{p} are independent of k. This is the analogue of the property of reactionlessness, but now seen as a property of the emitting component rather than of the receiving component. The quantity $\mu_{jk}(x)$ is the intrinsic rate at which emissions $j \to k$ occur in state x. Indiscriminacy implies that this rate is independent of k (so that flow between components is effectively governed entirely by the routing rule $\lambda_{jk}(n)$), as is the associated internal transition intensity Λ_{jk}^{p}.

Finally, we shall require the property of *partial routing balance*: that there exist a function $h(n|\lambda)$ satisfying

$$\sum_{k}\left[h(n + e_k|\lambda)\lambda_{kj}(n + e_k) - h(n + e_j|\lambda)\lambda_{jk}(n + e_j)\right] = 0 \tag{8}$$

for all j and all routing rules λ in the permitted class $\mathscr{L}$. This was a property we took for granted in the case of constant routing intensities (when it was satisfied with $h(n|\lambda) = \prod_{j} w_j^{n_j}$), but it now appears as one of the significant properties required.

Theorem 10.1 *Consider the system specified by the external and internal transition intensities (1) and (2). Suppose all routes balanced by Φ (cf. (3), (4)), all components indiscriminate and that the routing intensity λ satisfies the partial routing balance equation (8). Then:*

(i) *The system has equilibrium distribution*

$$\pi(x) \propto \Phi(x)h(n|\lambda) \tag{9}$$

on any given irreducible set.

(ii) *The system is weakly coupled, and shows partial balance over the non-progressive transitions; also over the progressive plus external transitions associated with a given component.*

Proof As in Theorem 4.1, it is enough to prove that expression (9) satisfies the partial balance relations asserted in (ii). By hypothesis, it satisfies (3), and so balances the non-progressive transitions. The balance over remaining transitions associated with component j can be written

$$\sum_k [h(n - e_j + e_k)\lambda_{kj}(n - e_j + e_k)S_{kj}\Phi - h(n)\lambda_{jk}(n)(\mu_{jk}\Phi - \Delta_{jk}^p)] = 0 \qquad (10)$$

where

$$\Delta_{jk}^p(x) = [\mathscr{X}, x](\Phi\Lambda_{jk}^p). \qquad (11)$$

For brevity, the x-arguments are suppressed in (11), as is also the λ-argument of h. In virtue of the balance relation (5) we can write (10) as

$$\sum_k (\mu_{jk}\Phi - \Delta_{jk}^p)[h(n - e_j + e_k)\lambda_{kj}(n - e_j + e_k) - h(n)\lambda_{jk}(n)] = 0 \qquad (12)$$

Indiscriminacy implies μ_{jk} and Δ_{jk}^p independent of k, so that (12) reduces to (8), satisfied by hypothesis. All assertions then follow. ■

So, the only property lost under our greatly relaxed assumptions is the product form of the distribution (9). The partial reversibility asserted in the theorems of section 6 is also retained. Let us define a transformation R on a transition intensity $\theta(x, x')$ by

$$R\theta(x, x') = \theta(x', x)\Phi(x')/\Phi(x) \qquad (13)$$

Theorem 10.2 *The system of Theorem 10.1 retains its structure under time-reversal (i.e. the transition intensities are of the form* (1), (2), *routes are balanced, components indiscriminate and the routing intensity partially balanced). Specifically, the intrinsic rates suffer the transformations*

$$\Lambda_{jk}^{p*} = R\Lambda_{kj}^p$$

$$\Lambda^{0*} = R\Lambda^0 \qquad (14)$$

$$\kappa_{jk}^* = R\kappa_{kj}$$

with the implication $\Phi^* = \Phi$, and the routing rates suffer the transformation

$$\lambda_{jk}^*(n) = \lambda_{kj}(n - e_j + e_k)h(n - e_j + e_k|\lambda)/h(n|\lambda) \qquad (15)$$

with the implications $h^* = h$ and $\lambda_j^*(n) = \lambda_j(n)$.

Proof This is again a matter of verification. The balance relations (3) and (4) transform into their respective time-reverses, and we see from (5) that balance plus indiscriminacy implies that $S_{kj}\Phi$ is independent of k, and this implies indiscriminacy of the time-reversed system. ∎

However, the input and output streams $0 \to j$ and $j \to 0$ will be independent constant-rate Poisson only under rather special circumstances.

Theorem 10.3 *The input streams of the system of Theorem 10.1 are independent (conditional on x) Poisson with rates $\lambda_{0j}(n)\mu_0(x)$ and the output streams are independent (conditional on x) Poisson with rates $\lambda_{0j}^*(n)\mu_0(x)$. Here $\mu_0(x)$ is the common value of $\mu_{0j}(x) = \sum_{x'} \kappa_{0j}(x, x')$ $(j = 1, 2, \ldots, m)$ and $\lambda_{0j}^*(n)$ equals $\lambda_{j0}(n + e_j)h(n + e_j|\lambda)/h(n|\lambda)$.*

Proof The first assertion is a matter of definition, and the second follows from the familiar argument and the easily-verifiable fact that $\sum_{x'} \kappa_{0j}^*(x, x') = \mu_0(x)$. ∎

So, inputs and outputs are accelerated by a common state-dependent factor $\mu_0(x)$. The rates of these streams are thus in general dependent on system state. They would be constant if, for example, λ were independent of n and μ_0 of x.

The independence statements of the theorem should perhaps be amplified. Conditional on process history at time $t-$ the various input events at time t are independent with the $x(t-)$-dependent intensities indicated. For the output streams this characterization holds in reverse time.

Exercises and comments

1. Verify that the conditions and conclusions of Theorem 10.1 do indeed reduce to those of Theorem 4.1 if one specializes the transition intensities (1) and (2) appropriately.

2. Suppose that for a given j and k it is true that $\lambda_{kj} = 0$ for all λ in $\mathscr{L}$. That is, the kj route is never used. Then Theorem 10.1 remains true without the balance condition (5) for the kj route.

3. Deduce the similar relaxation of the indiscriminacy condition (cf. Exercise 4.5).

4. For the simplest example which does not reduce to one of those already considered, suppose that if $x \to x'$ by addition of a unit to component j then x' is unique, and so can be written unambiguously as $x + e_j$. If we choose

$$\kappa_{jk}(x + e_j, x + e_k) = \Phi(x)/\Phi(x + e_j) \tag{16}$$

then this satisfies the balance and indiscriminacy conditions.

This model is then a slight generalization of that of section 9.3. A more substantial one is to suppose uniqueness of x' in a transition $x \to x'$ in that the unit is inserted into a particular place in a 'society' (cf. section 5) and to then assume a balance relation (16) for the more closely specified transitions of this type.

CHAPTER 11

Weak coupling

1. WEAK COUPLING

As explained in section 10.1, our contention is that weak coupling is a significant property of all the models of Chapters 5, 6, 9 and 10, in which one significant mode of interaction between components is the transfer of units of some kind. By weak coupling we mean that the equilibrium distribution of system state x shows the factorization

$$\pi(x\,|\,\lambda) = \Phi(x)h(n(x), \lambda) \tag{1}$$

where the parameter λ describes the routing of units and the statistic $n = n(x)$ describes the distribution of units over components. As noted in Theorem 10.1.1, this assumption is equivalent to the assumption that the distribution of x conditional upon n is independent of λ in equilibrium. That is, that the effect of the routing of units between components upon the equilibrium distribution of system state x is summed up by its effect on the distribution of n.

To be frank, whether this hypothesis is either necessary or sufficient for our purposes is not yet clear. Our purpose may be said to be to gain understanding of the equilibrium properties of a system of linked components. The Jackson network represents a class of systems for which equilibrium properties are simple and explicit. The properties which seem relevant in that they are observed for at least some Jackson networks may be listed:

(1) Product form
(2) Partial balance
(3) Quasi-reversibility
(4) Conservation of Poisson flows
(5) Weak coupling
(6) Constancy of 'unit potential' in that $h(n, \lambda)$ has the form $\prod_j w_j^{n_j}$.

In the case of functionally independent components we know from section 10.8 that the *component* condition of quasi-reversibility is enough, in that it implies properties (1)–(4) for the *system*. It does not imply weak coupling, which consequently emerges as a condition in some ways unnecessarily strong.

Nevertheless, the property of weak coupling does seem a natural one. It is stated for the system and, as we shall see in this chapter, can be stated for systems in which there is no mention of components, units or transport. In this more general version it appears as the concept of *imbedding* a given Markov process in a more general one; the natural precursor of next chapter's study of *insensitivity*.

Notions of weak coupling arise in quite other contexts. For example, when nations trade, interaction is represented by relatively few parameters: prices of goods and currency exchange rates. In other words, if availability of goods at the quoted prices is assumed, then a country need know no more about its trading partners, and is affected by its trading partners only in so far as these parameters change. This is a model of optimization rather than of statistical distribution, and resources (i.e. goods) are consumable rather than conserved, but there are obvious parallels.

2. WEAK COUPLING IN A GENERAL SYSTEM

One can develop the notion of weak coupling without reference to components, units or flows. Let us suppose simply that the state space $\mathscr{X}$ of a Markov system can be decomposed into sets $\mathscr{X}(n)$ $(n \in \mathscr{N})$. If n is termed the *count*, then we shall refer to these sets as *count classes*.

The count $n(x)$ of a state x is the label n of the class to which x belongs. *Internal* transitions are those within a class; *external* transitions are those between classes. We shall, however, allow the possibility that some external transitions allow state to return immediately to the same class, so it is not always true that count n changes in an external transition.

This reduces to the conception of section 10.1 if we suppose that n represents the count $\{n_j\}$ of units over components. The external transitions are just those in which there is a transfer of a unit. The external transitions in which n does not change correspond to the 'shunting' of a component. However, we are not bound to this interpretation: $\{\mathscr{X}(n)\}$ can be an arbitrary decomposition of $\mathscr{X}$.

With this more general understanding we now make assumptions about transition intensities in close analogue to those of section 10.10. Let us suppose that an external transition $x \to x'$ has intensity

$$\Lambda(x, x') = \lambda(n(x), n(x'))\kappa(x, x') \tag{1}$$

Here $\kappa(x, x')$ is an 'intrinsic' rate for the transition $x \to x'$, fixed for the system. The other factor $\lambda(n, n')$ is a 'routing rate' for the transition $n \to n'$. The routing matrix $\lambda = (\lambda(n, n'))$ is allowed to take values in a set $\mathscr{L}$ of permitted routings.

If one wishes to emphasize that the transition $x \to x'$ is one in which $n \to n'$ then one can write relation (1) rather as

$$\Lambda_{nn'}(x, x') = \lambda(n, n')\kappa_{nn'}(x, x') \tag{2}$$

where

$$\kappa_{nn'}(x, x') = \begin{cases} \kappa(x, x') & n(x) = n, n(x') = n' \\ 0 & \text{otherwise} \end{cases} \tag{3}$$

This corresponds to the usage of section 10.10, in which Λ_{jk}, κ_{jk} were used for external transitions in which $n \to n - e_j + e_k$. In general, if we are speaking of *external* transitions $x \to x'$ we shall assume n, n' are corresponding, so that $n(x) = n$, $n(x') = n'$.

Following again the pattern of (10.10.2), we shall suppose that a transition $x \to x'$ internal to $\mathcal{X}(n)$ has intensity

$$\Lambda(x, x') = \sum_{n'} \lambda(n, n') \Lambda_{nn'}^{P}(x, x') + \Lambda^0(x, x') \tag{4}$$

One might think that internal transitions should not be affected by routing, and Λ^0 indeed represents those transitions which are not, the 'non-progressive' transitions. However, we have already seen in sections 9.6 and 10.3 that we must allow for the possibility that a change of routing which produces an increased exodus from a component (or class) may stimulate an increased rate of associated internal transitions, the 'progressive' transitions. This is then the reason for the form (4); the term $\Lambda_{nn'}^{P}$ is to be regarded as representing those internal transitions associated with the external transitions for which $n \to n'$. Note that in this case $n(x) = n(x') = n$, since the transition is internal.

We now make a formal general definition: the system defined by the decomposition $\mathcal{X} = \{\mathcal{X}(n)\}$ and transition rules (2), (4) is *weakly coupling* if its equilibrium distribution has the form (1.1) for all λ in $\mathcal{L}$. The factor $\Phi(x)$ is regarded as unique (to within a normalization) in that all lack of determination of $\pi(x|\lambda)$ can be attributed to lack of determination of $h(n, \lambda)$.

The definition of $\kappa_{nn'}(x, x')$ can be changed by a factor dependent on (n, n'), this factor being absorbed in a redefinition of $\lambda(n, n')$ (and so of $\mathcal{L}$). Let us then assume henceforth that κ is so normalized that

$$\sum_x \sum_{x'} \Phi(x) \kappa_{nn'}(x, x') = 1 \tag{5}$$

Theorem 2.1 *Assume the structure* (2), (4), *the weak coupling condition* (1.1) *and the normalization* (5). *Then* $h(n, \lambda)$ *satisfies the balance equation*

$$\sum_{n'} [h(n', \lambda) \lambda(n', n) - h(n, \lambda) \lambda(n, n')] = 0 \qquad (n' \in \mathcal{N}) \tag{6}$$

and the system is reducible only in so far as the routing is reducible.

Proof Substituting expression (1.1) into the equilibrium equation we deduce the equation

$$h(n) \sum_{n'} \lambda(n, n') [\mathcal{X}, x] (\Phi \Lambda_{nn'}^{P}) + h(n) [\mathcal{X}, x] (\Phi \Lambda^0)$$
$$+ \sum_{n'} [h(n') \lambda(n', n) \sum_{x'} \Phi(x') \kappa_{n'n}(x', x) - h(n) \lambda(n, n') \Phi(x) \sum_{x'} \kappa_{nn'}(x, x')] = 0 \tag{7}$$

in which, for simplicity, we have written $h(n, \lambda)$ simply as $h(n)$. Summing equation (7) over x in $\mathscr{X}(n)$ and appealing to (5) we deduce (6). By hypothesis, the indeterminacies of π (corresponding to reducibility of the system) are all attributable to indeterminacies in h. There will be such indeterminacies if and only if the normalized solution of (6) is non-unique. Such non-uniqueness is equivalent to reducibility of the routing λ.

3. BALANCE AND INDISCRIMINACY CONDITIONS

We assume for this chapter and the next the structure specified by the count classes $\mathscr{X}(n)$ and the transition intensities (2.2), (2.4). It is then rather easy to show that, in some cases at least, certain class-balance relations must be satisfied if the system is to be weakly coupling.

Theorem 3.1 Class balance conditions. *Suppose the routing rates $\lambda(n, n')$ can adopt any non-negative value. Then it is necessary for weak coupling that Φ should balance internal transitions*

$$[\mathscr{X}, x](\Phi \Lambda^0) = 0 \tag{1}$$

and also satisfy the balance conditions

$$[\mathscr{X}(n'), x](\Phi \kappa) + [\mathscr{X}(n), x](\Phi \Lambda_{nn'}^{\mathrm{P}}) = 0 \qquad (x \in \mathscr{X}(n); \, n, n' \in \mathscr{N}) \tag{2}$$

Proof Set $\lambda(n, n') = 0$ for all n, n'. Then relation (1) follows from (2.7). Now assume $\lambda(n, n') = \lambda(n', n) > 0$ for given n, n', but assume λ zero on all other routings. From (2.7) and (2.6) we then deduce that $h(n) = h(n')$ for the given n, n', and

$$\sum_{x'} \Phi(x') \kappa_{n'n}(x', x) - \Phi(x) \sum_{x'} \kappa_{nn'}(x, x') + [\mathscr{X}(n), x](\Phi \Lambda_{nn'}^{\mathrm{P}}) = 0 \tag{3}$$

This is the relation more compactly expressed by (2). ∎

It is convenient to rewrite relation (3) as

$$v_{n'n}(x) - \mu_{nn'}(x)\Phi(x) + [\mathscr{X}(n), x](\Phi \Lambda_{nn'}^{\mathrm{P}}) = 0 \tag{4}$$

where

$$v_{n'n}(x) = \sum_{x'} \Phi(x') \kappa_{n'n}(x', x) \tag{5}$$

$$\mu_{nn'}(x) = \sum_{x'} \kappa_{nn'}(x, x') \tag{6}$$

and to note that, in virtue of the normalization (2.5), $v_{n'n}(x)$ is a distribution over $x(n)$:

$$\sum_{x \in \mathscr{X}(n)} v_{n'n}(x) = 1 \qquad (n, n' \in \mathscr{N}) \tag{7}$$

Relations (4), (7) have an interesting interpretation. The $\Phi(x)$ satisfying (4) for given n, n' can be identified as the expected time spent in state x during a single sojourn in $\mathscr{X}(n)$ for a process in which $\mathscr{X}(n)$ is entered at x with probability $v_{n'n}(x)$, makes transitions $x \rightarrow x'$ within $\mathscr{X}(n)$ with intensity $\Lambda_{nn'}^{p}(x, x')$ and leaves $\mathscr{X}(n)$ from state x with intensity $\mu_{nn'}(x)$ (cf. section 3.4). The requirement of balance is that $\Phi(x)$ should sustain this interpretation for all n', plus of course balancing non-progressive transitions (cf. (1)).

This interpretation is to be regarded as describing events in an 'auxiliary process' rather than in the actual system. In the actual system these events occur as in the auxiliary process but accelerated by a time factor $\lambda(n, n')$, in that $\Lambda_{nn'}^{p}$ is multiplied by this factor in (2.4) and the actual exit rate to $\mathscr{X}(n')$ from x of $\mathscr{X}(n)$ is $\lambda(n, n')\mu_{nn'}(x)$ by (2.2).

The routings λ may well be subject to some kind of constraint. A constraint that we shall find natural in particular applications is that they obey some kind of partial balance condition, in that the solution h of (2.6) also satisfies

$$\sum_{n' \in C_i(n)} [h(n', \lambda)\lambda(n', n) - h(n, \lambda)\lambda(n, n')] = 0 \tag{8}$$

where, for given n, the sets $C_i(n)$ $(i = 1, 2, \ldots, I(n))$ constitute a decomposition of $\mathscr{N}$.

Relations (8) can be more compactly expressed

$$[C_i(n), n] (h\lambda) = 0 \quad (i = 1, 2, \ldots, I(n) \ n \in \mathscr{N}) \tag{9}$$

Theorem 3.2 *Suppose the routing λ subject only to the partial balance conditions (9). Then it is necessary and sufficient for weak coupling that the balance equations (1), (2) should be satisfied and that, for every n, the quantity $v_{n'n}(x)$ defined by (5) should depend upon n' only through i.*

The last condition is a generalized indiscriminacy condition (cf. section 10.10). It is to be interpreted as saying that, for any given n and i, the quantity $v_{nn'}(x)$ does not vary with n' as n' varies within $C_i(n)$.

Proof The two trial routings employed in the proof of Theorem 3.1 are consistent with conditions (9). The balance conditions (1), (2) are consequently necessary. Substituting for $\mu_{nn'}(x)\Phi(x)$ from (3) into (2.7) we reduce the total balance equation (2.7) to

$$\sum_{n'} v_{n'n} [h(n')\lambda(n', n) - h(n)\lambda(n, n')] = 0 \tag{10}$$

This relation can hold for all h, λ consistent with (9) if and only if $v_{n'n}$ has a constant value as n' varies within $C_i(n)$, for any n, i. ∎

The quantity $v_{n'n}(x)$ can be viewed as something like an intrinsic rate of transition from $\mathscr{X}(n')$ to a state x of $\mathscr{X}(n)$, averaged over the states x' of $\mathscr{X}(n')$.

The condition on $v_{n'n}$ is that this should show a restricted dependence on n', indeed a kind of indiscriminacy condition. In view of (4) the combination $\mu_{nn'}(x)\Phi(x) - [\mathscr{X}(n), x](\Phi\Lambda_{nn'}^{p})$ must show the same kind of restricted dependence, and it seems reasonable to demand that $\mu_{nn'}$ and $\Lambda_{nn'}^{p}$ should show the same restricted dependence separately, in that they should depend upon n' only through i. This slight strengthening of the indiscriminacy condition is simplifying rather than necessary.

In Theorem 12.5.1 we give a rephrased version of Theorem 11.3.2 with this strengthened condition. This yields a criterion for weak coupling which illuminates a whole sequence of interesting special cases.

4. SPECIALIZATION TO THE JACKSON NETWORK

One might regard as characteristic of the particular case of a Jackson network that n can be identified with $\{n_j\}$, the array of unit-counts at the nodes, that the only possible external transitions $n \to n'$ are those of the form $n \to n - e_j + e_k$ for some j, k, and that the routing rate $\lambda(n, n')$ satisfies the partial balance conditions (10.10.8). The sufficient conditions for weak coupling stated in Theorem 3.3 then reduce the general structure to exactly that postulated in section 10.10. The fact that some transitions $n \to n'$ are not possible merely means that the corresponding balance conditions can be trivially satisfied, by supposing $\kappa_{nn'}$ and $\Lambda_{nn'}^{p}$ equal to zero for such transitions. We give the details in section 12.8.

If we specialize to the functionally independent case then we also impose, among other things, the condition that $\lambda_{jk}(n) := \lambda(n, n - e_j + e_k)$ should be independent of n. It is not then clear that $\mathscr{L}$ is then large enough that the conclusions of Theorem 3.2 hold. That is, that the balance and indiscriminacy conditions of that theorem (now reducing to those of Theorem 10.4.1) are necessary as well as sufficient for weak coupling. However, necessity has been proven in Whittle (1984a). Essentially, although one does not have the same rich possibility of varying λ, one can demand that, for a closed system, relation (1.1) should hold on all reducible sets $\sum_{j} n_j = N$ in a form independent of N, apart from a normalization factor. Variation of N then leads to a demonstration of necessity.

5. IMBEDDING AND INSENSITIVITY

The more general definition of weak coupling given in section 2 has real point in a much wider context than that of network flows. In particular, it gives a natural formulation of the concept of *insensitivity* which has been so much discussed in recent years. To develop this we must add one further condition to that of weak coupling.

Suppose we now adopt the Φ-normalization

$$\sum_{\mathcal{X}(n)} \Phi(x) = 1 \qquad (n \in \mathcal{N}) \tag{1}$$

of Theorem 10.1.1. This is consistent with the normalization (2.5), which normalizes κ for a given Φ. With this normalization $h(n, \lambda)$ is identifiable as the equilibrium distribution $\pi(n|\lambda)$ of n, and $\Phi(x)$ as the equilibrium distribution $\pi(x|n)$ of x conditional upon n. To emphasize this, let us write the weak coupling relation (1.1) as

$$\begin{aligned} \pi(x|\lambda) &= \Phi(x)\pi(n|\lambda) \\ &= \pi(x|n)\pi(n|\lambda) \end{aligned} \tag{2}$$

Note that the additional normalization (1) can be accommodated only if one is willing to redefine the routings $\lambda(n, n')$ by an (n, n')-dependent factor, and so redefine $\mathcal{L}$. If one is not willing to redefine $\mathcal{L}$, then normalization (1) represents a new constraint on the system.

One can now regard the *x-process* with transition intensities $\Lambda(x, x')$ determined by (2.1), (2.4) as an 'enrichment' of the *n-process*, defined as the process with state variable $n \in \mathcal{N}$ and transition intensity $\lambda(n, n')$. The enrichment is far from arbitrary: the random variable n has the same equilibrium statistics for both the x-process and for the simple n-process. By Theorem 2.1 and normalization (1) the *equilibrium distribution* $\pi(n|\lambda)$ is the same for both cases. From these results and normalization (2.5) it also follows that the *probability flux* $\mathcal{X}(n) \to \mathcal{X}(n')$ in equilibrium is

$$\sum_x \sum_{x'} \Phi(x)\pi(n|\lambda)\lambda(n, n')\kappa_{nn'}(x, x') = \pi(n|\lambda)\lambda(n, n') \tag{3}$$

and is thus the same as the flux $n \to n'$ for the *n*-process.

We regard the *n*-process as a crude model of a system, the *x*-process as a more detailed model. If the two are related as we have described (i.e. $\Lambda(x, x')$ derived from $\lambda(n, n')$ by (2.1) and (2.4), equilibrium distributions related by (2), and normalizations (1), (2.5) achievable without redefinition of $\mathcal{L}$), then we shall say that the *x*-process *imbeds* the *n*-process.

Imbedding implies that the equilibrium *n*-statistics are the same for the *x*-process (in both distribution and flux) as they were for the original *n*-process. Furthermore, this agreement holds identically for the class of *n*-processes specified by $\mathcal{L}$. We can thus say that *n*-statistics are *insensitive* to the difference in specification between the two models.

The imbedding *x*-process is so strongly constrained that interesting imbeddings may simply not exist. However, it will turn out, by exploitation of Theorems 3.2 and 3.3, that, the stronger the balance conditions satisfied by the *n*-process, the weaker the conditions that need be demanded of the imbedding *x*-process, and the more interesting the possible imbeddings. Such assertions correspond exactly to the 'insensitivity theorems' of which there are so many scattered examples, and which this formulation would seem to unify. We shall make these statements exact in the next chapter.

CHAPTER 12

Insensitivity

1. INSENSITIVITY: CONCEPT AND EXAMPLES

By 'insensitivity' is meant simply that some distributional aspect of a stochastic process is invariant under some variations of specification of that process. We have already seen a number of examples of insensitivity in equilibrium statistics. We saw in section 3.5 that the equilibrium distribution over states in a semi-Markov process depends only upon transition probabilities and expected lifetimes in states. It is thus insensitive to the distributions of these lifetimes, subject at least to their being independent and having prescribed expectations.

We saw in Theorem 5.7.2 that the equilibrium distribution of numbers of molecules formed from constituents of restricted abundance has the same constrained Poisson form whether we took a cellular or spatially aggregated model. That is, this distribution is insensitive to the degree of refinement of spatial description. Again, we saw in Theorem 6.1.1 that a component adopts the same Gibbs distribution in equilibrium whether it is exchanging energy quanta directly with other components, exchanging them with a 'heat-bath' of random local quantum density, or by exchanges with a heat-bath of fixed quantum density.

The constrained Poisson distribution arises again in the Erlang distribution of Exercise 4.7.3 and in its generalized version, the circuit-switched network of Exercise 4.7.4. The Erlang model is famous as one of the first examples of insensitivity: the equilibrium distribution of the number of busy channels depends upon the call length distribution only through its mean value. This distribution can even be made channel- and caller-specific if the mean value is held fixed and the right independence assumptions made. We shall prove this and similar results for the circuit-switched networks in section 7. See also Ex. 5.4.2.

Finally, in section 10.5 we encountered an example of insensitivity with a new feature. The notion of a sojourn time (for a unit entering a node) was replaced by that of a *service requirement*. If the service requirement were worked off at a constant rate then the two concepts would be identical. However, the service requirement is worked off at a random rate, this rate depending on numbers currently at that and other nodes and also upon queue discipline. The assertion of Theorem 10.5.2 and Corollary 10.5.1 is that the equilibrium distribution of the numbers at the nodes is insensitive to the distributions of service requirements, subject again to prescription of mean value and independence.

264

The first systematic studies of insensitivity were made by Matthes (1962) and co-workers (see for example, Koenig, Matthes and Nawrotzki (1967)) who saw the relationship of the concept to some partial balance property. In fact, they put conditions in terms of a 'local balance' property of a certain semi-Markov process. This local balance property becomes just a partial balance property if one achieves a Markov rather than a semi-Markov description. These studies have been continued, notably by Schassberger (1977; 1978a, b, c; 1979). The approach taken here is that set out in Whittle (1986) which sees insensitivity as the possibility of imbedding a class of Markov processes in a more general class. The material of sections 3 to 6 indicates the very direct relation between insensitivity and partial balance deduced in Whittle (1985a). A rather different approach is that taken by Hordijk and van Dijk (1983a, b) and Hordijk (1984).

2. INSERTION OF AN AUXILIARY PROCESS

Many insensitivity results are proved by the insertion of an auxiliary process; a particular way of expanding some set of states of a given process to a more complex set of 'sub-states'. We have already seen examples of this in section 3.5 and section 10.5(b). Such a procedure is nothing but a special case of imbedding, and we shall see it as such in the next section. However, it is so important both as a special case and as a technique that we should define it exactly.

The notion of an auxiliary process (λ, μ, v) has already been defined in section 3.4. This is a set $\mathscr{J}$ of states j, entered at j with probability $v(j)$, suffering internal transitions $j \to k$ with intensity $\lambda(j, k)$, and transition out of $\mathscr{J}$ from j with intensity $\mu(j)$. The expected time $u(j)$ spent in state j during a single sojourn in $\mathscr{J}$ satisfies the equations

$$v(j) - \mu(j)u(j) + \sum_k [u(k)\lambda(k, j) - u(j)\lambda(j, k)] = 0 \qquad (1)$$

and so also

$$1 = \sum_j v(j) = \sum_j \mu(j)u(j) \qquad (2)$$

We know (Theorem 3.4.2) that the duration τ of a single sojourn in $\mathscr{J}$ can be given a virtually arbitrary distribution on $\mathbb{R}_+$ by appropriate choice of (λ, μ, v). We shall assume the normalization $E(\tau) = 1$, which then implies that (λ, μ, v) is constrained by the condition

$$\sum_j u(j) = 1 \qquad (3)$$

Consider now a Markov process with state variable $n \in \mathscr{N}$ and transition intensity $\theta(n, n')$. Let A be a given subset of $\mathscr{N}$. We shall then say that this n-process is modified by 'insertion of the auxiliary process (λ, μ, v) in A' if we modify it to the process with state variable

$$x = \begin{cases} (n, j) & (n \in A) \\ n & (n \in \bar{A}) \end{cases} \tag{4}$$

and transition intensity

$$\Lambda(x, x') = \begin{cases} \theta(n, n') & (n, n' \in \bar{A}) \\ \theta(n, n')v(j') & (n \in \bar{A}, n' \in A) \\ \theta(n, n') & (j = j'; n, n' \in A) \\ \theta(n, \bar{A})\lambda(j, j') & (n = n' \in A) \\ \theta(n, n')\mu(j) & (n \in A, n' \in \bar{A}) \end{cases} \tag{5}$$

Here

$$\theta(n, \bar{A}) := \sum_{n' \in \bar{A}} \theta(n, n') \tag{6}$$

and any transition not listed in (5) has zero intensity.

Verbally, the transitions $n \to n'$ proceed at the previous rates $\theta(n, n')$ for n, n' both in A or both in $\bar{A}$. When n is in A the transitions of the auxiliary process follow the rates λ, μ accelerated by an n-dependent factor $\theta(n, \bar{A})$. More specifically, as n enters A then the set $\mathscr{J}$ is simultaneously entered at j with probability $v(j)$; (see the second relation of (5)). Transitions $j \to j'$ and $j \to \bar{\mathscr{J}}$ then have intensity $\theta(n, \bar{A})\lambda(j, j')$ and $\theta(n, \bar{A})\mu(j)$ (see the fourth and fifth relations). In the latter case one has the simultaneous transition $n \to n' \in \bar{A}$ with probability $\theta(n, n')/\theta(n, \bar{A})$.

If the equilibrium distribution of n is unaffected by the insertion into A of an arbitrary auxiliary process (with the normalization (3) understood) then we shall say that the equilibrium distribution of n is *insensitive* to such a modification. We shall see in section 5 that the necessary and sufficient condition for such insensitivity is that the n-process should show partial balance in A.

For the original n-process the event 'exit from A' has intensity $\theta(n, \bar{A})$. In the language of section 3.4 one can say that the 'nominal sojourn time τ in A' or the 'work-load τ to be processed in A' is exponentially distributed with unit mean, but worked off at rate $\theta(n, \bar{A})$. The effect of introducing an auxiliary process is to modify the distribution of τ, retaining unit mean, and retaining also the feature that τ is worked off at rate $\theta(n, \bar{A})$. Since, by Theorem 3.4.2, one can realize any distribution of τ, insensitivity of the n-distribution to insertion of an arbitrary auxiliary process implies insensitivity with respect to the distribution of τ (subject always to the understanding that τ has unit mean and is worked off at rate $\theta(n, \bar{A})$).

3. IMBEDDING AND INSENSITIVITY

As already observed, many insensitivity results are proved by showing that appropriate insertion of an auxiliary process does not affect equilibrium

n-statistics. This technique is natural and effective, but is often regarded as no more than a device; one which regrettably introduces irrelevancies and yields desired conclusions only by a limit argument. We shall on the contrary regard it as the prototype of an insensitivity assertion. The introduction of an auxiliary process is nothing but a form of imbedding, as defined in section 11.5. We shall now formalize the approach suggested in that section.

We take as starting point a 'simple' given Markov process with state variable $n \in \mathcal{N}$ and transition intensity $\theta(n, n')$. The equilibrium distribution $\pi(n)$ then satisfies the balance equation

$$[\mathcal{N}, n](\pi\theta) = 0 \qquad (n \in \mathcal{N}) \tag{1}$$

We often have in mind a set of such processes, generated by letting θ vary in a set Θ. If dependence upon variable θ needs emphasis we shall write $\pi(n)$ as $\pi(n \mid \theta)$.

Suppose this process (the *n-process*) now augmented to one with a more detailed description $x \in \mathcal{X}$, the *x-process*. The random variable n is still defined as a function $n(x)$ of x. The class of x for which $n(x) = n$ is denoted $\mathcal{X}(n)$, and is the analogue in the x-process of the simple point value n. As in section 11.2 we shall refer to these classes as the *count-classes*.

We derive the transition rules of the x-process from those of the n-process by setting

$$\Lambda(x, x') = \theta(n, n')\kappa(x, x') \qquad (x \in \mathcal{X}(n), x' \in \mathcal{X}(n')) \tag{2}$$

for external transitions (between count-classes) and

$$\Lambda(x, x') = \sum_{n'} \theta(n, n')\Lambda_{nn'}^{\text{P}}(x, x') + \Lambda^0(x, x') \qquad (x, x' \in \mathcal{X}(n)) \tag{3}$$

for internal transitions (within count-classes). In other words, we follow the rules (2.2), (2.4) for weak coupling except that the routing rate λ is replaced by the actual transition intensity θ of the n-process. We now regard θ as the quantity which parametrizes the possible n-processes. As in section 11.2, we regard κ as an 'intrinsic' external transition rate specified by the augmentation procedure, $\Lambda_{nn'}^{\text{P}}$ as an intensity for 'progressive' internal transitions associated with the external transition $n \to n'$, and Λ^0 an intensity for 'non-progressive' internal transitions. We shall often find it convenient to define

$$\kappa_{nn'}(x, x') = \begin{cases} \kappa(x, x') & x \in \mathcal{X}(n), x' \in \mathcal{X}(n') \\ 0 & \text{otherwise} \end{cases} \tag{4}$$

For our purposes the augmentation of the n-process to the x-process cannot be arbitrary; it must be one that preserves the equilbrium n-statistics if the x-process is indeed to be regarded as a more structured version of the n-process. We shall say that the x-process *imbeds* the n-process if, in equilibrium and for all θ in Θ (i) the distribution of n agrees with that for the n-process (ii) the probability fluxes $\mathcal{X}(n) \to \mathcal{X}(n')$ agree with the corresponding fluxes $n \to n'$ for the n-process, and (iii) the distribution of x conditional on n is independent of θ.

Theorem 3.1   *The imbedding conditions (i)–(iii) are equivalent to the statements that the equilibrium distribution of the x-process can be represented*

$$\pi(x|\theta) = \Phi(x)\pi(n|\theta) \tag{5}$$

where $\pi(n|\theta)$ is the equilibrium distribution of the n-process, and $\Phi(x)$ (identifiable with $\pi(x|n(x), \theta) = \pi(x|n(x)))$ satisfies

$$\sum_{\mathscr{X}(n)} \Phi(x) = 1 \tag{6}$$

$$\sum_x \sum_{x'} \Phi(x)\kappa_{nn'}(x, x') = 1 \qquad (n, n' \in \mathscr{N}) \tag{7}$$

Proof   Relations (5), (6) are certainly equivalent to the statements (i), (iii). In virtue of form (2) for the external transition intensity relation (7) is then equivalent to (ii) (cf. (11.5.3)).

The intensities (2), (3) are certainly consistent with (1) and (5): the balance equations

$$[\mathscr{X}, \mathscr{X}(n)](\pi\Lambda) = 0$$

with expression (5) inserted for $\pi(x|\theta)$ reduce just to (1) (cf. Theorem 11.2.1). ∎

The structure of imbedding is then just that of weak coupling with the difference that n-transition intensities θ replace routing intensities λ, and that there is the added condition that $h(n, \theta)$ be identifiable with $\pi(n|\theta)$, the equilibrium distribution of the n-process. This additional condition makes relations (6), (7) conditions rather than normalizations.

Note that some imbeddings are trivial; e.g. the modification of n to $x = (n, y)$, where y is the state variable of a Markov process physically and probabilistically independent of the n-process. Others are immediate but meaningful: e.g. the passage from a model with N statistically identical but unidentified molecules to one with identified molecules. Both of these examples have the feature that the variable $n(x)$ of the x-process is itself Markov. This is not in general the case.

If the n-process can be imbedded we regard this as an insensitivity assertion: that n-statistics are insensitive to the difference in specification between the n-process and the x-process. It is properties (i) and (ii) that express this. Property (iii) may be regarded as a less natural demand. It requires that, in a certain sense, the augmentation be θ-independent, transitions internal to the $\mathscr{X}(n)$ being affected by θ only by the differential acceleration implicit in formula (3). A more relaxed augmentation may be possible (cf. the quasireversible nodes of section 10.8) but this one seems to constitute at least a natural first choice.

We come now to the most significant assertion: the statement of sufficient (and partially necessary) conditions that an x-process constitute an imbedding. The theorem implies that, the more partial balance the n-process shows, the weaker the conditions on the imbedding process and the greater the insensitivity.

Theorem 3.2 *Consider the x-process constructed from the given n-process by the rules (2), (3). Suppose the n-process such that $\pi(n|\theta)$ satisfies the partial balance equations*

$$[C_i(n), n](\pi\theta) = 0 \qquad \begin{pmatrix} i = 1, 2, \ldots, I(n) \\ n \in \mathcal{N} \end{pmatrix} \qquad (8)$$

for all θ in Θ, where the sets $C_i(n)$ constitute a decomposition of $\mathcal{N}$ for given n. Suppose that one can find $\Phi(x)$ satisfying (6) and

$$[\mathcal{X}, x](\Phi\Lambda^0) = 0 \qquad (9)$$

$$\sum_{\mathcal{X}(n)} v_i(x) = 1 \qquad (10)$$

$$v_i(x) - \mu_i(x)\Phi(x) + [\mathcal{X}(n), x](\Phi\Lambda_i^p) = 0 \qquad \begin{pmatrix} x \in \mathcal{X}(n) \\ i = 1, 2, \ldots, I(n) \\ n \in \mathcal{N} \end{pmatrix} \qquad (11)$$

Here

$$\left. \begin{aligned} v_i(x) &= \sum_{x'} \Phi(x')\kappa_{n'n}(x', x) \\ \mu_i(x) &= \sum_{x'} \kappa_{nn'}(x, x') \\ \Lambda_i^p(x, x') &= \Lambda_{nn'}^p(x, x') \end{aligned} \right\} \qquad \begin{pmatrix} x \in \mathcal{X}(n); \; n' \in C_i(n) \\ i = 1, 2, \ldots, I(n); \; n \in \mathcal{N} \end{pmatrix} \qquad (12)$$

the quantities on the right thus being required to depend upon n' only through i. Then the x-process has equilibrium distribution (5) and imbeds the n-process.

Proof This is just a statement of Theorem 11.3.2 with θ replacing λ, and the condition that $v_{n'n}$ should depend upon n' only through i strengthened to the condition that this should hold for $v_{n'n}$, $\mu_{nn'}$ and $\Lambda_{nn'}^p$. It is argued in section 11.3 that this is a natural strengthening, although we can then only assert that the conditions of Theorem 3.2 are sufficient for imbedding, rather than also necessary. Relation (10) is equivalent to condition (7). ■

Relation (11) again exhibits $\Phi(x)$ as the expected time spent in x during a sojourn in $\mathcal{X}(n)$ under the rules of an auxiliary process for which the entrance probability, transition intensities and exit intensity are specified by v_i, Λ_i^p and μ_i. These events in the auxiliary process are accelerated by a factor $\theta(n, C_i(n))$ in the actual x-process, in that expression (3) for the internal transition intensity now reduces to

$$\Lambda(x, x') = \sum_i \theta(n, C_i(n))\Lambda_i^p(x, x') + \Lambda^0(x, x') \qquad (13)$$

and, by (2), the transition $x \to \sum_{C_i(n)} \mathcal{X}(n')$ has intensity $\theta(n, C_i(n))\mu_i(x)$ for $x \in \mathcal{X}(n)$.

The consequence of the balance condition (8) for the n-process is that the number of conditions placed upon the quantities $\Lambda^P_{nn'}$, $\mu_{nn'}$ and $v_{n'n}$ (essentially the (λ, μ, v) of the effective auxiliary processes) is reduced. This is the indication that every partial balance property satisfied by the n-process implies some degree of insensitivity.

Our programme will now be to follow through the consequences of Theorem 3.2 under progressively stronger balance conditions on the n-process.

4. INSENSITIVITY WITHOUT PARTIAL BALANCE

Suppose we assume nothing of the n-process but that it has a unique equilibrium distribution satisfying the total balance equation (3.1). Then (3.8) holds only for $C_i(n) = \mathcal{N}$ and relations (3.10–3.12) reduce to

$$v(x) - \mu(x)\Phi(x) + [\mathcal{X}(n), x](\Phi\Lambda^P) = 0 \tag{1}$$

Here

$$\left.\begin{aligned}v(x) &= \sum_{x'} \Phi(x')\kappa_{n'n}(x', x) \\[2mm] \mu(x) &= \sum_{x'} \kappa_{nn'}(x, x')\end{aligned}\right\} \qquad \left(\begin{aligned}&x \in \mathcal{X}(n) \\ &n, n' \in \mathcal{N}\end{aligned}\right) \tag{2}$$

must be independent of n' and satisfy

$$\sum_{\mathcal{X}(n)} v(x) = \sum_{\mathcal{X}(n)} \Phi(x) = 1 \qquad (n \in \mathcal{N}) \tag{3}$$

Expression (3.3) for the internal transition intensity reduces to

$$\Lambda(x, x') = \theta(n)\Lambda^P(x, x') + \Lambda^0(x, x') \qquad (x, x' \in \mathcal{X}(n)) \tag{4}$$

where

$$\theta(n) := \theta(n, \mathcal{N}) = \sum_{n'} \theta(n, n') \tag{5}$$

Relations (1)–(5) are very near the specification of an independent auxiliary process in each state of the n-process used in the construction of Theorem 3.5.1. In fact they reduce exactly to this construction if we set

$$\Lambda^0 = 0 \tag{6}$$

and specify $\kappa(x, x')$ by

$$\kappa(x, x') = \mu(x)v(x') \tag{7}$$

for given μ, v consistent with (3). Relations (2) are then automatically satisfied; our imbedding process reduces to the construction of Theorem 3.5.1, and application of Theorem 3.2 leads, in this special case, to a reproof of Theorem 3.5.1. That is, to the assertion that $\pi(n)$ is insensitive to the distribution of sojourn times in the states n, provided only these are independent with respective expectations $\theta(n)^{-1}$.

The essence of a semi-Markov process is the regeneration rule (7): that on passage from x in $\mathscr{X}(n)$ to x' in $\mathscr{X}(n')$ there is a fresh start, in that $P(x'|x, n') = P(x'|n')$. The conditions (1)–(5) certainly do not imply conditions as strong as (6), (7), and so do not imply that the imbedding process should reduce to a semi-Markov process. It is possible that any further imbedding than that supplied by the semi-Markov process is trivial, in the sense explained in section 3. Whether this is so or not remains a matter for further investigation.

5. SIMPLE PARTIAL BALANCE

Suppose the n-process shows partial balance in a set A of $\mathscr{N}$, so that

$$[A, n](\pi\theta) = 0 \qquad (n \in A)$$

One can then assert that balance relations (2.4) hold for C equal to A or $\bar{A}$ for n in A, but only for $C = \mathscr{N}$ for n in $\bar{A}$.

A convenient way of writing conditions (3.12) in this case is then to set

$$\sum_{x'} \Phi(x')\kappa_{n'n}(x', x) = \begin{cases} v_A(x) & (n \in A, n' \in \bar{A}) \\ v(x) & (\text{otherwise}) \end{cases} \tag{1}$$

$$\sum_{x'} \kappa_{nn'}(x, x') = \begin{cases} \mu_A(x) & (n \in A, n' \in \bar{A}) \\ \mu(x) & (\text{otherwise}) \end{cases} \tag{2}$$

$$\Lambda = \begin{cases} \theta(n, A)\Lambda^P + \theta(n, \bar{A})\Lambda_A^P + \Lambda^0 & (n \in A) \\ \theta(n)\Lambda^P + \Lambda^0 & (n \in \bar{A}) \end{cases} \tag{3}$$

The balance relations (3.11) then become

$$\left. \begin{aligned} v(x) - \mu(x)\Phi(x) + [\mathscr{X}(n), x](\Phi\Lambda^P) &= 0 & (n \in \mathscr{N}) \\ v_A(x) - \mu_A(x)\Phi(x) + [\mathscr{X}(n), x](\Phi\Lambda_A^P) &= 0 & (n \in A) \end{aligned} \right\} \tag{4}$$

with the subsidiary requirements

$$\left. \begin{aligned} \sum_{\mathscr{X}(n)} v(x) = \sum_{\mathscr{X}(n)} \Phi(x) &= 1 & (n \in \mathscr{N}) \\ \sum_{\mathscr{X}(n)} v_A(x) &= 1 & (n \in A) \end{aligned} \right\} \tag{5}$$

If we can find a structure which fulfils relations (1)–(5) we shall have determined on imbedding x-process. The construction of section 4 is certainly one such, with $\mu_A = \mu$, $v_A = v$ and $\Lambda_A^P = \Lambda^P$, but it does not exploit the partial balance. A construction which exploits just the partial balance is one that inserts an auxiliary process in A, following the prescription of section 2. One can easily verify that this specification constitutes an imbedding of the n-process with

$$\kappa(x, x') = \begin{cases} 1 & (n, n' \in \bar{A}) \\ \delta(j, j') & (n, n' \in A) \\ v(j') & (n \in \bar{A}, n' \in A) \\ \mu(j) & (n \in A, n' \in \bar{A}) \end{cases} \tag{6}$$

and $\Lambda^P = \Lambda^0 = 0, \Lambda_A^P(x, x') = \lambda(j, j')$. Equations (1)–(5) are all satisfied if we make the identifications

$$\left. \begin{aligned} v(x) = \Phi(x) &= \begin{cases} u(j) & (n \in A) \\ 1 & (n \in \bar{A}) \end{cases} \\ v_A(x) &= v(j) \\ \mu(x) &= 1 \\ \mu_A(x) &= \mu(j) \end{aligned} \right\} \tag{7}$$

Theorem 5.1 *The equilibrium distribution of n is insensitive to the insertion of an arbitrary auxiliary process in A (i.e. to the distribution of nominal sojourn time τ in A, subject to $E(\tau) = 1$) if and only if the n-process shows partial balance over A.*

Proof We have just proved sufficiency. However, in proving necessity we also obtain a direct demonstration of sufficiency.

Let $\bar{\pi}(x)$ be the equilibrium distribution of the x-process specified by (2.5). Suppose that this indeed has the form

$$\bar{\pi}(x) = \begin{cases} \pi(n)u(j) & (n \in A) \\ \pi(n) & (n \in \bar{A}) \end{cases} \tag{8}$$

Appealing to (2.1)–(2.5) and (8) we find that the balance equations for $\bar{\pi}$ reduce to

$$\left. \begin{aligned} [\mathcal{N}, n](\pi\theta) &= 0 & (n \in \mathcal{N}) \\ (v(j) - \mu(j))[A, n](\pi\theta) &= 0 & (n \in A) \end{aligned} \right\} \tag{9}$$

Partial balance over A consequently implies that expression (8) is the equilibrium distribution, and that the distribution of n is indeed insensitive to insertion of an auxiliary process in A.

Now, insensitivity does not in itself imply that (8) holds; it implies only that

$$\left. \begin{aligned} \sum_j \bar{\pi}(n, j) &= \pi(n) & (n \in A) \\ \bar{\pi}(n) &= \pi(n) & (n \in \bar{A}) \end{aligned} \right\} \tag{10}$$

Flux balance between $\mathcal{X}(n)$ and $\mathcal{X}$ under $\bar{\pi}$ yields the relation

$$\sum_j \mu(j)\bar{\pi}(n, j) = \pi(n) \qquad (n \in A) \tag{11}$$

If we consider an auxiliary process for which $\mathscr{J}$ contains just two elements and $\mu(1) \neq \mu(2)$ then (10), (11) and (2.2), (2.3) imply that $\bar{\pi}$ must indeed have the form (8). Equations (9) are thus necessary for insensitivity, and partial balance over A necessary for insensitivity if we can choose the auxiliary process so that $v(j) - \mu(j) \not\equiv 0$. The choice (employed by Schassberger and Kelly) $v(1) = v(2) = 1/2$ $\lambda(j, k) = 0$, for which $u(j) = (2\mu(j))^{-1}$, will serve. ∎

So, there are particular imbeddings which demonstrate insensitivity of $\pi(n)$ with respect to distribution of nominal sojourn time in states, and nominal sojourn time in A. However, there must also be a hybrid version, its specification latent in conditions (1)–(5), which simultaneously implies both of these insensitivities. The auxiliary processes cannot be independently specified for such an imbedding, because exit from A must also imply simultaneous exit from some n of A. There must be a natural imbedding which takes account of this. It has yet to be found, but is indeed latent in conditions (1)–(5).

6. JOINT PARTIAL BALANCE AND INSENSITIVITY

Suppose that the n-process shows joint partial balance over the sets A_1, $A_2, \ldots, A_d$ of $\mathscr{N}$. The concept of joint partial balance was defined and discussed in section 3.1, where we also introduced the notation

$$A_{EF} = \left(\bigcap_F A_i \right) \left(\bigcap_{E-F} \bar{A}_i \right) \qquad (F \subset E) \tag{1}$$

for E a subset of $D = \{1, 2, \ldots, d\}$. We shall find it convenient to write A_{DE} simply as A_E. That is,

$$A_E = \left(\bigcap_E A_i \right) \left(\bigcap_{\bar{E}} \bar{A}_i \right) \tag{2}$$

and the statement that n lies in A_E then implies that n lies in exactly those A_i for which $i \in E$.

If n lies in A_E then, by Theorem 3.1.1, the A_{EF} for $F \subset E$ constitute a decomposition of $\mathscr{N}$ and A_E is partially balanced against each of the A_{EF}. This is at least part of the point of the notion of joint partial balance: that under it $\mathscr{N}$ decomposes into disjoint sets A_{EF}, each of which partially balances A_E.

For n in A_E we can take the sets $C_i(n)$ of (3.8) for varying i as the sets A_{EF} for varying $F \subset E$. We assume that $\Lambda^p_{nn'}$ reduces to being dependent on n' only through E, F for n in A_E, so that, corresponding to (3.13), we can write the internal transition intensity as

$$\Lambda(x, x') = \sum_{F \subset E} \theta(n, A_{EF}) \Lambda^p_{EF}(x, x') + \Lambda^0(x, x') \tag{3}$$

for $n(x)$ in A_E. The relations analogous to (3.11), (3.12) then become

$$v_{EF}(x) - \mu_{EF}(x)\Phi(x) + [\mathscr{X}(n), x](\Phi\Lambda_{EF}^P) = 0 \qquad (x \in \mathscr{X}(n), n \in A_E) \qquad (4)$$

$$v_{EF}(x) = \sum_{x'} \Phi(x')\kappa_{n'n}(x', x)$$

$$\mu_{EF}(x) = \sum_{x'} \kappa_{nn'}(x, x') \qquad\qquad (n \in A_E, n' \in A_{EF}) \qquad (5)$$

$$\sum_{\mathscr{X}(n)} v_{EF}(x) = 1 \qquad (n \in A_E) \qquad (6)$$

We can then restate Theorem 3.2 for this special case as follows.

Theorem 6.1 *Suppose the n-process shows joint partial balance over the sets* $A_1, A_2, \ldots, A_d$ *of* $\mathscr{N}$. *Then the x-process has the equilibrium distribution* (3.5) *and imbeds the n-process if* Φ *satisfies the normalization condition* (3.6), *the* Λ^0 *balance condition* (3.9), *and conditions* (4)–(6) *for* $F \subset E \subset D$.

As formerly, relations (5) constitute not merely definitions of v_{EF} and μ_{EF}, but also indiscriminacy conditions on κ.

However, the interest is to find particular imbeddings. The easiest case is that in which we again make the hypothesis of Theorem 3.1.2, that there is no transition $n \to n'$ under which more than one A_i is entered or more than one A_i is vacated. This hypothesis allows independent specifications of auxiliary processes in the different A_i.

Theorem 6.2 *Suppose the n-process shows partial balance on each of the sets* A_1, $A_2, \ldots, A_d$ *and that there is no transition under which more than one* A_i *is entered or more than one* A_i *is vacated. Then* $\pi(n)$ *is insensitive to insertion of arbitrary auxiliary processes in the sets* $A_1, A_2, \ldots, A_d$. *That is, it is insensitive to the distributions of nominal sojourn times in* $A_1, A_2, \ldots, A_d$ *if these are independent with unit mean and the nominal sojourn time in* A_i *is worked off at rate* $\theta(n, \bar{A}_i)$.

Proof Insertion of the auxiliary processes indeed constitutes an imbedding under the conditions stated. The variable and parameters of the ith auxiliary process will be distinguished by a subscript i; e.g. $v(j)$ is modified to $v_i(j_i)$.

It is useful to introduce the notation

$$\delta_E(j, j') = \prod_{i \in E} \delta(j_i, j_i')$$

Introduction of the d independent auxiliary processes is now equivalent to the specification for an imbedding: $\Lambda^0 = 0$, and, for n in A_E,

$$\Lambda_{nn'}^P(x, x') = \begin{cases} \lambda_i(j_i, j_i')\delta_{E-i}(j, j') & (n' \in A_{E-i} \text{ or } A_{E-i+m}) \\ 0 & (\text{otherwise}) \end{cases} \qquad (7)$$

$$\kappa(x, x') = \begin{cases} \delta_E(j, j') & (n' \in A_E) \\ \delta_E(j, j')v_m(j_m) & (n' \in A_{E+m}, m \notin E) \\ \delta_{E-i}(j, j')\mu_i(j_i) & (n' \in A_{E-i}, i \in E) \\ \delta_{E-i}(j, j')\mu_i(j_i)v_m(j_m) & (n' \in A_{E-i+m}, i \in E, m \notin E) \end{cases} \tag{8}$$

In virtue of the entrance/exit hypotheses these are the only cases that need be considered.

We then find that all the conditions of Theorem 5.1 are satisfied with, for n in A_E,

$$\Phi(x) = \prod_{i \in E} u_i(j_i) \tag{9}$$

$$\sum_{x'} \Phi(x')\kappa_{n'n}(x', x) = \begin{cases} \Phi(x) & (n' \in A_E, A_{E+m}) \\ \Phi(x)v_i(j_i)/u_i(j_i) & (n' \in A_{E-i}, A_{E-i+m}) \end{cases} \tag{10}$$

$$\sum_{x'} \kappa_{nn'}(x, x') = \begin{cases} 1 & (n' \in A_E, A_{E+m}) \\ \mu_i(j_i) & (n' \in A_{E-i}, A_{E-i+m}) \end{cases} \tag{11}$$

Expressions (7), (10), (11) effectively define Λ_{EF}^P (for $F = E - i$), v_{EF} and μ_{EF} (for $F = E, E - i$), respectively; the values of F indicated being the only ones which occur.

7. EXAMPLES

It is Theorem 6.2 which finds ready application. Consider, for example, the Erlang example of Exercise 4.7.3. The most basic description of this is simply the number of busy channels. This is the variable of a reversible Markov process. A more adequate description (itself an imbedding of the first process, but one which is more than formal in that it conveys additional physical information) is that which lists the busy channels. This process, with state variable n, say, is again reversible. So, if we define A_i as the set of states for this process under which channel i is busy, then there is certainly partial balance within each of the sets $A_1, A_2, \ldots, A_m$. These sets obey moreover the entrance/exit hypothesis of Theorem 6.2, since multiple events have zero intensity.

The hypotheses of Theorem 6.2 are then satisfied with $\theta(n, \bar{A}_i) = \mu$. The conclusion of that theorem then implies that the equilibrium statistics of n are unchanged if call length has an arbitrary distribution of mean μ^{-1}. This distribution may vary between channels, although we continue to suppose the lengths of different calls independent.

The same argument can be applied to the circuit-switching example of Exercise 4.7.4, a generalization of the Erlang example, with the notion of a simple channel now being replaced by that of the circuit between a prescribed pair of nodes.

Note that we are not utilizing the full reversibility of the n-process; merely the fact that the sets $A_1, A_2, \ldots, A_m$ are jointly partially balanced and satisfy the entrance/exit hypothesis. For a fully reversible process there must be a yet stronger imbedding, since the condition that $\nu_{n'n}$, $\Lambda_{nn'}^p$ and $\mu_{nn'}$ should depend upon n' only through i is then no constraint at all.

The reason for the insensitivity of the models of section 5.7 can be seen in various ways. That the more refined model can be aggregated (over spatial cells) is a consequence of Poisson statistics. The possibility of the reverse (i.e. that the aggregated model can be imbedded in more refined models) can again be seen as a consequence of the detailed balance of the absorption/emission of a given quantum by a given component. Occupation statistics for the components are then unchanged if the wanderings of a quantum outside the component are given more structure. That is, if an auxiliary process is inserted which represents the migration of the quantum over spatial cells.

Exercises and comments

1. Suppose, for the Erlang example, that there are a number of caller types, all however having the same values of call rate λ and service rate μ. Take as description n of the process a listing of which channels are occupied by which caller-types, and suppose the allocation of incoming calls to free channels independent of channel label or caller-type. Show that call-length distributions may be of arbitrary channel/type-specific form (of mean μ^{-1}) without affecting equilibrium n-statistics.

8. JACKSON NETWORKS

As a final special class of cases, let us suppose that the n-process represents a Jackson network of unstructured nodes. We shall interpret this statement as implying that n is the vector of node-counts $\{n_j\}$, that the only transitions are of the form $n \rightarrow n - e_j + e_k$ for varying j, k, and that in equilibrium the process shows flux balance for the transitions $n \rightleftharpoons C_j(n)$ $(j = 1, 2, \ldots, m)$ where

$$C_j(n) = \{n - e_j + e_k; k = 1, 2, \ldots, m\} \qquad (1)$$

If we can demonstrate imbeddings of this process then we shall have achieved two goals. First, we shall have demonstrated an insensitivity of the simple Jackson network. Secondly, we shall have constructed a more complicated model, which we may regard as a Jackson network with structured nodes, and known equilibrium distribution.

Applying Theorem 3.2 with the $C_j(n)$ given by (1) we deduce just the structured network of section 10.10 with the modifications that routing intensity $\lambda_{jk}(n)$ is replaced by the transition intensity $\theta(n, n - e_j + e_k)$, and the identification $h(n, \lambda) = \pi(n \mid \theta)$ and normalization (3.6) are also required. These differences occur because in section 10.10 we were looking simply for weak coupling, and not for an imbedding.

The fact that n is partially balanced against the sets $C_j(n)$ given by (1) implies that the n-process shows partial balance on sets of constant n_j. Insertion of an auxiliary process into these sets corresponds exactly to the construction of section 10.4: of structured nodes which are functionally independent, balanced and reactionless.

The sets of constant n_j are each partially balanced, but do not satisfy the entrance/exit condition of Theorem 3.1.2, because a migration induces passage out of two such sets and into two such sets. However, the sets A_{ij} for which the identified unit i occupies node j ($i = 1, 2, \ldots, N; j = 1, 2, \ldots, m$) do satisfy this condition. They are also individually partially balanced if we require, as well as route partial balance, the symmetry assumptions of section 10.5 for passage of a given unit in and out of a given node. One can then insert an arbitrary auxiliary process into each A_{ij}. It is because of this that we could obtain the insensitivity assertions of section 10.5, and that it is indeed also possible to insert an auxiliary process into sets of constant n_j.

A view of the process in which state is described by giving the locations of units rather than the numbers at nodes is closely related to Hordijk's concept of a 'job-mark' process (see Hordijk and van Dijk, 1983; Hordijk, 1984). Here the units are jobs, each marked by its current location in the system which is also an indication of the stage it has reached in processing. The 'entrance-exit' condition is equivalent to the assumption that only the job changes mark in any given event. Kelly (1979) refers to such processes as 'spatial' processes—rather a special use of the term, but again conveying the notion that only one unit moves at a time.

The notion of several and changing types of units can rather easily be incorporated, by letting the nodes of the model of section 10.10 describe a node/type combination. However, ringing of changes can become an empty exercise, and section 10.8 probably remains the treatment of the multi-type case which best combines conciseness and generality.

PART IV

Bonding models; polymerization and random graphs

In Chapter 7 we considered the statistics of molecules constructed from elements in prescribed abundance, assuming however that only a finite number of types of molecule were possible. If bonding can go on indefinitely, so that molecules of indefinite size and complexity can be built up, then one is considering the actual situation for organic and polymer molecules. It has the new feature that various types of critical behaviour (phase change) can occur.

However, polymer molecules constitute only one example of such behaviour. We are in general interested in the formation (and dissolution) of clusters, aggregates or complexes which have at least the possibility of becoming indefinitely large and richly linked. Such processes occur at all kinds of levels above the molecular. For example, the neutralization of antigens by antibodies occurs by the formation of cross-links, and so of clusters which in the end are so large as to become inactive. At a higher level still, one can think of human social and economic institutions as created by the formation of bonds between individuals. This example demonstrates a new feature of great significance. In such institutions individuals become differentiated, in that they adopt different roles. There is an interaction between the bonds an individual forms and the role he adopts, and it is this interaction which one expects will produce a collectively-working structure.

We shall discuss some of these other applications in Chapter 17, but before that use largely the polymer terminology, as being standard in the best-developed application.

Most of the work on what one might call clustering or aggregation processes is to be found in the polymer literature, followed, in more recent years, by the random graph literature. We give a brief survey in section 13.8.

CHAPTER 13

Polymerization; the simplest models

1. POLYMERS, GRAPHS, GELATION

We consider processes in which basic particles ('units') can associate to form clusters of indefinite size, which can also dissociate, and we are interested in the ensuing equilibrium distribution of the sizes and type of cluster. Such processes, which one could variously term clustering, aggregation or polymerization processes, occur in a variety of contexts. The best-developed applications are those of polymer chemistry, in which one considers the chained and branched structures which can be built up by the formation of bonds between basic chemical building blocks. However, there are many other applications, e.g. the study of antigen–antibody reactions (Goldberg, 1952, 1953), of the dynamic behaviour of social groupings (Wasserman, 1978) and of socioeconomic structure generally (Whittle, 1977a). We shall nevertheless employ the nomenclature of polymer chemistry, as an established vocabulary for the best-developed application.

One may say that the study of interactions between given units, and of consequent structures, embraces the whole study of particulate matter. However, the polymerization literature makes one drastically simplifying assumption relative to this general case: the absence of a spatial metric structure. There is no notion of distance or dimensionality; the only relationship between units lies in the presence or absence of a mutual bond. It is true that one can begin to build in a rudimentary spatial element, by allowing units to lie in various regions, and permitting bonds only between regions considered spatially adjacent. Nevertheless, there is, at least in the simplest models, no notion of distance, 'action at a distance', or of a field which mediates interaction. Bonding models are thus a very special case of interaction models, but they retain enough structure that they can produce some interesting effects.

The chemical kinetic models of Chapter 7 were in effect bonding models, because we regarded a molecule as a 'cluster' with internal interactions (unspecified) which held it together, and we assumed no interaction between the separated molecules. The only interaction occurred upon 'encounter' in space, when there was the possibility of reaction; i.e. of bond formation. We shall now assume very much the statistics of those molecular models, but there are three new points which arise.

(i) In Chapter 7 we presumed that there were only a finite number of types of molecule. We shall now suppose that, even with only a single element, models of an indefinite size and variety of structure can be built up. This indeed occurs in polymers and organic molecules, where a few basic types of unit can combine to form an unlimited range of macromolecules.

(ii) With this increased variety of structure, there must be a definite description of structure at some level of detail.

(iii) The reaction rates were regarded as given in Chapter 7, consistent with reversibility. Determination of these rates can now be part of the problem; there are so many ways in which structures can combine or divide that it is a real combinatorial task to enumerate them. We shall in fact circumvent this problem rather than solve it, but to do so requires development of a new viewpoint.

The molecules of Chapter 7 were made up of atoms of different elements. We shall now use the neutral word 'units' rather than 'atoms'. This is because in the polymer context the units may themselves be molecules (effectively fixed in form) as well as atoms. In other contexts they may be cells, bacteria, antibodies or human beings.

A variety of structures can be built up from even a single type of unit. The simple unit ◯ is the *monomer*. A cluster of two units is the *dimer*, which necessarily has

the form ◯——◯ if directed, multiple, or self-bonds are not allowed.

With this restriction there are two types of *trimer*, the linear form

◯——◯——◯ and the cyclic form (triangle figure) . In general, the

simple chain ◯——◯——◯– – – – – –◯ is referred to as a

linear polymer. If any unit is bonded to more than two others then one has a *branched* polymer

As we saw in the case of the *trimer*, there is also the possibility of *cycle* or *ring-formation*

One may also allow the possibility of directed bonding if a bond has an asymmetric character (e.g. is initiated by one of the units, is a relation between a superior and a subordinate, or is a path for transmission of some flow in a definite direction). One may also wish to allow multiple bonds and self-bonds , directed or undirected.

We shall largely use the polymer terminology, but it is clear from these diagrams that the polymer is a graph, and we can also use the terminology of graph theory (see section 1.15). In fact, a collection of polymers made up from N units is a graph on N nodes (Fig. 13.1.1).

The nodes of the graph represent units, and are supposed distinguishable, i.e. labelled. The nodes of the graphs may be *coloured* if different types of unit are

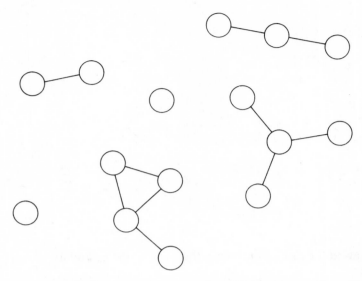

Fig. 13.1.1 A collection of polymers made up from N units as a graph on N nodes.

possible. The arcs of the graph represent bonds. If the arcs are directed then they indicate a directional property of the bond; perhaps that of being *initiated* by the unit from which the arc emerges. Multiple bonding and self-bonding are not excluded in general. The connected components of the graphs represent just the molecules or *polymers*.

It should perhaps be emphasized that the graphs we are now considering are quite distinct from the network graphs of Part III. Those graphs specified a fixed routing scheme between nodes. The graphs we now consider describe the randomly changing connections between units. Nor is the component of a graph anything to do with the components of Part III. These latter were the processors situated at the nodes of a processing network. The components we now consider are just the clusters themselves.

The striking critical effect both observed and predicted for polymers is that of *gelation* or the *sol/gel transition*. Depending upon the parameters of the process, the matter available may be in the *sol* state (in which there are no very large molecules) or the *gel* state (in which most of the matter present has aggregated into a single giant complex). As one changes parameters in such a way as to favour association over dissociation there may be a transition from the sol state to the gel state, this transition being discontinuous in the thermodynamic limit.

The various historical approaches to a model of polymerization listed in section 8 have, despite their differences, produced a remarkably consistent theory of gelation. Mathematicians have also studied the effect in the random graph literature (see again section 8). The prescription adopted most often in this literature is that only single undirected arcs between distinct nodes may occur, and occur independently (for distinct node-pairs) with constant probability p. This may seem like the simplest prescription, but is not; that palm should be awarded to the Poisson model of section 6. On the other hand, it is simpler than physical realism would demand; as a prescription of the effective potential energy of a configuration it is simple in the extreme (see section 5). However, it does exhibit a gelation effect. If we set $p = a/N$ then for N large there is a critical value $\bar{a}$ of a. For $a < \bar{a}$ all polymers (connected components) are bounded in size, with a probability that tends to unity with increasing N. For $a > \bar{a}$ the probability approaches unity that most of the N nodes are connected; i.e. belong to a single polymer molecule, whose size must then be of order N.

Suppose the different types of polymer molecule can be indexed by r, taking values in a set of types $\mathscr{R}$. We shall term a molecule of type r an r-mer and, as in Chapter 7, denote it by A_r. A unit of type α will be denoted e_α.

What one means by 'type' corresponds to the level of description adopted. The fullest statistical description would specify which units comprised the molecule and their complete pattern of connection. The fullest structural description would not distinguish between units of the same type, but would still specify the bond-pattern in full. However, one will adopt the lowest level of description at which the model is still Markov. This means, in most cases, the minimal description needed to determine the configurational energy of the molecule. At such a level a description r thus covers many *isomers*: molecules whose structures differ, but only in aspects considered irrelevant.

If n_r is the number of r-mers the assumption is then that $n = \{n_r\}$ is the state variable of a continuous-time Markov process. Despite the lack of any spatial concept in the model, we do need the idea of volume. We shall assume then that all reactions are confined to a region of volume V, the region itself also often being referred to as V. The concentration of r-mers is then

$$c_r = \frac{n_r}{V}.$$

2. DETERMINISTIC KINETIC MODELS

The models we shall consider are essentially identical with those of Chapter 7. They will be more special, in that, for simplicity, we shall consider only certain types of reaction. On the other hand, we allow an infinite variety of polymers. Also, we shall assume that only the abundance of units is restricted, not of energy. The system is thus in a heat-bath of a definite temperature, and the process is assumed reversible rather than transition-reversible. As we noted in the last section, the reaction rates at a certain level of description may have to be derived from a model at a higher level of description.

The only reactions we shall consider are those of change of internal state of a polymer,

$$A_r \rightleftharpoons A_s \tag{1}$$

and association/dissociation,

$$A_r + A_s \rightleftharpoons A_u \tag{2}$$

Let us suppose, consistently with (7.1.11), that reaction (1) has forward rate $\kappa_{rs}c_r$ and backward rate $\kappa'_{rs}c_s = \kappa_{sr}c_s$ where c_r is the concentration of A_r. That is, these are the numbers of reactions per unit time and volume. The corresponding rates for reaction (2) are supposed to be $\lambda_{rsu}c_rc_s$ and $\lambda'_{rsu}c_u$. The kinetic equations for the concentrations can then be written

$$\dot{c}_v = \sum_r \sum_s \kappa_{rs}c_r(\delta_{sv} - \delta_{rv}) + \sum_{r \leqslant s} \sum_u (\lambda_{rsu}c_rc_s - \lambda'_{rsu}c_u)(\delta_{uv} - \delta_{rv} - \delta_{sv}) \tag{3}$$

a special case of (7.1.9). The constraint $r \leqslant s$ in the summation is to imply that a given pair of values r, s shall appear only once in the summation.

The deterministic detailed balance conditions for reactions (1), (2) are

$$\kappa_{rs}c_r = \kappa_{sr}c_r \tag{4}$$

$$\lambda_{rsu}c_rc_s = \lambda'_{rsu}c_u \tag{5}$$

If these have a solution $c_r = \gamma_r$ then this is a stationary solution; that they *should* have a solution is a deterministic reversibility condition.

Suppose, as in Chapter 7, that the units are immutable, that an A_r contains $a_{\alpha r}$ α-units, and that total unit numbers are conserved in the reactions (1), (2). Then, as in (7.1.15), if $c_r = \gamma_r$ solves (4), (5) then so does

$$c_r = \gamma_r \prod_\alpha w_\alpha^{a_{\alpha r}} \tag{6}$$

for arbitrary w_α. The parameters w_α are to be determined by the conditions

$$\sum_r a_{\alpha r}n_r = N_\alpha \tag{7}$$

or

$$\sum_r \gamma_r a_{\alpha r} \prod_\beta w_\beta^{a_{\beta r}} = \rho_\alpha \qquad (\alpha = 1, 2, \ldots, p) \tag{8}$$

where $\rho_\alpha = N_\alpha/V$ is the prescribed total density of α-units.

It follows by the methods of section 7.2 that expression (6) gives the equilibrium solution on an irreducible set, at least if the abundance conditions (7) can be satisfied. Under irreducibility conditions on the process equations (8) have a unique solution for the w_α, and expression (6) gives them the unique equilibrium solution for the prescribed abundances.

In the case when $\mathcal{R}$ was finite equations (8) always had a solution, at least if free units were permitted (see section 7.2). However, if $\mathscr{R}$ is infinite then they may not. The trouble is not as simple as a divergence of the infinite series in (8), but is indeed related to the occurrence of polymers of infinite size. This was the point discussed in section 6.3; see also Exercises 1 and 2.

Conditions (8) can be expressed rather more compactly if we define a generating function

$$G(w) = \sum_r \gamma_r \prod_\alpha w_\alpha^{a_{\alpha r}} \tag{9}$$

The abundance conditions (8) can then be written in the form

$$w_\alpha \frac{\partial G(w)}{\partial w_\alpha} = \rho_\alpha \qquad (\alpha = 1, 2, \ldots, p) \tag{10}$$

These equations should determine the parameters w of the equilibrium solution (6).

Exercises and comments

1. Consider a case in which there is only a single type of unit, and the size of a cluster is an adequate description. We effectively then make the identification $r = a_r$, and $G(w)$ reduces to

$$G(w) = \sum_{r=1}^{\infty} \gamma_r w^r$$

Suppose that $\gamma_r = \kappa \sigma^r$. (This implies that the configuration energy of an r-mer contains a term simply for the existence of the molecule plus a term proportional to the size of the molecule. Exercise 3.1 gives a mechanism that would produce such a dependence.) Then $G(w) = \kappa(1 - \sigma w)^{-1} - \kappa$, and the determining relation (10) for w becomes

$$\frac{\kappa \sigma w}{(1 - \sigma w)^2} = \rho$$

This has a solution for any ρ, increasing with ρ. Although $G(w)$ diverges at $w_0 = \sigma^{-1}$, the value of w determined by (11) is always smaller than w_0. So, although the distribution $c_r = \gamma_r w^r$ shifts its mass to the heavier molecules as ρ increases, the sum $\Sigma r c_r$ remains finite, indicating that molecules remain finite.

2. Suppose now instead that

$$\gamma_r = \frac{1.3.5 \ldots (2r - 3)}{2^{r+1} r(r!)} \sigma^r \qquad (r \geqslant 1) \tag{12}$$

This has been so chosen that

$$w \frac{\partial G}{\partial w} = 1 - \sqrt{1 - \sigma w}$$

and equation (10) for w becomes

$$\sqrt{1 - \sigma w} = 1 - \rho \tag{13}$$

This possesses a solution for w if and only if $\rho \leqslant 1$. For $\rho = 1$, when $w = \sigma^{-1}$, the infinite sum for $\partial G / \partial w$ still converges, but has reached its radius of convergence. For $\rho > 1$ the analysis becomes meaningless. In fact 'gelation' has occurred, and the available mass is concentrated largely in indefinitely large molecules.

3. For the γ_r of (12) one finds that

$$\gamma_r \sim \text{const.} \ \sigma^r r^{-5/2}$$

for large r. One feels then that large molecules are marginally less likely in this case than in the case $\gamma_r = \kappa \sigma^r$ of Exercise 1, and it seems paradoxical that gelation should occur in the first case but not in the second. The explanation lies in the fact that a randomly chosen unit lies in a molecule of size r with probability proportional to rc_r (in the case when size is a sufficient description) so that the expectation of polymer size on this basis is then

$$\frac{\Sigma r^2 c_r}{\Sigma r c_r} = \frac{\Sigma r^2 \gamma_r w^r}{\Sigma r \gamma_r w^r}$$

For the example of Exercise 1 the two series in this expression converge or diverge together. For the example of Exercise 2 there is a value of w for which $\Sigma r \gamma_r w^r$ converges (so that unit density ρ is finite) but $\Sigma r^2 \gamma_r w^r$ diverges, indicating that the average size of the molecule within which a randomly-chosen unit finds itself is infinite.

Of course, to construct an example by specifying γ_r is artificial. In later sections we shall effectively deduce γ_r from more primitive assumptions, and shall then find that the phenomenon of gelation is the rule rather than the exception.

3. STOCHASTIC KINETIC MODELS

As the last section was in fact a special case of section 7.1, so is this in fact a special case of section 7.3.

We shall stochasticize the model of the last section by supposing that reaction (2.1) has total forward intensity $\kappa_{rs} n_{rs}$, that reaction (2.2) has total forward intensity $\lambda_{rsu} n_r (n_s - \delta_{rs}) / V$ and total backward intensity $\lambda'_{rsu} n_u$. That is, these are the intensities for a single molecular event of the type indicated in the whole volume V of the reaction vessel. The reason why these should be regarded as the appropriate stochastic versions of the deterministic rates was explained in Theorem 7.3.1.

The deterministic kinetic equation (2.3) now has the stochastic analogue

$$\frac{\partial \Pi}{\partial t} = \sum_r \sum_s \kappa_{rs} (z_s - z_r) \frac{\partial \Pi}{\partial z_r} + \sum_{r \leqslant s} \sum_u (z_u - z_r z_s) \left[\frac{\lambda_{rsu}}{V} \frac{\partial^2 \Pi}{\partial z_r \partial z_s} - \lambda'_{rsu} \frac{\partial \Pi}{\partial z_u} \right] \tag{1}$$

where $\Pi(z) = E \prod_r z_r^{n_r}$ is the p.g.f. of molecular abundances n_r (cf. (7.3.8)).

If the detailed balance equations (2.4), (2.5) have a solution $c_r = \gamma_r$, then equation (1) has a stationary solution

$$\Pi(z) \propto \exp\left[V \sum_r \gamma_r z_r \right] \qquad (2)$$

If element abundances are conserved so that the detailed balance equations admit the more general solution (2.6) then equation (1) correspondingly admits the more general solution

$$\Pi(z) \propto \exp V \sum_r z_r \gamma_r \prod_\alpha w_\alpha^{a_{\alpha r}} \qquad (3)$$

with arbitrary w_α.

The deterministic reversibility conditions (2.4), (2.5) are indeed equivalent to the assumption of stochastic reversibility (see Lemma 7.3.1). Solution (3) represents the n_r as independently Poisson distributed, with expectation values determined by the parameters w_α. Alternatively, one can regard the w_α as marker variables for the numbers of α-units, and so deduce (cf. Theorem 7.3.2).

Theorem 3.1 *Assume the reversibility and conservation conditions which make (3) a stationary solution of (1) for any w, and assume the process irreducible for given N_α ($\alpha = 1, 2, \ldots, p$). Then the p.g.f. of the n_r under the abundance constraints*

$$\sum_r a_{\alpha r} n_r = N_\alpha \qquad (\alpha = 1, 2, \ldots, p) \qquad (4)$$

of a closed system is proportional to the coefficient of $\prod_\alpha w_\alpha^{N_\alpha}$ in the power series expansion of expression (3).

Note that solution (3) may be purely formal, in the sense that the infinite sum in the exponent may not converge on $|z_r| = 1$ ($r \in \mathcal{R}$) for any value of w. Theorem 3.1 nevertheless remains valid; the determination of the constrained p.g.f. depends upon the values of γ_r only for r such that $a_{\alpha r} \leqslant N_\alpha$ for all α. On the other hand again, it is the behaviour of the infinite series which determines the behaviour of the process in the thermodynamics limit, as we have seen in Exercise 2.2.

With solution (3) and Theorem 3.1 the problem may seem to be essentially solved. This is very far from the case. We have not yet considered at what level the polymers are to be described. At any level other than the most detailed description the label r corresponds to a whole class of *isomers*, and the coefficient γ_r is to be interpreted as

$$\gamma_r = \omega_r e^{-U_r/kT} \qquad (4)$$

where U_r is the potential energy of an A_r (i.e. of a polymer in the class r) and ω_r is the number of isomers of A_r (i.e. the number of members of this class).

The determination of the coefficient ω_r is a combinatorial problem, as is then also the determination of the rates κ, λ. For example, the rate λ'_{rsu} will depend upon the number of ways in which an A_u can split into an A_r and an A_s, and to determine this is a very substantial combinatorial problem.

The way to circumvent this situation is that taken in section 5. We shall revert to a complete description, and then regard the problem as one of deducing coarse-description statistics.

This calculation is not trivial or uninteresting; it is the essential problem of 'statistical integration' that one must face when going from a micro-level specification to macro-level conclusions.

The natural level of description is fixed for one by the form of the energy function U_r. The energy of a polymer will depend upon parameters of the model. Two polymers will be regarded as isomers (and so as effectively indistinguishable) if their energies are the same identically in these parameters. We shall see how this principle works out in Chapter 14.

Exercises and comments

1. *Linear polymers.* Suppose again, as in Exercises 2.1 and 2.2, that size is a sufficient description, and suppose that A_r is necessarily a linear polymer of size r. Suppose that the reactions $A_r + A_s \rightleftharpoons A_{r+s}$ take place at rates $\lambda n_r (n_s - \delta_{rs})/V$ and μn_{r+s} in the forward and backward directions respectively.

Show that the model is reversible and that

$$\gamma_r = \frac{\mu}{\lambda}$$

This is of the form considered in Exercise 2.1, and we know that it can never lead to gelation. Linear polymers do not gel, because there is only one polymer of a given size.

One might have expected the dissociation rate to differ by a factor of two according as $r \neq s$ or $r = s$, because there seem to be two bond-breaks which yield the reaction $A_{r+s} \rightarrow A_r + A_s$ in the first case. However, there is indeed only one, because we are considering a dissociation into *identified* r-mers.

2. *Linear and loop polymers.* Suppose that polymers can also form simple loops. Let A_r and $A_{r'}$ be the linear and loop polymers of size r respectively ($r = 1, 2, 3 \ldots$). We suppose the association/dissociation rates of the last section plus the linear/loop transition $A_r \rightleftharpoons A_{r'}$ at intensities κn_r and $v r n_{r'}$ respectively. The factor r in the second intensity reflects the fact that there are r bond-breaks which will open the loop to linear form. Show that again one has reversibility, with

$$\gamma_r = \frac{\mu}{\lambda}, \qquad \gamma_{r'} = \frac{\mu \kappa}{\lambda v} \frac{1}{r}$$

Note that this model still does not show gelation, in that there is a finite w solving

$\sum_r (\gamma_r + \gamma_{r'})w^r = \rho$ for every finite ρ. There are still too few structures of size r that gelation is possible.

3. *The sweeping of bacteria.* Suppose α can take two values, the units for α equal to 1 or 2 being identified as antibodies and bacteria respectively. Suppose that clusters are labelled simply by $r = (r_1, r_2)$, where r_α is the number of α-units in the cluster and that the only values possible are $(0, 1)$ and $(1, j)$ $(j = 0, 1, 2, \ldots)$. That is, free bacteria or an antibody to which j bacteria are attached. Suppose, consistent with this, that the only reaction is

$$A_{1j} + A_{01} \rightleftharpoons A_{1,j+1} \qquad (j = 0, 1, 2, \ldots)$$

with total forward and backward reaction rates $n_{1j}n_{01}H_{j+1}/V$ and $n_{1,j+1}H_j$. Show that the detailed balance equations are solved by

$$\gamma_{01} = 1$$
$$\gamma_{1j} = H_j \qquad (j = 0, 1, 2, \ldots)$$

For the particular case when the forward and backward rates are $\mu\sigma(f-j)_+ n_{1j}n_{01}/V$ and $\mu(j+1)$, where f is a positive integer, we can take

$$\gamma_{1j} = \binom{f}{j}\sigma^j$$

This model represents a situation in which the only bonding consists in attachment of bacteria to antibodies. The term H_j represents the 'intrinsic' probability of an A_{1j}. The final special form corresponds to the case when an antibody has f independent attachment sites (is an 'f-functional unit', in polymer parlance).

In this latter case we have

$$G(w) = w_1(1 + \sigma w_2)^f + w_2$$

Show that the abundance equations (2.10) have a unique positive solution for all ρ. The absence of gelation is of course due to the impossibility of infinite structure. If one allows antibodies also to bond to each other then gelation is possible, and it is indeed by the formation of large complexes that bacteria are neutralized.

4. CRITICALITY

If, as one varies parameters of the model, there is a discontinuous transition from the sol to the gel state, then the values of the parameters at which this occurs are referred to as 'critical'. There is naturally considerable interest in determining whether this or other phase transitions can occur, and in locating the critical points. We shall find a relatively easy way of doing this when we revert to a complete description, but we might as well note how criticality manifests itself in the solution as we have it.

From the evaluation of $\Pi(z)$ for prescribed N_α given by Theorem 3.1 we deduce, for example, that

$$E(n_r) = V\gamma_r \frac{I\left[\prod_\alpha w_r^{a_{\alpha r}}\right]}{I[1]} \tag{1}$$

Here

$$I[\phi(r)] := \oint \ldots \oint \phi(r) e^{VG(w)} \prod_\alpha \frac{\mathrm{d}w_\alpha}{w_\alpha^{N_\alpha+1}} \tag{2}$$

where $G(w)$ has the definition (2.9) and the integration contour in (2) is the complex one $|w_\alpha| = \delta \,(\alpha = 1, 2, \ldots, p)$, the constant δ being chosen small enough that $G(w)$ has no singularities within this contour.

In the thermodynamic limit (i.e. when N_α and V become infinite with fixed density $\rho_\alpha = N_\alpha/V$, for all α) evaluation (1) then becomes

$$E(n_r) = V\gamma_r \prod_\alpha \bar{w}_\alpha^{a_{\alpha r}} \tag{3}$$

where $\bar{w}$ is the saddle-point of the function $G(w) - \Sigma \rho_\alpha \log w_\alpha$. This saddle-point, if it exists, is located by the stationarity conditions

$$\frac{\partial G(w)}{\partial w_\alpha} - \frac{\rho_\alpha}{w_\alpha} = 0 \qquad (\alpha = 1, 2, \ldots, p) \tag{4}$$

These are just the familiar abundance conditions (2.10) of the deterministic case. One expects that the relevant saddle-point is just the positive real solution $\bar{w}$ of these equations.

Suppose one considers the prescribed unit-densities as parameters of the model which may be varied (although certainly not the only parameters). It is readily verified that $\bar{w}$ increases with increasing ρ. It may be that, as ρ thus increases, there comes a point where equations (4) fail to have a solution. This is usually because $\bar{w}$ has moved on to a value which is a branch point of the function $G(w)$. We have already seen an example in Exercise 2.2, and Exercise 2.1 seemed to demonstrate that it was a branch point which was needed rather than a simple pole.

This is indeed how criticality manifests itself, but we shall give a more directly physical treatment of the matter in section 6. See also Exercise 2.

Exercises and comments

1. In looking for a most probable configuration we maximize $\sum_r [c_r(1 + \log\gamma_r) - c_r \log c_r]$ with respect to c subject to

$$\sum_r a_{\alpha r} c_r = \rho_\alpha \qquad (\alpha = 1, 2, \ldots, p)$$

The dual problem is just the minimization of $G(w) - \Sigma \rho_\alpha \log w_\alpha$ with respect to w; the location of the saddle-point $\bar{w}$. These extremal problems, familiar from Chapter 5 can change in character now that $\mathcal{R}$ is infinite. If the equilibrium state is the gel state then the primal problem will have an infinite solution and the dual problem no solution (cf. section 6.3).

2. The consideration of moments in Exercise 2.3 would lead us to conclude that we have reached criticality when $\bar{w}$ is such that the matrix of second derivatives G_{ww} is infinite, even if the vector of first derivatives G_w exists (both evaluated at $\bar{w}$). That is, even if $G(w) - \Sigma \rho_\alpha \log w_\alpha$ has a minimum at $\bar{w}$, it also shows infinite curvature in some direction.

5. EQUILIBRIUM IN THE COMPLETE DESCRIPTION

Let us henceforth consider the case of only a single type of unit; we shall return to the multi-type case in Chapters 16 and 17.

The complete description of the configuration $\mathscr{C}$ of N given units is then the bond-graph on N identifiable nodes. Let us label these nodes $a = 1, 2, \ldots, N$; we shall often denote a typical pair of nodes by a, b. Despite the fact that we shall often not distinguish physically between an ab bond and a ba bond, it turns out to be mathematically natural to make a distinction, initially at least, and so to regard the graph $\mathscr{C}$ as a *directed* graph. Specification of $\mathscr{C}$ is equivalent to specification of $\{s_{ab}; a, b = 1, 2, \ldots, N\}$, where s_{ab} is the number of bonds from unit a to unit b. If one likes, s_{ab} is the number of bonds between these two units which are *initiated* by unit a.

For a given configuration $\mathscr{C}$ we shall denote the total number of bonds formed by B. This can be written

$$B = \sum_a \sum_b s_{ab} \tag{1}$$

A component of $\mathscr{C}$ is a sub-graph whose nodes are all connected (at some remove), but which have no connection with the other nodes of $\mathscr{C}$. The components are then just the *polymers*; we shall denote their number by C. One has

$$B + C \geqslant N \tag{3}$$

with equality if and only if the components of $\mathscr{C}$ are all trees. (See Exercise 1.)

We now require a model which determines the equilibrium distribution of $\mathscr{C}$; from this distribution we hope then to determine the equilibrium characteristics of the process at a grosser level. Now, to specify a reversible dynamic model is virtually to specify an equilibrium distribution; in constructing the first one must have a form for the second in mind. We shall then take specification of the equilibrium distribution as the starting point, regarding this as the specification of the model in at least its equilibrium characteristics. This is not as arbitrary as it may seem; the distribution has to be seen as a Gibbs distribution for the configuration. Nor should it be thought that one has by-passed the essential difficulties of the problem by starting with a 'solution', any more than that prescription of an equiprobable distribution in phase-space in Chapter 5 implied that the real problems were then solved. The real problem is the deduction of bulk behaviour, i.e. of phenomena at a low but significant level of prescription. To specify a complete-description distribution is like specifying a differential equation or a transition probability; one still has to integrate (over time or over statistical variation) to deduce useful conclusions.

The distribution we shall postulate is one of the general form

$$P_N(\mathscr{C}) \propto Q_N(\mathscr{C}) := V^N \left[\prod_a \prod_b \frac{(h/2)^{s_{ab}}}{s_{ab}!} \right] \Phi(s)$$

$$= V^N \left[\frac{(h/2)^B}{s_{ab}!} \right] \Phi(s) \tag{4}$$

Here V is the volume of the region within which the N units are confined, h is a parameter which we shall regard as inversely proportional to volume

$$h = \frac{1}{\kappa V} \tag{5}$$

and $\Phi(s)$ is a function of $s = \{s_{ab}\}$ whose form we shall vary. Note that s and $\mathscr{C}$ are equivalent full descriptions of the configuration, for given N.

Let us consider the meaning of the factors of $Q_N(\mathscr{C})$ in order. The term V^N is constant, but becomes significant later, when we consider migration of polymers between sub-volumes.

The product of terms $(h/2)^{s_{ab}}/s_{ab}!$ would, on its own, state that the s_{ab} are independent Poisson variables, with common expectation $h/2$. This is, in a sense, our basic statistical assumption—that such a distribution would hold were it not for constraints and configuration-dependent energy effects. Now, we see the reason for choosing h as in (5): if N and V increase subject to constant unit density

$$\rho = N/V \tag{6}$$

then one will wish the total expected number of bonds initiated by a given unit to have a finite limit. We have in fact fixed this expectation as

$$Nh/2 = N/2\kappa V = \rho/2\kappa \tag{7}$$

One thinks of this basic Poisson distribution as being 'coloured' by the factor

$$\Phi(s) = e^{-U(s)/kT} \tag{8}$$

where $U(s)$ represents the potential energy of the configuration s. The quantity h must itself be related to the potential energy of a bond, but there will in general be other interactive effects, represented by $U(s)$.

The Poisson factor cannot be absorbed in $\Phi(s)$, because it represents itself a kind of zeroth-order model which incorporates two definite effects not appearing in Φ. First of all, the $1/V$ dependence of h reflects the volume-dependence of encounter rate between a given pair of units. In a model with more detailed spatial structure this would be represented differently (a unit would bond only with units in its neighbourhood, cf. section 5.7), but for the moment we omit spatial location from the description.

The second feature of the Poisson factor is the occurrence of the combinatorial factors $(s_{ab}!)^{-1}$. These occur because the bonds are potentially infinite in number and are not identified individually. In fact, the Poisson statistics are very nearly what one would get if the volume containing the N units were immersed in an infinite 'bond-bath', with bonds immigrating, emigrating, attaching and detaching independently (see Exercises 4 and 5).

No other combinatorial terms occur in the description at this level, unless we include other fundamental variations (e.g. that a unit can form a bond in more than one way—see Whittle, 1965b).

Distribution (4) of course allows both multiple bonding and self-bonding, but the occurrence of these becomes negligible in the thermodynamic limit, as h tends to zero. In general, there are a number of effects which one would be inclined to exclude on physical preconceptions, but which the mathematics make it natural to include. In such cases, one does well to follow the mathematics.

Exercises and comments

1. A tree is a connected graph with the minimal number of arcs; i.e. removal of an arc would create two components. It can thus have no cycles. Suppose a tree of n units requires $b(n)$ bonds. Breaking of one bond creates two new trees, whence we deduce that $b(n_1) + b(n_2) + 1 = b(n_1 + n_2)$, and $b(n) = n - 1$. From this it follows that, in the notation of the text, $B = N - C$ for a graph whose components are all trees. More generally, (3) holds for any graph, with equality if and only if the graph has no cycles, i.e. if all its components are trees.

2. Let $s + e_{ab}$ denote the value of s if s_{ab} is increased by one, and $\Lambda(s, s + e_{ab})$ the intensity of the transition $s \to s + e_{ab}$ (i.e. creation of an additional ab bond), etc. Then distribution (4) would be the equilibrium distribution for a reversible dynamic model for which

$$\frac{\Lambda(s, s + e_{ab})}{\Lambda(s + e_{ab}, s)} = \frac{\Phi(s + e_{ab})}{\Phi(s)} \left(\frac{1}{2\kappa V} \right) \left(\frac{1}{s_{ab} + 1} \right)$$

The Φ-ratio represents the ratio of Gibbs factors, the term in V^{-1} the rate of ab encounter in bond formation, and the $(s_{ab} + 1)$ term the fact that there are this many bonds to choose from in bond dissolution.

3. Consider the model for which

$$\Lambda(s, s + e_{ab}) = \frac{v}{2V}$$

$$\Lambda(s, s - e_{ab}) = \mu s_{ab}$$

This has an equilibrium distribution of the Poisson form

$$P(s) \propto \prod_{a,b} \frac{(h/2)^{s_{ab}}}{s_{ab}!}$$

with $h = v/\mu V$. The assumption is that a given pair a,b has intensity $v/2V$ of acquiring a directed bond; all bonds have independent intensity μ of dissolution.

4. *The bond-bath.* A more explicit mechanism for the reaction of Example 3 is to assume that bonds may exist on their own, having one or both ends free. Let a free end be interpreted as bound to a dummy node $a = 0$, and let s be the full set $\{s_{ab};\ a,b = 0,1,2,\ldots,N\}$. Then s_{00} is the number of free bonds, s_{a0} the number of bonds from a with female end free, and s_{0b} the number of bonds to b with male end free. Let us assume the transition intensities $\Lambda(s,s')$ specified by

$s' - s$	$\Lambda(s,s')$
e_{00}	vV
$-e_{00}$	μs_{00}
$e_{a0} - e_{00}$	$\lambda_m s_{00}/V$
$e_{0b} - e_{00}$	$\lambda_f s_{00}/V$
$-e_{a0} + e_{00}$	$\mu_m s_{a0}$
$-e_{0b} + e_{00}$	$\mu_f s_{0b}$
$e_{ab} - e_{a0}$	$\lambda_f s_{a0}/V$
$e_{ab} - e_b$	$\lambda_m s_{0b}/V$
$-e_{ab} + e_{a0}$	$\mu_f s_{ab}$
$-e_{ab} + e_{0b}$	$\mu_m s_{ab}$

The first two transitions represent immigration or emigration of free bonds, the next four represent gain or loss of the first attachment of a bond, the final four represent gain or loss of the second attachment. Show that in equilibrium the s_{ab} are independent Poisson with

$$E(s_{ab}) = \begin{cases} \dfrac{vV}{\mu} & (a = b = 0) \\[2ex] \dfrac{\lambda_m}{\mu_m} \dfrac{v}{\mu} & (b = 0,\ a > 0) \\[2ex] \dfrac{\lambda_f}{\mu_f} \dfrac{v}{\mu} & (a = 0,\ b > 0) \\[2ex] \dfrac{\lambda_m \lambda_f}{\mu_m \mu_f} \dfrac{v}{\mu} \dfrac{1}{V} & (a,b > 0) \end{cases}$$

6. GELATION IN THE POISSON MODEL

We shall refer to distribution (5.4) as a 'model', since, although not specifying a dynamic model, it is sufficient as a starting point for the study of equilibrium behaviour. We shall refer to the case when $\Phi(s) \equiv 1$ and

$$P_N(\mathscr{C}) \propto Q_N(\mathscr{C}) = V^N \prod_{\alpha s} \prod \frac{(h/2)^{s_{ab}}}{s_{ab}} \tag{1}$$

as the *Poisson model*.

In the next chapter we shall find ways of deducing the polymer statistics and hence studying gelation. However, if one is interested only in testing whether the process is in the gel state or not, then there is a very quick and intuitive method of doing so.

Let us suppose that, rather than having N units in a region of volume V, we have $2N$ units in two compartments each of volume V. We shall make the additional assumption, that

$$Q_{2N}(\mathscr{C}) = Q_{N_1}(\mathscr{C}_1) Q_{N_2}(\mathscr{C}_2) \qquad (N_1 + N_2 = 2N) \tag{2}$$

where $\mathscr{C}_i$ is a configuration of N_i units in compartment i ($i = 1, 2$) and the individual factors both have the form (1).

The fact that we write $\mathscr{C}$ as $(\mathscr{C}_1, \mathscr{C}_2)$ implies that we allow no bonding between units in different compartments: $\mathscr{C}_1$ and $\mathscr{C}_2$ are separated graphs. The fact that we are allowing N_1 and N_2 to vary subject to $N_1 + N_2 = 2N$ indicates that we are allowing free migration between the two compartments. The migrations must be of complete polymers, however, since only within-compartment bonding is permitted.

Model (2) is an extension of model (1), with the most modest intimation of spatial structure. It turns out to be consistent, however, in that the two models gel under the same conditions (with reservations; see sections 15.4, 15.5).

Let us define

$$Q_N = \sum_{\mathscr{C}} Q_N(\mathscr{C}) \tag{3}$$

where the sum is over all configuration of N units. For the Poisson case (1) with $h = (\kappa V)^{-1}$ we have

$$Q_N = V^N e^{N^2/2\kappa V} \tag{4}$$

The probability that the $2N$ units distribute themselves in a given way between the two compartments, regardless of configuration within compartments, is then proportional to $Q_{N_1} Q_{N_2}$. The units are still identified. If we sum over all permutations of units consistent with the N_1, N_2 allocation we obtain

$$\dot{P}(N_1, N_2) \propto \frac{Q_{N_1} Q_{N_2}}{N_1! N_2!} \tag{5}$$

where $P(\cdot)$ is used to represent the equilibrium distribution of whichever quantity is indicated. Let us write (N_1, N_2) as $(N + n, N - n)$, so that n represents the imbalance in distribution between the two compartments. By (4), (5) this has the distribution

$$P(n) \propto \frac{\exp(n^2/\kappa V)}{(N + n)! \, (N - n)!} \qquad (-N \leqslant n \leqslant N) \qquad (6)$$

Theorem 6.1 *Distribution* (5) *has a single maximum at* $n = 0$ *or symmetric maxima each side of* $n = 0$ *according as* $\rho < \kappa$ *or* $\rho > \kappa$.

More specifically, in the thermodynamic limit the random variable n/V *has a distribution in which it takes values* $\pm \delta$ *with equal probability, where* $\delta = 0$ *for* $\rho < \kappa$ *and* $\delta > 0$ *for* $\rho > \kappa$.

Proof The distribution $P(n)$ is plainly symmetric, and one can test whether $n = 0$ is a local minimum or maximum of the distribution by comparing $P(0)$ with $P(1)$. This yields the criteria of the first assertion.

In fact, if one sets $\delta = n/V$ then for large V we have $V^{-1} \log P(n) \sim$ const $+ f(\delta)$ where

$$f(\delta) := \frac{\delta^2}{\kappa} - (\rho + \delta) \log(\rho + \delta) - (\rho - \delta) \log(\rho - \delta)$$

This expression has turning points at the roots of

$$2\delta = \kappa \log\left(\frac{\rho + \delta}{\rho - \delta}\right)$$

For $\rho < \kappa$ this has only the solution $\delta = 0$ (a maximum of $f(\delta)$) and for $\rho > \kappa$ has roots at 0 and at non-zero values $\pm \delta_0$ (respectively a minimum and symmetrically-placed maxima of $f(\delta)$). Given the dependence of $P(n)$ upon V and $f(\delta)$, the conclusions of the theorem follow. ■

The interpretation is then that there is a critical density $\rho_c = \kappa$. If $\rho < \rho_c$ there is statistical equidistribution of matter between the two compartments, but if $\rho > \rho_c$ then matter tends to 'lump' in one compartment or the other. This lumping becomes total in the thermodynamic limit (in that all but a vanishing fraction concentrates in one of the compartments).

One construes this 'lumping' as the gel-state; matter is in a single complex which can only be in one compartment or the other. The value ρ_c is the critical density of gelation.

Pictorially, the situation is as in Fig. 13.6.1. As often, a critical transition is marked by a splitting (of a peak in a distribution, of an eigenvalue, of a spectral line). Symmetry is retained after gelation, in that the gel mass is equally likely to reside in one compartment or the other. On the other hand, symmetry is broken in that equidistribution is lost; the mass will reside in only one of the compartments at a given time.

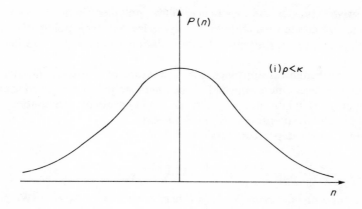

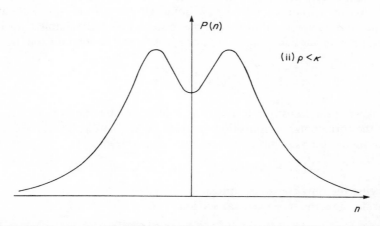

Fig. 13.6.1 The occurrence of gelation in the Poisson model. The variable n is the difference in the numbers of units in the two compartments. For density ρ less than the critical value κ matter is equidistributed between the two compartments, as indicated in (i). For ρ greater than κ matter tends to lump itself in one compartment or the other, as indicated in (ii).

7. EQUILIBRIUM SOLUTIONS AS TIME-DEPENDENT SOLUTIONS

Flory's original approach (see section 8) was not to consider the equilibrium regime of a process of association and dissociation, but rather to consider the time-dependent version of a purely associative process. That is, a process in which bonds form but do not break. He then deduced the occurrence of gelation at a time when the degree of bonding had reached the critical value.

This view expresses the experience of the polymer chemist: when physical conditions are created which favour polymerization, then polymerization proceeds progressively, and, after some time, gelation is observed as a rather definite event.

Flory's mathematical approach was a mixture of statistical mechanics and heuristics. However, one is impelled to ask whether equilibrium distributions such as (6.1) or (5.4) might not be seen as time-dependent distributions for an appropriate stochastic process of pure bonding.

Consider a time-dependent version of (6.1):

$$Q(s, t) = V^N \prod_a \prod_b \frac{(\lambda t/2V)^{s_{ab}}}{s_{ab}!} \tag{1}$$

Here the bonding parameter $h = 1/\kappa V$ has been replaced by $\lambda t/V$. We have also dropped the N subscript for simplicity.

Distribution (1) implies that the s_{ab} are independent Poisson variables with expectation $\lambda t/2V$ at time t. This is exactly the distribution that would hold for a process, completely dissociated at $t = 0$, in which for any a, b there is a fixed probability intensity $\lambda/2V$ that a new bond will form from unit a to unit b, independent of the previous history of the process. This is the simplest stochastic bonding process one can imagine, and one sees from Theorem 6.1 that it gels at time

$$t = \frac{1}{\lambda \rho}$$

Let us write the transition in which s_{ab} is increased by one as $s \to s + e_{ab}$, and denote the corresponding transition intensity as $\Lambda(s, s + e_{ab})$. Then the dynamic Poisson model we have just considered is characterized by

$$\Lambda(s, s + e_{ab}) = \lambda/2V \qquad (a, b = 1, 2, \ldots, N)$$

these being the only transitions possible.

Consider now the more general version

$$Q(s, t) = \left[V^N \prod_a \prod_b \frac{(\lambda t/2V)^{s_{ab}}}{s_{ab}!} \right] \Phi(s) \tag{2}$$

corresponding to (5.4). This can indeed be seen as the time-dependent solution, either exactly or in the thermodynamic limit, of a natural stochastic bonding process.

Theorem 7.1 *Consider the pure bonding process for which*

$$\Lambda(s, s + e_{ab}) = \frac{\lambda}{2V} \frac{\Phi(s + e_{ab})}{\Phi(s)} \tag{3}$$

and let $P(s, t)$ denote the distribution of s at time t for this process, given complete dissociation at time $t = 0$. Let

$$P^*(s, t) = \frac{Q(s, t)}{\sum_s Q(s, t)} \tag{4}$$

denote the normalized version of expression (2). Then P, P^* respectively satisfy

$$\frac{\partial P(s, t)}{\partial t} = \sum_a \sum_b P(s - e_{ab}, t) \Lambda(s - e_{ab}, s) - \Lambda(s) P(s, t) \tag{5}$$

$$\frac{\partial P^*(s, t)}{\partial t} = \sum_a \sum_b P^*(s - e_{ab}, t) \Lambda(s - e_{ab}, s) - \Lambda^*(t) P^*(s, t) \tag{6}$$

where

$$\Lambda(s) := \sum_a \sum_b \Lambda(s, s + e_{ab}) = \frac{\lambda}{2V \Phi(s)} \sum_a \sum_b \Phi(s + e_{ab}) \tag{7}$$

$$\Lambda^*(t) := \sum_s P^*(s, t) \Lambda(s) \tag{8}$$

If for all t $\Lambda(s)$ is constant (in s) for all s such that $P^*(s, t) > 0$ then $P(s, t) = P^*(s, t)$.

Proof This is essentially an application of Theorem 3.6.1. Equations (5), (7) constitute the usual Kolmogorov forward equation for the Markov process specified by (3). One readily verifies that expression (2) satisfies

$$\frac{\partial Q(s, t)}{\partial t} = \sum_a \sum_b Q(s - e_{ab}, t) \Lambda(s - e_{ab}, s) \tag{6}$$

and hence that its normalized version satisfies (6). The two equations (5), (6) evidently agree under the condition stated, and only under an almost-everywhere version of it. More explicitly, suppose we have established that $P(s, \tau) = P^*(s, \tau)$ for $\tau \leqslant t$. To establish equality of rate of change we then require, by (5) and (6), that

$$(\Lambda(s) - \Lambda^*(t)) P^*(s, t) = 0$$

which is equivalent to the condition of the theorem. ∎

The condition is quite a restrictive one. Unless there are configurations which are actually forbidden then $P^*(s, t) > 0$ for all s and all $t > 0$, and the condition of the theorem would require that

$$\bar{\Lambda}(s) := \frac{1}{N} \Lambda(s) \propto \frac{1}{N^2} \sum_a \sum_b \Phi(s + e_{ab})/\Phi(s)$$

be independent of s (see Exercise 1). However, one can imagine that, under appropriate conditions on Φ, the random variable $\bar{\Lambda}(s)$ would converge (in the appropriate probabilistic sense) to a constant in the thermodynamic limit under the distribution $P^*(s, t)$ for fixed t. That is, that in the thermodynamic limit the condition of the theorem would be satisfied and P, P^* would be equal in some acceptable asymptotic sense. The argument is plausible, although closer analysis is obviously required to justify it.

Exercises and comments

1. Show that the conditions of the theorem are satisfied if $\Phi(s)$ can be represented

$$\Phi(s) = \int \prod_a \prod_b \lambda_{ab}(\zeta)^{s_{ab}} \phi(d\zeta)$$

where $\sum_a \sum_b \lambda_{ab}(\zeta)$ is independent of ζ for all ζ occurring in the integral. Interpret.

8. LITERATURE REFERENCES

The literature on polymers is of course enormous, but we are concerned only with the literature on what one might term bonding statistics. The classic pioneering work on polymer statistics is due to Flory (1953, and references there quoted) and Stockmayer (1943, 1944). Both of these authors assumed a model of pure association whose parameters changed deterministically with time, and whose statistics were assumed to adapt immediately to the corresponding Gibbs equilibrium distribution. An alternative approach due to Gordon (1962) and Good (1963) saw the mechanism as that of a branching process. Both of these approaches require, for mathematical tractability, that the polymers all have trees as their bond-graphs.

The approach adopted in these chapters (13–17) is that developed by the author over several years. It is the only one which constructs a Markov model of the mechanism and determines its equilibrium behaviour. It is also the only one which dispenses with the restriction to trees, with indeed an actual simplification in treatment. However, reassuringly, all approaches seem to use mathematics which is essentially similar, and to produce common statistics for common cases.

The study of random graphs was begun by Erdös and co-authors (see Erdös and Rényi, 1960; Erdös and Spencer, 1974) and now has a large literature (see Bollobas, 1985). This study at least began in ignorance of existing work on polymers, but inevitably soon also found interest in critical effects; abstract analogues of gelation.

The material of sections 2–4 of this chapter was first presented in Whittle (1965a), of section 5 in Whittle (1981), of section 6 in Whittle (1980a, b) and of section 7 in Whittle (1984c).

CHAPTER 14

Polymer and unit statistics

1. THE FUNDAMENTAL LEMMA

For this and the next chapter we shall keep to the case of a single type of unit.

The following combinatorial lemma plays a role, implicit or explicit, in a variety of contexts (see, for example, Percus, 1971, p. 52). It could be said to explain occurrences of Poisson statistics in an open process. We shall quote it in a form suitable for present purposes.

Lemma 1.1 *Suppose that to each graph $\mathscr{C}$ on an arbitrary number of labelled nodes can be attached a weight $W(\mathscr{C})$ with the properties (i) that $W(\mathscr{C})$ is invariant under permutation of the nodes, and (ii) that $W(\mathscr{C}) = W(\mathscr{C}_1)W(\mathscr{C}_2)$ if $\mathscr{C}$ can be decomposed into mutually unconnected graphs $\mathscr{C}_1$, $\mathscr{C}_2$. Then*

$$\sum_{\mathscr{C}} W(\mathscr{C})/N(\mathscr{C})! = \exp\left[\sum_i W(\mathscr{C}^{(i)})\right] \tag{1}$$

where $\sum_{\mathscr{C}}$ covers all graphs, $N(\mathscr{C})$ is the number of nodes in $\mathscr{C}$, the term for $N(\mathscr{C}) = 0$ is taken as unity and $\mathscr{C}^{(i)}$ $(i = 1, 2, \dots)$ are the distinct connected graphs.

Proof Set $W(\mathscr{C}^{(i)}) = W_i$, as that, by (ii),

$$W(\mathscr{C}) = \prod_i W_i^{m_i(\mathscr{C})}$$

where $m_i(\mathscr{C})$ is the number of times $\mathscr{C}^{(i)}$ occurs as a component of $\mathscr{C}$. We have then

$$Q_N := \sum_{\mathscr{C}}^N W(\mathscr{C}) = \sum_{\mathscr{C}}^N \prod_i W_i^{m_i(\mathscr{C})}$$

$$= N! \sum_m^N \prod_i \frac{W_i^{m_i}}{m_i!} \tag{2}$$

where Σ^N denotes a sum consistent with a total of N nodes. The final equality in (2) follows from (i): by summing $W(\mathscr{C})$ over all permutations of nodes consistent with prescribed values $m_i(\mathscr{C}) = m_i$ $(i = 1, 2, \dots)$. There are $N!$ permutations, but only $N!/\prod_i m_i!$ are distinct, because any individual identification of the m_i components of type i is arbitrary.

303

It follows then from (2) that

$$\sum_{\mathscr{C}} W(\mathscr{C})/N(\mathscr{C})! = \sum_N Q_N/N! = \exp\left(\sum_i W_i\right), \qquad (3)$$

equivalent to (1). ∎

Since the $\mathscr{C}^{(i)}$ are to be identified with polymers, the usefulness of the assertion is plain. However, only at the highest level of description will one regard each $\mathscr{C}^{(i)}$ as a distinct type of polymer. A polymer A_r in general corresponds to a whole class of isomers, of those $\mathscr{C}^{(i)}$ which are regarded as equivalent at the level of description r. The quantity r labels equivalence classes at the level of description adopted. Let us then refer to it as the *description*, and denote the description of $\mathscr{C}^{(i)}$ by $r(i)$. A minimal level of description would seem to be the size of the polymer, so let us suppose that all $\mathscr{C}^{(i)}$ of description r contain the same number of units (nodes) a_r.

Theorem 1.1 *Let $P_N(\mathscr{C})$ denote the probability of configuration $\mathscr{C}$ on N nodes, and let $\Pi_N(z) = E\left(\prod_r z_r^{n_r}\right)$ be the joint p.g.f. of the numbers n_r of r-mers in this configuration. Suppose that*

$$P_N(\mathscr{C}) \propto Q_N(\mathscr{C}) := \prod_i F_i^{m_i(\mathscr{C})} \qquad (4)$$

where $m_i(\mathscr{C})$ is the number of components of $\mathscr{C}$ of the form $\mathscr{C}^{(i)}$. Then $\Pi_N(z)$ is proportional to the coefficient of w^N in the power series expansion of

$$\exp\left[\sum_i F_i z_{r(i)} w^{a_{r(i)}}\right] = \exp\left[\sum_r \Gamma_r z_r w^{a_r}\right] \qquad (5)$$

where Γ_r is the sum of F_i over all i such that $r(i) = r$.

Proof Define, for $N(\mathscr{C}) = N$,

$$W(\mathscr{C}) = Q_N(\mathscr{C}) w^N \prod_r z_r^{n_r}$$

$$= \prod_i W_i^{m_i(\mathscr{C})} \qquad (6)$$

where

$$W_i = F_i z_{r(i)} w^{a_{r(i)}} \qquad (7)$$

$\Pi_N(z)$ is proportional to the coefficient of w^N in $\sum_{\mathscr{C}} W(\mathscr{C})/N(\mathscr{C})!$; by (6), (7) and Lemma 1.1 this sum has the evaluation (5). ∎

The assumption behind the form of distribution (4) is that it is a Gibbs distribution which expresses absence of interaction between components of the graph (i.e. between polymers) in that the energy of the total configuration is the sum of energies associated with individual components. Of course, there is dynamic interaction, in that components (molecules) join and split, but this is a weak coupling which leads to independence in open equilibrium statistics. Formula (5) expresses the familiar independent Poisson statistics of the open process.

A comparison of (5) and (13.3.2) leads one to expect the identification

$$\Gamma_r = V\gamma_r.$$

This is a point to which we shall return in sections 5 and 7.

If one adopts the natural level of description then one should not need the explicit marker variables z_r. There should then be parameters of the energy function which can serve as effective marker variables. Otherwise expressed, Γ_r should be determinable from the sum $\sum_i F_i = \sum_r \Gamma_r$, in that it should be identifiable with a term of a particular form in this sum. We shall see examples of this in the next few sections.

2. THE FIRST-SHELL MODEL

The Poisson model (13.6.1) is indeed of the form (1.4). It is also the simplest model of any interest, and interesting enough that we shall spend all of Chapter 17 studying multi-type versions of it. However, physical realism demands that we move on to models showing more structure and interaction.

A natural next step up is to assume that

$$P_N(\mathscr{C}) \propto Q_N(\mathscr{C}) = V^N \left[\prod_a \prod_b \frac{(h/2)^{S_{ab}}}{S_{ab}!} \right] \left[\prod_j H_j^{N_j} \right] v^{B+C-N} \tag{1}$$

where we recall (see section 13.5) that B and C represent respectively the number of bonds and of polymers in the configuration of N units. The first bracket represents the Poisson statistics of (13.6.1), with h still having the volume-dependent evaluation $(\kappa V)^{-1}$. We define N_j as the number of units which have formed exactly j bonds (in either direction); let us call such a unit a u_j. The second bracket introduces then a dependence on the number of bonds which units form. For example, we have considered already in Exercise 13.3.3 a case where a unit had f equivalent and independent attachment sites available, and it proved natural then to choose

$$H_j = \binom{f}{j} \sigma^j$$

The binomial factor is combinatorial, corresponding to a summation over several ways of making j bonds. In general H_j will also contain a term reflecting the energy of the j-bond configuration. For the Poisson model, the intensity of formation of a new bond by a unit was independent of the number of bonds it had already formed; this is no longer so.

Such an effect is termed a 'first-shell' effect, in that it is an effect dependent on a unit's immediate neighbours (i.e. on the units to which it is directly bound). Second-shell effects would depend upon neighbours at one remove, etc.

The final term, v^{B+C-N}, reflects a difference in the probability of intramolecular and intermolecular bond formation. Suppose a new bond were added. Then this final factor would change by a factor v if the bond were one internal to an existing polymer (when $B+C-N$ would increase by one), but not change at all if the bond were to join distinct existing polymers (when $B+C-N$ would not change).

If v were set equal to zero, then intramolecular bond formation would be prohibited altogether. The only permitted configurations would be those for which $B+C=N$, and all the components would be trees (Exercise 13.5.1). That is, all the polymers would be purely branching, with no cycles.

The distribution (1) is indeed of the form (1.4) with

$$
\begin{aligned}
F &= v\left(\frac{V}{v}\right)^{R}(vh)^{L}\prod_{j}H_{j}^{r_{j}} \\
&= v\left(\frac{v}{V}\right)^{L-R}(2\kappa)^{-L}\prod_{j}H_{j}^{r_{j}}
\end{aligned}
\tag{2}
$$

where R is the number of units in the component, L the number of bonds, and r_{j} the number of $u_{j}(j=0, 1, 2, \ldots)$.

The statistics R, L and $r = \{r_{0}, r_{1}, r_{2}, \ldots\}$ are all dependent upon the form of the particular polymer (component), and the natural level of description is that for which these quantities have prescribed values. Then (V/v) is a marker variable for the number of units, vh for the number of bonds, H_{j} for the number of u_{j}, and v for the number of components.

In fact, prescription of the r_{j} implies prescription of R and L, for

$$
\begin{aligned}
R &= \sum_{j}r_{j} \\
L &= \tfrac{1}{2}\sum_{j}jr_{j}
\end{aligned}
\tag{3}
$$

(the half occurring because the sum counts bonds twice—once from each end). So, for model (1) the statistic $r = \{r_{0}, r_{1}, r_{2}, \ldots\}$ indeed constitutes the natural level of description of a polymer A_{r}.

3. UNIT STATISTICS FOR THE FIRST-SHELL MODEL

The simplest task, before examining polymer structure, is to determine the distribution of the obvious random variables in expression (2.1), the N_{j}. This is what we mean by 'unit statistics'. In particular, by evaluating $E(N_{j})/N$ we determine the distribution of the numbers of bonds which units form. Note that, corresponding to formulae (2.3)

$$
N = \sum_{j}N_{j}
$$

$$B = \tfrac{1}{2}\sum_j jN_j \tag{1}$$

We have already observed that the H_j, as well as being parameters of the physical model, serve as marker variables for the N_j. Specifically, regard the sum $Q_N = \sum_{\mathscr{C}} Q_N(\mathscr{C})$ as a function

$$Q_N = Q_N[H] = Q_N[\{H_j\}]$$

of the H_j. Then $P(\{N_j\})$ is proportional to the term in $\prod_j H_j^{N_j}$ in the expansion of $Q_N[H]$. Otherwise expressed:

$$E\left(\prod_j \zeta_j^{N_j}\right) = \frac{Q_N[\{H_j\zeta_j\}]}{Q_N[\{H_j\}]} \tag{2}$$

This has as consequence, for example, that

$$EN_j = H_j \frac{\partial \log Q_N}{\partial H_j} \tag{3}$$

For simplicity, we shall investigate matters first for the case of equal intra- and inter-molecular reaction rates, $v = 1$, and return to the more general case in section 4.

Theorem 3.1 Evaluation of unit statistics for the first-shell model. *Assume the form (2.1) for the unnormalized equilibrium distribution $Q_N(\mathscr{C})$, but with $v = 1$. Then the sum $Q_N = Q_N[H]$ has the evaluation*

$$Q_N = V^N \sqrt{\frac{\kappa V}{2\pi}} \int_{-\infty}^{\infty} H(\xi)^N e^{-\kappa V \xi^2 / 2} \, d\xi \tag{4}$$

where

$$H(\xi) := \sum_{j=0}^{\infty} \frac{H_j \xi^j}{j!} \tag{5}$$

Proof Suppose that

$$H_j = \int x^j m\,(dx) \tag{6}$$

for some signed measure m on the real line. Then

$$Q_N = V^N \int \cdots \int \sum_s \left[\prod_a \prod_b \frac{(hx_a x_b/2)^{s_{ab}}}{s_{ab}!} \right] \left[\prod_a m\,(dx_a) \right]$$

$$= V^N \int \cdots \int \exp\left[\tfrac{1}{2} h \left(\sum_a x_a \right)^2 \right] \prod_a m\,(dx_a) \tag{7}$$

Appeal now to the identity

$$\exp\left(\tfrac{1}{2}h\sum{}^2\right) = \frac{1}{\sqrt{2\pi h}} \int_{-\infty}^{\infty} \exp\left[\xi\sum - \xi^2/2h\right] d\xi \tag{8}$$

Setting $\sum = \sum_a x_a$ in (8) and substituting from (8) into (7) we deduce expression (4) with

$$H(\xi) = \int e^{x\xi} m\,(dx) = \sum_0^\infty \frac{H_j \xi^j}{j!}$$

(Recall also that $h = (\kappa V)^{-1}$.) ∎

It is useful to appeal to a formal representation (6), but (4) is valid as an identity in the H_j whether or not the representation holds at some recognized level of regularity.

Corollary 3.1 *The unit totals N_j have the joint distribution*

$$P(\{N_j\}) \propto \frac{(2B)!}{B!\,(2\kappa V)^B} \prod_j \frac{(H_j/j!)^{N_j}}{N_j!} \qquad \left(\sum_j N_j = N\right) \tag{9}$$

Proof Since $P(\{N_j\})$ is the term in $\prod_j H_j^{N_j}$ in expression (4) we deduce that

$$P(\{N_j\}) \propto \prod_j \frac{(H_j/j!)^{N_j}}{N_j!} \int \xi^{\Sigma jN_j} e^{-\kappa V \xi^2/2} d\xi \tag{10}$$

Now, $\Sigma jN_j = 2B$, and so we derive (9) from (10) by standard integration formulae. ∎

Expression (9) is always correct, in that it is always proportional to the sum of expression (2.1) over all $\mathscr{C}$ consistent with prescribed $\{N_j\}$. However, it is possible that the distribution is not summable. Even though the N_j are all finite (for prescribed N) the label j can be infinite, and it is possible that the distribution has positive mass on infinite j. That is, that the tendency to bonding is so extreme that some units have formed infinitely many bonds. The following condition is sufficient to exclude this possibility.

Theorem 3.2 *The distribution (9) will be summable for given N, and so j finite with probability one, if $\log(H(\xi))$ is of less than quadratic growth in ξ at infinity. That is, if*

$$\lim_{|\xi| \to \infty} \left[\frac{\log(H(\xi))}{|\xi|^2} \right] = 0 \tag{11}$$

Proof The sum $\sum P(\{N_j\})$ and the integral (4) representing it will converge or diverge together. Condition (11) is sufficient to ensure convergence of the integral, because it ensures that for arbitrarily small positive β

$$|H(\xi)| < e^{\beta \xi^2/2}$$

for large enough $|\xi|$. ∎

Condition (11) implies a very mild constraint on the rate of growth of H_j with increasing j (and so on the extent to which existence of bonds encourages formation of further bonds). For $\log H(\xi)$ to equal $\frac{1}{2}\beta\xi^2$ precisely would imply that

$$H_{2j} = \frac{(2j)!\,(\beta/2)^j}{j!} \sim j^j \left(\frac{2\beta}{e}\right)^j$$

That is, the configuration energy of a u_j is of order $j \log j$, which is more than one would normally expect on physical grounds.

Corollary 3.2 *Suppose $\log H(\xi)$ of less than quadratic growth at infinity, and let $\bar{\xi}$ be the value of ξ maximizing $\rho \log H(\xi) - \kappa\xi^2/2$. Then the p.g.f. of the number of bonds j formed by a unit is*

$$E(\zeta^j) = \frac{H(\bar{\xi}\zeta)}{H(\bar{\xi})} \tag{11}$$

in the thermodynamic limit.

Proof We have, by (3), (4)

$$E(N_j) \propto \int \frac{H_j \xi^j}{j!} H(\xi)^{N-1} e^{-\kappa V \xi^2/2} \sim \text{const.}\,\frac{H_j \bar{\xi}^j}{j!}$$

whence the assertion follows.

The special case of the Poisson model (13.6.1) corresponds to $H_j = 1$, or

$$H(\xi) = e^\xi \tag{12}$$

The model with

$$H_j = \sigma^j \binom{f}{j} \tag{13}$$

for integral f corresponds, as we have seen in Exercise 13.3.3, to the assumption that a unit has f sites at which it can form single bonds, attachment at different sites being equiprobable and independent. This is referred to in the literature as the *equireactive f-functional* unit. We see that for this model

$$H(\xi) = (1 + \sigma\xi)^f \tag{14}$$

Exercises and comments

1. Consider the Poisson model. It follows from (12) and Corollary 3.2 that the number of bonds a unit forms is Poisson distributed with expectation $\bar{\xi} = \rho/\kappa$.

This is indeed no more than was assumed in the original model. The total number of bonds a unit forms is $\sum_b (s_{ab} + s_{ba})$, which is Poisson distributed with parameter $2N(2\kappa V)^{-1} = \rho/\kappa$.

2. Consider the f-functional model. We see from (14) and Corollary 3.2 that the number of bonds a unit forms is binomially distributed: f independent trials with a bond probability on each of

$$p = \frac{\sigma \bar{\xi}}{1 + \sigma \bar{\xi}} \tag{15}$$

Here p is determined by

$$\frac{p}{(1-p)^2} = \frac{\rho f \sigma^2}{\kappa} \tag{16}$$

We shall see (Exercise 4.2 and Theorem 8.1) that the process gels when p becomes as large as $(f-1)^{-1}$. This corresponds to a critical density of

$$\rho = \frac{\kappa(f-1)}{\sigma^2 f(f-2)^2} \tag{17}$$

3. Consider the model for which $H(\xi) = \cosh(\xi)$. Since this has an expansion in powers of ξ^2, units are permitted to have only an even number of bonds. The value $\bar{\xi}$ maximizing

$$\rho \log \cosh(\xi) - \kappa \xi^2/2$$

is zero for $\rho \leqslant \kappa$, positive for $\rho > \kappa$. That is, virtually no bonds form until ρ reaches a critical value, κ. The process also gels at the same point; see Corollary 4.1.

One might think that bonding was impossible for this model, because the fact that H_1 is zero implies that a unit cannot form a single bond, and so there is no reaction path by which it can form the permitted even number of bonds. If we considered rather a model with

$$H(\xi) = \tfrac{1}{2}[(1+\varepsilon)e^{\xi} + (1-\varepsilon)e^{-\xi}]$$

then a reaction path would exist for ε positive. However, for ε small enough the function $\rho \log H - \kappa \xi^2/2$ again displays the phenomenon that $\bar{\xi}$ jumps discontinuously from a small value to a larger one as ρ increases through a critical value. In other words, the behaviour for $\varepsilon = 0$ is identical with the limit behaviour for $\varepsilon \downarrow 0$.

4. Suppose the bond direction does matter, so that $\prod_j H_j^{N_j}$ in (2.1) is replaced by $\prod_j \prod_k H_{jk}^{N_{jk}}$, where j and k are respectively the numbers of outgoing and incoming bonds for a given unit. Generalize the treatment of this section to that case.

4. GELATION

We shall move on to polymer statistics in the next section. However, now that we have expression (3.4) for Q_N, we can apply the argument of section 13.6 to deduce very quickly a criterion for gelation. Let us write the integrand of (3.4) as $\exp VJ(\xi)$, so that

$$J(\xi) = \rho \log H(\xi) - \kappa \xi^2 / 2 \qquad (1)$$

The quantity $\bar{\xi}$ is then determined as the root of $J' = 0$, or of

$$\rho \frac{H'}{H} = \kappa \xi \qquad (2)$$

Here the prime indicates differentiation with respect to ξ, and all functions of ξ are evaluated at $\bar{\xi}$. Note that

$$J'' = \rho \frac{H''}{H} - \rho \left(\frac{H'}{H} \right)^2 - \kappa \qquad (3)$$

Theorem 4.1 *Consider the first-shell model in the case $v = 1$, and suppose that $\log H(\xi)$ is of less than quadratic growth at infinity. The criterion that matter should be in the sol state is that the value $\bar{\xi}$ maximizing $J(\xi)$ (and so satisfying (2)) should satisfy*

$$\rho \frac{H''}{H} < \kappa \qquad (4)$$

Gelation occurs when equality holds in (4).

Proof Define

$$I(\phi) = \int \phi(\xi) \exp[VJ(\xi)] \, d\xi \qquad (5)$$

One finds then that

$$\frac{I(\phi)}{I(1)} = \phi + \tfrac{1}{2} v \phi'' + o(V^{-1}) \qquad (6)$$

where ϕ, ϕ'' are evaluated at $\bar{\xi}$, and

$$v = -\frac{1}{VJ''} \qquad (7)$$

(see Exercise 1). From (6) it follows that

$$\frac{Q_{N-1}Q_{N+1}}{Q_N^2} = \frac{I(H)I(H^{-1})}{I(1)^2} = 1 + v \left(\frac{H'}{H} \right)^2 + o(V^{-1}). \qquad (8)$$

Now, by the argument of section 13.6, the criterion that matter should be in the sol state is that

$$\frac{Q_{N-1}Q_{N+1}}{Q_N^2} < 1 + \frac{1}{N}$$

or

$$v\left(\frac{H'}{H}\right)^2 < \frac{1}{N}$$

for large V. Substituting for v in (9) from (7), (3) we deduce the criterion (4).

In the next chapter we shall improve this argument by clarifying its physical basis and by conducting the argument *under* the integral of (3.4). This then puts matters directly in terms of the properties of the function $J(\xi)$, and gives a much more elegant and powerful treatment.

We saw from the 'cosh' model of Exercise 3.3 that, if $H_1 = 0$, so that a unit is constrained to have no bonds or more than one, then there is a transition point in $\bar{\xi}$. Specifically, as ρ increases, there is a value at which $\bar{\xi}$ jumps from zero to a positive value. We now see that this transition point coincides with the gel point.

Corollary 4.1 *Suppose that $H_1 = 0$, so that, as ρ increases, there is a value at which $\bar{\xi}$ jumps from zero to a positive value. This transition point is also the gel point.*

Proof We have $J'(0) = 0$, and $\xi = 0$ is a maximum of J as long as $J''(0) < 0$. This is exactly criterion (4), since J'' is given by (3), and $H'(0) = 0$. ∎

One can say that, for those processes in which $\bar{\xi}$ itself shows no transition at the gel point, the unit itself would not be aware from 'local' information that gelation had taken place. Local information would tell the unit that bonding was infinitesimally stronger, i.e. that he had infinitesimally more neighbours, but not that this infinitesimal increase has suddenly given him infinitely many neighbours at some remove.

For the models considered in Corollary 4.1 there is also a local change. The taboo on having a single neighbour means that there is a sudden jump from no neighbours to several (immediate) neighbours; this jump also coincides with the acquisition of infinitely many neighbours at some remove.

In Chapter 17 we shall encounter a much more natural version of the same phenomenon. In that chapter we allow units to have 'roles', which are not fixed. At gelation there is in general a discontinuous change also in the distribution of roles; both one's own and one's neighbours'.

Exercise and comments

1. The expression $I(\phi)/I(1)$ can, for large V, be regarded as the expectation of $\phi(\xi)$ on the assumption that ξ is normally distributed with mean $\bar{\xi}$ and variance v given by (7). Hence expression (6). A more careful analysis shows that the remainder is indeed $o(V^{-1})$.

2. Consider the f-functional model of Exercise 3.2, with H given by (3.14). Confirm the critical density (3.17) from criterion (4).

5. POLYMER STATISTICS FOR THE FIRST-SHELL MODEL

With Theorem 1.1 and evaluation (3.4) we can now deduce a central identity which formally yields the full polymer statistics. Recall from (1.5) the quantities $\Gamma_r w^{a_r}$, identifiable as $E(n_r)$ for an open process. Define the function

$$\Gamma(H, w) = \sum_r \Gamma_r w^{a_r} \qquad (1)$$

We adopt the description $r = (r_0, r_1, r_2, \dots)$ natural for the general first-shell model. The coefficient Γ_r is then determined as the term in $\prod_j (wH_j)^{r_j}$ in the expansion of this function, necessarily $a_r = \sum_j r_j$.

Theorem 5.1 *Consider the first-shell model (2.1) and suppose $\log H(\xi)$ of less than quadratic growth at infinity. Then the generating function (1) has the evaluation*

$$\exp(\Gamma(H, w)) = \left[\sqrt{\frac{\kappa V}{2\pi v}} \int_{-\infty}^{\infty} \exp\left[\frac{V}{v}(wH(\xi) - \kappa\xi^2/2) \right] d\xi \right]^v \qquad (2)$$

This result does indeed determine polymer statistics for, as we have just seen, Γ_r is determinable from $\Gamma(H, w)$ and, by Theorem 1.1, $\Pi_N(z)$ is proportional to the coefficient of w^N in the expansion of $\exp(\Sigma_r \Gamma_r w^{a_r} z_r)$.

Proof Consider first the case $v = 1$. Evaluation (2) for that case then follows from

$$\exp(\Gamma(H, w)) = \sum_N Q_N w^N / N!$$

(cf. (1.3) with Q_N evaluated by (3.4). But we see from (2.2) that the effect of changing from $v = 1$ to a general value of v is to multiply Γ by v, and modify V to V/v. The general form (2) thus follows from that for $v = 1$. ∎

The growth conditions on H are made so that the integral (3.4) will be convergent, and the process have a proper equilibrium for prescribed finite N. Note, however, that the integral (2) will always diverge if $H_j > 0$ for some $j > 2$; i.e. if branching is possible in the polymers. This implies that the series defining $\Gamma(H, w)$ will diverge under the same circumstances, if $w > 0$. The divergence is immaterial; expression (2) defines $\Gamma(H, w)$ as a power series, term by term, and Theorem 1.1 then determines the p.g.f. $\Pi_N(z)$ unambiguously and rigorously in terms of the earlier coefficients of this series.

Corollary 5.1 *The expectation parameter Γ_r has the evaluation*

$$\Gamma_r = \omega_r V (v/V)^{L+1-R} (2\kappa)^{-L} \prod_j H_j^{r_j} \qquad (3)$$

where ω_r is purely combinatorial, $R = a_r = \sum_j r_j$ is the number of units in an r-mer

and $L = \frac{1}{2}\sum_j jr_j$ the number of bonds.

Evaluation (3) follows from the fact that Γ_r is a sum of terms (2.2) over all isomers *i* corresponding to the description *r*. Indeed, ω_r is just the number of such isomers. Alternatively, (3) follows from (2) by a rescaling of the variable of integration, ξ. ∎

The analysis of section 1 would lead us to identify Γ_r with the $V\gamma_r$ of section 13.3, and so to expect it to be of order V. Under hypothesis (2.1) this is the case only under quite limited conditions.

Corollary 5.2 *The expected abundance (3) is, in its dependence on V, of order V if $L + 1 - R = 0$, i.e. if the polymer is a tree. For fixed v it is of smaller order in V for other polymers. It is of order V for all polymers if v is itself made proportional to V.*

That is, tree-polymers have an abundance of order V while other polymers (let us call them *ring-polymers*) have an abundance of lower order, the order decreasing with the amount of cyclization of the polymer. One can say that there are so many ring-polymers that the abundance of any given one must be low. However, their total effect may be significant. Indeed, it is because of the multitude of contributions from ring-polymers that integral (2) and series (1) diverge (if branching is possible and w non-zero). We shall see in the next section that $\Gamma(H, w)$ converges if we consider the contribution to it from tree-polymers alone, so long as w does not exceed a value which is related to the phenomenon of gelation.

If one sets $v = 0$ then one forbids cyclization altogether, and all polymers are trees. If one takes v positive and proportional to V then one is implying that the chance of bonding for two units in the same molecule is of the order of V times that for units in different molecules, other things being equal. But this is indeed what one would expect: that a pair of units held in proximity should have encounter rate of the order of V times that for a random pair, effects such as steric hindrance apart (i.e. that there is some geometric constraint on the structure). Indeed, this view is consistent with the approach of section 13.3, where it was assumed that association of a given pair of polymers took place at a rate of order V^{-1}, whereas an internal transition within a given polymer took place at a rate independent of volume. The difficulty is that a first-shell model cannot represent how the bonding rate between two units in the same polymer depends upon the size or more than local structure of the polymer.

Observe that Theorem 5.1 effectively solves the combinatorial problem discussed at the end of section 2. The combinatorial factor occurring in γ_r or C_r is just the factor ω_r of (3), counting the numbers of isomers of an A_r, and this is implicitly determined by the assertions of the theorem. Note also that, by specifying and then reducing the full-description equilibrium distribution, we have indeed avoided the combinatorial problem mentioned in section 13.3: that of determining the rates λ, λ', κ, κ' occurring in a reduced model.

In fact, explicit determination of ω_r still seems difficult, although we do find the general form for tree-polymers in section 7. However, it is more important that the conclusions of Theorem 5.1 help to determine the general bulk character of the polymerization process, and of critical effects in particular.

6. NOTHING BUT TREES: THE CASE $v = 0$

The case $v = 0$, when only tree-polymers are permitted, is interesting as being the classical case, as allowing a treatment which is in respects simpler, and in having interesting alternative interpretations (see Exercise 7.1). The evaluation of the generating function $\Gamma(H, w)$ is also an interesting limit version of the evaluation (5.2).

Theorem 6.1 *Consider the first-shell model in the case $v = 0$, when only tree-polymers can occur, and suppose $\log H(\xi)$ of less than quadratic growth at infinity. The generating function $\Gamma(H, w)$ then has the evaluation*

$$\Gamma(H, w) = V \max_{\xi} \left[wH(\xi) - \kappa \xi^2 / 2 \right] \tag{1}$$

Here the maximum is the positive local maximum of the bracket which exists for small enough positive w, and which tends to $\xi = 0$ as w tends to zero. The value of w at which this maximum disappears is the critical value of this parameter, at which the process gels.

Proof The result seems to be an easy enough formal consequence of (5.2), and so it would be if the integral of that expression were convergent. However, since the integral is divergent in all cases which permit branching, an alternative proof must be found. This must use the fact that, in writing down the divergent integral (5.2), one is essentially specifying rules by which the coefficients in a power series expansion are to be calculated.

Let us define

$$M(H, V) = \sqrt{\frac{\kappa V}{2\pi}} \int_{-\infty}^{\infty} \exp \left\{ V \left[wH(\xi) - \kappa \xi^2 / 2 \right] \right\} d\xi \tag{2}$$

which is expression (5.2) in the case $v = 1$. Let us also define the operators

$$T = \sum_j j H_j \frac{\partial}{\partial H_j}$$

$$U = \sum_j H_{j+1} \frac{\partial}{\partial H_j} \tag{3}$$

and note that

$$TH(\xi) = \xi H'(\xi)$$
$$UH(\xi) = H'(\xi) \tag{4}$$

where $H' = \partial H / \partial \xi$. Note then that M satisfies the equation

$$\kappa V T M = U^2 M \tag{5}$$

because, from (2), (4),

$$U^2 M - \kappa V T M = \sqrt{\frac{\kappa V}{2\pi}} \int [(VwH')^2 + VwH'' - \kappa V^2 w\xi H'] \, (\exp) \, d\xi$$

$$= \sqrt{\frac{\kappa V}{2\pi}} \int \frac{\partial}{\partial \xi} [VwH' \, (\exp)] \, d\xi \tag{6}$$

where (exp) denotes the exponential integrand of (2). These formal calculations are correct in that the relations of (6) do express correct relationships between coefficients in power series in w. Expression (6) is to be regarded as zero in that the coefficient of w^N in the integral of (6) is proportional to

$$\int\limits_{-\infty}^{\infty} \frac{\partial}{\partial \xi} [H^{N-1} H' e^{-\kappa V \xi^2 / 2}] \, d\xi$$

which is certainly zero if $\log H$ is of less than quadratic growth.

For the case of general v we now have, from (5.2), that

$$e^\Gamma = M(H, V/v)^v \tag{7}$$

From (5), (7) we then find that

$$\kappa V T \Gamma = (U\Gamma)^2 + vU^2 \Gamma$$

so that, in the limit $v = 0$, Γ satisfies

$$\kappa V T \Gamma = (U\Gamma)^2 \tag{8}$$

We again emphasize that, even if M or Γ are divergent as generating functions, equations (5), (7) and (8) correctly express relations between coefficients in those generating functions.

We now verify that (1) is the appropriate solution of (8). Denote expression (1) by Γ^*. Because the bracket is stationary with respect to ξ we have

$$\frac{\partial \Gamma^*}{\partial H_j} = \frac{Vw\xi^j}{j!}$$

and so

$$T\Gamma^* = Vw\xi H'$$
$$U\Gamma^* = VwH'$$

where ξ has in all cases the maximizing evaluation of (1). Thus

$$(U\Gamma^*)^2 - \kappa VT\Gamma^* = V^2 w H'(wH' - \kappa\xi)$$

This is zero, by the stationarity condition of (1), so expression (1) indeed solves equation (8). It is also the correct solution. To see this, note that we are looking for a solution which is a power series in the H_j, beginning with a term in H_0, which we may as well take as being wVH_0, where w is the one arbitrary parameter of the solution. Equation (7) has only one such solution, because it expresses coefficients of terms of a given weight $N = \sum_j jN_j$ in terms of coefficients of lower weight. But expression (1) is just such a solution, for w small enough.

The final assertion of the theorem stems only from the observation that some kind of singularity must be reached when w is large enough that evaluation (1) fails. In fact, this value w is a branch-point of the function $\Gamma(H, w)$ for given w, and we confirm by an alternative argument in the next section that for w to reach the branch-point is to bring about gelation. The argument is in fact that $\partial^2 \Gamma/\partial w^2$ becomes infinite at the critical value of w, and we know from Exercise 13.2.3 that this implies that the expected size of the polymer containing a randomly-chosen unit is infinite. ∎

Returning to the treatment of section 13.3 we see that we can identify $\Gamma(H, w)$ with $VG(w) = V\sum_r \gamma_r w^R$, where $V\gamma_r w^R$ is the expected value of n_r in an open process, the parameter w being determined by unit abundance. We thus deduce from expression (1).

Corollary 6.1 *Consider the first-shell model in the case $v = 0$, when only tree-polymers can occur, and suppose $\log H(\xi)$ of less than quadratic growth at infinity. Then one has the evaluation*

$$G(w) := \sum_r \gamma_r w^R = \max_{\xi} \left[wH(\xi) - \kappa\xi^2/2 \right] \tag{9}$$

For the closed process we know from section 13.4 that we have the evaluation $E(n_r) \sim V\gamma_r \bar{w}^R$ in the thermodynamic limit, where $\bar{w}$ is the root of

$$w\frac{\partial G(w)}{\partial w} = \rho \tag{10}$$

That is, the value of w minimizing $G(w) - \rho \log w$. Indeed, let the values of w and ξ determined by (10) and the stationarity condition of (9) be jointly denoted $\bar{w}, \bar{\xi}$.

Theorem 6.2 *The quantities $\bar{w}, \bar{\xi}$ are determinea by the pair of equations*

$$wH' = \kappa\xi$$
$$wH = \rho \tag{11}$$

The quantity $\bar{\xi}$ is identical with the root $\bar{\xi}$ of (4.2), and gelation occurs when first the inequality

$$\rho\left(\frac{H''}{H}\right) < \kappa \tag{12}$$

familiar from (4.4) is violated. That is, the criticality condition for the first-shell model is the same whether ν is 0 or 1.

Proof Relations (11) follow from (9), (10) and from the stationarity condition for ξ in (9). Eliminating w from equations (11) we deduce equation (4.2), and so the identity of the definitions of $\bar{\xi}$. Criticality will be reached when the local maximum in the bracket of (9) fails to exist. This occurs when a maximum and a minimum coalesce, and the second derivative becomes zero. That is, when the inequality

$$wH'' - \kappa < 0 \tag{13}$$

is first violated. Eliminating w from (11), (13) we see that (13) is indeed equivalent to the inequality (12), which is just (4.4). ■

It is of course the final conclusion which is the important one: that the criticality condition is the same whether ν is 0 or 1. This inspires the conjecture that the critical value ρ_c of ρ is independent of ν. We shall see in the next chapter that this is indeed the case, but also that a second and earlier type of transition appears as ν is increased.

When we speak of 'earlier' we mean 'for smaller ρ'; we should indeed develop a feeling for the dependence of $\bar{\xi}$, $\bar{w}$ upon ρ. We shall see in section 9 that, below criticality, the quantities $\bar{\xi}$, $\bar{w}$ and ρ all increase from zero together. The behaviour above criticality differs, interestingly and significantly.

Criticality has many manifestations, and we shall see in the next section that one is that G_{ww} becomes infinite, as does some polymer.

Note from (11), (13) that the subcriticality condition can be written purely in terms of ξ:

$$\frac{\xi H''}{H} < 1 \tag{14}$$

where ξ is of course understood to have the value $\bar{\xi}$.

Exercises and comments

1. Suppose that $H_j = 0$ for $j > 2$, so that branching can never occur, and only linear and loop polymers can be formed. Then criterion (14) becomes

$$\frac{H_2\xi}{H_1 + H_2\xi} < 1$$

This is always satisfied for $\xi \geqslant 0$, with the implication that gelation never occurs. This is of course a conclusion already reached, in the exercises of section 13.3.

7. TREE STATISTICS FOR THE FIRST-SHELL MODEL

With the evaluation (6.9) we are in a position to determine polymer statistics more explicitly. All theorems are stated for the first-shell model with $v = 0$, assuming $\log H(\xi)$ of less than quadratic growth at infinity. They can in fact be extended to provide information on statistics for the tree polymers even in the case $v > 0$, but this is a line we shall not follow.

Recall that we denote the solutions of (6.11) by $\bar{w}$, $\bar{\xi}$. It will often be understood that functions occurring in the formulae are evaluated at $(\bar{w}, \bar{\xi})$. We shall henceforth find it more convenient to denote the size of an A_r by R rather than a_r. If we write (6.9) as

$$G(H, w) = \max_{\xi} F(H, \xi, w) \tag{1}$$

then we shall appeal repeatedly to the general formulae

$$G_w = F_w \tag{2}$$

$$G_{ww} = F_{ww} - F_{w\xi} F_{\xi\xi}^{-1} F_{\xi w} \tag{3}$$

where ξ is given the maximizing value of (1) wherever it occurs. Formulae (2), (3) are valid even for the case of vector w, ξ; the subscripts indicate partial derivatives, following the conventions of section 1.10.

Theorem 7.1 *Below criticality and in the thermodynamic limit the probability distribution of the number of bonds a randomly chosen unit has formed is*

$$P(j) = \frac{1}{H(\bar{\xi})} \frac{H_j \bar{\xi}^j}{j!} \tag{4}$$

The expected number of bonds formed per unit volume is $\frac{1}{2} \kappa \bar{\xi}^2$.

Proof In the thermodynamic limit we have

$$E(n_r/V) = \gamma_r \bar{w}^R$$

(see (13.2.3)) so that

$$E(N_j/V) = \sum_r r_j \gamma_r \bar{w}^R$$

$$= H_j \frac{\partial}{\partial H_j} G(H, \bar{w})$$

$$= \bar{w} \frac{H_j \bar{\xi}^j}{j!} \tag{5}$$

whence (4) follows. The expected number of bonds per unit volume is

$$\frac{1}{2V} E\left(\sum_j jN_j\right) = \frac{1}{2} \bar{w} \bar{\xi} H' = \frac{1}{2} \kappa \bar{\xi}^2 \tag{6}$$

by appeal to (5) and (6.11), respectively.

We again have agreement with the case $v = 1$; distribution (4) is just that asserted in Corollary 3.2. ■

Theorem 7.2 *The expected size of the polymer within which a randomly chosen unit finds itself is*

$$E(R) = 1 + \frac{w(H')^2}{H(\kappa - wH'')}$$

$$= 1 + \frac{\rho(H')^2}{H(\kappa H - \rho H'')} \tag{7}$$

where all evaluations are at $\bar{w}$, $\bar{\xi}$.

Proof By the argument of Exercise 13.2.3 we have

$$E(R) = \frac{\sum_r R^2 \gamma_r \bar{w}^R}{\sum_r R \gamma_r \bar{w}^R} = 1 + \frac{w^2 G_{ww}}{w G_w}$$

Applying the rules (2), (3) we then find that

$$G_w = H$$

$$G_{ww} = \frac{-(H')^2}{wH'' - \kappa}$$

whence expressions (7) follow. ■

Expression (7) confirms that violation of condition (6.12) indeed marks the onset of gelation, in that expected polymer size then becomes infinite.

Regarding R as the random variable 'size of the polymer within which a randomly chosen unit lies' we can in fact deduce a good deal more about its distribution, from the fact that w is a marker variable for R. Consider another random variable R': the number of units connected by some path to a given end of a randomly chosen bond.

Theorem 7.3 *Consider polymer statistics for a fixed value $\bar{w}$ of w (corresponding either to an open process or to a closed process in the thermodynamic limit). Let $\bar{\xi}$ be the solution of*

$$\bar{w} H'(\bar{\xi}) = \kappa \bar{\xi} \tag{8}$$

and ξ^* the solution of

$$\bar{w}\zeta H'(\xi^*) = \kappa \xi^* \tag{9}$$

Then

$$U(\zeta) := E(\zeta^R) = \frac{\zeta H(\xi^*)}{H(\bar{\xi})} \tag{10}$$

and

$$V(\zeta) := E(\zeta^R) = \frac{\zeta H'(\xi^*)}{H'(\bar{\xi})} = \frac{\xi^*}{\bar{\xi}} \tag{11}$$

Proof By the usual reasoning we have

$$E(\zeta^R) = \frac{\sum_r R\gamma_r (\bar{w}\zeta)^R}{\sum_r R\gamma_r \bar{w}^R} = \frac{(G_w)_{w = \bar{w}\zeta}}{(G_w)_{w = \bar{w}}} \tag{12}$$

Since $G_w = wH$ then (10) follows.

We have now

$$V(\zeta)^2 = \frac{\sum_r (R-1)\gamma_r (\bar{w}\zeta)^R}{\sum_r (R-1)\gamma_r \bar{w}^R} \tag{13}$$

Both sides are identifiable as the probability generating function of the size of a polymer within which a randomly chosen bond lies. Since there are $R - 1$ bonds in an r-mer then $(R-1)$ replaces the factor R of (12). On the other hand this expectation is

$$E(\zeta^{R'+R''}) = V(\zeta)^2$$

where R'' is the number of units connected to the other side of the bond; we appeal to the fact that R' and R'' are independent but have the same distribution.

Now, by the argument of Theorem 7.1 we have

$$\sum_r (R-1)\gamma_r w^R = \tfrac{1}{2}\kappa\xi^2$$

We thus deduce from (13) that

$$V(\zeta)^2 = \left(\frac{\xi^*}{\bar{\xi}}\right)^2 \tag{14}$$

Taking the square root (necessarily positive, for real positive ζ) of each side of (14) we deduce the outer equality of (11). The final equality follows from (8), (9). ∎

In Exercise 1 we shall see that these assertions have a significant interpretation in terms of a branching process model.

Finally, we can deduce the γ_r themselves by appeal to the Lagrange expansion of an implicitly defined function. Suppose w a free parameter, and suppose we wish to evaluate a function $M(\xi)$, with ξ determined implicitly by (1) (in fact, by (6.9)). This then has the evaluation

$$M(\bar{\xi}) = \frac{1}{2\pi i} \oint M(\xi) F_{\xi\xi} F_{\xi}^{-1} \, d\xi \tag{15}$$

where the contour of integration includes the pole of the integrand at $\bar{\xi}$, but no other singularities of the integrand. The $F_{\xi\xi}$ factor is included to make the residue at this pole just $M(\bar{\xi})$. We can write (15)

$$
\begin{aligned}
M(\bar{\xi}) &= \frac{1}{2\pi i} \oint \frac{M(\xi)\left(1-\dfrac{w}{\kappa}H''\right)}{\left(1-\dfrac{w}{\kappa}\dfrac{H'}{\xi}\right)} \frac{d\xi}{\xi} \\
&= \frac{1}{2\pi i} \oint M(\xi)\left(1-\frac{w}{\kappa}H''\right) \sum_{k=0}^{\infty} \left(\frac{wH'}{\kappa\xi}\right)^k \frac{d\xi}{\xi} \\
&= \sum_{k=0}^{\infty} \{\text{coefficient of } \xi^k \text{ in } [M(\xi)(1-wH''/\kappa)(wH'/\kappa)^k]\}
\end{aligned}
\tag{16}
$$

We can now determine γ_r explicitly and, in so doing, determine the combinatorial term ω_r in (5.3), at least for the tree-polymers.

Theorem 7.4 *The factor γ_r in $E(n_r)$ has the evaluation*

$$
\gamma_r = \begin{cases} H_0 & (R=1) \\ \left(\dfrac{1}{\kappa}\right)^{R-1} \dfrac{(R-2)!}{\prod_j r_j!} \prod_{j>0}\left[\dfrac{H_j}{(j-1)!}\right]^{r_j} & (R>1) \end{cases}
\tag{17}
$$

if $\sum_j jr_j = 2(R-1)$, and is zero otherwise.

Proof We can determine γ_r as the term in $\prod_j H_j^{r_j}$ in $\sum_r \gamma_r$ and so can as well set $w=1$. We know also from Corollary 5.1 that κ must occur as $\kappa^{-L}=\kappa^{1-R}$, and so can as well normalize κ to unity. It follows from (6.9) that

$$
\sum_r R\gamma_r = H(\xi)
$$

where ξ is the root of

$$
H'(\xi) = \xi
$$

Appealing to evaluation (16) we then have

$$
R\gamma_r = \text{the term in } \prod_j H_j^{r_j} \text{ in } \Sigma
$$

where

$$
\Sigma := \sum_{k=0}^{\infty} \{\text{coefficient of } \xi^k \text{ in } H(1-H'')(H')^k\}
\tag{18}
$$

Let us set

$$H'(\xi) = \sum_1^\infty c_j \xi^{j-1}$$

where then

$$c_j = \frac{H_j}{(j-1)!}$$

We have

$$\text{coefficient of } \xi^{k+s} \prod_j c_j^{r_j} \text{ in } (H')^k = \begin{cases} \dfrac{R!}{\prod_j r_j!} & \text{if} \begin{cases} k = R := \sum_j r_j \\ \sum_j j r_j = 2R - s \end{cases} \\ 0 & \text{otherwise} \end{cases}$$

It thus follows that

$$\text{coefficient of } \xi^s \prod_j c_j^{r_j} \text{ in } \sum_k (1 - H'')(H'/\xi)^k = \frac{(R-1)!}{\prod_j r_j!}\left[R - \sum_i (i-1)\, r_i\right]$$

$$= \frac{(R-1)!\, s}{\prod_j r_j!} \qquad \text{if} \begin{cases} \sum_j r_j = R \\ \sum_j j r_j = 2R - s \end{cases}$$

and is otherwise zero. Finally, we have then

$$\text{coefficient of } \xi^0 \prod_j c_j^{r_j} \text{ in } H \sum_k (1 - H'')(H'/\xi)^k$$

$$= \frac{(R-2)!}{\prod_j r_j!}\left[\frac{\Sigma r_j}{s}\, s\right]$$

$$= R\frac{(R-2)!}{\prod_j r_j!} \qquad \text{if} \begin{cases} \sum_j r_j = R \\ \sum_j j r_j = 2(R-1) \end{cases} \tag{19}$$

and is otherwise zero. Multiplying this expression by $R^{-1}\prod_{j>1} c_j^{r_j}$, we deduce expression (17) for γ_r in the case $R > 1$. The expression is not valid in the case $R = 1$ because expression (19) is derived from terms for $k = R - 1, k = R - 2$ in (18). In the case $R = 1$ there will only be a contribution from the term for $k = 0$, which is readily verified to give $\gamma_r = H_0$. ■

Of course, a factor in H_0 can occur only if $R = 1$, and must then occur to first order—this is the γ for the monomer form.

Exercises and comments

1. *The branching process interpretation.* Watson (1958) viewed the formation of polymers as a branching process, a model that Gordon (1962) and Good (1963) formulated independently and in much more analytic detail. A unit is regarded as an individual; one of his bonds represents descent from his father, the others represent paternity of his sons. The more bonds, the more sons, and the larger will be the progeny over all generations from a single initial ancestor. This total progeny, connected by relationship, is regarded as a polymer, connected by bonds. Whether the probability of infinite progeny is zero or positive is regarded as an indication of the sol state or gel state respectively.

It is not clear how this mechanism can be reconciled with the actual polymerization mechanism, but the mathematics agrees (at least when polymers are restricted to being trees, as is necessary for the branching process).

Let $\Pi_0(z)$ be the p.g.f. of the total number j of bonds a unit forms. Let $\Pi_1(z)$ be the p.g.f. of the further number of bonds formed by a unit which has already at least one bond (i.e. of $j - 1$ conditional on $j > 0$). Then

$$\Pi_1(z) = \frac{\Pi_0'(z)}{\Pi_0'(1)} \tag{20}$$

and it is $\Pi_1(z)$ which is the p.g.f. of the number of sons. Then the expected total progeny will be infinite if the expected number of sons $\Pi_1'(1)$ exceeds unity (see, for example, Grimmett and Stirzaker, 1982, p. 93). The p.g.f. $V(\zeta)$ of total progeny from a single individual (himself included) in fact satisfies the functional equation

$$V(\zeta) = \zeta \Pi_1(V(\zeta)) \tag{21}$$

and this will have a proper solution only if $\Pi_1'(1) < 1$.

Now $V(\zeta)$ is the p.g.f. of progeny from individuals with a father: it is analogous to the random variable R' of the text. However, if one chooses an individual at random to contribute the root of the tree then the total population with which he is related includes relations through his father as well as his actual offspring, and has p.g.f.

$$U(\zeta) = \zeta \Pi_0(V(\zeta)) \tag{22}$$

Now, for the tree polymerization process we know from Theorem 7.1 that

$$\Pi_0(z) = \frac{H(\bar{\xi}z)}{H(\bar{\xi})} \tag{23}$$

and, so, by (20), that

$$\Pi_1(z) = \frac{H'(\bar{\xi}z)}{H'(\bar{\xi})} \tag{24}$$

Show that the $U(\zeta)$, $V(\zeta)$ of Theorem 7.3 in fact satisfy relations (21), (22) with Π_0, Π_1 given by (23), (24), and so that the mathematics of our analysis is consistent with that for a branching-process model. We see from (24) that the condition $\Pi_1'(1) < 1$ for non-criticality is just (6.14).

8. SOME SPECIAL TREE-POLYMER PROCESSES

We can make the calculations of the last section explicit for some special processes. Consider first the Poisson model (in the case $v = 0$) for which

$$H(\xi) = e^{\xi} \tag{1}$$

Theorem 8.1 *Consider the Poisson model with only tree-polymers permitted. The model still gels at the critical density $\rho = \kappa$. The natural level of description of a polymer is its size R; the expected number of R-mers per unit volume below criticality is*

$$\gamma_R \bar{w}^R = \kappa \frac{R^{R-2}}{R!} \left(\frac{\rho}{\kappa} e^{-\rho/\kappa} \right)^R \qquad (R \geqslant 1) \tag{2}$$

Proof We know from Theorem 6.2 that the critical density is just that determined for the case $v = 1$ in section 13.6. Equations (6.1) yield

$$\begin{aligned} \bar{\xi} &= \rho/\kappa \\ \bar{w} &= \rho\, e^{-\rho/\kappa} \end{aligned} \tag{3}$$

By the same argument as that used in Theorem 7.6 we have $R\gamma_R = $ coefficient of $\zeta^0 w^{R-1}$ in

$$\begin{aligned} \sum_{k=0}^{\infty} e^{\xi} \left(1 - \frac{w}{\kappa} e^{\xi} \right) \left(\frac{w e^{\xi}}{\kappa \zeta} \right)^k \\ = \frac{1}{\kappa^{R-1}} \left[\frac{R^{R-1}}{(R-1)!} - \frac{R^{R-2}}{(R-2)!} \right] \\ = \frac{1}{\kappa^{R-1}} \frac{R^{R-2}}{(R-1)!} \end{aligned} \tag{4}$$

if $R > 1$. If $R = 1$ then the term in R^{R-2} in the penultimate expression is missing, but the final expression is correct. From (3), (4) we deduce expression (2). ∎

The series $\sum_R u^R R^{R-2}/R!$ has radius of convergence $u = 1$; at this value it has a branch point rather than a simple pole. Note that $(\rho/\kappa) \exp(-\rho/\kappa)$ increases to unity as ρ increases to the critical value κ, and thereafter decreases. That is, distribution (2) again becomes proper for $\rho > \kappa$. We shall see in the next section that it then gives the distribution in the *sol fraction*; the material which is in polymers of finite size.

A classic example is the *equi-reactive f-functional* case

$$H(\xi) = (1 + \sigma\xi)^f$$

already considered in Exercise 13.3.3 and section 3. We now consider this in the case $v = 0$.

Theorem 8.2 *Consider the model of equi-reactive f-functional units in the tree-restricted case. The parameters $\bar{w}$ and $\bar{\xi}$ are determined by*

$$w(1 + \sigma\xi)^f = \rho \tag{5}$$

$$\sigma f w (1 + \sigma\xi)^{f-1} = \kappa\xi \tag{6}$$

and the critical density is

$$\rho_c = \frac{\kappa(f-1)}{\sigma^2 f(f-2)^2} \tag{7}$$

The natural level of description of a polymer is its size R, and the expected number of R-mers per unit volume is

$$\gamma_R \bar{w}^R = \left(\frac{\kappa}{\sigma^2}\right) \frac{(Rf - R)!}{R!\,(Rf - 2R + 2)!} \left(\frac{\bar{w}\sigma^2 f}{\kappa}\right)^R \tag{8}$$

Proof Assertions (5)–(7) again follow from (6.11) and the criterion (6.12). To deduce the distribution (8), note that

$R\gamma_R =$ coefficient of $\xi^0 w^{R-1}$ in

$$\sum_{k=0}^{\infty} (1 + \sigma\xi)^f \left(1 - \frac{w\sigma^2}{\kappa} f(f-1)(1+\sigma\xi)^{f-2}\right)\left[\frac{w\sigma f}{\kappa\xi}(1+\sigma\xi)^{f-1}\right]^k$$

$$= \left(\frac{\sigma^2 f}{\kappa}\right)^{R-1}\left[\binom{Rf - R + 1}{R - 1} - (f-1)\binom{Rf - R}{R - 2}\right]$$

$$= \left(\frac{\sigma^2 f}{\kappa}\right)^{R-1}\frac{(Rf - R)!\,f}{(R-1)!\,(Rf - 2R + 2)!}$$

whence expression (8) follows. ∎

These formulae can be interpreted rather more graphically with a reparametrization. We know from Theorem 7.1 that the number of bonds found on a randomly chosen unit has the binomial distribution

$$P(j) = \binom{f}{j} p^j q^{f-j}$$

when the bonding probability $p = 1 - q$ has the expression

$$p = \frac{\sigma\bar{\xi}}{1 + \sigma\bar{\xi}}$$

We then have the evaluation

$$\frac{\bar{w}\sigma^2 f}{\kappa} = pq^{f-2}$$

for the combination occurring in the distribution (8), and the subcriticality condition (6.14) becomes simply

$$p < \frac{1}{f-1} \tag{9}$$

Condition (9) accords with the branching interpretation of Exercise 7.1: $(f-1)p$ is just the expected number of sons in that interpretation. Equations (5), (6) transform to the equation

$$pq^{-2} = \sigma^2 f\rho/\kappa$$

for p.

9. BEYOND CRITICALITY; STATISTICS OF THE SOL FRACTION

We note the minor but significant distinctions: bonding is possible if $H_j > 0$ for some $j > 0$, multiple bonding is possible if $H_j > 1$ for some $j > 1$, branching is possible if $H_j > 0$ for some $j > 2$. Simple bonding is possible if $H_1 > 0$.

Let us rewrite equations (6.11) to express ρ and w parametrically in terms of ξ:

$$\rho = \rho(\xi) = \frac{\kappa\xi H(\xi)}{H'(\xi)} \tag{1}$$

$$w = w(\xi) = \frac{\kappa\xi}{H'(\xi)} \tag{2}$$

Lemma 9.1 *The equation*

$$\xi H''(\xi) = H'(\xi) \tag{3}$$

has a unique positive root if branching is possible; no positive root otherwise.

Proof Equation (3) can be written

$$H_1 = \sum_{j=2}^{\infty} (j-1)\frac{H_{j+1}}{j!}\xi^j$$

The value of the series is zero if branching is impossible; otherwise it increases monotonically from 0 to $+\infty$ as ξ does. ∎

We shall define ξ_C as being the root of (3) if branching is possible, and $+\infty$ otherwise.

Theorem 9.1

 (i) *As ξ increases from 0 to ξ_c then w increases monotonically from 0 to $w_c := w(\xi_c)$. As ξ increases from ξ_c to $+\infty$ then w decreases from w_c to 0.*

 (ii) *Suppose simple bonding possible. As ξ increases from 0 to $+\infty$ then $\rho(\xi)$ increases monotonically from 0; it increases to $+\infty$ if $\log H(\xi)$ is of less than quadratic growth at infinity.*

(iii) *The critical value ρ_c of density is just $\rho(\xi_c)$.*

The critical value referred is that we know to be common to the cases $v = 0$, $v = 1$.

Proof Assertion (i) follows from the facts that $w(\xi)$ is certainly increasing initially and that equation (3) is the condition for a turning point in $w(\xi)$. This equation we know has only a simple positive root, at ξ_c.

Alternatively, there is a graphical view of the matter which makes the w/ξ relation immediate. Note that the function $H'(\xi)$ is convex, strictly so if branching is possible. The value of constant w which makes the graph of $\kappa\xi/w$ intersect with the graph of $H'(\xi)$ at a given value of ξ then evidently behaves as indicated in Fig. 14.9.1. Otherwise expressed, if $w < w_c$ these two graphs intersect at two values of ξ, one less than ξ_c and one greater; if $w = w_c$ there is a tangency at ξ_c; if $w > w_c$ there is no intersection.

For assertion (ii), it is again clear from (1) that if $H_1 > 0$ then $\rho(\xi)$ increases with ξ initially. The condition for a turning point in this expression is

$$HH' + \xi(H')^2 - (HH'') = 0$$

or

$$\sum_j \sum_k (j-k)^2 \frac{H_j H_k}{j! k!} \xi^{j+k-1} = 0$$

which has no positive root. The function $\rho(\xi)$ is thus monotone increasing. Now, since $\log H(\xi)$ is completely monotone and of less than quadratic growth at infinity, then H'/H is completely monotone and of less than linear growth at infinity. This implies that $\rho(+\infty) = +\infty$.

Assertion (iii) follows simply from the fact that (3) is the criticality condition expressed in terms of ξ; see (6.1). ∎

Note that, for given ρ, the value of ξ satisfying (1) is the value $\bar{\xi}$ maximizing

$$J(\xi) = \rho \log H(\xi) - \kappa \xi^2 / 2 \tag{4}$$

familiar from section 4.

Note also that we can write

$$w = \rho / H(\xi) \tag{5}$$

The value

$$w_c = w(\xi_c) = \max_{\xi \geqslant 0} w(\xi)$$

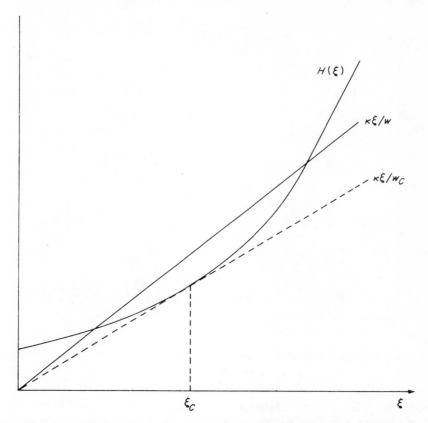

Fig. 14.9.1 A graphical representation of the relation between the parameters w and ξ and of criticality.

is just the value of w at which the straight line $\kappa\xi/w$ is tangential to the graph of $H'(\xi)$; the graphs would not intersect for larger w. The value w_C is a branch-point of the function $G(w) = \Sigma\gamma_r w^R$; the function is simply not defined for larger w.

Under the assumptions of Lemma 9.1 we can write w as a function $w[\rho]$ of density which behaves as in Fig. 14.9.2: it increases monotonically from 0 to w_C as ρ increases from 0 to ρ_C, and decreases monotonically to zero thereafter.

Theorem 9.2 *Suppose the assumptions of Theorem 9.1 satisfied, and that $\rho > \rho_C$. Then, at least for $v = 1$, the sol-fraction has the same statistics as does the process at the subcritical density ρ', where ρ' is the smaller root of $w[\rho'] = w[\rho]$.*

By 'sol-fraction' we mean that fraction of the matter which is in polymers of finite size.

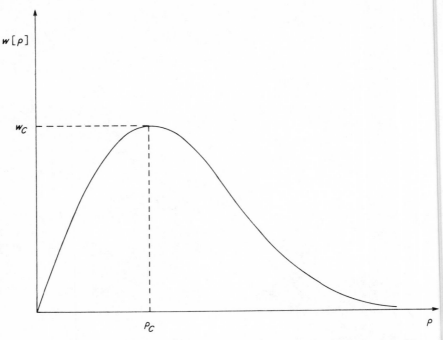

Fig. 14.9.2 The dependence of the abundance parameter w of the sol-fraction distribution upon density ρ.

Proof   We shall have proved the result if we can show that, for any finite sub-set $\mathcal{R}'$ of $\mathcal{R}$ and for finite $n_r \, (r \in \mathcal{R}')$

$$P(n_r; r \in \mathcal{R}') \propto \prod_{r \in \mathcal{R}'} [\Gamma_r w[\rho]^R / n_r!]^{n_r} \tag{6}$$

in the thermodynamic limit. That is, that the numbers n_r of the finite polymers are independent Poisson variables with expectations $\Gamma_r w[\rho]^R = \Gamma_r w[\rho']^R$.

By appealing to Theorem 5.1 in the case $v = 1$ we deduce that

$$P(n_r; r \in \mathcal{R}') \propto \left[\prod_{\mathcal{R}'} \Gamma_r^{n_r} / n_r! \right] I_m \tag{7}$$

where $m = \sum_{\mathcal{R}'} n_r$ and

$$I_m := \text{coefficient of } w^N \text{ in } (w^m \exp{(\Gamma(H, w))}$$

$$\propto \frac{V^{N-m}}{(N-m)!} \int_{-\infty}^{\infty} H(\xi)^{N-m} e^{-\kappa V \xi^2 / 2} d\xi \tag{8}$$

For finite m we have then in the thermodynamic limit

$$I_m \propto [\rho/H(\bar{\xi})]^m$$

where $\bar{\xi}$ is the value maximizing $J(\xi)$. By equation (5) and the comment before it we have then

$$I_m \propto (w[\rho])^m$$

which with (7) establishes (6). ∎

Of course, this distribution accounts for only $V\rho'$ of the total mass. The remainder, $V(\rho - \rho')$, is in the gel state, and concentrated on r of infinite R. The theorem holds also in the case $v = 0$; we indicate the reasoning in Exercises 2–5.

Exercises and comments

1. The case when simple bonding is impossible can be viewed as a limit case $H_1 \downarrow 0$. Alternatively, we can adopt the point of view which we shall see in the next chapter to be the profitable one, and parametrize in terms of w rather than ξ.

We see from Fig. 14.9.3 that ξ remains equal to zero as w increases from zero to $w_C = \kappa/H_2$. It follows from (5) that ρ follows the linear rule $\rho = wH_0$. As w then

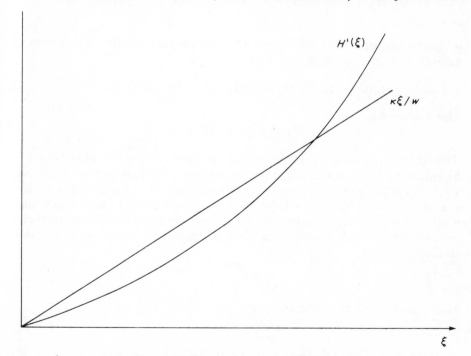

Fig. 14.9.3 The w/ξ relation and criticality in the case when simple bonds are forbidden.

decreases from w_C then matters follow much the same supercritical course as in other cases. Note that the sol fraction consists purely of monomers, both below and above criticality. Indeed, for such processes $\Gamma_r = 0$ if $0 < R < \infty$.

2. Let $\Pi_1(z)$ be the p.g.f. of the number of sons in a branching process, as in Exercise 7.1. Let p_{nj} be the probability that an initial ancestor has j offspring in the nth generation, and let

$$M_n(z) = \sum_j p_{nj} z^j$$

be the corresponding p.g.f. Show that $M_0(z) = z$ and that

$$M_n(z) = M_{n-1}(\Pi_1(z)) = \Pi_1(M_{n-1}(z)) \qquad (n = 1, 2, 3, \dots).$$

3. Continuing the same example, let ε_n be the probability that the line is extinct by generation n. Show that $\varepsilon_0 = 0$ and

$$\varepsilon_n = \Pi_1(\varepsilon_{n-1}) \qquad (n = 1, 2, 3, \dots)$$

and that ε_n increases to the smaller root of

$$\varepsilon = \Pi_1(\varepsilon)$$

as n increases.

4. Continuing still, let $\bar{M}_n(z)$ be the p.g.f. of the progeny *conditional* on ultimate extinction of the line. Show that $\bar{M}_0(z) = z$ and

$$\bar{M}_n(z) = M_n(\varepsilon z)/M_n(\varepsilon) = M_n(\varepsilon z) \qquad (n = 1, 2, 3, \dots)$$

This implies that

$$\bar{M}_n(z) = \bar{M}_{n-1}(\Pi_1(\varepsilon z))$$

The effect of requiring ultimate extinction is then to replace $\Pi_1(z)$ by $\Pi_1(\varepsilon z)$.

5. Returning to the tree-polymer process, consider the analogues of $U(\zeta)$ and $V(\zeta)$ modified by the conditioning that the random bond or unit one chooses lies in a *finite* molecule. This is a demand that one considers the sol-fraction only, and is equivalent to the condition of extinction imposed in Exercise 4. Show that these are evaluated as before except that $H(\xi)$ is replaced by $H(\varepsilon\xi)$, where ε is the smaller root of

$$\varepsilon = \frac{H'(\varepsilon\bar{\xi})}{H'(\bar{\xi})}$$

Thus $\bar{\xi}$ and $\varepsilon\bar{\xi}$ are the two roots ξ of

$$\bar{w}H'(\xi) = \kappa\xi$$

Hence show that Theorem 9.2 also holds in the case $\nu = 0$.

10. ATTRIBUTIONS

Stockmayer (1943, 1944) studied tree-polymers of equireactive f-functional units and deduced size distributions such as (8.8) and conditions for criticality. Gordon (1962) gave a branching process formulation and Good (1963) used this to deduce sol fraction statistics. Good also (1960) deduced Stockmayer's distributional results by a systematic appeal to the Lagrange expansion of an implicitly determined function: essentially relation (7.16). Whittle (1965a) formulated a Markov model for a general first-shell (but tree-restricted) model and deduced identity (6.9), criticality conditions and the polymer distribution (7.17). The extension to the non-tree case was achieved in Whittle (1980a–c); this work included in particular the argument of section 4. This treatment was much simplified in Whittle (1981) from which the material of sections 1, 2, 3 and 5 is taken. The material of Theorem 7.3 and Exercise 7.1 (establishing in detail the relation between the mathematics of a polymerization and a branching process model) is believed to be largely new, as also is the material of section 9, on sol fraction statistics.

CHAPTER 15

Compartmental statistics

1. A COMPARTMENTAL FORMULATION

Our standard formulation has been to consider N units in a volume V with free bonding between any of them. However, already in sections 13.6 and 14.4 we found it useful to consider a 'two compartment' model with $2N$ units distributed freely over two identical compartments each of volume V, although with the restriction that bonding could occur only between units in the same compartment. That is, units and polymers could migrate, but had to place themselves totally in one compartment or another.

This is an extended formulation which we shall follow up in this chapter. It has many advantages. For one thing, it gives a rudimentary spatial structure to a model sadly deficient in anything such. For another, it gives a very quick test for criticality, as we have seen in sections 13.6 and 14.4. Indeed, at the moment it is our only test for criticality in the case $v = 1$. It gives a vivid picture: gelation reveals itself in the fact that matter tends to concentrate itself in one of several identical compartments. The statistics of the compartments not containing the gel-fraction (all but one) indicate how the sol-fraction behaves above criticality.

The other advantages seem to be accidental technical ones, although there is doubtless no accident. In the case $v = 1$ it is actually easier to calculate the statistics of size distribution if one allows several compartments, as we shall see in section 6. Finally the case $v = m$ (m integral) and one compartment of volume V has a correspondence with the case of m compartments of volume V/m and $v = 1$, as we shall see in section 5.

The analysis of this chapter is that given in Whittle (1980a).

Let us assume that we have indeed m compartments, labelled $i = 1, 2, \ldots, m$. Since we assume no cross-compartmental bonding, the total configuration $\mathscr{C}$ of N identified units is just $\mathscr{C} = (\mathscr{C}_1, \mathscr{C}_2, \ldots, \mathscr{C}_m)$; the set of compartmental configurations. As in sections 13.6 and 14.4 we shall assume that

$$P(\mathscr{C}) \propto \prod_i Q_i(\mathscr{C}_i) \tag{1}$$

where $Q_i(\mathscr{C}_i)$ is an i-dependent form of the Q already postulated in (14.2.1):

$$Q_i(\mathscr{C}_i) = V_i^{N_i} \Phi_i(s_i) \prod_{a,b}^i [(h_i/2)^{s_{ab}}/s_{ab}!] \tag{2}$$

Here $N_i = N(\mathscr{C}_i)$ is the number of units in compartment i, the product $\prod\limits_{a,b}^{i}$ is over pairs of units in compartment i, and s_i is the bond configuration in compartment i. Quantities h, V, etc. are as before, with an additional i subscript and we shall again assume h_i of order V_i^{-1}:

$$h_i = (\kappa_i V_i)^{-1} \tag{3}$$

We assume distribution (1) constrained only by

$$\sum_i N_i = N \tag{4}$$

so that free migration of polymers is allowed.

Note that we are using an ambiguous notation: N_i for number in compartment i, N_j for number of u_j and (later) N_α for number of units of type α. So long as we have a general subscript (i,j or α) in place there is no ambiguity, and we shall use multiple subscripts when multiple classification is necessary.

Very often we shall assume compartments identical, so that V_i, κ_i, Φ_i are independent of i. However, there is no need to suppose this as yet.

As in (13.6.5), the distribution of compartment totals is given by

$$P(\{N_i\}) \propto \prod_i (Q_{iN_i}/N_i!) \tag{5}$$

subject to (4), where Q_{iN_i} is the sum of $Q_i(\mathscr{C}_i)$ over all configurations of N_i units. These sums integrate over configurations within compartments and the factorial terms in (5) come from an integration over permutations of units.

2. MANIFESTATIONS OF CRITICALITY

Criticality can manifest itself in a number of ways, some of which can be much easier to test for than others. The cast-iron indication is usually taken to be the evaluation of $E(R)$; the expected size of the polymer within which a randomly chosen unit lies. When this is infinite (i.e. tends to infinity in the thermodynamic limit), then gelation has occurred. Hitherto, we have been able to make this calculation only in the case $v = 0$ (Theorem 14.7.2).

Another indication is obtained by testing for departure from spatial equidistribution. That is the touchstone we shall apply systematically in this chapter. It is a test we have already applied to the first-shell model with $v = 1$ in section 14.4, when it gave a very easy and rapid indication. We shall now deepen this calculation, and also the analysis of section 14.9. It will transpire in section 6 that the critical point thus determined agrees with that revealed by polymer size statistics. It will also transpire, in section 5, that this analysis extends to other values of v, but that there is more to gelation than we have hitherto realized.

A third possible symptom of criticality is the obtrusion of a singularity into the mathematical analysis. For example, we have seen in section 13.4 that a saddle-point may move on to a singularity, and in Theorem 14.6.1 that an implicit determination of a function may fail. When such things happen one knows that *some* critical transition has occurred, although closer analysis may be needed to determine what.

Sometimes critical effects are apparent at the level of unit statistics, as for the 'cosh' model of Exercise 14.3.3; see also the discussion after Corollary 14.4.1. This seldom happens for models with fixed type-abundances. However, if units are free to change some of their characteristics, such as type (Chapter 17) then they in general do so revealingly at criticality.

3. COMPARTMENTAL STATISTICS FOR THE FIRST-SHELL MODEL ($v = 1$)

One can, of course, deduce complete polymer statistics for the compartmented first-shell model as in Theorem 14.5.1. That is, that in a hypothetical open version of the model all types of polymer are Poisson distributed with expectations whose generating functions are determined by a formula of type (14.5.2). The conclusion is straightforward and indicated in Exercise 1.

However, one of the points of a compartmental formulation is to exhibit critical effects by a study of spatial statistics alone. In the case $v = 1$ this study is both easy and informative, and has implications for the model with other values of v. We shall simply use formula (1.5) for compartmental statistics $\{N_i\}$, with evaluation (14.3.4) for the Q_{iN_i}. That is,

$$P(\{N_i\}) \propto \prod_i \int_{-\infty}^{\infty} \frac{[V_i H_i(\xi_i)]^{N_i}}{N_i!} e^{-\kappa_i V_i \xi_i^2/2} d\xi_i \tag{1}$$

The p.g.f. of the $\{N_i\}$ for prescribed $N = \sum_i N_i$ is then

$$E\left(\prod_i \eta_i^{N_i}\right) \propto \int \cdots \int \left[\sum_i f_i H_i(\xi_i)\eta_i\right]^N \exp\left[-\frac{V}{2}\sum_i f_i \kappa_i \xi_i^2\right] \prod_i d\xi_i \tag{2}$$

where

$$V = \sum_i V_i \tag{3}$$

is the total volume and

$$f_i = V_i/V \tag{4}$$

is the fraction of volume constituted by the ith compartment. Let us denote $(\xi_1, \xi_2, \ldots, \xi_m)$ by ξ, define the function

$$J(\xi) = \rho \log\left(\sum_i f_i H_i(\xi_i)\right) - \frac{1}{2}\sum_i f_i \kappa_i \xi_i^2 \tag{5}$$

and define the operation on a function

$$\mathscr{I}[\phi(\xi)] = \int \ldots \int e^{\nu J(\xi)} \phi(\xi) \, d\xi_1 \ldots d\xi_m \tag{6}$$

It follows then from (2) that, for example,

$$E(N_i) = N \mathscr{I}[\pi_i(\xi)]/\mathscr{I}[1] \tag{7}$$

where

$$\pi_i(\xi) = \frac{f_i H_i(\xi_i)}{\sum_k f_k H_k(\xi_k)} \tag{8}$$

More generally

$$E\left[\prod_i N_i^{(k_i)}\right] = N^{(k)} \mathscr{I}\left[\prod_i \pi_i(\xi)^{k_i}\right]\Big/ \mathscr{I}[1] \tag{9}$$

where $k = \sum_i k_i$.

The point of these observations is that the direct study of the nature and location of the maxima of $P(\{N_i\})$ can be replaced by the corresponding study for $J(\xi)$; simpler, in that it is more explicit. However, this will be true only if integrals such as (2) are convergent and $J(\xi)$ is maximal for finite ξ. Convergence is irrelevant as long as integrals are only a means for expressing identities between coefficients, but becomes essential if the actual magnitude of the integral is important.

We assume then that $\log H_i(\xi_i)$ is of less than quadratic growth at infinity for all i. This ensures that integral (2) is convergent, and that $J(\xi)$ has its global maximum at a finite value of ξ. We may also assume the maximizing ξ non-negative, with one improbable but significant exception.

Lemma 3.1 $J(\xi) \geqslant J(T\xi)$ for $\xi \geqslant 0$, where T is an operator changing the sign of the ξ_i for some fixed set of i. Equality will hold in this inequality if and only if the corresponding H_i are even functions, in which case a maximum of $J(\xi)$ at $\bar{\xi}$ is matched by an equivalent one at $T\bar{\xi}$. These images of maxima in the positive orthant will not affect the value of $\mathscr{I}[\phi(\xi)]/\mathscr{I}[1]$ if $\phi(T\xi) = \phi(\xi)$; i.e. if $\phi(\xi)$ has the same symmetries as $J(\xi)$.

Proof All assertions follow from the first inequality, which is a simple consequence of $H_{ij} \geqslant 0$. Note that the inequality of the Lemma will be strict if $H_{i1} > 0$, i.e. if single bonds are possible in all compartments. ∎

Define now

$$p_i = N_i/N$$

and denote the vector $\{p_i\}$ by p. This describes then the distribution of units over compartments, and is itself random. We shall correspondingly denote by $\pi(\xi)$ the vector with elements $\pi_i(\xi)$.

Theorem 3.1 *Suppose that $J(\xi)$ attains its global maximum at several values ξ^u ($u = 1, 2, \ldots$). Then in the thermodynamic limit the compartmental distribution p takes the values $\pi(\xi^u)$ with probabilities α_u ($u = 1, 2, \ldots$) where*

$$\alpha_u \propto |-J_{\xi\xi}(\xi^u)|^{-1/2} \tag{10}$$

Proof By $|-J_{\xi\xi}(\xi^u)|$ we mean the determinant of the negative Hessian $-J_{\xi\xi}$ evaluated at ξ^u.

The assertion follows immediately from asymptotic evaluation of the integrals in (9). In the thermodynamic limit relation (9) becomes, for fixed $\{k_i\}$,

$$E\left[\prod_i p_i^{k_i}\right] = \sum_u \alpha_u \prod_i \left[\pi_i(\xi^u)^{k_i}\right] \tag{11}$$

That is, p converges in distribution to the random vector that takes value $\pi(\xi^u)$ with probability α_u ($u = 1, 2, 3, \ldots$), which is the assertion of the theorem. ∎

The significance of the theorem is that the nature of the maxima of $J(\xi)$ determines the character of the compartmental distribution, and a change in this nature heralds a phase change. For example, with m identical compartments, we expect in the sol state that equidistribution will hold, and that p can only take the value $(\frac{1}{m}, \frac{1}{m}, \ldots, \frac{1}{m})$. In the gel state, we expect matter to lump into one compartment, all compartments being equally likely to receive the gel mass. That is, p can take the values $(c, d, d, \ldots, d)$, $(d, c, d, \ldots, d)$, $\ldots$, $(d, d, d, \ldots, c)$, and these each have probability m^{-1}. A passage from the first case to the second would correspond to the splitting of the maximum of $J(\xi)$ into m distinct and equivalent maxima. There is a simultaneous analogous splitting of the maxima of $P(\{N_i\})$, but, we repeat, it is easier to study the effect directly on $J(\xi)$.

In the ensuing treatment we shall find it convenient to assume that $H_i'(0) > 0$ and that $H_i'(\xi_i)$ is strictly convex for $\xi_i \geqslant 0$ for all i. (Here and subsequently we use H', H'', $\ldots$ to denote derivatives when there is no ambiguity as to the relevant variable.) The first assumption implies that $H_{i1} > 0$, i.e. that single-bonding is possible. The second implies that $H_{ij} > 0$ for some $j > 2$, so that branching is possible. We can easily discuss the cases excluded later, as limit versions of that treated, and shall do so. However, they show special features which it is tiresome to have to distinguish during the main discussion. Note that $H_{i1} > 0$ will certainly have the implication that J attains its global maximum only in the positive orthant.

If a value of ξ maximizes $J(\xi)$ globally we shall refer to it as a *stable* solution; if it maximizes $J(\xi)$ locally we shall refer to it as a *locally stable* solution; if it is locally stable but not stable we shall refer to it as *metastable*. These characterizations are supported by the fact that maxima of J correspond to maxima of the same character of $P(\{N_i\})$.

Exercises and comments

1. In an open version of the process the numbers n_{ir} of r-mers in compartment i are independent Poisson variables. $E(n_{ir})$ is the term in $\prod_j H_{ij}^{r_j}$ in the expansion of $\Gamma_i(H_i, w)$, where Γ_i is an i-dependent version of the evaluation (14.5.2). The value of w is common to all compartments, however, because $N = \sum_i N_i$ is the only invariant if migration is free, and w marks N.

4. CRITICALITY, AS REVEALED BY COMPARTMENTAL STATISTICS

Theorem 4.1 *Consider the compartmentalized first-shell model in the case $v = 1$, and suppose $\log H_i(\xi_i)$ of less than quadratic growth at infinity for all i. The necessary and sufficient conditions for a value ξ to maximize $J(\xi)$ locally are that it should satisfy*

$$w H_i'(\xi_i) = \kappa_i \xi_i \qquad (i = 1, 2, \ldots, m) \tag{1}$$

where

$$w := \frac{\rho}{\bar{H}} := \frac{\rho}{\sum_i f_i H_i(\xi_i)} \tag{2}$$

and that either (i) the smaller root of (1) be taken for all i or (ii) the smaller root of (1) be taken for all but one value of i, and

$$S := 1 + \sum_i \frac{f_i (H_i'/\bar{H})^2}{(\kappa_i/\rho) - (H_i''/\bar{H})} \leqslant 0 \tag{3}$$

It is natural to identify the two cases with the regimes below and above criticality respectively. This inference will be justified by the fuller characterization of Theorem 4.2. However, it is convenient to anticipate, and refer to the two cases as 'subcritical' and 'supercritical' respectively. The use of w to denote expression (2) is appropriate, cf. (14.9.5).

Proof Equations (1) are readily verified as the stationarity conditions for J. Since J is differentiable, has all its maxima at finite ξ, and ξ is unconstrained, then all maxima must be stationary points. Equations (1) thus possess solutions corresponding to these maxima, at least.

For given i and fixed w equation (1) has solutions corresponding to intersection of the straight line $\kappa_i \xi_i$ with the graph of $w H_i'(\xi_i)$, which is convex for positive ξ (cf. Fig. 14.9.1). There are thus two solutions (possibly coincident) or none. Since solutions exist, there are two. The equation system (1) may then have many solutions; we must determine which of these indeed correspond to local maxima of J.

The negative Hessian

$$K = -J_{\xi\xi}$$

must be non-negative definite at $\bar{\xi}$ if $\bar{\xi}$ is to be the position of a local maximum. We readily verify that

$$K_{ij} = B_i\delta_{ij} + A_iA_j \tag{4}$$

where

$$A_i = wf_iH_i'/\sqrt{\rho}$$

$$B_i = f_i(\kappa_i - wH_i'')$$

The eigenvalues v of the matrix (4) are the m roots of the equation

$$\left[\prod_i (v - B_i)\right]\left[1 - \sum_i \frac{A_i^2}{v - B_i}\right] = 0$$

All roots of this equation are real, and we require that they be non-negative.

One readily verifies graphically (and a more general result is proved in Theorem 17.1.2) that the eigenvalues are non-negative iff either (i) $B_i \geqslant 0$ for all i, or (ii) $B_i \geqslant 0$ for all i but one, and

$$1 + \sum_i A_i^2/B_i \leqslant 0. \tag{5}$$

The expression on the left is just the quantity S, defined in (3). Note that $S \geqslant 0$ in case (i).

Now, $B_i > 0$ or $\kappa_i > wH_i''$, corresponds to the fact that the line $\kappa_i\xi_i$ crosses the graph of $wH_i'(\xi_i)$ from below; i.e. to the fact that we are dealing with the smaller root of (1). The condition $B_i < 0$ correspondingly implies that $\bar{\xi}_i$ must be the larger root, while $B_i = 0$ corresponds to a double root. The statements (i), (ii) thus imply the final conditions of the theorem. ∎

It is useful to know how the solutions $\bar{\xi}_i$ of (1) vary with ρ. However, it is in fact more natural to take w as parametrizing the solution. Differentiating (1) with respect to w we see that

$$\frac{\partial\bar{\xi}_i}{\partial w} = f_iH_i'/B_i \tag{6}$$

The density ρ is determined in terms of w by (2); from this relation and (6) we find that

$$\frac{\partial\rho}{\partial w} = \bar{H}S = \rho S/w \tag{7}$$

whence the significance of S emerges. From these equations we deduce the following result.

Lemma 4.1

(i) *For a subcritical solution ρ and $\bar{\xi}_i$ increase with w.*

(ii) *For a locally stable supercritical solution ρ and the exceptional (upper) $\bar{\xi}_i$ decrease with w; the other roots $\bar{\xi}_i$ increase with w.*

The sequence of equilibrium solutions generated as ρ increases from 0 to ∞ is now roughly as follows. We let w increase from zero and generate the unique subcritical solution (corresponding also to ρ increasing from zero). Such a solution exists until w becomes so large that the two solutions of (1) become coincident for some i; this point marks criticality. After criticality we let w decrease and follow the path of a supercritical solution. That is, we follow the greater (and increasing) root ξ_i of (1) for one of the compartments i for which criticality occurred, and follow the smaller (and decreasing) root for all others. This will still be a path of increasing ρ if the solution is locally stable, so that $S \leqslant 0$.

This corresponds exactly to the sequence of events observed in Theorem 14.9.1. Moreover, the fact that above criticality one takes the greater solution of (1) in one of the compartments and the smaller one in all the others corresponds exactly to the statement of Theorem 14.9.2. That is, that the gel-fraction settles in one of the compartments and that the sol-fraction in the other compartments has the statistics of a subcritical process at the conjugate density.

However, this proposed sequence of events may have to be modified in two ways if we seek the stable (rather than only locally stable) solution. First, there may be several supercritical solutions, and it may be that one of these 'overtakes' another as global maximizer as ρ increases through some value. Secondly, if for a given supercritical solution the graph of ρ against w is not monotonic then there is a forbidden range of w for which $S > 0$. It is forbidden, because the corresponding values of ξ are not stable. At some density there must then be a discontinuity in the critical solution, as the appropriate w jumps the forbidden interval. This phenomenon can affect the situation just below criticality: the subcritical solution there is always locally stable, but may only be metastable.

We can make the following general observations on overtakings.

Lemma 4.2 *If $\bar{\xi}^{(1)}, \bar{\xi}^{(2)}$ are two stationary points of J, and $\bar{\xi}^{(2)}$ overtakes $\bar{\xi}^{(1)}$ at density ρ_0, then $\bar{H}^{(2)} \geqslant \bar{H}^{(1)}$ and $w^{(2)} \leqslant w^{(1)}$ at ρ_0.*

Here by 'overtaking' we mean

$$J\left(\xi^{(1)}\right) - J\left(\xi^{(2)}\right) \quad \begin{cases} > 0 & (\rho < \rho_0) \\ < 0 & (\rho > \rho_0) \end{cases} \tag{8}$$

and the superscript notation is self-explanatory.

Proof At ρ_0 we have

$$\frac{\partial J^{(1)}}{\partial \rho} \leqslant \frac{\partial J^{(2)}}{\partial \rho}$$

Now, since $J(\bar{\xi})$ is stationary with respect to $\bar{\xi}$ we have

$$\frac{\partial J}{\partial \rho} = \log \bar{H} \qquad (9)$$

From these last two relations we deduce that $\bar{H}$ increases at the transition. Thus w decreases, since $\rho \, (= w\bar{H})$ varies only infinitesimally. ■

Suppose now that for some supercritical solution the graph of ρ against w has the non-monotonic form illustrated in Fig. 15.4.1 (necessarily supercritical, because $\partial \rho / \partial w > 0$ for a subcritical solution). This solution is locally stable if $S \leqslant 0$, i.e. if $\partial \rho / \partial w \leqslant 0$, and so for $w \leqslant w_B$ and $w \geqslant w_C$, but perhaps only metastable.

We shall denote the value of $J(\bar{\xi})$ corresponding to w_A as J_A, etc., and write the varying value of ρ along the solution path as $\rho(w)$.

Lemma 4.3   *For the situation of Fig. 15.4.1 $J_A > J_C$ and $J_B < J_D$. For some value ρ_0 of ρ strictly between $\rho(w_B)$ and $\rho(w_C)$ the solution on the AB arc overtakes the solution on the CD arc.*

That is, if one chooses the globally maximizing solution *on this path*, then, as ρ varies, the solution follows the heavy line of Fig. 15.4.1(b).

Proof   It follows from (9) that, on the path parametrized by w, the quantities J and ρ increase and decrease together. Thus $J(\bar{\xi})$ decreases continuously on the arcs AB and CD, and the final assertion of the lemma follows from the two inequalities asserted.

We have now

$$J_C - J_A = \int_{w_A}^{w_C} \frac{\partial J(\bar{\xi})}{\partial w} \, dw = \int \frac{\partial J}{\partial \rho} \, d\rho(w) = \int_{w_A}^{w_C} \log \bar{H} \, d\rho(w)$$

where in the integral ρ decreases from $\rho(w_A)$ to $\rho(w_B)$ and then increases to $\rho(w_C) = \rho(w_A)$, so that $d\rho(w)$ is negative on AB and positive on BC. Let w_E and w_F be the abscissae on AB and BC for a prescribed value of ρ. Since $w_E < w_F$ we have $\bar{H}_E > \bar{H}_F$, and the contribution to the integral from these two points will be proportional to $[-\log \bar{H}_E + \log \bar{H}_F]|d\rho| < 0$. The integral is thus negative, and we have established the first inequality. The second follows by a similar argument. ■

Let us now denote the values of $\rho(w)$ for the subcritical solution and the supercritical solution continuous with it at criticality by $\rho_1(w)$ and $\rho_2(w)$ respectively; corresponding values of $\bar{H}$ will be denoted $\bar{H}_1(w)$ and $\bar{H}_2(w)$, and the values of J for sub- and super-critical solutions at w_A by J_{1A} and J_{2A}, respectively. Since plainly $\bar{H}_1(w) \leqslant \bar{H}_2(w)$ and $\rho = w\bar{H}$ then

$$\rho_1(w) \leqslant \rho_2(w)$$

with equality only at criticality.

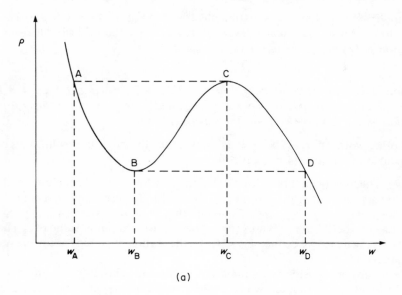

(a)

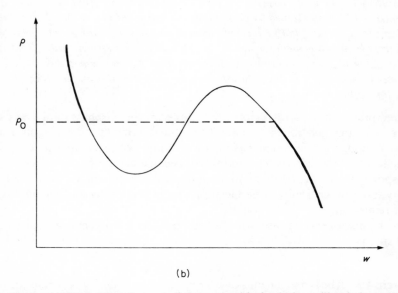

(b)

Fig. 15.4.1   (a) Illustrating the part of the supercritical solution will only be metastable if the graph of ρ against w is not monotonic. (b) The stable solution leaps from one descending part of the $\rho(w)$ curve to another at a transition density ρ_0.

Suppose now we have the situation illustrated in Fig. 15.4.2, where $\rho_2(w)$ is not monotonic. More particularly, as w decreases from the critical value w_C the function $\rho_2(w)$ first decreases, and then increases.

Lemma 4.4 *For the situation of Fig. 15.4.2 $J_{2A} > J_{2C} = J_{1C}$ and $J_{2B} < J_{1D}$. For some value ρ_M of ρ strictly between $\rho_1(w_D)$ and $\rho_1(w_C) = \rho_C$ the supercritical solution on the AB arc overtakes the subcritical solution.*

In other words, if $\rho_2(w)$ behaves as shown, then the subcritical solution is only metastable in an interval $\rho_M < \rho \leqslant \rho_C$.

Proof The first inequality follows as before, which implies all conclusions of the second part except that $\rho_M > \rho_1(w_D)$. To establish the second inequality and this fact, we note that $\rho_2(w) > \rho_1(w)$ implies that $w_E < w_F$, where w_E and w_F are abscissae on the BC and CD arcs corresponding to a common value of ρ. Hence $\bar{H}_{2E} > \bar{H}_{1F}$, whence we deduce $J_{2B} < J_{1D}$ by the argument employed for the proof of Lemma 4.3.

We are now in a position to summarize conclusions. The basically simple point, that criticality is the point at which equipartition becomes unstable, has become interestingly complicated by the possibility of metastability.

Theorem 4.2 *As ρ increases from zero then w increases from zero, and the $\bar{\xi}$ maximizing J is that obtained by taking the smaller root of (1) for all i. This solution is at least locally stable until ρ has increased to the value ρ_C for which the two roots of (1) first become coincident for some i, although possibly only metastable from some smaller value ρ_M. As ρ continues to increase from ρ_M then w decreases (perhaps discontinuously) and a stable solution must be such that the smaller root of (1) is taken for all but one value of i.*

The interpretation is that all regions are occupied by the sol-phase until ρ becomes large enough that gelation takes place. The gel-mass is contained in just one single region, although the favoured region could change with changing ρ.

The fact that the subcritical solution can become metastable before criticality corresponds to the fact that random migration between regions can raise local density above the criticality value. Gelation then begins locally, and, having begun, is self-reinforcing.

For the homogeneous case, when all compartments are identical, one can be more specific.

Theorem 4.3 The homogeneous case. *For ρ sufficiently small J is maximized if all ξ_i are equal to the smaller root of*

$$wH'(\xi) = \kappa\xi \tag{10}$$

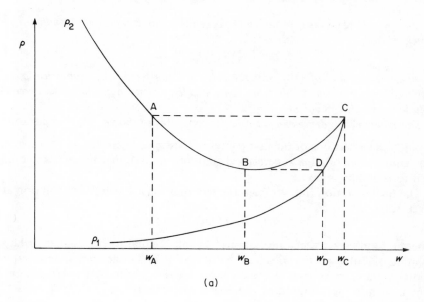

(a)

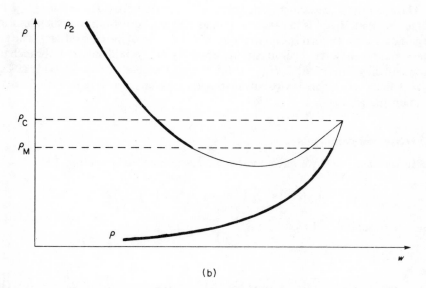

(b)

Fig. 15.4.2   (a) Illustrating that part of the subcritical solution just below criticality may only be metastable. (b) The stable solution leaps from the subcritical branch to a descending part of the supercritical branch at a transition density ρ_M somewhat less than the critical density ρ_0.

where $\rho = wH(\xi)$. *This root and w increase with ρ until ρ attains the value ρ_C for which*

$$H''(\xi)/H(\xi) = \kappa/\rho \tag{11}$$

when the two roots of (10) coalesce. This solution is locally stable for $\rho \leqslant \rho_C$ although possibly metastable from some smaller density ρ_M. The m supercritical solutions are all equivalent.

The value of ρ_M is decreasing in m, and is certainly less than ρ_C for m large enough.

Proof The assertions of the first paragraph are specializations of Theorem 4.2. The equivalence of all m supercritical solutions implies that the gel can equally well form in any region.

The final assertion follows from the fact that for a supercritical solution at a given value of w

$$S = 1 + m^{-1}[(m-1)c_1 + c_2]$$

where c_1 and c_2 are the values of mA_i^2/B_i for ξ_i equal to the smaller or larger solution of (10) respectively. Since the values of c_1 and c_2 depend upon w but not upon m, and c_1 is positive, one can make S positive by choosing m large enough. One is then in the situation of Fig. 15.4.2, implying that $\rho_M < \rho_C$. ∎

Effectively, the violence of 'gelational collapse' increases with m; see Theorem 5.2.

The exceptional cases mentioned can now be dealt with. If $H_{ij} = 0$ for $j > 2$ for some i, so that $H_i'(\xi_i)$ is linear, then (1) has only a single solution, which limiting arguments show to correspond to the smaller solution of the general case. The gel-mass could then never occur in region i, and polymers would have to unbranch the moment they entered it! If $H_{i1} = 0$ for some i the smaller root of (1) is always $\xi_i = 0$. In this case there is virtually no polymerization in those regions that do not contain the gel-mass.

Exercise and comments

1. In the case $v = 0$ we have, by the same argument as in section 14.6,

$$\Gamma(H,w) = \sum_i V_i \max_{\xi_i} [wH_i(\xi_i) - \kappa_i\xi_i^2/2] \tag{12}$$

Here w is determined by

$$w\frac{\partial\Gamma}{\partial w} = N \tag{13}$$

i.e. by

$$w\sum_i f_i H_i(\bar{\xi}_i) = \rho \tag{14}$$

where $\bar{\xi}_i$ is the extremizing value of (12). In view of (14) it is natural to make the identification (2).

5. THE CASE OF GENERAL v

Suppose that we have m identical compartments, each of volume V/m, and that we consider polymer statistics of the whole system, without regard to compartmental distribution. It follows then from formula (1.1) that Theorem 14.5.1 will hold with, in the case $v = 1$,

$$\exp\left[\Gamma\left(H,w\right)\right] = \left[\sqrt{\frac{\kappa V}{2\pi m}} \int_{-\infty}^{\infty} \exp\left[\frac{V}{m}\left(wH\left(\xi\right) - \kappa\xi^2/2\right)\right] d\xi\right]^m \tag{1}$$

From comparison of (1) and formula (14.5.2) we deduce

Theorem 5.1 *The polymer and unit statistics for the case of a first-shell model with $v = m$ (an integer) in a single compartment of volume V are identical with those of the same model with $v = 1$ in m identical compartments each of volume V/m.*

This gives a picture of the effect of increasing v through values $1, 2, 3, \ldots$. It is as though a unit can bond only to those other units sharing a cell of size V/v with it, but also that the bonding rate is increased from $(2\kappa V)^{-1}$ to $v(2\kappa V)^{-1}$, i.e. increased by a factor of v.

Corollary 5.1 *The critical density of gelation ρ_C for a given first-shell model is independent of v for $v = 0, 1, 2, \ldots$.*

Proof We have already seen (Theorem 4.3) that ρ_C is independent of m in the case $v=1$, and equal to the value for $v=0$. The assertions then follow from Theorem 5.1. ∎

It is plausible that the assertion holds for all v: this would for example be true if ρ_C were monotonic in v. However, to treat this point would take us further than we wish.

One might see the result as following from the fact that the statistics of tree-polymers are independent of v, at least for a given value of the parameter w, and that other polymers have abundances of a lower order in V (see Corollary 14.5.1).

However, it also appeared from Theorem 4.3 that, for m sufficiently large (and so for v sufficiently large), the subcritical statistics might only be metastable even for values of ρ below the critical value ρ_C. One may say that, even if given non-tree structures have lower abundance than tree-structures, there are a great many of them.

The observations of Theorem 4.3 on this point can be strengthened.

Theorem 5.2 *Suppose that $\log H\left(\xi\right)$ is of less than quadratic growth at infinity, and that branching is possible. Then, for any given value of density ρ, the supercritical solution overtakes the subcritical solution if v is chosen large enough.*

Proof By virtue of Theorem 5.1 we can discuss the problem in terms of equilibrium distribution over v identical compartments if v is integral. Let P_1 be the probability of statistical equidistribution (i.e. the sum of $P(\{N_i\})$ over its local maximum at exact equidistribution) and P_2 the probability of concentration in a single compartment. If we can demonstrate that for any ρ we can find a v which ensures that

$$(P_2/P_1)^{1/N} \geqslant C > 1 \tag{1}$$

for all large enough N then we shall have proved the assertion of the theorem.
The probability of exact equidistribution is, by (1.5) and (14.3.4),

$$P(N_i = N/v; \; i = 1, 2, \ldots, v)$$

$$= k \left[\frac{(V/v)^{N/v}}{(N/v)!} \int\limits_{-\infty}^{\infty} H(\xi)^{N/v} \exp\left[-\kappa V \xi^2/2v\right] d\xi \right]^v \tag{2}$$

where k is a normalization factor. In integrating $P(\{N_i\})$ around exact equidistribution we multiply expression (2) by a term of order $N^{v/2}$. One will then find that

$$P_1^{1/N} \sim k^{1/N} (e/\rho) \max_{\xi} \left[H(\xi) e^{-\kappa \xi^2/2\rho} \right]$$

$$= k^{1/N}(e/\rho)H(\xi_1)e^{-\kappa \xi_1^2/2\rho} \tag{3}$$

say. The probability of complete concentration in compartment 1 is, analogously to (2),

$$P_2 = P(N_1 = N) = k \frac{(V/v)^N}{N!} \int\limits_{-\infty}^{\infty} H(\xi)^N \exp\left[-\kappa V \xi^2/2v\right] d\xi$$

whence

$$P_2^{1/N} \sim k^{1/N}(e/v\rho) \max_{\xi} \left[H(\xi) e^{-\kappa \xi^2/2v\rho} \right]$$

$$\geqslant k^{1/N}(e/v\rho)H(\xi_1\sqrt{v})e^{-\kappa \xi_1^2/2\rho} \tag{4}$$

From (3), (4) we then deduce that

$$\left(\frac{P_2}{P_1} \right)^{1/N} \geqslant \frac{H(\xi_1\sqrt{v})}{vH(\xi_1)} \tag{5}$$

for N large enough. But, if branching is possible, then $H_j > 0$ for some $j > 2$, and $H(\xi_1\sqrt{v})$ increases at least as $v^{3/2}$ with increasing v. We can then make the ratio (5) exceed any prescribed value by choosing v large enough. ∎

We saw from Corollary 14.5.2 that possible structures all had abundance of order V exactly only in the cases $v = 0$ or v of order V, the second assumption corresponding to the idea that an encounter between a given pair of units is of the order of V times more probable if they are in the same molecule than if they are not. However, we see from Theorem 5.2 that the second assumption will also lead to the phenomenon of gelation under all conditions. That is, intra-molecular bonding is so easy that the subsequent rich bonding is virtually irreversible. What is needed to escape from this dilemma is a condition that restricts the possibility of intra-molecular bonding. For example, units which are indirectly bonded at some small remove may be physically unable to get close enough to form a direct bond; units which are indirectly bonded at some large remove may be no more likely to encounter each other than if they move in different molecules.

6. POLYMER SIZE STATISTICS

We have not calculated quantities such as expected polymer size $E(R)$ for any case except that for $v = 0$ (Theorem 14.7.2), because there are technical difficulties in doing so. These difficulties are now much eased in the compartmental version of the model.

Let us for simplicity consider the homogeneous case, with a total of N units distributed over a volume V divided into m identical compartments of volume V/m. We also assume $v = 1$ for the moment. It follows then, as in Theorem 14.5.1, that the p.g.f. $\Pi(z) = E\left(\prod_r z_r^{n_r}\right)$ of polymer abundances has the form

$$\Pi(z) = C_N \exp\left(\sum_r \Gamma_r z_r w^R\right) \tag{1}$$

where C_N denotes the operation 'extraction of the coefficient of w^N' and

$$\Gamma(H,w) = \sum_r \Gamma_r w^R \tag{2}$$

has the evaluation (5.1). We can write this evaluation as

$$e^\Gamma = I^m \tag{3}$$

where

$$I := \sqrt{\frac{\overline{\kappa V}}{2\pi m}} \int_{-\infty}^{\infty} \exp\left[\frac{V}{m}(wH/\xi) - \kappa\xi^2/2)\right] d\xi$$

It follows from (1), (3) that

$$E(n_r) \propto C_N(\Gamma_r w^R I^m) \tag{4}$$

and so that

$$E\left[\sum_r n_r R^k\right] \propto C_N\left[I^m \left(w \frac{\partial}{\partial w}\right)^k \Gamma\right]$$

We thus deduce

Theorem 6.1 *For an m-compartment first-shell model with $v = 1$*

$$E\left[\sum_r n_r R^k\right] = m C_N \left[I^m \left(w \frac{\partial}{\partial w}\right)^k \log I\right] \bigg/ C_N(I^m) \tag{5}$$

It is simplest if we set $w = e^\alpha$, so that $w \dfrac{\partial}{\partial w} = \dfrac{\partial}{\partial \alpha}$, and denote differentials of I with respect to α by $I_\alpha, I_{\alpha\alpha}$, etc. So, the average polymer size

$$\bar{R} = \frac{\sum\limits_r R^2 \Gamma_r w^R}{\sum\limits_r R \Gamma_r w^R}$$

will now have the expression

$$\bar{R} = \frac{C_N(\Gamma_{\alpha\alpha} I^m)}{C_N(\Gamma_\alpha I^m)} = \frac{C_N(I_{\alpha\alpha} I^{m-1} - I_\alpha^2 I^{m-2})}{C_N(I_\alpha I^{m-1})} \tag{6}$$

We can rather easily evaluate all the expressions in (6) in the thermodynamic limit provided that I does not occur to a negative power, i.e. if $m \geq 2$. Correspondingly, we can easily evaluate expression (5) in the limit provided that $m \geq k$.

We shall use the averaging operator $\mathscr{I}$ defined in (3.6), and use H_i to denote $H(\xi_i)$.

Theorem 6.2 *The average polymer size has the expression*

$$\bar{R} = 1 + \frac{m(N-1)}{2} \mathscr{I}\left[\left(\frac{H_1 - H_2}{\Sigma H_k}\right)^2\right] \bigg/ \mathscr{I}(1) \tag{7}$$

with asymptotic evaluations

$$\bar{R} \sim 1 + \frac{\rho[H'(\omega)/H(\omega)]^2}{\kappa - \rho H''(\omega)/H(\omega)} \tag{8}$$

below criticality and

$$\bar{R} \sim N \left[\frac{H(\omega_1) - H(\omega_2)}{H(\omega_1) + (m-1)H(\omega_2)}\right]^2 \tag{9}$$

above criticality. Here $\xi = (\omega, \omega, \ldots, \omega)$ and $\xi = (\omega_1, \omega_2, \ldots, \omega_2)$ are subcritical and supercritical solutions of (4.1) respectively.

Proof Differentiation under the $\mathscr{I}$ integral leads to the evaluations

$$C_N(I_\alpha I^{m-1}) \propto N \mathscr{I}\left(\frac{H_1}{\Sigma H_k}\right)$$

$$C_N(I_\alpha^2 I^{m-2}) \propto N(N-1) \mathscr{I}\left(\frac{H_1 H_2}{(\Sigma H_k)^2}\right)$$

$$C_N(I_{\alpha\alpha} I^{m-1}) \propto N \mathscr{I}\left(\frac{H_1}{\Sigma H_k}\right) + N(N-1) \mathscr{I}\left(\left(\frac{H_1}{\Sigma H_k}\right)^2\right)$$

whence evaluation (7) follows. The supercritical evaluation (9) is a fairly obvious consequence. In the subcritical case (7) becomes approximately

$$\bar{R} \sim 1 + \frac{m(N-1)}{2} \left[\frac{H'(\omega)}{H(\omega)} \right]^2 \frac{\mathscr{I}(\xi_1 - \xi_2)^2}{\mathscr{I}(1)}$$

We can regard this last ratio of $\mathscr{I}$ integrals as an 'expectation' of $(\xi_1 - \xi_2)^2$ on the assumption that $\xi_1, \xi_2, \ldots, \xi_m$ are normally distributed all with mean ω and covariance matrix $(NK)^{-1}$. From this we deduce that

$$\frac{\mathscr{I}(\xi_1 - \xi_2)^2}{\mathscr{I}(1)} \sim \frac{2}{VB} = \frac{2m\rho}{N(\kappa - \rho H''/H)}$$

whence evaluation (7) follows. ∎

Evaluation (8) is the familiar one for the case when intrapolymer reaction is forbidden; that we recover it again is just another indication of that fact that statistics below criticality are independent of m, at least up to the point of metastability.

For the Poisson case $H = e^\xi$ the subcritical evaluation (8) reduces to $\kappa/(\kappa - \rho)$, and (9) reduces to

$$\bar{R} \sim N \left[\frac{\omega_1 - \omega_2}{\omega_1 + (m-1)\omega_2} \right]^2 \tag{10}$$

where

$$\omega_1 e^{-\omega_1} = \omega_2 e^{-\omega_2}$$
$$\omega_1 + (m-1)\omega_2 = m\rho/\kappa$$

and $\omega_1 > \omega_2$. For ρ large we have $\omega_1 \sim m\rho/\kappa$, $\omega_2 \sim 0$ so that (10) implies $\bar{R} \sim N$, as one would expect.

7. DEGREE OF BONDING AND CYCLIZATION

If we modify $H(\xi)$ to $wH(\beta\xi)$, then, just as w is the marker variable for the number of units $N = \Sigma\Sigma N_{ij}$ then so is β the marker variable for the number of bond-ends $2B = \Sigma\Sigma_j N_{ij}$. We have

$$E(B) = \frac{C_N(mI_\beta I^{m-1})}{2C_N(I^m)}$$

$$= Nm\mathscr{I}(\xi_1 H_1'/\Sigma H_k)/2\mathscr{I}(1) \tag{1}$$

this being the expectation for a randomly chosen polymer. The expectation for the polymer containing a randomly chosen unit is, analogously to (6),

$$\bar{B} = \frac{C_N(I_{\alpha\beta}I^{m-1} - I_\alpha I_\beta I^{m-2})}{2C_N(I_\alpha I^{m-1})} \tag{2}$$

where β can be put equal to unity once the differentiations are performed. Expression (1) has the obvious approximate evaluations

$$E(B) \sim \frac{N \omega H'(\omega)}{2 H(\omega)}$$

and

$$E(B) \sim \frac{N(\omega_1 H(\omega_1) + (m-1)\omega_2 H(\omega_2))}{2(H(\omega_1) + (m-1)H(\omega_2))}$$

below and above criticality respectively.

Analogously to (6.7) we find the expression

$$\bar{B} = E(B/N) + \frac{m(N-1)}{4} \mathscr{I} \left[\frac{(H_1 - H_2)(\xi_1 H_1' - \xi_2 H_2')}{(\Sigma H_k)^2} \right] \Bigg/ \mathscr{I}(1) \qquad (3)$$

The evaluation of $\bar{B}$ is best expressed in terms of the measure of degree of cyclization

$$Z = (\bar{B} - \bar{R} + 1)/\bar{B}$$

By methods similar to those used in deriving expressions (6.8), (6.9), we find that

$$Z \sim 0 \qquad (4)$$

below criticality, and

$$Z \sim \frac{\omega_1 H_1'(\omega_1) - \omega_2 H_2'(\omega_2) - 2(H(\omega_1) - H(\omega_2))}{\omega_1 H'(\omega_1) - \omega_2 H'(\omega_2)} \qquad (5)$$

above criticality. In the Poisson case expression (5) reduces to

$$Z \sim \frac{\omega_1 + \omega_2 - 2}{\omega_1 + \omega_2}$$

which tends to unity for large ρ.

Relation (4) is true independently of m, and hence of v, for $\rho < \rho_M$.

Note that the subcritical evaluations (6.8) and (4) are independent of m, at least for $m \geqslant 2$, and so independent of v for $v \geqslant 2$. One expects them to be indeed totally independent of v. However, the value of m (and so v) plainly affects supercritical statistics and, as we know (Theorem 5.2), affects also the density ρ_M at which the completely sol form becomes only metastable.

CHAPTER 16

Multi-type models

1. FORMULATION

Most bonding models suggested by real life do indeed require several types of unit. In the chemical context of Chapter 7 molecules were built up from atoms of several elements, of prescribed abundances. In the polymer context, polymer molecules are built up from several types of basic building block. These constitute elements in the effective sense, if not in the strict chemical sense, in that they may be assumed immutable, and prescribed in abundance. In biological contexts, the neutralization of antigens by antibodies is achieved by the formation of large antigen–antibody clusters. In socioeconomic models the individuals which enter into a community or organization will have differing socioeconomic roles. In fact, they may be able to change these roles, so this is a case where 'elements' can 'mutate'. That is, an individual's role and his environment (as constituted by the bonds he has formed) interact; role adapts to environment and conversely. This adaptation is a new and significant effect, which we shall study specifically in the next chapter.

The treatment we shall give in this chapter is to some extent purely a formal generalization of that of Chapters 13 and 14. However, in trying to achieve the natural generalization one is led by the mathematics to concepts which are a distinct advance on those for the single-type case.

We shall index unit type by $\alpha = 1, 2, \ldots, p$ and speak of a unit of type α as an α-unit. The type (role, colouring) of unit a will be denoted by $c(a)$. The total number of α-units in the system will be denoted N_α, so that the total number of units of any type is

$$N = \sum_\alpha N_\alpha \tag{1}$$

(As noted in section 15.1, we hope to keep the meanings of N_j, N_α and N_i distinct by use of a generic subscript, always using a Greek subscript for type. There will be no discussion of compartmental statistics in this chapter.)

If the system is closed and all types are immutable then all the N_α are invariant. If types are freely mutable in that any assignment of types to identified units can change to any other by some sequence of possible mutations then N is the only invariant in a closed system. We shall refer to these two extreme cases as the *immutable* and *mutable* cases respectively. There is plainly a whole range of intermediate cases, whose treatment will be largely clear from that of the two extreme cases.

The configuration $\mathscr{C}$ on N units is now described by the type $c(a)$ of unit a and the number of bonds s_{ab} initiated from a to b ($a, b = 1, 2, \ldots, N$). This is a specification of the corresponding graph by the 'colouring' of its nodes and the pattern of its directed arcs. We shall then write $\mathscr{C} = (c, s)$, where c is the colouring function $c(\cdot)$ and s the array of bond totals $\{s_{ab}\}$ for the N nodes of the graph.

The natural generalization of the expression (13.5.4) for the assumed form of the complete-description equilibrium distribution is

$$P(\mathscr{C}) \propto Q(\mathscr{C}) := V^N \left(\prod_\alpha \sigma_\alpha^{N_\alpha} \right) \left(\prod_a \prod_b \frac{\varepsilon_{ab}^{s_{ab}}}{s_{ab}!} \right) \Phi(c, s) \tag{2}$$

where

$$\varepsilon_{ab} = \tfrac{1}{2} h_{c(a)c(b)} \tag{3}$$

and h has the volume-dependence

$$h_{\alpha\beta} = \frac{1}{\kappa_{\alpha\beta} V} \tag{4}$$

The term $\prod_\alpha \sigma_\alpha^{N_\alpha}$ will be constant in the immutable case. In the mutable case it reflects the fact that the different types may have different intrinsic probabilities ('intrinsic' meaning: in the absence of interaction effects). Indeed, if this were the only term occurring in (2) then a summation of (2) over permutations of units would demonstrate the N_α as being multinomially or independent Poisson distributed in the closed and open cases respectively, with $E(N_\alpha)$ proportional to σ_α. One can relate σ_α to the potential energy of role α.

The second bracket in (2) again represents, on its own, an assumption that the s_{ab} are independent Poisson variables although with parameters depending upon the types of the relevant unit pairs, a and b. One can regard $\kappa_{\alpha\beta}^{-1}$ as measuring the intrinsic strength of an $\alpha\beta$ bond. It is not necessarily true that $\kappa_{\alpha\beta} = \kappa_{\beta\alpha}$, although we shall in general assume this symmetry.

The term $\Phi(c, s)$ represents, as before, interaction effects in the configuration. Note that we could write

$$\prod_a \prod_b \varepsilon_{ab}^{s_{ab}} = \prod_\alpha \prod_\beta \left(\tfrac{1}{2} h_{\alpha\beta} \right)^{B_{\alpha\beta}} \tag{5}$$

where $B_{\alpha\beta}$ is the number of bonds (in the whole configuration of N units) initiated by an α-unit to a β-unit. This is not the same as the number of $\alpha\beta$ bonds which is

$$B_{\alpha\beta}^* = \begin{cases} B_{\alpha\alpha} & \alpha = \beta \\ B_{\alpha\beta} + B_{\beta\alpha} & \alpha \neq \beta \end{cases} \tag{6}$$

The total number of bonds is

$$B = \sum_\alpha \sum_\beta B_{\alpha\beta} \tag{7}$$

For the first shell model the idea of a u_j now generalizes to that of a $u_{\alpha j}$, an α-unit with j_β undirected bonds to β-units ($\beta = 1, 2, \ldots, p$). The subscript j now stands for the p-vector of integers $(j_1, j_2, \ldots, j_p)$. For the first shell model one will assume, corresponding to (14.2.1), that

$$\Phi(c, s) = \left(\prod_\alpha \prod_j H_{\alpha j}^{N_{\alpha j}} \right) v^{B + C - N} \tag{8}$$

where $N_{\alpha j}$ is the number of $u_{\alpha j}$ and C is, as ever, the number of components in the graph (polymers in the configuration). As in section 14.2 $H_{\alpha j}$ measures the preference for the local configuration of a $u_{\alpha j}$ and v measures the preference for intra- over inter-polymer bonds. One could bring bond-direction into specification of a $u_{\alpha j}$ but there seem to be enough variations to consider already.

Note the identities

$$\sum_j N_{\alpha j} = N_\alpha \tag{9}$$

$$\sum_j (N_{\alpha j} j_\beta - N_{\beta j} j_\alpha) = 0 \tag{10}$$

The second follows from the fact that both sums evaluate $B_{\alpha\beta}^*$, the total number of undirected $\alpha\beta$ bonds.

2. UNIT STATISTICS

This section is the multi-type analogue of section 14.3. Let us define $Q_{N.}$ as the sum of $Q(\mathscr{C})$ as defined in (1.2) over all s for the N units of prescribed types. In the mutable case let us define Q_N as the sum of $Q(\mathscr{C})$ over all (c, s) for the prescribed N units.

We assume that $Q(\mathscr{C})$ is invariant under a permutation of the identities of units. The sum $Q_{N.}$ can then be a function only of the N_α ($\alpha = 1, 2, \ldots, p$). For the mutable case we then immediately derive an expression for the joint distribution of the N_α:

$$P(\{N_\alpha\}) \propto \frac{Q_{N.}}{\prod_\alpha N_\alpha!} \tag{1}$$

This is completely analogous to expression (15.1.5) for the joint distribution of compartmental totals, the factorial terms coming again from summation of $Q_{N.}$ over permutation of units of like type. Indeed, in the mutable case criticality can show itself in the distribution of type-totals N_α, just as it did in the distribution of compartment-totals N_i. However, distribution (1) is more general than (15.1.5), because units of different types may bond, whereas units in different compartments may not.

Note that, in the mutable case

$$\frac{Q_N}{N!} = \sum \frac{Q_{N.}}{\prod_\alpha N_\alpha!} \tag{2}$$

where the sum is over all N_α ($\alpha = 1, 2, \ldots, p$) consistent with prescribed N.

In both the mutable and immutable cases expression (1) also determines the joint distribution of the $N_{\alpha j}$, in that $P(\{N_{\alpha j}\})$ is proportional to the term in $\prod\prod_j H_{\alpha j}^{N_{\alpha j}}$ in the expansion of expression (1).

We wish to determine the evaluation of $Q_{N.}$ in the first shell case analogous to expression (14.3.4). It turns out that the integration over ζ which occurred in that expression has to be generalized in a somewhat surprising fashion.

Define ζ as the array of p^2 variables $\{\zeta_{\alpha\beta}; \alpha, \beta = 1, 2, \ldots, p\}$ and define the generating functions

$$H_\alpha(\zeta) = \sum_j H_{\alpha j} \prod_\beta \zeta_{\alpha\beta}^{j_\beta}/j_\beta! \tag{3}$$

If $\phi(\zeta)$ is any function of ζ then define the p^2-fold integral

$$\int \phi(\zeta)d\eta = \int_{-\infty}^{\infty} \cdots \int_{-\infty}^{\infty} \phi(\zeta)\prod_\alpha\prod_\beta d\eta_{\alpha\beta}, \tag{4}$$

where the integration is along the infinite real axis for each $\eta_{\alpha\beta}$, and ζ is determined in terms of η by

$$\zeta_{\alpha\beta} = \begin{cases} \dfrac{1}{\sqrt{2}}(\eta_{\alpha\beta} + i\eta_{\beta\alpha}) & (\beta < \alpha) \\[2mm] \eta_{\alpha\alpha} & (\beta = \alpha) \\[2mm] \dfrac{1}{\sqrt{2}}(\eta_{\alpha\beta} - i\eta_{\beta\alpha}) & (\beta > \alpha) \end{cases} \tag{5}$$

Theorem 2.1 *Consider the first shell model defined by the equilibrium distribution* (1.2), (1.8) *with* $v = 1$. *Then the sum* $Q_{N.}$ *has the evaluation*

$$Q_{N.} = V^N\left(\prod_\alpha \sigma_\alpha^{N_\alpha}\right)\left(\prod_\alpha\prod_\beta \frac{\kappa_{\alpha\beta}V}{2\pi}\right)^{1/2} \int\left(\prod_\alpha H_\alpha(\zeta)^{N_\alpha}\right)\exp\left(-\frac{V}{2}\sum_\alpha\sum_\beta \kappa_{\alpha\beta}\zeta_{\alpha\beta}\zeta_{\beta\alpha}\right)d\eta \tag{6}$$

Proof   Assume the representation

$$H_{\alpha j} = \int \left(\prod_\beta x_\beta^{j_\beta} \right) m_\alpha \, (\mathrm{d}x) \tag{7}$$

Then, corresponding to (14.3.7), one has

$$Q_{N.} = V^N \left(\prod_\alpha \sigma_\alpha^{N_\alpha} \right) I \tag{8}$$

where

$$I = \int \dots \int \exp \left[\tfrac{1}{2} \sum_a \sum_b h_{c(a)c(b)} x_{a,\,c(b)} x_{b,\,c(a)} \right] \prod_a m_{c(a)} \, (\mathrm{d}x_a) \tag{9}$$

and $x_{a\beta}$ is the βth component of x_a. The integral (9) can be written

$$I = \int \dots \int \exp \left[\tfrac{1}{2} \sum_\alpha \sum_\beta h_{\alpha\beta} \Sigma_{\alpha\beta} \Sigma_{\beta\alpha} \right] \prod_a m_{c(a)} \, (\mathrm{d}x_a) \tag{10}$$

where

$$\Sigma_{\alpha\beta} := \sum_{\substack{a: \\ c(a) = \alpha}} x_{\alpha\beta} \tag{11}$$

Appeal now to the identity (14.3.8) with $\Sigma = \Sigma_{\alpha\alpha}$, and to the identity

$$\exp[h_{\alpha\beta}\Sigma_{\alpha\beta}\Sigma_{\beta\alpha}]$$

$$= \frac{1}{2\pi h_{\alpha\beta}} \int\limits_{-\infty}^{\infty}\!\!\int \exp\left[\xi_{\alpha\beta}\Sigma_{\alpha\beta} + \xi_{\beta\alpha}\Sigma_{\beta\alpha} - \xi_{\alpha\beta}\xi_{\beta\alpha}/h_{\alpha\beta}\right] \mathrm{d}\eta_{\alpha\beta}\, \mathrm{d}\eta_{\beta\alpha} \tag{12}$$

for $\alpha \neq \beta$ (cf. Exercise 1). Inserting these in (10) we deduce that

$$I = \int \mathrm{d}\eta \exp \left[\sum_a \sum_\beta \xi_{c(a)\beta} x_{a\beta} \right] e^S \prod_a m_{c(a)} \, (\mathrm{d}x_a) \tag{13}$$

where S is the exponent in the integrand of (6). Expression (6) now follows from (13) by appeal to (7), (8). ■

We deduce from (2), (6) the expression

$$Q_N = V^N \left(\prod_\alpha \prod_\beta \frac{\kappa_{\alpha\beta} V}{2\pi} \right)^{1/2} \int \left[\sum_\alpha \sigma_\alpha H_\alpha(\xi) \right]^N \exp\left[-\frac{V}{2} \sum_\alpha \sum_\beta \kappa_{\alpha\beta} \xi_{\alpha\beta} \xi_{\beta\alpha} \right] \mathrm{d}\eta \tag{14}$$

This can be regarded as a p.g.f. for the N_α with the σ_α as marker variables, and as a p.g.f. for the $N_{\alpha j}$ with the $H_{\alpha j}$ as marker variables, etc. It is similar to the expression (15.3.2) for the compartmental model, with the classification now being by types rather than by compartments. However, there are now cross-variables $\xi_{\alpha\beta}$

corresponding to the fact that bonding is possible between different types, as was not possible between different compartments.

Exercises and comments

1. Assume, as is true, that identity (14.3.8) holds for complex Σ. Show that identity (12) then follows.

2. Note that the form of integration prescribed by (4), (5) enforces the constraints (1.10). The probability of prescribed $N_{\alpha j}$ is proportional to the coefficient of

$$\prod_{\alpha} \prod_{j} H_{\alpha j}^{N_{\alpha j}}$$ in expression (6), and so proportional to

$$\int \prod_{\alpha} \prod_{j} \left(\prod_{\beta} \xi_{\alpha\beta}^{j_\beta} \right)^{N_{\alpha j}} e^S \, d\eta = \int \left(\prod_{\alpha} \prod_{\beta} \xi_{\alpha\beta}^{B_{\alpha\beta}^*} \right) e^S \, d\eta \tag{15}$$

where S is the exponent in the integrand of (6) and

$$B_{\alpha\beta}^* = \sum_{j} N_{\alpha j} j_\beta$$

Suppose that $\alpha > \beta$. The part of the second integral in (15) which involves $B_{\alpha\beta}^*$ and $B_{\beta\alpha}^*$ can then be factored out as

$$\iint \xi_{\alpha\beta}^{B_{\alpha\beta}^*} \xi_{\beta\alpha}^{B_{\beta\alpha}^*} e^{-V\kappa_{\alpha\beta}\xi_{\alpha\beta}\xi_{\beta\alpha}} \, d\eta_{\alpha\beta} \, d\eta_{\beta\alpha}.$$

Transforming to polar coordinates in real η-space we see that this integral is zero unless

$$B_{\alpha\beta}^* = B_{\beta\alpha}^*$$

which is identity (1.10).

3. POLYMER STATISTICS

The natural description of the polymer for the general first-shell model is in terms of $r_{\alpha j}$, the number of $u_{\alpha j}$ for varying α and j. That is, $r = \{r_{\alpha j}, \alpha = 1, 2, \ldots, p; j\}$. We shall follow a uniform convention that in the immutable case w denotes the p-vector $(w_1, w_2, \ldots, w_p)$ and in the mutable case it denotes the scalar common value $w = w_1 = w_2 = \ldots = w_p$.

The fundamental Lemma 14.1.1 has an obvious multi-colour version (see Exercise 1). Coupling this with the evaluation (2.4) we deduce immediately the following evaluation of polymer statistics for the multi-type case, in analogue with Theorem 14.5.1.

Theorem 3.1 *Define the quantity* $\Gamma(H, \omega)$ *by*

$$e^{\Gamma} = \left[\left[\prod_{\alpha} \prod_{\beta} \left(\frac{\kappa_{\alpha\beta} V}{2\pi v} \right)^{1/2} \right] \int \exp\left[\frac{V}{v} \left(\sum_{\alpha} w_{\alpha} \sigma_{\alpha} H_{\alpha}(\xi) - \tfrac{1}{2} \sum_{\alpha} \sum_{\beta} \kappa_{\alpha\beta} \xi_{\alpha\beta} \xi_{\beta\alpha} \right) \right] d\eta \right]^{v} \tag{1}$$

and set

$$\Gamma = \Sigma \Gamma_r, \tag{2}$$

where Γ_r *is the term in* $\prod_{\alpha} \prod_{j} H_{\alpha j}^{r_{\alpha j}}$ *in the power series expansion of* Γ. *Define then*

$$\chi(w, z) = \exp\left(\sum_r \Gamma_r z_r \right) \tag{3}$$

and $\Pi(z) = E\left(\prod_r z_r^{n_r} \right)$, *the p.g.f. of polymer numbers. Then in the immutable case*

$\Pi(z)$ *is proportional to the coefficient of* $\prod_{\alpha} w_{\alpha}^{N_{\alpha}}$ *in the expansion of* $\chi(w, z)$ *in the powers of the* w_{α}. *In the mutable case* $\Pi(z)$ *is proportional to the coefficient of* w^N.

Expression (3) is the familiar independent Poisson p.g.f. of a hypothetical or actual open version of the model (grand canonical ensemble). The novelty of the theorem compared with previous versions is the evaluation (1) of $\Gamma = \Sigma \Gamma_r$, involving as it does integration over complex ξ-values. This does seem to represent something quite fundamental in the pair-bonding of units of different types.

Exercises and comments

1. In the multi-colour version of the fundamental Lemma 14.1.1 condition (i) will be modified to the requirement that $W(\mathscr{C})$ be invariant under permutations of nodes *of the same colour*. Show that conclusion (14.1.1) then becomes modified to

$$\sum_{\mathscr{C}} \frac{W(\mathscr{C})}{\prod_{\alpha} N_{\alpha}(\mathscr{C})!} = \exp\left[\sum_i W(\mathscr{C}^{(i)}) \right]$$

4. CRITICALITY; MULTI-COLOURED TREES

Despite its restrictiveness (which can have profound consequences; see section 17.1) the case $v = 0$ is a useful one to consider, because of its explicitness and quick indication of criticality. We give the following multi-type version of Theorem 14.6.1.

Theorem 4.1 *For the first-shell model the generating function $\Gamma(H, w)$ has the evaluation*

$$\Gamma(H, w) = V \operatorname*{stat}_{\xi} \left[\sum_{\alpha} w_{\alpha} \sigma_{\alpha} H_{\alpha}(\xi) - \tfrac{1}{2} \sum_{\alpha} \sum_{\beta} \kappa_{\alpha\beta} \xi_{\alpha\beta} \xi_{\beta\alpha} \right] \tag{1}$$

in the case $v = 0$. Here 'stat' indicates that the value of ξ must be a stationary point of the bracket, and one seeks that stationary point which tends to zero as all w_{α} tend to zero.

The proof is an analogue of that of Theorem 14.6.1 (see Whittle, 1965b). The character of the stationary point is no longer that of a local maximum; we return to this point in section 7.

The stationarity point is located by the conditions

$$w_{\alpha} \sigma_{\alpha} \frac{\partial H_{\alpha}}{\partial \xi_{\alpha\beta}} = \kappa_{\alpha\beta} \xi_{\beta\alpha} \qquad (\alpha, \beta = 1, 2, \ldots, p) \tag{2}$$

As in section 14.6, the expected number of r-mers for the closed model is determined by

$$E(n_r) = \Gamma_r \prod_{\alpha} w_{\alpha}^{a_{\alpha r}} \tag{3}$$

in the subcritical case. Here the w_{α} are in general distinct in the immutable case and have values determined by

$$w_{\alpha} \sigma_{\alpha} H_{\alpha} = \rho_{\alpha} \qquad (\alpha = 1, 2, \ldots, p) \tag{4}$$

where $\rho_{\alpha} = N_{\alpha}/V$ is the prescribed density of α-units. In the mutable case all w_{α} have a common value w determined by

$$w \Sigma \sigma_{\alpha} H_{\alpha} = \rho \tag{5}$$

where $\rho = N/V$ is the prescribed total density.

The criteria for criticality can now be summarized as follows

Theorem 4.2
(i) *The set $\mathscr{W}$ of real w for which equations (2) possess solutions in $\xi \geqslant 0$ constitutes the region in which $\Gamma(H, w)$ is defined as a function of real w.*
(ii) *As ρ (and so w) increases from zero then criticality occurs when first the vector $(w_1, w_2, \ldots, w_p)$ determined by (2) and (4) (immutable case) or (2) and (5) (mutable case) reaches the boundary of $\mathscr{W}$. This is signalled by the fact that the $p^2 \times p^2$ matrix $-F_{\xi\xi}$ with $(\alpha\beta, \lambda\mu)^{th}$ element*

$$-\frac{\partial^2 F}{\partial \xi_{\alpha\beta} \partial \xi_{\lambda\mu}} = \kappa_{\alpha\beta} \delta_{\alpha\mu} \delta_{\beta\lambda} - \delta_{\alpha\lambda} w_{\alpha} \sigma_{\alpha} \frac{\partial^2 H_{\alpha}}{\partial \xi_{\alpha\beta} \partial \xi_{\lambda\mu}} \tag{6}$$

becomes singular.

Proof The use of F is a return to the notation of (14.6.9) and (14.7.1); we are writing relation (1) as

$$G(w) = \operatorname*{stat}_{\xi} F(\xi, w) \tag{7}$$

where

$$G(w) = \frac{1}{V} \Gamma(H, w) = \sum_r \gamma_r \prod_\alpha w_\alpha^{a_{\alpha r}}$$

We have the relations

$$G_w = F_w \tag{8}$$

$$G_{ww} = F_{ww} - F_{w\xi} F_{\xi\xi}^{-1} F_{\xi w} \tag{9}$$

$$\xi_w = -F_{\xi\xi}^{-1} F_{\xi w} \tag{10}$$

(see Exercise 1), where ξ has the evaluation determined by (7) for a given w and we have used the conventions of section 1.10 for vectors and matrices of derivatives. That w reaches the boundary of $\mathcal{W}$ is signalled by non-finiteness of ξ_w; we see from (10) that this corresponds to singularity of $F_{\xi\xi}$. We see from (9) that singularity of $F_{\xi\xi}$ also implies non-finiteness of G_{ww}, indicating (see Exercise 2) the occurrence of infinite polymers. ∎

One would expect the multi-type analogue of Theorem 14.4.1, stating that the conditions for criticality are the same in the cases $v = 0, 1$. Indeed, one would expect the multi-type analogues of the material of Chapter 15 and Theorem 14.9.2, stating that the conditions for criticality are the same in the cases $v = 0, 1, 2, \ldots$ and that, past gelation, sol-fraction statistics are the same as at a 'conjugate density' below gelation. Such assertions must again be based upon studies of the function

$$J(\xi) = \begin{cases} \sum_\alpha \rho_\alpha \log\left(\sigma_\alpha H_\alpha(\xi)\right) - \frac{1}{2} \sum_\alpha \sum_\beta \kappa_{\alpha\beta} \xi_{\alpha\beta} \xi_{\beta\alpha} \\ \\ \rho \log\left[\sum_\alpha \sigma_\alpha H_\alpha(\xi)\right] - \frac{1}{2} \sum_\alpha \sum_\beta \kappa_{\alpha\beta} \xi_{\alpha\beta} \xi_{\beta\alpha} \end{cases} \tag{11}$$

the two forms being the definitions appropriate for the immutable and mutable cases respectively. The integrands in the evaluations (2.6) and (2.14) are equal to $\exp(VJ)$, and in the single-type case it was the maximizing value $\bar{\xi}$ of J which was significant. The conditions (2) and (4) or (5) are indeed stationarity conditions for J. However, the stationary point they now locate is not necessarily a simple maximum; we follow up this point in section 7. However, the following assertion can be made.

Theorem 4.3
(i) *The relations (2) and (4)/(5) locate a stationary point $\bar{\xi}$ of $J(\xi)$.*
(ii) *Regard $F(\xi, w)$ and $J(\xi)$ as functions of η rather than ξ, by application of the transformation (2.5). Then, below criticality, the matrices $F_{\eta\eta}$ and $J_{\eta\eta}$ are negative definite at $\bar{\xi}$.*

By '(4)/(5)' is meant a reference to relation (4) for the immutable case and relation (5) for the mutable case.

Proof Assertion (i) is a matter of immediate verification. For assertion (ii), consider the quadratic form

$$x^\mathsf{T} F_{\xi\xi} x = \sum_\alpha \sum_\beta \sum_\lambda \sum_\mu \frac{\partial^2 F}{\partial \xi_{\alpha\beta} \partial \xi_{\lambda\mu}} x_{\alpha\beta} x_{\lambda\mu}$$

In the case of small density, when w is small, this reduces to

$$x^\mathsf{T} F_{\xi\xi} x = -\sum_\alpha \sum_\beta \kappa_{\alpha\beta} x_{\alpha\beta} x_{\beta\alpha}$$

$$= -\sum_\alpha \sum_\beta \kappa_{\alpha\beta} y_{\alpha\beta}^2$$

where y is related to x as η is to ξ (see (2.5)). The form is thus negative definite in y, which is equivalent to saying that the matrix $F_{\eta\eta}$ is negative definite. It will remain negative definite as parameters change until $F_{\eta\eta}$ (and so $F_{\xi\xi}$) becomes singular, which occurs exactly at criticality.

We shall prove the assertion on J only for the immutable case, which is a sufficient pattern for the mutable. One finds that

$$\frac{\partial^2}{\partial \xi_{\alpha\beta} \partial \xi_{\lambda\mu}} (J - F) = -\frac{\rho_\alpha \sigma_\alpha}{H_\alpha^2} \frac{\partial H_\alpha}{\partial \xi_{\alpha\beta}} \frac{\partial H_\alpha}{\partial \xi_{\lambda\mu}}$$

$$= -\delta_{\alpha\lambda} (\rho_\alpha \sigma_\alpha)^{-1} \kappa_{\alpha\beta} \xi_{\beta\alpha} \kappa_{\lambda\mu} \xi_{\mu\lambda}$$

so that

$$x^\mathsf{T} (J_{\xi\xi} - F_{\xi\xi}) x = -\sum_\alpha (\rho_\alpha \sigma_\alpha)^{-1} \left(\sum_\beta \kappa_{\alpha\beta} x_{\alpha\beta} \xi_{\beta\alpha} \right)^2$$

whence the assertion on $J_{\eta\eta}$ follows. ∎

The fact that one has to examine a $p^2 \times p^2$ matrix to establish criticality seems a surprising escalation in difficulty. However, we shall see in the next section that this can be reduced to consideration of a $p \times p$ matrix in the Poisson case.

Exercises and comments

1. In the implicit definition (7) of $G(w)$ the variable ξ is determined in terms $\xi(w)$ of w by the relation $F_\xi = 0$. Hence deduce the relationships $G_w = F_w$ and

$$G_{ww} = F_{ww} + 2F_{w\xi}\xi_w + \xi_w^\mathsf{T} F_{\xi\xi} \xi_w$$

$$F_{\xi w} + F_{\xi\xi}\xi_w = 0$$

(at $\xi = \xi(w)$) and deduce relations (8)–(10).

2. In the mutable case we have, as in Theorem 14.7.2, that

$$\bar{R} = 1 + \frac{wG_{ww}}{G_w}$$

where $\bar{R}$ is the expected size of the polymer within which a randomly chosen unit lies. For the immutable case, let $\bar{R}_{\alpha\beta}$ be the expected number of β-units in the polymer within which a randomly chosen α-unit lies. Show that

$$\bar{R}_{\alpha\beta} = \delta_{\alpha\beta} + w_\beta \left(\frac{\partial^2 G}{\partial w_\alpha \partial w_\beta} \right) \Big/ \left(\frac{\partial G}{\partial w_\alpha} \right)$$

5. CRITICALITY FOR THE POISSON CASE

We regard the Poisson case as the first shell case for which

$$H_\alpha(\xi) = \exp\left(\sum_\beta \phi_{\alpha\beta} \xi_{\alpha\beta} \right) \tag{1}$$

We usually assume $v = 1$, but can take (1) to be the sole criterion, so that we can talk about the Poisson case for various values of v.

An $\alpha\beta$ bond becomes more favoured as the parameter $\phi_{\alpha\beta}$ increases. This parameter would seem to be redundant when we already have the $\kappa_{\alpha\beta}$ parameter. However, it is simplest to leave both in: the effective combination of them will emerge.

The criticality condition for the Poisson polymerization process turns out to be much simpler than the general condition of Theorem 4.2 (ii). We consider the immutable case now, and leave the mutable case for the next chapter.

Theorem 5.1 *Consider the Poisson polymerization process with immutable types and prescribed unit densities ρ_α ($\alpha = 1, 2, \ldots, p$) in the tree-restricted case $v = 0$. Then the process is sub-critical if and only if the $p \times p$ matrix*

$$M := (\rho_\alpha^{-1} \delta_{\alpha\beta} - \psi_{\alpha\beta}) \tag{2}$$

is positive-definite, where

$$\psi_{\alpha\beta} = \frac{\phi_{\alpha\beta} \phi_{\beta\alpha}}{\kappa_{\alpha\beta}} \tag{3}$$

Proof The parameters w of (4.1) are determined by relation (4.4), where H_α has the ξ-argument solving the stationarity condition (4.2). The matrix element (4.6) now has the evaluation

$$-F_{\alpha\beta,\,\lambda\mu} = \kappa_{\alpha\beta} \delta_{\alpha\mu} \delta_{\beta\lambda} - \delta_{\alpha\lambda} w_\alpha \sigma_\alpha H_\alpha \phi_{\alpha\beta} \phi_{\alpha\mu}$$
$$= \kappa_{\alpha\beta} \delta_{\alpha\mu} \delta_{\beta\lambda} - \delta_{\alpha\lambda} \rho_\alpha \phi_{\alpha\beta} \phi_{\alpha\mu}$$

The process has reached criticality when this matrix is singular, so that the equations

$$\sum_\lambda \sum_\mu (\kappa_{\alpha\beta}\delta_{\alpha\mu}\delta_{\beta\lambda} - \delta_{\alpha\lambda}\rho_\alpha\phi_{\alpha\beta}\phi_{\alpha\mu})x_{\lambda\mu} = 0$$

possess a non-zero solution x. These relations reduce to

$$\kappa_{\alpha\beta}x_{\beta\alpha} = \rho_\alpha\phi_{\alpha\beta}\sum_\mu \phi_{\alpha\mu}x_{\alpha\mu} \tag{4}$$

If we define

$$y_\alpha = \rho_\alpha\sum_\mu \phi_{\alpha\mu}x_{\alpha\mu}$$

then we find by appropriate summation in (4) that

$$\rho_\beta^{-1}y_\beta = \sum_\alpha \psi_{\alpha\beta}y_\alpha \tag{5}$$

For this to have a solution is to say that the matrix M defined in (2) is singular, so singularity of M is the condition for criticality. The process is certainly subcritical if all the ρ_α are small enough, when M is indeed positive definite. The process must have passed through criticality if and only if M is not positive definite. ∎

In fact, because one has the simple evaluation (6) of $Q_{N.}$ in this Poisson case with $v = 1$ one can reapply the compartmental argument of section 13.6 to deduce very rapidly.

Theorem 5.2 *The criticality criterion of Theorem 5.1 holds also in the case $v = 1$.*

Proof Assuming $v = 1$ and H_α of the form (1) one deduces readily from (1.2), (1.8) that

$$Q_{N.} = V^N\left(\prod_\alpha \sigma_\alpha^{N_\alpha}\right)\exp\left[\frac{1}{2V}\sum_\alpha\sum_\beta \psi_{\alpha\beta}N_\alpha N_\beta\right] \tag{6}$$

If we define

$$P(N.) = \frac{Q_{N.}}{\prod_\alpha N_\alpha!} \tag{7}$$

then the distribution of units of prescribed total abundance $2N.$ over two identical compartments of volume V is then $\prod_1^2 P(N_{i.})$ subject to

$$\sum_1^2 N_{i.} = 2N.$$

where the i indicates a compartment label. We find then from expressions (6), (7) that, for small vector Δ,

$$\log\left[\frac{P(N.+V\Delta)P(N.-V\Delta)}{P(N.)^2}\right] = -V\Delta^{\mathsf{T}}M\Delta + o(\Delta^2) \tag{8}$$

where M is just the matrix (2). The condition that M be positive definite is then the necessary and sufficient condition that spatial equidistribution be a stable configuration, which we take as a criterion of subcriticality. ■

One gets a feeling for various cases by considering the various extreme possibilities in the case $p = 2$, for which the critical boundary $|M| = 0$ in ρ-space becomes

$$\frac{1}{\rho_1\rho_2} - \frac{\psi_{11}}{\rho_2} - \frac{\psi_{22}}{\rho_1} + (\psi_{11}\psi_{22} - \psi_{12}^2) = 0 \tag{9}$$

Case (i) $\psi_{12} = 0$. In this case (9) reduces to

$$(\rho_1 - \psi_{11}^{-1})(\rho_2 - \psi_{22}^{-1}) = 0$$

so that the process is no longer totally in the sol state if ρ_α exceeds $\psi_{\alpha\alpha}^{-1}$ for any α. This is, of course, the case when the two types of units do not interact at all, and

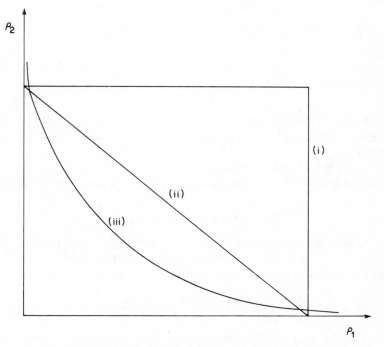

Fig. 16.5.1 Critical boundaries in ρ-space in the Cases (i)–(iii).

the process splits into two separate one-type processes, with critical densities $\rho_\alpha = \psi_{\alpha\alpha}^{-1}$ ($\alpha = 1, 2$). The example also illustrates that, if the process decomposes in this way, then one can have one component of matter in the sol state and another in the gel state.

Case (ii). $\psi_{12} = \sqrt{\psi_{11}\psi_{22}}$. In this case the critical locus (9) reduces to a straight line

$$\psi_{11}\rho_1 + \psi_{22}\rho_2 = 1.$$

This is the case when there is a certain substitutability of units, in that the bond configurations (a), (b) in Fig. 16.5.2 have the same 'energy'.

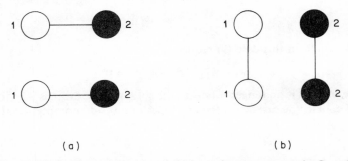

(a)   (b)

Fig. 16.5.2 Configurations (a) and (b) have the same energy in Case (ii).

Case (iii). $\psi_{11} = \psi_{22} = 0$. In this case the critical locus (9) reduces to a rectangular hyberpola

$$\rho_1\rho_2 = \psi_{12}^{-2}.$$

This is the case when 12 bonds are the only possible ones, so that polymers require both types of unit to form. As ρ_1 decreases to zero the critical value of ρ_2 increases to infinity.

Note the form of the contours in Fig. 16.5.1, changing from concavity to convexity as inter-type attraction increases relative to intra-type attraction.

6. EQUI-REACTIVE SITES

The concept of the f-functional unit, i.e. a unit with f independent and equi-reactive sites, is a classic polymer model introduced in Exercise 13.3.3 and Section 14.3. It has several multi-type generalizations. For instance, if one sets

$$H_\alpha(\xi) = \left(1 + \sum_\beta \phi_{\alpha\beta}\xi_{\alpha\beta}\right)^{f_\alpha} \tag{1}$$

one is saying that an α-unit has f_α independent sites, which can form bonds of any type, but β units being preferred in proportion to $\phi_{\alpha\beta}$. If, on the other hand, one sets

$$H_\alpha(\xi) = \prod_\beta (1 + \phi_{\alpha\beta}\xi_{\alpha\beta})^{f_{\alpha\beta}} \tag{2}$$

one is saying that an α-unit has $f_{\alpha\beta}$ sites for bonding to β-units only, with a propensity for such a bond proportional to $\phi_{\alpha\beta}$ ($\beta = 1, 2, \ldots, p$). There are plainly intermediate possibilities, and these may be differently chosen for different unit types.

Let us for specificity consider case (1), and write this as

$$H_\alpha = K_\alpha^{f_\alpha}$$

One finds then, as in the last section, that the $\xi_{\alpha\beta}$ are determined by

$$\frac{f_\alpha \rho_\alpha \phi_{\alpha\beta}}{K_\alpha} = \kappa_{\alpha\beta}\xi_{\beta\alpha} \tag{3}$$

and that the process is wholly in the sol-state just so long as the matrix

$$\tilde{M} = \left(\frac{K_\alpha^2 \delta_{\alpha\beta}}{\rho_\alpha f_\alpha (f_\alpha - 1)} - \psi_{\alpha\beta} \right) \tag{4}$$

is positive definite. Here $\psi_{\alpha\beta}$ still has the definition (5.3).

The degrees of bonding $\xi_{\alpha\beta}$ are determined only implicitly by (3), although one can, as in section 14.8, illuminate the situation by introducing the quantities

$$p_{\alpha\beta} = \frac{\phi_{\alpha\beta}\xi_{\alpha\beta}}{1 + \sum_\beta \phi_{\alpha\beta}\xi_{\alpha\beta}}. \tag{5}$$

Expression (5) can be interpreted as the probability that a given site on an α-unit bonds, and to a β-unit.

For example, consider the two-type case with

$$H_1 = (1 + \xi_{12})^{f_1},$$
$$H_2 = (1 + \xi_{21})^{f_2}. \tag{6}$$

For this only 12 bonds are possible and the model is both of type (1) and type (2). It has some importance for, for example, the antigen–antibody reaction, since bonding occurs effectively only between antigen and antibody.

Theorem 6.1 *For the 'antigen–antibody' model (6) the condition for criticality is*

$$p_{12}p_{21} < \frac{1}{(f_1 - 1)(f_2 - 1)}. \tag{7}$$

Here p_{12} and p_{21}, the probabilities of bonding for a site on a 1-unit or a 2-unit *respectively, are determined in terms of the unit densities by*

$$\rho_1 f_1 q_{12} q_{21} = \kappa p_{21}$$
$$\rho_2 f_2 q_{12} q_{21} = \kappa p_{12}$$

(8)

where we have set $q_{\alpha\beta} = 1 - p_{\alpha\beta}$, $\kappa_{12} = \kappa$.

Proof Since units are immutable we can as well set $\sigma_1 = \sigma_2 = 1$. Let us also simply set $\kappa_{12} = \kappa$. The conditions (4.2), (4.2) determining ξ, w then reduce to

$$\left. \begin{array}{l} w_1 f_1 (1 + \xi_{12})^{f_1 - 1} = \kappa \xi_{21} \\ w_2 f_2 (1 + \xi_{21})^{f_2 - 1} = \kappa \xi_{12} \end{array} \right\}$$

(9)

$$\left. \begin{array}{l} w_1 (1 + \xi_{12})^{f_1} = \rho_1 \\ w_2 (1 + \xi_{21})^{f_2} = \rho_2 \end{array} \right\}$$

(10)

and definitions (5) become

$$p_{12} = \frac{\xi_{12}}{1 + \xi_{12}}, \quad p_{21} = \frac{\xi_{21}}{1 + \xi_{21}}$$

(11)

The condition that $\tilde{M}$ should be positive definite is readily found to reduce to

$$\frac{(1 + \xi_{12})^2 (1 + \xi_{21})^2}{\rho_1 \rho_2 f_1 f_2 (f_1 - 1)(f_2 - 1)} > \frac{1}{\kappa^2}$$

(12)

We can easily eliminate w from relations (9), (10) and solve for ξ in terms of the ps from (11). With these steps we find that (12) reduces to inequality (7) in the $p\alpha\beta$, the relation between p and ρ being expressed by (8). ∎

One can derive formulae for the actual γ_r, i.e. the expected distribution of r-mers, as in sections 14.7 and 14.8. One must appeal to a multi-variable form of the Lagrange expansion due to Good (1960).

Theorem 6.2 (Good) *If* $\zeta_j = u_j / f_j(u)$, *where* $f_j(u)$ *is analytic in the neighbourhood of the origin and* $f_j(0) \neq 0$ ($j = 1, 2, \ldots, n$), *and* $F(u)$ *is meromorphic in the neighbourhood of the origin, then the coefficient of* $\prod_j \zeta_j^{m_j}$ *in the expansion of*

$F[u(\zeta)]$ *is equal to that of* $\prod_j u_j^{m_j}$ *in the expansion of*

$$F(u) \left[\prod_j f_j(u)^{m_j} \right] \det \left[\delta_{jk} - \frac{u_j}{f_j(u)} \frac{\partial f_k(u)}{\partial u_j} \right]$$

(13)

Let us apply this theorem to the antigen–antibody model (6) for which then

$$G(w) = \underset{\xi}{\mathrm{stat}} \left[w_1 (1 + \xi_{12})^{f_1} + w_2 (1 + \xi_{21})^{f_2} - \kappa \xi_{12} \xi_{21} \right]$$

$$= \sum_r \sum_s \gamma_{rs} w_1^r w_2^s$$

(14)

say. Here we have again set $\sigma_1 = \sigma_2 = 1$ and $\kappa_{12} = \kappa$. The natural level of description of a cluster is (r, s); the number of units of each type in the cluster.

Theorem 6.3 *For the 'antigen–antibody' model* (6) *the quantity* $E(n_{rs})$ $= V \gamma_{rs} w_1^r w_2^s$ *has the alternative expressions*

$$E(n_{rs}) = c_{rs} \left(\frac{f_1 w_1}{\kappa} \right)^r \left(\frac{f_2 w_2}{\kappa} \right)^s$$

$$= c_{rs} \left(\frac{p_{21} q_{12}^{f_1 - 1}}{q_{21}} \right)^r \left(\frac{p_{12} q_{21}^{f_2 - 1}}{q_{12}} \right)^s \tag{15}$$

where

$$c_{rs} = V \kappa \frac{(f_1 r - f_1)! \, (f_2 s - f_2)!}{r! s! (f_1 r - r - s + 1)! (f_2 s - r - s + 1)!} \tag{16}$$

Proof As in the proof of Theorem 14.7.4 we have

$$\sum_r \sum_s r \gamma_{rs} w_1^r w_2^s = w_1 (1 + \bar{\xi}_{12})^{f_1} \tag{17}$$

where $\bar{\xi}$ is the value of ξ determined by equations (9). Thus, by Good's theorem, $r\gamma_{rs}$ is the coefficient of $\xi_{21}^{r-1} \xi_{12}^{s}$ in the expansion of

$$\frac{f_1^{r-1} f_2^s (1 + \xi_{12})^{f_1}}{\kappa^{r+s-1}} \left[1 - \frac{(f_1 - 1)(f_2 - 1) \xi_{12} \xi_{21}}{(1 + \xi_{12})(1 + \xi_{21})} \right] (1 + \xi_{12})^{(r-1)(f_1 - 1)} (1 + \xi_{21})^{s(f_2 - 1)}$$

One finds with some reduction that this yields the first expression of (15). The second follows if we use (9), (11) to express w in terms of p. The parameters p are of course determined in terms of the prescribed densities ρ by (8). ∎

Exercises and comments

1. Deduce the analogues of (3), (4) for the case when H_α is of the form (2) for all α.

7. ASYMPTOTICS FOR THE CASE $v = 1$

One could continue to consider criticality of size statistics from a compartmental formulation of the case $v = 1$, as in the last chapter.

The only point we shall make here is that asymptotic evaluation of the integrals concerned now presents quite new features. Consider only the single compartment case, for which we derive unit statistics from the integral (2.6) for Q_N. We can write this as

$$Q_N \propto \int \exp (V J (\xi)) \mathrm{d}\eta \tag{1}$$

where J has the first evaluation of (4.11). We are interested in evaluating integral (1) for large V. In the single-type case this meant we were interested in the values of ξ maximizing J. However, the situation is less simple in the multi-type case, because ξ now follows a complex integration path. Presumably we have to seek an appropriate hybrid of maximizing and saddle-point values.

The situation is illuminated by considering the Poisson case (5.1) for which we have the exact evaluation (5.6) for Q_{N}. One can verify that this exact evaluation is that which would be yielded by the following asymptotic procedure: to replace the complex transformation (2.5) in the integral (1) by the real transformation

$$
\xi_{\alpha\beta} = \begin{cases} \dfrac{1}{\sqrt{2}}(\zeta_{\alpha\beta} + \zeta_{\beta\alpha}) & (\beta < \alpha) \\[2mm] \zeta_{\alpha\beta} & (\beta = \alpha) \\[2mm] \dfrac{1}{\sqrt{2}}(\zeta_{\alpha\beta} - \zeta_{\beta\alpha}) & (\beta > \alpha) \end{cases} \tag{2}
$$

and to seek for the $\zeta_{\alpha\beta}$ values that maximize $J(\xi)$ for $\alpha \geqslant \beta$ and minimize it for $\alpha < \beta$. It is at least suggestive to note that, for the simplest non-trivial case, the interaction of two types of unit ($p = 2$), one deduces in this way three 'space-like' axes (positive-definiteness of form) and one 'time-like' axis (negative-definiteness of form).

This suggested min-max evaluation is consistent with the complex integration path required in Theorem 2.1. Essentially we are integrating $\zeta_{\alpha\beta}$ along the real axis for $\alpha \geqslant \beta$ and along the imaginary axis for $\alpha < \beta$. In the first case we would look for a maximizing value; in the second for a saddle-point. But, if the saddle-point lies on the real $\zeta_{\alpha\beta}$ axis it will be minimizing in the direction normal to the integration path, i.e. along the real axis.

Attributions

The material of sections 1–5 and section 7 is due to Whittle (1965b, 1980c and 1981). For the material of section 6: Stockmayer (1943, 1944) deduced size-distributions for the multi-type equi-reactive site case by direct combinatorial arguments; Good (1960) showed how appeal to the multi-variable Lagrange expansion provided a systematic method of derivation.

CHAPTER 17

Role-adaptation; new critical effects

1. ROLE-ADAPTIVE MODELS

If one allows mutation of types in a multi-type bonding process then one can represent, for example, adaptation to various roles by individuals in a model of a society or an economy. One allows an individual to choose and even change his role; this choice will affect and be affected by his circumstances, i.e. by the bonds he has formed with other members of the society. In other words, one is allowing a two-way interaction between type and bonding, and our models are capable of representing this.

Another example might be the differentiation of biological cells. Cells begin in an undifferentiated state; as the interactions which lead to organization develop, so do cells acquire a function (nerve, bone, liver . . .) and adapt to that function. The situation is again then one of organization developing by an interplay between inter-cell interactions (bonding) and acquisition of function (role).

Our models are capable of representing this interplay, and constitute a stage (albeit an early one) in the development of models for self-organizing systems.

A crude characterization of the state of a role-adaptive model might be the distribution of roles: the expected numbers $E(N_\alpha)$ of individuals in the various roles. A transition in structure at critical parameter values will usually be reflected by a transition in this distribution, just as it was by the compartmental distribution in Chapter 15. However, role adaptation allows much richer possibilities than does simple compartmental migration, and we now hope for evidence of some interesting degree of organization.

An early attempt by the author (Whittle, 1977a) to demonstrate changes in the role-distribution at criticality failed, for an interesting reason. At that time there was a theory only for the case $v = 0$, when clusters are restricted to being trees. If two clusters combine then one might regard this as analogous to the merger of two companies. The essence of such a merger in the real world is then that there is an internal reorganization of roles, links, etc. as the new organization seeks its way to an efficient and internally consistent structure. However, if companies are represented by trees, such reorganization can scarcely occur. New internal bonds are forbidden, and the loss of a bond means the immediate division of the company into two separate structures.

371

However, the possibility of allowing intra-cluster bonding indeed led to the demonstration of a role-shift at criticality (see Whittle, 1980c, 1982a,b from which the material of this chapter is largely drawn; the material of section 3 is new).

This is, however, a study which has only just begun. In this chapter we consider the simplest set of models: Poisson bonding models with $v = 1$ and unrestricted mutation of types. These allow treatment by simple means, but already show new effects.

To emphasize the impermanence of type we shall now consistently use the term 'role', which also better suits the rather different applications we now have in mind. In these applications we shall also use the word 'individual' rather than 'unit', and rather than 'gelation', shall speak of condensation or 'integration'.

2. ROLE-DISTRIBUTION FOR THE POISSON MODEL

We shall take the Poisson model with $v = 1$ as being expressed by the equilibrium distribution (16.1.2) with the specialization $\Phi \equiv 1$. That is

$$P_N(\mathscr{C}) \propto Q_N(\mathscr{C}) = V^N e^{\sum_\alpha \gamma_\alpha N_\alpha} \prod_a \prod_b \frac{\varepsilon_{ab}^{s_{ab}}}{s_{ab}!} \tag{1}$$

where

$$\varepsilon_{ab} = \frac{\psi_{c(a)c(b)}}{2V} \tag{2}$$

Here $c(a)$ is the colour of node a; the role of individual a. It is convenient to replace σ_α by e^{γ_α}; these quantities represent the intrinsic attractiveness of role α. Likewise it is convenient to replace $\kappa_{\alpha\beta}^{-1}$ by $\psi_{\alpha\beta}$; this measures the intrinsic strength of bonding between individuals in roles α and β. We assume complete mutability of role, so that the configuration $\mathscr{C}$ describes both bonding pattern $\{s_{ab}\}$ and role allocation $\{c(a)\}$. All values of these are possible for the given N individuals.

We shall again write $N.$ for $\{N_\alpha\}$, the array of numbers N_α in the various roles. Correspondingly ρ will denote total density N/V, and $\rho.$ the vector of role-densities $\rho_\alpha = N_\alpha/V$.

We shall suppose the values of $\psi_{\alpha\beta}$ and γ_α finite in magnitude. This will in particular imply that all roles are represented; i.e. that $\rho_\alpha > 0$ for all α with probability one in the thermodynamic limit.

Theorem 2.1   *The distribution of role-totals for model* (1) *is*

$$P(N.) \propto V^N \left(\prod_\alpha \frac{e^{\gamma_\alpha N_\alpha}}{N_\alpha!} \right) \exp\left(\frac{1}{2V} \sum_\alpha \sum_\beta \psi_{\alpha\beta} N_\alpha N_\beta \right), \tag{3}$$

constrained only by

$$\sum_\alpha N_\alpha = N \tag{4}$$

Proof Summing over all the s_{ab} in (1) we convert the final double product in (1) to the exponential term of (3). Summing over all permutations of individuals we acquire the factorial terms $\left(\sum_\alpha N_\alpha! \right)^{-1}$, and so deduce (3). ▪

The first bracket of (3) would indicate the N_α as being multinomially distributed (i.e. independently Poisson but constrained by (3)). The exponential term strongly modifies this, however, and reflects the interactions between individuals. We have averaged over bonding patterns, but the effect of bonding is manifest in this term.

We look for the effective regime by looking for the $N.$ which maximizes $P(N.)$ globally. This regime will overwhelm all others (i.e. those corresponding to other local maxima) in the thermodynamic limit. However, rather than maximizing $P(N.)$ subject to constraint (4) it is better to maximize

$$\log P(N.) - \lambda \sum_\alpha N_\alpha \tag{5}$$

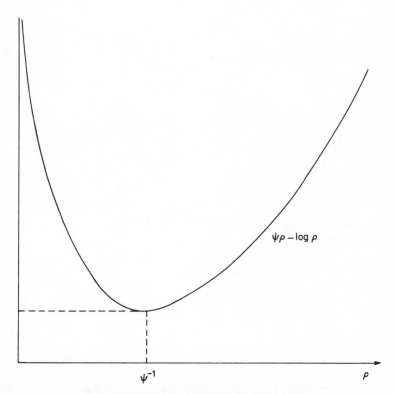

Fig. 17.2.1 Graph of the function $\psi\rho - \log\rho$.

freely in $\mathbb{R}^p$, where λ is a Lagrangian multiplier. The parametrization of the problem in terms of λ rather than of ρ leads to a nice picture of events, best appreciated when one has looked at the simplest case of all: that of a single role. In this case the quantity to be maximized is

$$\frac{1}{V}\log P(N.) - \lambda\rho \sim \frac{\psi\rho^2}{2} - \rho\log\rho - (\lambda - \gamma - 1)\rho \qquad (6)$$

with exact equality in the thermodynamic limit. The condition that expression (6) should be stationary with respect to ρ is

$$\psi\rho - \log\rho + \gamma = \lambda \qquad (7)$$

Now, the graph of $\psi\rho - \log\rho$ is as in Fig. 17.2.1. As ρ increases from zero the function first decreases monotonically from $+\infty$ to a minimum of $1 + \log\psi$ at $\rho = \rho_C = \dfrac{1}{\psi}$, after which it increases monotonically to $+\infty$ again. Equation (7)

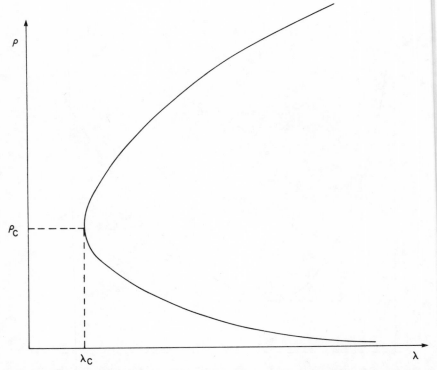

Fig. 17.2.2 The two branches of the function inverse to $\psi\rho - \log\rho$ correspond to sub-critical and super-critical regimes.

thus has two solutions for ρ or none according as λ is greater or less than $\lambda_c := 1 + \log \psi + \gamma$. The graph of the ρ solutions as a function of λ is thus as in Fig. 17.2.2. The solution locus has two branches, meeting at the point (λ_c, ρ_c). One is on the lower branch or the upper branch according as $\rho < \rho_c$ or $\rho > \rho_c$.

In fact, the lower branch is a locus of solutions (i.e. of most probable values) corresponding to the sub-critical or *uncondensed* phase; the upper branch corresponds to solutions in the super-critical or *condensed* phase. One passes from one branch to the other at the density ρ_c, which is exactly the critical density. We shall now justify this interpretation, and give its multi-role version.

By 'subcriticality' we shall mean exactly that spatial equidistribution is at least a metastable state, i.e. that a compartmental distribution shows at least a local maximum at a configuration in which all compartments are of identical constitution. As we shall see in section 5, it will be a feature of the role-adaptive model that this state may indeed be no more than metastable, in that the subcritical solution is overtaken by a supercritical solution at a density smaller than the critical density ρ_c. This is a very significant effect.

Theorem 2.2

(i) *The stationarity conditions locating a maximum of expression (5) are, in the thermodynamic limit,*

$$\sum_\beta \psi_{\alpha\beta}\rho_\beta - \log \rho_\alpha + \gamma_\alpha = \lambda \qquad (\alpha = 1, 2, \ldots, p) \qquad (8)$$

(ii) *Equations (8) have in general several solutions; we classify them as* lower branch *or* upper branch *according as the $p \times p$ matrix*

$$M := (\rho_\alpha^{-1}\delta_{\alpha\beta} - \psi_{\alpha\beta}) \qquad (9)$$

is positive definite or not. Lower and upper branch solutions correspond to uncondensed and condensed phases respectively.

(iii) *A necessary condition for acceptability of an upper branch solution is that the quantity*

$$\frac{\partial \rho}{\partial \lambda} = -\mathbf{1}^\mathsf{T} M^{-1} \mathbf{1} \qquad (10)$$

be positive and that the matrix M have a single negative eigenvalue.

(iv) *A condition equivalent to (iii) is that the matrix*

$$\tilde{M} := M - \mathbf{1}(\mathbf{1}^\mathsf{T} M^{-1} \mathbf{1})^{-1}\mathbf{1}^\mathsf{T} \qquad (11)$$

be non-negative definite.

Proof It follows from (3) that we have the expression in the thermodynamic limit

$$F(\rho.) := \lim V^{-1} \log P(N.)$$

$$= \text{constant} + \tfrac{1}{2}\sum_\alpha \sum_\beta \psi_{\alpha\beta}\rho_\alpha\rho_\beta - \sum_\alpha \rho_\alpha \log \rho_\alpha + \sum_\alpha (1+\gamma_\alpha)\rho_\alpha \tag{12}$$

The stationarity conditions $\partial F/\partial \rho_\alpha = \lambda$ for the Lagrangian form (5) then yield (8), and we find also that

$$\left(\frac{\partial^2 F}{\partial \rho_\alpha \partial \rho_\beta}\right) = -M \tag{13}$$

Assertion (i) is then immediate, and assertion (ii) follows from the analysis of Theorem 16.5.2.

It follows immediately from (8) that

$$M\frac{\partial \rho.}{\partial \lambda} = -1 \tag{14}$$

and so that

$$\frac{\partial \rho}{\partial \lambda} = \mathbf{1}^\mathsf{T}\frac{\partial \rho.}{\partial \lambda} = -\mathbf{1}^\mathsf{T} M^{-1}\mathbf{1} \tag{15}$$

as asserted in (10). Let us set $\rho. = \bar{\rho}. + \Delta$. Then if $\bar{\rho}.$ indeed locates a local maximum of $F(\rho.)$ subject to prescribed ρ then it must be that $\Delta^\mathsf{T} M\Delta \geqslant 0$ for all Δ such that $\Delta^\mathsf{T}\mathbf{1} = 0$. Let us decompose an arbitrary perturbation Δ into components Δ_i and Δ_0 in and out of the plane $\Delta^\mathsf{T}\mathbf{1} = 0$:

$$\Delta = \Delta_i + \Delta_0 \tag{16}$$

where

$$\Delta_0 = \frac{(\Delta^\mathsf{T}\mathbf{1})(M^{-1}\mathbf{1})}{\mathbf{1}^\mathsf{T} M^{-1}\mathbf{1}} \tag{17}$$

We find then that

$$\Delta^\mathsf{T} M\Delta = \Delta_i^\mathsf{T} M\Delta_i + \Delta_0^\mathsf{T} M\Delta_0$$
$$= \Delta^\mathsf{T}\tilde{M}\Delta + (\Delta^\mathsf{T}\mathbf{1})^2(\mathbf{1}^\mathsf{T} M^{-1}\mathbf{1})^{-1} \tag{18}$$

The particular decomposition (16) was chosen so that Δ_i would be a permissible perturbation, i.e. $\Delta_i^\mathsf{T}\mathbf{1} = 0$, and also so that $\Delta_i^\mathsf{T} M\Delta_0 = 0$, so that $\Delta^\mathsf{T} M\Delta$ decomposes into separate quadratic forms, as in (18). Since $\Delta_i^\mathsf{T}\mathbf{1} = 0$ we must have $\Delta_i^\mathsf{T} M\Delta_i \geqslant 0$ or $\Delta^\mathsf{T}\tilde{M}\Delta \geqslant 0$ for all Δ. That is, $\tilde{M}$ must be positive definite under all circumstances. If $\mathbf{1}^\mathsf{T} M^{-1}\mathbf{1} > 0$ then M is positive definite and we are in the lower branch case. We can then be in the upper branch case only if $\mathbf{1}^\mathsf{T} M^{-1}\mathbf{1} < 0$, and the decomposition (18) then implies that M has a single negative eigenvalue. ∎

There is only one uncondensed solution; that obtained by following the branch starting from $\rho. = 0$ at $\lambda = +\infty$. However, there can be a multitude of condensed (upper branch) solutions; one can decide between these only by testing which gives the global maximum of $P(N.)$. We follow up this point in the next section.

3. EQUILIBRIUM FAR FROM CRITICALITY

Consider the equilibrium equations (2.8) for the ρ_α when λ is large, so that conditions are far from criticality. The single-role version (2.7) makes clear that either ρ is of order λ or that $-\log \rho$ is of order λ, i.e. ρ is of order $e^{-\lambda}$. This exactly describes the behaviour on the two branches for large λ. Consider now the multi-role case; let us assume that ρ_α is of order λ for α in a set A, and that $-\log \rho_\alpha$ is of order λ in the complementary set $\bar{A}$. We shall term A the *basis* of the solution of (2.8), for it is essentially the set of roles which are positively represented, for large λ. If A is empty then ρ is itself exponentially small in λ and we are on the unique subcritical branch. If A is non-empty then ρ is itself of order λ and we are on a super-critical branch. The question is then: which choice of A corresponds to the dominant regime, i.e. to the value of ρ. globally maximizing $P(N.)$?

Theorem 3.1 *Suppose ρ large and A the set of α for which ρ_α is of order ρ. Then*

(i) *The basis A corresponding to the dominant regime is that which minimizes*

$$D_A := \mathbf{1}_A^\mathsf{T} \psi_{AA}^{-1} \mathbf{1}_A \tag{1}$$

subject to

$$\psi_{AA}^{-1} \mathbf{1}_A > 0 \tag{2}$$

$$\psi_{\bar{A}A} \psi_{AA}^{-1} \mathbf{1}_A \leqslant \mathbf{1}_{\bar{A}} \tag{3}$$

and then

$$\rho_A \sim (\rho/D_A) \psi_{AA}^{-1} \mathbf{1}_A \tag{4}$$

(ii) *This dominant basis has the additional property that ψ_{AA} has just one positive eigenvalue. Equivalently, the matrix $\psi_{AA} - D_A^{-1} \mathbf{1}_A \mathbf{1}_A^\mathsf{T}$ is non-positive definite.*

Here we have used the notation for sub-vectors and sub-matrices defined in section 1.10. Thus, for example, ρ_A is the vector $\{\rho_\alpha; \alpha \in A\}$.

Proof Equations (2.8) for α in A yield

$$\rho_A = \lambda \psi_{AA}^{-1} \mathbf{1}_A \tag{5}$$

if we retain only terms of order λ. Expression (5) must be positive, whence we deduce (2). The requirement that the equations (2.8) for α in $\bar{A}$ have a solution for the remaining ρ_α yields condition (3). Normalization of (5) by $\mathbf{1}_A^\mathsf{T} \rho_A = \rho$ yields

$$\lambda = D_A^{-1} \rho$$

and determination (4).

Inserting (5) into $\log P(N.)$ we find the dominant term to be

$$\frac{\lambda^2 V}{2} D_A = \frac{V \rho^2}{2 D_A}$$

whence it follows that the global maximum corresponds to the admissible A for which D_A is minimal.

To prove assertion (ii), note from the definition (2.9) that, with an appropriate reordering of roles

$$M = \begin{pmatrix} M_{AA} & M_{A\bar{A}} \\ M_{\bar{A}A} & M_{\bar{A}\bar{A}} \end{pmatrix} \sim \begin{pmatrix} -\psi_{AA} & -\psi_{A\bar{A}} \\ -\psi_{\bar{A}A} & -\psi_{\bar{A}\bar{A}} + \Lambda \end{pmatrix}$$

Here Λ is a diagonal matrix with entries ρ_α^{-1} $(\alpha \in \bar{A})$ on its diagonal. It is consequently positive and exponentially large in λ. Thus $\xi' M \xi > 0$ if $\xi_{\bar{A}} \neq 0$, and so M has exactly as many negative eigenvalues as does $-\psi_{AA}$. Moreover, it is readily verified that $-M^{-1} \sim \begin{pmatrix} \psi_{AA}^{-1} & \cdot \\ \cdot & \end{pmatrix}$. Assertions (ii) thus transfer from the corresponding assertions for M of Theorem 2.2. ■

Exercises and comments

1. One can ask to what extent the properties asserted in (i) of the theorem imply those asserted in (ii), and conversely. There is at least one implication. We know from (ii) that if A is the basis of the dominant regime then

$$\xi^{\mathsf{T}} (\psi_{AA} - D_A^{-1} \mathbf{1}_A \mathbf{1}_A^{\mathsf{T}}) \xi \leq 0 \tag{6}$$

Let C be a subset of A and choose

$$\xi = \begin{pmatrix} \psi_{CC}^{-1} \mathbf{1}_C \\ \cdot \end{pmatrix}$$

Note that inequality (6) then implies that

$$D_C - D_A^{-1} D_C^2 \leq 0$$

which implies either that $D_C < 0$ or $D_C > D_A$. The negative definiteness condition (6) thus implies that no subset of A is possible as a basis for the dominant regime.

4. THE LATTICELESS ISING MODEL

Suppose that there are two possible roles, which we shall regard as 'orientation'. These are indicated by $\alpha = +, -$, and are supposed symmetric in that

$$\begin{aligned} \gamma_+ &= \gamma_- = 0 \\ \psi_{++} &= \psi_{--} = \psi_1 \\ \psi_{+-} &= \psi_{-+} = \psi_2 \end{aligned} \tag{1}$$

say. This choice is motivated by the Ising model of ferromagnetism, in which molecules are arranged at the nodes of a lattice $\mathbb{Z}^d$. The molecules have magnetic spin which is quantized either 'up' or 'down'; hence the $\pm$ notation. There is an interaction between neighbours, with a pairwise potential which takes one value for a pair of like-oriented spins, another value for a pair of unlike-oriented spins. As one changes parameters in such a way as to favour like orientation there comes the transition point of 'alignment' or magnetization, when there is a general tendency for the molecules to line up together one way or the other. (See, for example, Reichl, 1980, pp. 280–292.)

To calculate the exact statistics of the Ising model on $\mathbb{Z}^d$ has been a standing problem for decades. The case $d = 1$ (when there is no phase transition) is trivial: the case $d = 2$ was solved by Onsager and several alternative treatments have since been developed. The cases $d \geqslant 3$ remain unsolved.

The specification (1) amounts to the specification of a 'latticeless' Ising model in which bonds are created at random rather than existing between spatial neighbours only. However, the bonding pattern interacts with the orientation pattern. The symmetry expressed in (1) implies that one need only distinguish 'like-like' and 'like-unlike' bonds (see Exercise 3).

The parameter which varied in the classical Ising model is temperature T, affecting bond strengths by the usual formula

$$\psi_{\alpha\beta} = e^{-U_{\alpha\beta}/kT}$$

where $U_{\alpha\beta}$ is the potential energy of an $\alpha\beta$ bond. For this latticeless Ising model we shall rather take molecular density ρ as the independent parameter.

Theorem 4.1 *The latticeless Ising model undergoes condensation as the density ρ increases through*

$$\rho_C = \frac{2}{\psi_1 + \psi_2} \tag{2}$$

If $\psi_1 > \psi_2$ then there is a further phase transition: the model undergoes alignment as the density ρ increases through

$$\rho_A = \frac{2}{\psi_1 - \psi_2} \tag{3}$$

Proof The lower branch solution is one of equidistribution: $\rho_+ = \rho_- = \frac{1}{2}\rho$. This branch will terminate and condensation occur when the matrix

$$M = \begin{pmatrix} 2\rho^{-1} - \psi_1 & -\psi_2 \\ -\psi_2 & 2\rho^{-1} - \psi_1 \end{pmatrix} \tag{4}$$

becomes singular. The determinant $|M|$ is zero for ρ equal to ρ_C or ρ_A, and $\rho_A < 0$ or $\rho_A > \rho_C$ according as $\psi_1 < \psi_2$ or $\psi_1 > \psi_2$. The condensation transition thus takes place at $\rho = \rho_C$.

The equipartition solution remains valid on the upper branch for all ρ if $\psi_1 < \psi_2$, or until density ρ_A is reached if $\psi_1 > \psi_2$. That is, matter is equidistributed between the two orientations, even though it is not equidistributed over space for $\rho > \rho_C$. In the case $\psi_1 > \psi_2$ the matrix (4) develops *two* negative eigenvalues, unacceptably, when ρ reaches ρ_A. The global maximum of $P(N.)$ is then attained at values for which ρ_+ and ρ_- are unequal, corresponding to the emergence of a net alignment in the assembly. ■

Alignment ('magnetization') is best exhibited by examining the distribution

$$P(N.) \propto \frac{\exp\{V^{-1}[\psi_1(N_+^2 + N_-^2) + 2\psi_2 N_+ N_-]\}}{N_+! N_-!} \qquad (N_+ + N_- = N) \quad (5)$$

directly. The stationarity condition (2.8) is

$$(\psi_1 - \psi_2)(2\rho_+ - \rho) = \log\left(\frac{\rho_+}{\rho - \rho_+}\right) \tag{6}$$

This has a solution $\rho_+ = \frac{1}{2}\rho$. For $\psi_1 < \psi_2$ and for $\psi_1 > \psi_2$, $\rho < \rho_A$ the distribution (5) for N_+ has a single maximum at $N_+ = N/2$, corresponding to the root $\rho_+ = \frac{1}{2}\rho$ of (6), and to statistical equipartition between the two orientations. However, for $\psi_1 > \psi_2$ and $\rho > \rho_A$ this maximum divides into two separate maxima, symmetrically located about a local *minimum* at $N/2$, just as in Fig. 13.6.1. This corresponds to the emergence of two more real solutions of (6). As for the deviation from spatial equipartition in section 13.6, the distribution is symmetric, but breaks symmetry. That is, equipartition fails, but a deviation from equipartition in one way or the other is equally likely.

This direct examination of $P(N.)$ exhibits the magnetization transition very clearly. Note, however, that it reveals no sign of the preceding condensation transition.

Note also that both transitions are 'continuous' in that the $\{\rho_\alpha\}$ solution (and all 'local' statistics) are continuous through the transition. For example, as ρ increases through ρ_A the single maximum of $P(N_+)$ divides continuously into two maxima, i.e. the equipartition solution bifurcates continuously into the two asymmetric solutions. This is to some extent because of the symmetry of the model in the $\pm$ orientations (see Exercise 3 and the next section).

Exercises and comments

1. Our 'latticeless' Ising model is very similar in outcome to the Bragg-Williams approximation to the classic model (see Reichl, 1980, p. 285). This approximation consists in assuming the proportion of $++$ neighbours to be equal to $(\rho_+/\rho)^2$ and then leads in fact to the distribution (5). The Bragg-Williams approximation will naturally not exhibit a condensation point, based as it is on a fixed-lattice model.

2. Confirm that the solution pattern described by Theorem 4.1 agrees for large ρ with that characterized by Theorem 3.1(i).

3. Examine behaviour of the model in the case $\gamma_+ > \gamma_-$. This partial loss of symmetry corresponds to the presence of an external magnetic field, biasing the assembly in favour of a + orientation.

5. THE FARMER-TRADER MODEL

This is again a two-role model. However, in distinction to the Ising model, there is asymmetry between the roles.

We suppose that the units correspond to individuals in a society, and that the labelling $\alpha = 1, 2$ correspond to the roles of farmer or of trader respectively. We suppose γ_1 considerably larger than γ_2, corresponding to the fact that, if population is so sparse that human contacts are few, then it is much more reasonable to take the role of farmer than that of trader. It will be to someone's advantage to assume the role of trader only when population density is high enough to make human contact sufficiently frequent. We shall assume that $\psi_{12} = \psi_{21} > 0$ (corresponding to the possibility of client–agent relations between farmer and trader) and that $\psi_{22} > 0$ (corresponding to the possibility of mercantile relations between traders). However, we shall assume $\psi_{11} = 0$; farmers are too similar for any direct economic relationship to be advantageous.

A direct examination of $P(N_1) = P(N_1, N - N_1)$ reveals the possibility of a *discontinuous* transition in this case, essentially because of the asymmetry between the roles. Instead of the splitting of a maximum of this distribution into two (with increasing ρ, say) one can have the sequence of events illustrated in Fig. 17.5.1. That is, $P(N_1)$ can have a single well-defined maximum at low density. As density increases a subsidiary local maximum can appear at a completely separated value of N_1/N. As density increases further this subsidiary maximum can grow and overtake the first maximum. With appropriate choices of parameters this overtaking indeed occurs for the farmer–trader model, and corresponds to a discontinuous transition in the thermodynamic limit.

This direct study is illuminating, but does not show how changes in role-distribution correlate with the event of condensation, which one might term *economic integration* in this context. The following theorem summarizes assertions evident from numerical study rather than mathematical proof.

Theorem 5.1 *For the farmer–trader model the upper branch of the ρ/λ relation need not be monotonic. It is then possible that an upper branch solution can overtake the lower branch solution at a population density lower than the critical condensation density ρ_C. This overtaking corresponds to a transition from the uncondensed to a condensed state at which there is a discontinuity in role-distribution. Otherwise expressed, the model can show a discontinuous transition*

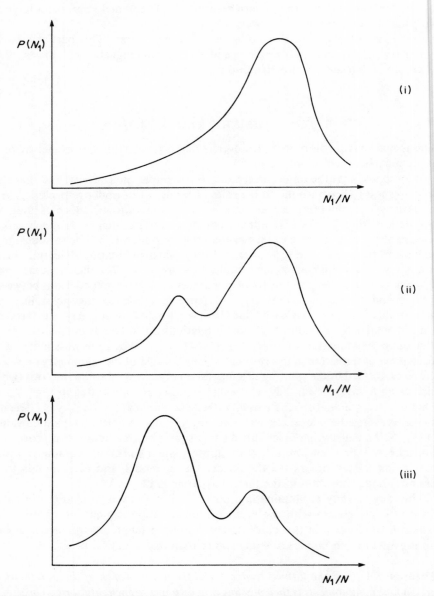

Fig. 17.5.1 A situation in which a new maximum appears and 'overtakes' the old one, giving a discontinuous transition in the thermodynamic limit. This is to be compared with the situation of Fig. 13.6.1, in which an existing maximum splits; this gives a continuous transition.

which can be characterized simultaneously as an economic integration and as a shift from a rural to a mercantile economy.

These assertions follow from the observed form of ρ/λ curve when one solves equations (2.8). For some values of parameters both branches are monotonic, as in Fig. 17.5.2. In these cases, λ changes continuously with increasing ρ, even though the condensation point, and the corresponding solution $\{\rho_\alpha\}$ also in fact changes continuously. For others the situation is more that presented in Fig. 17.5.3. The upper branch is not monotonic, so that there can be several values of λ corresponding to a given value of ρ. Not all of these correspond to acceptable solutions; stability demands that $\partial\rho/\partial\lambda$ be positive on the upper branch (cf. Theorem 2.2(iii)). However, with appropriate choice of parameters the farmer–trader model does indeed demonstrate such multiple λ-solutions at values of ρ less than ρ_C. Even more: there can be a value $\rho_T < \rho_C$ at which the upper branch solution overtakes that on the lower branch. In that case, at $\rho = \rho_T$ there is a transition from the lower branch to the upper (implying passage to the integrated phase) and also a discontinuity in λ, implying a simultaneous discontinuity in the actual role-distribution $\{\rho_\alpha\}$.

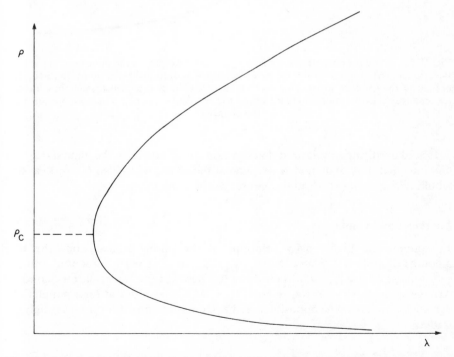

Fig. 17.5.2 A situation in which the transition to the integrated ('condensed') state at the critical density ρ_C is continuous.

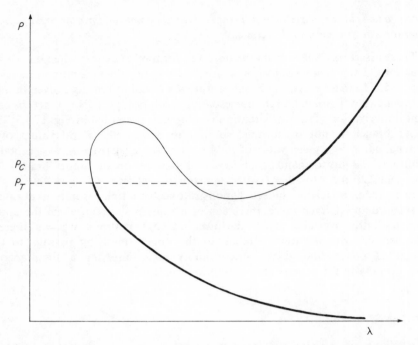

Fig. 17.5.3 A situation where the transition is discontinuous. At the density ρ_T (less than the critical density ρ_C) the solution point jumps discontinuously to a point on the upper branch of the solution locus. This corresponds to a simultaneous economic integration (the change of regime) and shift from a rural to a mercantile economy (the discontinuity of the change).

It need hardly be emphasized that the model is schematic in the highest degree, and intended only to demonstrate some of the effects which can occur. Realism would demand a host of other features in the model.

Exercises and comments

1. Examine the high-density solutions of the farmer–trader model by the methods of section 3. Show that if $\psi_{22} < \psi_{12}$ then $A = (1, 2)$ for the optimal solution, while if $\psi_{22} > \psi_{12}$ then $A = (2)$. That is, if the trader–farmer bond is stronger than the trader–trader bond then both roles occur at high population densities, but that, in the opposite case, the population consists overwhelmingly of traders.

PART V

Spatial models; random fields

This part is rather distinct from the others, in that the concepts of reversibility and weak coupling play no great role. However, enough of our systems have had their components spatially distributed that there is reason to consider the spatial aspect. Also, many of the features of spatial processes which were at least once regarded as somewhat puzzling make sense when one sees the process as a spatio-temporal process, perhaps in its equilibrium state. Finally, the phenomenon of stochastic invariance under some group of transformations natural to the context, which we have found so pervasive, strikingly pervades again the theory of spatial processes.

CHAPTER 18

Random fields

1. THE SPATIAL FORMALISM: SPACE AS A GRAPH

A change of context often forces a change in notation, as mutually inconsistent traditions meet. We have used 'x' rather consistently to denote a generic random variable, and, in particular, the process or state variable. We shall now replace it for this purpose by ξ, and use x to denote spatial coordinate: position in space. Both uses of x are well established, but the second is earlier, and so has precedence. The symbol $\mathscr{X}$ will now denote the set of possible spatial sites rather than state space.

We have considered *temporal processes* $\{\xi(t)\}$: random functions of time. We now wish to consider *spatio-temporal processes* $\{\xi(x, t)\}$, or sometimes purely *spatial processes* $\{\xi(x)\}$. One does not have to seek far to find quantities whose variation in space is of equal interest to their variation in time: meteorological variables, the fertility of a field, the density and nature of plant cover, particle density. The space and time coordinates can be on a par, in that they take values in sets of comparable cardinality, and the process may show stochastic invariance under transformations of space as well as of time. On the other hand, it is along the time axis that one has the 'causal flow', at least in those models for which relativistic effects are absent or negligible.

A spatial process (i.e. a random function of x, or of x and t) is often referred to as a *random field*. The term is not entirely a good one, since the physical concept of a field conveys more than just the idea of a quantity defined throughout space. It often implies also that this quantity mediates interaction between physical objects imbedded in that space. However, the use of the term in the weaker sense now has currency.

One begins by thinking of the space as a conventional physical space, a Euclidean space of d dimensions, say, so that $x \in \mathbb{R}^d$. However, for many purposes the aspects of distance and dimension play no role, and it is natural at some level to regard the sites x at which the field is defined as the nodes of a graph. The relationship between two sites x, x' is that of *neighbourhood* of some degree (depending upon the pattern of arcs connecting them in the graph) rather than that of relative orientation and mutual distance.

As stated, one begins indeed by regarding x as denoting position in physical space, $\mathbb{R}^d$. However, it is natural, in crystallographic and other contexts, to consider the *discrete* case, in which the sites are the points of a lattice L, rather than of a continuous space. The cubic lattice $\mathbb{Z}^d$ is the obvious example. Now, $\mathbb{Z}^d$ is embedded in $\mathbb{R}^d$, so that distances and vector distances between points in the lattice are defined. However, for many purposes these turn out to be irrelevant. What is relevant is the implicit assumption that the 2^d sites $x \pm e_j$ ($j = 1, 2, \ldots, d$) are regarded as *neighbours* of x. (Here e_j is the vector corresponding to a unit displacement along the jth axis.) It is this neighbourhood relationship rather than a distance relationship which is important for many purposes.

One is thus simplifying the lattice L to a graph Γ on the set of prescribed nodes $\mathscr{X}$. The nodes x, x' are connected by an arc in Γ if and only if they are regarded as physical neighbours. Usually the neighbourhood relationship is regarded as symmetric, so that the arcs are undirected. It is true that the networks of Part III constituted examples of spatial systems described by a graph in which the nodes were indeed directed, because they allowed flows in definite directions. However, we scarcely consider systems with flow in this part and so shall consider the neighbourhood relationship symmetric.

To fix notation, then, we denote the set of possible spatial sites by $\mathscr{X}$. If this is endowed with a neighbourhood relation we shall denote the consequent graph on $\mathscr{X}$ by Γ. If it is further endowed with a distance measure, and so becomes a metric space, we shall denote the consequent lattice (which may be continuous) by L.

Note that the graph Γ is *fixed* and describes *spatial relationships*. It is not to be confused with the graphs representing polymers in Part IV, which were *random* and described *interactions*. The interactions are specified in the present case by the probabilistic structure, which has yet to be superimposed upon the spatial structure. That is, by specification of the joint distribution of field value at the different sites. We consider this probabilistic structure in the following sections.

In this chapter and the next we consider purely spatial processes, so that the field ξ is a function $\xi(x)$ of x alone. If Λ is a subset of the nodes of Γ then the partial description of the field $\{\xi(x); x \in \Lambda\}$ is sometimes written as ξ_Λ. The whole field, $\xi_{\mathscr{X}}$, will usually be written simply as ξ.

There are a few necessary definitions. The *complement* of Λ in $\mathscr{X}$, i.e. $\mathscr{X} \setminus \Lambda$, is written $\bar{\Lambda}$. The set consisting of Λ itself and all neighbours of members of Λ is called the *neighbourhood* of Λ, written $\mathscr{N}(\Lambda)$. The difference $\mathscr{N}(\Lambda) \setminus \Lambda$ is termed the *boundary* of Λ, written $\partial\Lambda$. That is, these are the nodes which are neighbours of members of Λ without themselves belonging to Λ; a subset of $\bar{\Lambda}$. In the case when Λ consists of the single site x we write $\mathscr{N}(\Lambda)$ and $\partial\Lambda$ as $\mathscr{N}(x)$ and ∂x respectively.

A *clique* is a set of nodes any pair of which are neighbours. A set consisting of a single node is regarded as eligible; i.e. a site is self-neighbour, for this purpose. Cliques tend to be smaller than one would think; see Exercise 1.

Note that what we have termed simply 'neighbours' are what physicists often refer to as 'nearest neighbours'. This reflects the fact that one can have neighbours of various degrees; see Exercise 2.

Exercises and comments

1. Suppose that L is the cubic lattice $\mathbb{Z}^d$, with the neighbours of x being the sites $x \pm e_j$ ($j = 1, 2, \ldots, d$). Note that the cliques are all of size one or two: single sites and pairs of neighbouring sites.

2. Let us define $\mathcal{N}^{(p)}(\Lambda)$ as $\mathcal{N}(\mathcal{N}^{(p-1)}(\Lambda))$ ($p = 1, 2, \ldots$) with $\mathcal{N}^{(0)}(\Lambda) = \Lambda$. Then $\mathcal{N}^{(p)}(\Lambda)$ is the set of sites which are neighbours of Λ at at most p steps. The set $\mathcal{N}^{(p)}(\Lambda) - \mathcal{N}^{(p-1)}(\Lambda)$ is the set of sites which are neighbours of Λ at exactly p steps (or, popularly, 'at $p - 1$ removes').

2. THE PROBABILISTIC FORMALISM: ORDERS OF INTERACTION

Let us assume, for simplicity, that the field variable can only take a discrete set of values at each site, so that probabilities $P(\xi_\Lambda)$ and conditional probabilities $P(\xi_{\Lambda_1} | \xi_{\Lambda_2})$ have their naïve definitions, at least for finite Λ.

Then, at least for finite Γ, the quantity $P(\xi)$ is well-defined. If $P(\xi)$ were an equilibrium Gibbs distribution then we would have

$$P(\xi) \propto e^{-V(\xi)/kT}$$

where $V(\xi)$ is the potential energy of configuration ξ, T is absolute temperature and k is Boltzmann's constant. Let us in any case simply define a normalized potential $U(\xi)$ by

$$P(\xi) = e^{-U(\xi)} \tag{1}$$

The formal point of this is that multiplicative decompositions of P correspond to additive decompositions of U, which seem more natural and physically meaningful.

The hope is, of course, that a meaning can be attached to $P(\xi)$ and $U(\xi)$ even in the case of infinite Γ and continuously varying ξ. Continuous variation of ξ will be taken into account by defining densities relative to an appropriate measure. There are two devices for coping with the fact that Γ may be infinite.

One is to regard the conditional probabilities

$$P(\xi_\Lambda | \xi_{\bar{\Lambda}}) = \frac{P(\xi)}{\sum_{\xi_\Lambda} P(\xi)} \tag{2}$$

for finite Λ as defining the statistics of the field. That is, one considers the field distribution inside the finite set Λ conditional on specification of field outside. The trouble with this is that these conditional probabilities obey a number of rather unevident consistency conditions, to the point that one can ensure consistency of a given specification only by deriving it from a specification of $P(\xi)$ itself, as in (2). The other point, and a most interesting one, is that there may be more than one $P(\xi)$ consistent with a self-consistent specification of $P(\xi_\Lambda | \xi_{\bar{\Lambda}})$ for varying finite Λ. This effect corresponds exactly to changes of state in physical systems (more exactly, to coexistence of different physical states) and has drawn great attention in recent years (see Preston (1974); Kindermann and Snell (1980) and particularly the papers by Dobrushin quoted in these).

Another, not unrelated, way of handling the case of infinite Γ is to assume that the ratio $P(\xi)/P(\xi')$ is well-defined if the fields ξ, ξ' differ only at a finite number of sites. To this end, one supposes that there is a 'ground-state' at each site x: $\xi(x) = 0$, say (see Exercise 1 for enumeration of the abuses of the term 'state' in this context). That is, the value 0 is a kind of reference value, corresponding to absence of whatever may be of interest—excitation, matter, life, infection, etc. Define the null field $\xi = 0$ as the field adopting the ground-state $\xi(x) = 0$ at all x. Then the field ξ is regarded as being built up from the null-field by modification at successive sites, and $P(\xi)/P(0)$ has a clear evaluation if ξ is non-null at only finitely many sites.

Define the operator $C_x(\xi)$ which modifies the value of field at site x to $\xi(x)$, for a prescribed ξ. That is, if ξ, η are two values of the field and $f(\xi)$ a function of field then $C_x(\xi)f(\eta)$ equals $f(\bar{\eta})$, where

$$\bar{\eta}(y) = \begin{cases} \eta(y) & y \neq x \\ \xi(y) & y = x \end{cases}$$

Theorem 2.1 *If ξ deviates from the null field at only finitely many sites then the potential $U(\xi)$ can be written*

$$U(\xi) = G + \sum_{x_1} G_{x_1}(\xi(x_1)) + \sum_{x_1 < x_2} \sum G_{x_1 x_2}(\xi(x_1), \xi(x_2))$$

$$+ \sum_{x_1 < x_2 < x_3} \sum \sum G_{x_1 x_2 x_3}(\xi(x_1), \xi(x_2), \xi(x_3)) + \dots \qquad (3)$$

where the terms G_{x_1}, $G_{x_1 x_2}$, ... are zero if any $\xi(x)$ argument takes the ground-state, and have the evaluations

$$G = U(0)$$
$$G_{x_1} = (C_{x_1} - 1)U(0) \qquad (4)$$
$$G_{x_1 x_2} = (C_{x_1} - 1)(C_{x_2} - 1)U(0)$$

etc.

The functions G and operators C of (4) have ξ arguments which have been suppressed for notational simplicity. The notations $x_1 < x_2 < x_3$, etc. are meant to indicate that a given set of values (x_1, x_2, x_3) occurs only once, permutations being counted as repetitions, and with each site occurring at most once in a set. In fact, (3) could be written more compactly as

$$U(\xi) = \sum_{\Lambda} G_{\Lambda}(\xi_{\Lambda})$$ (5)

The inverse relations (4) can correspondingly be written

$$G_{\Lambda}(\xi_{\Lambda}) = \prod_{x \in \Lambda} (C_x(\xi) - 1) U(0)$$

$$= \sum_{M \subset \Gamma} (-)^{|\Lambda - M|} U(\xi_{(M)})$$ (6)

where $|\Lambda - M|$ is the number of elements in $\Lambda - M$, and

$$\xi_{(M)}(x) = \begin{cases} \xi(x) & x \in M \\ 0 & x \notin M \end{cases}$$ (7)

The point of the successive sums in (3) is that they represent successive levels of interaction. That is, G_{x_1} is the change in potential $C_{x_1} U(0) - U(0)$ induced by modification of a null-field to $\xi(x_1)$ at the single site x_1; $G_{x_1 x_2}$ is the *pairwise interaction* between sites x_1 and x_2, in that it is the deviation of $C_{x_1 x_2} U(0)$ from $U(0) + [C_{x_1} U(0) - U(0)] + [C_{x_2} U(0) - U(0)]$. Correspondingly $G_{x_1 x_2 x_3}$ is the third-order interaction: the component of a three-site interaction not explained by a sum of lower-order interactions, etc.

Proof This is immediate. Setting

$$D_x = C_x - 1$$

we have

$$U(\xi) = \left[\prod_x C_x \right] U(0)$$

$$= \left[\prod_x (1 + D_x) \right] U(0)$$ (8)

Expanding expression (8) in powers of the D_x we derive the representation (3), (4). ■

This argument is correct if ξ equals the null field outside a finite set Λ_0 say, because then $D_x = 0$ for x outside Λ_0. However, representation (3) is formally correct for any ξ if interactions past a finite order m are zero, i.e. if $G_{\Lambda}(\xi_{\Lambda}) = 0$ for $|\Lambda| > m$. The representation is correct in that all contributions are accounted for in (3), even if the sum diverges, i.e. $P(\xi) = 0$. If interactions persist up to any order then there may be a deeper convergence problem.

If we make the convention that an interaction $G_{x_1 x_2 \ldots}(\xi(x_1), \xi(x_2), \ldots)$ is zero if any of the site values x_i are repeated, then we can write (3) rather as

$$U(\xi) = G + \sum_{x_1} G_{x_1} + \frac{1}{2!} \sum_{x_1} \sum_{x_2} G_{x_1 x_2} + \frac{1}{3!} \sum_{x_1} \sum_{x_2} \sum_{x_3} G_{x_1 x_2 x_3} + \cdots$$

where the summations are free. This is equivalent to *defining D_x to be nilpotent,*

$$D_x^j = 0 \qquad (j = 2, 3, \ldots)$$

and modifying (8) to

$$U(\xi) = e^{\sum_x D_x} U(0)$$

Finally, we shall need the following result, almost but not quite self-evident.

Theorem 2.2 *Suppose there is a factorization*

$$P(\xi) = P_1(\xi) P_2(\xi)$$

of $P(\xi)$ which separates $\xi(x_1)$ and $\xi(x_2)$, in that $P_1(\xi)$ is functionally independent of $\xi(x_2)$ and $P_2(\xi)$ functionally independent of $\xi(x_1)$. Then

$$G_\Lambda(\xi_\Lambda) = 0$$

for any Λ which contains both sites x_1 and x_2.

Proof We can achieve the representation

$$P_i(\xi) = \exp\left(-\sum_\Lambda G_\Lambda^{(i)}(\xi) \right)$$

for each factor, and it follows from the definition of G_Λ that

$$G_\Lambda = G_\Lambda^{(1)} + G_\Lambda^{(2)}$$

But $G_\Lambda^{(1)} = 0$ if Λ contains site x_2 and $G_\Lambda^{(2)} = 0$ if Λ contains site x_1, whence the assertion follows. ∎

Exercises and comments

1. We have already listed some of the uses and abuses of the term 'state' in Chapter 1, 'state' in the Markov sense and 'state of matter' being two clearly motivated but conflicting usages. The term 'ground-state' introduced in this section simply refers to a particular local value of field, and is comparable to our speaking in Chapter 10 of the 'state' of a component which is only part of the full system. The quite peculiar and confusing usage is to describe the *statistics* of a situation as a 'state'. For example, distribution (1) is by some referred to as a 'Gibbs state' with potential U, and the spatially Markov processes of the next section are referred to as a 'Markov state'. The usage presumably derives from the idea that statistics correspond to a physical regime, i.e. to a 'state of matter'.

2. Theorem 2.2 is a form of *Brook's theorem* (Brook, 1964) which states that if

$P(\xi)$ has several factorizations $\mathscr{F}_i$ then there exists a factorization $\mathscr{F}$ separating all arguments which are separated in any $\mathscr{F}$.

3. MARKOV RANDOM FIELDS

The notion of characterizing interaction as 'local' in some sense is a persistent one, with the aim of reducing interaction effects to their essentials. For temporal processes one expresses local interaction in time by requiring the Markov property. What is natural or possible as the spatial analogue of the Markov property was for a long time regarded as unclear by probabilists. Physicists of course found it natural to specify 'nearest neighbour interactions only' as an expression of local interaction. Whether these concepts coincide depends upon one's definition of the spatial Markov property (see Exercise 3). We now give the definition which has come to be regarded as right for purely spatial processes. However, note that a change to a spatio-temporal formulation might change one's ideas (see Chapter 20).

Definition 3.1 *The Markov field $\{\xi\}$ on Γ is Markov if for any set of sites Λ*

$$P(\xi_\Lambda | \xi_{\bar{\Lambda}}) = P(\xi_\Lambda | \xi_{\partial \Lambda}) \tag{1}$$

That is, it is only the field on the boundary of Λ, $\xi_{\partial\Lambda}$, which effectively conditions ξ_Λ. This seems a fair expression of the Markov property, and it does reduce to the conventional definition in the one-dimensional case, as we shall see.

We shall also see that Markov random fields do indeed exist, and can be characterized in terms of the form of $P(\xi)$.

Theorem 3.1 *The field $\{\xi\}$ on Γ is Markov if and only if the expressions G_Λ in the representation*

$$P(\xi) = \exp\left(-\sum_\Lambda G_\Lambda(\xi_\Lambda) \right) \tag{2}$$

are zero whenever Λ is not a clique.

Proof The condition is sufficient, because it implies for any Λ that

$$P(\xi_\Lambda | \xi_{\bar{\Lambda}}) := \frac{P(\xi)}{\sum_{\xi_\Lambda} P(\xi)}$$

is a function of $\xi(x)$ only for those x which belong to a clique of some member of Λ. This implies that the expression is functionally independent of $\xi(x)$ for $x \notin \mathscr{N}(\Lambda)$, which implies the Markov condition (1) (cf. Theorem 1.14.1).

To establish necessity, choose $\Lambda = x$ in (1). We then have

$$P(\xi) = P(\xi_{\bar{x}}) P(\xi_x | \xi_{\bar{x}})$$
$$= P(\xi_{\bar{x}}) Q(\xi_{\mathscr{N}(x)})$$

say. There thus exists a factorization of $P(\xi)$ which separates $\xi(x)$ from $\xi(x')$ if x' is not a neighbour of x. It thus follows from Theorem 2.2 that $G_\Lambda(\xi_\Lambda)$ is zero if Λ contains any pair of nodes which are not neighbours; i.e. if Λ is not a clique. ∎

Consider, for example, the one-dimensional case $L = \mathbb{Z}$, when x effectively runs through the signed integers, and the neighbours of x are $x \pm 1$. The Markov property (1) would then imply, for example, that

$$P(\xi(x)|\xi(y);\ y \neq x) = P(\xi(x)|\xi(x-1),\ \xi(x+1)),$$

which seems weaker than the conventional temporal Markov property. The conventional property is implied, however.

Corollary 3.1 *The Markov property* (1) *reduces to the conventional Markov property in the temporal case.*

Proof The only cliques are singletons x and pairs $(x-1, x)$, and Theorem 3.1 then states that the Markov property (1) implies a factorization

$$P(\xi) = \prod_x A_x(\xi(x))B_x(\xi(x-1),\ \xi(x))$$

and so a factorization

$$P(\xi(y);\ y \leqslant x) = \prod_{y \leqslant x} A_y^*(\xi(y))B_y^*(\xi(y-1),\ \xi(y))$$

say. This latter factorization implies that

$$P(\xi(x)|\xi(y);\ y < x) = A_x^*(\xi(x))B_x^*(\xi(x-1),\ \xi(x))$$

This, being a function of $\xi(x)$ and $\xi(x-1)$ alone, can be identified with $P(\xi(x)|\xi(x-1))$, by Theorem 1.14.1. This implies the conventional Markov property. ∎

Theorem 3.1 is celebrated in that it completely characterizes the spatial Markov property, whose very possibility had at times been doubted. The theorem is sometimes referred to as the Hammersley–Clifford theorem; these and several other authors have contributed. In fact, the theorem is an immediate consequence of Brook's theorem (see Exercise 2.2). Brook (1964) did not use the notion of a clique, and so stopped just short of stating the Hammersley–Clifford theorem in the explicit form of Theorem 3.1. He did, however, deduce Corollary 3.1 and multi-dimensional versions of it.

Exercises and comments

1. Note that one need only demand the Markov property (1) for the Λ consisting of single sites x. The necessity and sufficiency proofs of Theorem 3.1 then imply the property for general Λ.

2. Define $\partial^{(p)}\Lambda = \mathcal{N}^{(p)}(\Lambda) - \Lambda$, the boundary of Λ of *thickness p*. One can demand a pth order Markov property

$$P(\xi_\Lambda | \xi_{\bar{\Lambda}}) = P(\xi_\Lambda | \xi_{\partial^{(p)}\Lambda})$$

One can deduce an analogue of Theorem 3.1 for $p > 1$ simply by defining a new neighbour relationship $\mathcal{N}^*$ by $\mathcal{N}^*(x) = \mathcal{N}^{(p)}(x)$. That is, one is a 'starred' neighbour if one is a neighbour of degree not greater than p.

4. INTERACTIONS ON A TREE; CRITICAL EFFECTS

We shall now give an example of a field on an infinite graph which shows critical effects. The classic example of such a situation is the Ising model on $\mathbb{Z}^d$, whose solution for $d = 2$ was a mathematical challenge solved by L. Onsager (1944), but which remains unsolved for $d \geqslant 3$. However, the example of this section includes the case of the Ising model on a tree, and has a direct treatment which transparently demonstrates critical effects.

One begins by specifying the potentials, which implies specification of the conditional probabilities $P(\xi_\Lambda | \xi_{\bar{\Lambda}})$. It then turns out that, under certain circumstances, there is more than one set of field statistics corresponding to this specification.

We suppose that Γ is the infinite rooted tree with r branchings at each node. That is, r arcs emanate from the root, and $r + 1$ from any other node. We suppose a potential

$$U(\xi) = \sum_x u(\xi(x)) + \sum\sum_{x < y} v(\xi(x), \xi(y)) \qquad (1)$$

where the second sum is over neighbour pairs alone, so that the y are the 'sons' of x. We suppose that the point field $\xi(x)$ can adopt a finite set of values $j \in \mathcal{J}$, say.

The field is then Markov. The singleton potential $u(j)$ and neighbour-pair potential $v(j, k)$ are independent of position in the tree; they are not necessarily normalized relative to a null field as are the interaction potentials G_Λ of Theorem 2.1. The potential $v(j, k)$ need not be supposed to be symmetric, although it could be normalized to be so away from the root. We shall find it convenient to define

$$a_j = e^{-u(j)}, \qquad b_{jk} = e^{-v(j, k)}$$

Consider now the set of nodes Λ consisting of the root and the following $n - 1$ 'generations' of nodes. The corresponding graph is also a tree, with $N := (r^n - 1)/(r - 1)$ nodes. Let the root be labelled $x = 0$. We conjecture that, for large n,

$$\sum_{\xi_{\Lambda - 0}} \exp(-U_\Lambda(\xi_\Lambda)) \sim \pi_{\xi(0)} \lambda^N \qquad (2)$$

where $U_\Lambda(\xi_\Lambda)$ is the potential obtained by restricting the summations in (1) to $x, y \in \Lambda$.

Effectively, λ^N is the partition function for the partial field ξ_Λ, and π_j is proportional to the probability that $\xi(0) = j$, both of these being asymptotic evaluations for large n. This conjecture and some other points of detail are justified in the references cited at the end of the chapter.

Theorem 4.1 *The reduced partition function λ and the unnormalized probability distribution π of field at the root of Γ satisfy the non-linear eigenvalue system*

$$\lambda \pi_j = a_j \left(\sum_k b_{jk} \pi_k \right)^r \qquad (j \in \mathscr{J}) \tag{3}$$

Proof This follows by expressing the summation (2) for a tree of $n+1$ generations in terms of the summations for the r trees of n generations rooted in the first generation, and appeal to the conjectured relation (2). We then have

$$\pi_j \lambda^{rN+1} = a_j \sum_{k_1} \sum_{k_2} \cdots \sum_{k_r} \prod_{i=1}^{r} (b_{jk_i} \pi_{k_i} \lambda^N)$$

whence (3) follows. ■

One suspects that the non-linear system (3) can have more than one solution for λ and π, indicating that field statistics are indeed incompletely determined by specification of U. For an example, consider the particular case of the Ising model. In this the site state $\xi(x)$ can take two values, which we shall regard as 'orientations' and denote by $+$ and $-$. We shall choose $a_+ = a_- = 1$, $b_{++} = b_{--} = b_1$, and $b_{+-} = b_{-+} = b_2$. There is thus a tendency for neighbouring sites to adopt like or unlike orientations according as $b_1 > b_2$ or $b_1 < b_2$. These two cases are sometimes termed the 'attractive' and 'repulsive' cases respectively.

Equations (3) now become

$$\begin{aligned} \lambda \pi_+ &= (b_1 \pi_+ + b_2 \pi_-)^r \\ \lambda \pi_- &= (b_2 \pi_+ + b_1 \pi_-)^r \end{aligned} \tag{4}$$

Eliminating λ from equations (4) we deduce the equation

$$\theta = \left(\frac{b_1 \theta + b_2}{b_1 + b_2 \theta} \right)^r \tag{5}$$

for

$$\theta := \pi_+ / \pi_- \tag{6}$$

θ is a significant quantity; the ratio of probabilities that the root-orientation is $+$ or $-$ in an infinite tree. However, equation (5) may well have several solutions, indicating that prescription of the potential (1) does *not* uniquely determine the statistics of ξ.

Theorem 4.2 Criticality for the Ising model on a tree. *In the case*

$$r > \frac{b_1 + b_2}{b_1 - b_2} > 0$$

(necessarily the attractive case) equation (5) has real, positive solutions $\theta = 1$, $\phi^{\pm 1}$, where $\phi \neq 1$, and it is the solutions $\theta = \phi^{\pm 1}$ which are stable. In all other cases $\theta = 1$ is the only real positive root, corresponding to equiprobability of $+$ and $-$ orientations at the root of the infinite tree.

Proof Write (5) as $\theta = g(\theta)$. For the repulsive case, $b_1 < b_2$, the function $g(\theta)$ is monotone decreasing, with a graph as in Fig. 18.4.1. There is then only a single root, which is indeed $\theta = 1$.

In the attractive case, $b_1 > b_2$, the function $g(\theta)$ is monotone increasing, and one can have one of the situations of Figs 18.4.2 and 18.4.3. If $g'(1) < 1$, i.e.

$$r < \frac{b_1 + b_2}{b_1 - b_2}$$

then we are in the situation of Fig. 18.4.2, and there is again only the single solution $\theta = 1$, corresponding to equipartition. If, however, $g'(1) > 1$, so that the condition of the theorem holds, then we are in the situation of Fig. 18.4.3, and equation (5) has three solutions, at 1, ϕ and ϕ^{-1}, say. That at $\theta = 1$ is now unstable. The other two are stable, and correspond to a tipping of relative abundances towards either the value ϕ or its reciprocal ϕ^{-1}. ∎

By 'stability' of a root θ of $\theta = g(\theta)$ we mean that it is a possible limit point of the sequence θ_i generated by a recursion $\theta_{i+1} = g(\theta_i)$ for some interval of values of θ_0.

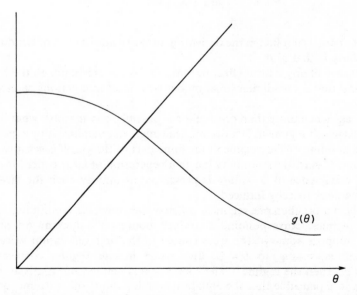

Fig. 18.4.1 The Ising model on a tree; absence of multiple solutions in the repulsive case.

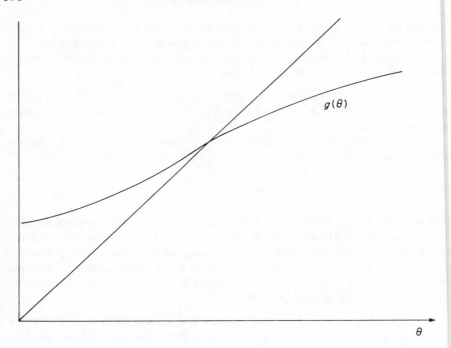

Fig. 18.4.2 The Ising model on a tree; absence of multiple solutions in the insufficiently attractive case.

Such a recursion is implicit in the derivation of the relations (3). The condition for such stability is that $g'(\theta) < 1$.

The absence of any critical effect in the case $r = 1$ corresponds to the frequent observation that a one-dimensional model with local interactions cannot show criticality.

In the case when more than one regime is possible, one may ask what it is that determines which regime in fact occurs. One might characterize it by a 'boundary condition at infinity'. The regime in the finite part of the graph is sensitive to the prescription of statistics at infinity (i.e. in a generation of large order). If a heavy (or light) occurrence of $+$ values is prescribed at infinity, then the 'heavy' (or 'light') regime is thereby induced.

Actually, if more than one regime is possible then there is a continuous range of possible regimes, corresponding to mixed boundary values at infinity. For example, suppose some of the trees rooted in the first generation were in the regime corresponding to $\theta = \phi$, and some in the regime corresponding to $\theta = \phi^{-1}$. Then the regime at the root $x = 0$ is neither, but a mixed version. However, it is plausible that the equations (3) determine the 'extreme' or 'pure' regimes. This point is studied by Zachary (1983).

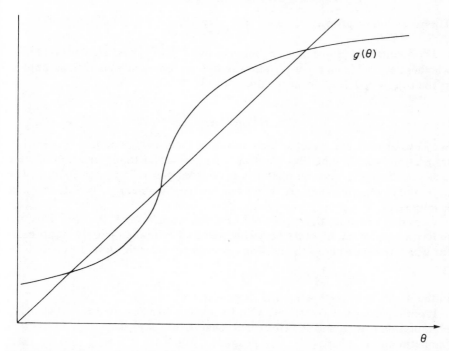

Fig. 18.4.3 The Ising model on a tree; the occurrence of multiple solutions if the condition of Theorem 4.2 is satisfied.

The tree model had its origins in work by Bethe (1939) who gave a first analysis; it was subsequently analysed in detail by Preston (1974) and Spitzer (1979). Later work has been done by Moore and Snell (1979) and Zachary (1983).

Exercises and comments

1. Note that equations (3) do not determine λ if $r > 1$. Indeed, for the particular case (4) it is only $\lambda \pi_+^{1-r}$ which is determined once θ is determined from (5).

2. An alternative approach to the recursive evaluation of the sum (2) adopted here is to note that, for the tree-model with a potential of form (1), one can write

$$P(\xi) = \pi_{\xi(0)} \prod_{0 \leqslant x < y} p_{\xi(x)\xi(y)} \tag{7}$$

where $x = 0$ is the root of the tree and a pair of nodes x, y for which $x < y$ is assumed to label a 'father/son' pair; i.e. a pair of neighbours for which x immediately precedes y in the tree. Here p_{jk} is the transition probability of a reversible Markov chain with state-space $\mathscr{I}$ and π_j is its equilibrium distribution.

The reversibility property implies that specification (7) is independent of the choice of root.

The quantities p_{jk} must now be chosen so that $P(\xi_x|\xi_{\bar{x}})$ has the same evaluation whether deduced from (1) or (7). Show that these consistency conditions, applied at the root $x = 0$, lead again to equations (3).

5. INVARIANCES

We have already encountered the concept of stochastic invariance under time translation (section 1.5), time reversal (Chapter 4) and transformation of state space (section 3.7), and the matter is considered generally in Appendix 1. The notion of stochastic invariance under some of the possible space transformations is important.

Consider a permutation σ of vertices which transforms Γ into $\sigma\Gamma$. This will induce a transformation of the field which we shall write as $\xi \to S\xi$. There are cases for which the statistics of the field do not change under this transformation, i.e.

$$\xi \sim S\xi \tag{1}$$

in the notation of section 1.6 and Appendix 1.

If one regards ξ as a vector with x^{th} element $\xi(x)$, then S is a matrix. In fact, it is a permutation matrix, and so orthogonal. That is, the $(jk)^{\text{th}}$ element of S is unity if the permutation σ takes site k to position j, otherwise it is zero. Clearly, then, interchange of the roles of j and k yields the inverse permutation, so that

$$S^{\mathsf{T}} = S^{-1}.$$

If the structure of the graph Γ on $\mathscr{X}$ and the probabilistic structure of ξ are at all closely related then a transformation leaving the statistics invariant will also leave the graph invariant:

$$\sigma\Gamma = \Gamma.$$

That is, the pattern of arcs is unchanged by the permutation of nodes. The transformations that leave Γ invariant form a group $\mathscr{G}$, so that if σ_1 and σ_2 belong to $\mathscr{G}$ then so do $\sigma_1\sigma_2$ and σ_1^{-1}. The transformations which leave field statistics invariant may constitute only a subset of the full group $\mathscr{G}$. If all invariances are in fact equivalences then the subset will be a subgroup of $\mathscr{G}$, and the induced transformations S will also form a group $\mathscr{S}$, with the equivalence

$$\xi \sim S\xi$$

holding for all S of $\mathscr{S}$.

If the nodes of Γ can indeed be identified with the points of a lattice L in $\mathbb{R}^d$ then the permutations σ of Γ are usually rigid transformations of the space which take L into itself.

For example, suppose L is $\mathbb{Z}^d$, and $\mathscr{G}$ is the group of translations along the coordinate axes (see Exercise 1). The property of invariance under all such translations is the property of *stationarity* (cf. section 1.5). If the lattice is not cubic then stationarity could in general hold only in weaker versions (see Exercise 2).

If L is $\mathbb{R}^d$ itself, then the group of translations is generated by the infinitesimal rather than the unit translations. Invariance of statistics under this group again implies full stationarity. Invariance under all rotations about a given axis implies *rotational symmetry* (in the statistical sense). Invariance with respect to all rigid motions (i.e. all translations and rotations) implies *isotropy*: identity of statistical structure under any choice of Cartesian axes.

Another possible lattice L would be the vertices of a regular polyhedron. The rotations about the centre of the polyhedron that take the polyhedron into itself (so that vertices are in the same positions in space, although permuted) form a group.

The simplest special case of this would be that in the plane, when L consists of the vertices of the regular n-gon, P_n. The transformations taking P_n into itself are the rotations about the centre of the n-gon by a multiple of $2\pi/n$. A rotation by just $2\pi/n$ amounts to the cyclic permutation τ of vertices which takes sites $(1, 2, 3, \ldots, n)$ to $(n, 1, 2, 3, \ldots, n-1)$. That is, under τ the node which was formerly at position j is now moved to position $j+1$, position $n+1$ being identified with position 1. A rotation of P_n through $2\pi r/n$ amounts to the application of τ^r to the set of vertices. We shall term this lattice the *circulant lattice of size n*, denoted P_n.

The matrix corresponding to the permutation τ is

$$
T = \begin{bmatrix}
\cdot & \cdot & \cdot & \ldots & \cdot & 1 \\
1 & \cdot & \cdot & \ldots & \cdot & \cdot \\
\cdot & 1 & \cdot & \ldots & \cdot & \cdot \\
\multicolumn{6}{c}{\cdots\cdots\cdots\cdots\cdots} \\
\cdot & \cdot & \cdot & \ldots & 1 & \cdot
\end{bmatrix}
\tag{2}
$$

That is, the $(jk)^{\text{th}}$ element is unity if $j - k = 1 \pmod{n}$, otherwise zero.

This cyclic permutation is the analogue of a *translation* for a finite array of sites. One effectively regards the sites as being arranged regularly on the circumference of a circle rather than on a line, so that the site displaced by a translation at one end of the line is inserted in the vacancy left at the other end of the line. One may regard it as a unit translation on an interval adapted to have an inverse. Statistical invariance under τ is then the natural version of stationarity on a finite one-dimensional space. From the postulated invariance

$$
\xi \sim T\xi
\tag{3}
$$

we deduce equivalence under all shifts

$$
\xi \sim T^j \xi
\tag{4}
$$

We can say that τ generates the whole Abelian group of cyclic shifts (τ^j). Since $\tau^n = 1$, or $T^n = I$, relation (4) is significant only for $j = 0, 1, 2, \ldots, n-1$.

The frequent use of 'periodic boundary conditions' or a 'toroidal domain' in spatial problems corresponds exactly to use of this idea. However, the idea is more than a subterfuge. One can regard the cyclic displacement as the proper analogue for a finite space of a translation for an infinite space. Further, as we shall see in the next section, the concept immediately generates the Fourier theory which is so inevitable in any context of translation invariance, but so much simpler for a finite space than for an infinite one.

Fourier theory, associated with translation invariance, is a special case of the rich algebraic consequences of invariance under a group of transformations. These algebraic consequences have been followed up by several authors, especially for the case of invariance of second-order statistical properties (see McLaren, 1963; Hannan, 1965).

There are spatial transformations other than the rigid ones which are of interest. One is the *reflections*, corresponding to a reversal of one or more coordinate axes. This cannot be viewed as a rigid motion unless one is prepared to immerse the space in a higher space.

Another non-rigid transformation is the *dilation*. A uniform dilation about the origin in $\mathbb{R}^d$ would transform x into λx, where λ is a positive scalar. A degree of invariance under dilations corresponds to a degree of scale invariance; see section 20.4.

Exercises and comments

1. The natural translations of the lattice $\mathbb{Z}^d$ are the translations $x \to x + \sum_{1}^{d} s_j e_j$, effected by the operator $\prod_j T_j^{s_j}$, for arbitrary integral s_j ($j = 1, 2, \ldots, d$). Here e_j is the unit displacement along the jth axis and T_j the operator which achieves this.

2. For the simplest possible non-cubic lattice, consider a one-dimensional lattice with sites at the integers $x = j$ ($j \in \mathbb{Z}$) and at the points $x = j + \delta$, where $0 < \delta < 1/2$. The only translations which take the lattice into itself are those which shift the lattice by an integral amount. Statistical invariance under these translations would still permit $\xi(j)$ and $\xi(j + \delta)$ to have an arbitrary joint distribution.

One could view this lattice as the direct product of $\mathbb{Z}$ and a two-point set. The integral translations of the lattice are then the natural translations of $\mathbb{Z}$.

3. The shift T^s by an arbitrary amount s in $\mathbb{R}$ can be written e^{sD}, where $D = \partial/\partial x$. We thus see the differential operator D as the *infinitesimal generator* of the shifts. Derive the analogues for arbitrary shifts and rotations in $\mathbb{R}^d$.

6. SECOND-ORDER PROCESSES; SECOND-ORDER STATIONARITY

Suppose that $\xi(x)$ takes values on the real line, so that $\xi \in \mathbb{R}^n$ if we regard ξ as the column vector $(\xi(x))$ and $|\mathcal{X}| = n$.

For some purposes one is content to study the first and second moments of ξ, rather than the whole distribution $P(\xi)$. That is, one concentrates attention on the *mean vector*

$$\mu := E(\xi) \tag{1}$$

and the *covariance matrix*

$$V := \text{cov}(\xi) = E[(\xi - \mu)(\xi - \mu)^{\mathsf{T}}] \tag{2}$$

A process for which these moments exist, and for which no more is necessarily specified, is a *second-order process*.

A Gaussian process (for which the elements of ξ are jointly normally distributed) is fully determined by the first and second moments, so that a second-order specification is a complete specification in the Gaussian case. However, even in other cases one can derive a theory of optimal interpolation and extrapolation of the field (in a linear least square sense) on the basis of moments (1) and (2) alone.

We have

Theorem 6.1 *The process $\{\xi\}$ is invariant to second order under the site-permutation σ if and only if*

$$\mu = S\mu \tag{3}$$
$$VS = SV \tag{4}$$

where S is the permutation matrix corresponding to σ. Then also $\xi \sim S\xi$ to second order.

Proof The transformed field $\xi' = S\xi$ has mean $S\mu$ and covariance matrix SVS^{T}, so we require (3) and

$$V = SVS^{\mathsf{T}} \tag{5}$$

for invariance. Relation (5) is equivalent to (4), since S is orthogonal. Relations (3), (4) imply the corresponding relations with S replaced by S^{-1} (which exists), hence the final assertion of the theorem. ∎

It is interesting that the second-order invariance condition (5) can be stated as the commutation condition (4).

We now return to the important special case of stationarity, restricting ourselves for simplicity to the one-dimensional finite-space version.

Theorem 6.2 *Suppose the field ξ defined on the sites $x = 1, 2, \ldots, n$ of a circulant lattice of size n, and suppose ξ second order stationary, so that*

$$\mu = T\mu \tag{6}$$
$$VT = TV \tag{7}$$

where T is the cyclic permutation matrix (5.2). Then $E[\xi(x)]$ is constant in x and $\text{cov}(\xi(x), \xi(x'))$ is a function of $x - x'$ alone, $v(x - x')$ say, with period n.

The covariance matrix $V = (v(x - x'))$ has eigenvalues (real and non-negative)

$$\phi_j = \sum_{r=0}^{n-1} v(r)\theta^{-jr} \tag{8}$$

where

$$\theta := e^{2\pi i/n} \tag{9}$$

The corresponding normalized right and left eigenvectors are $\zeta_j = n^{-1/2}(1, \theta^j, \theta^{2j}, \ldots, \theta^{(n-1)j})$ and $\bar{\zeta}_j = n^{-1/2}(1, \theta^{-j}, \theta^{-2j}, \ldots, \theta^{-(n-1)j})$. The spectral representation $V = \sum_j \phi_j \zeta_j \zeta_j^+$ thus becomes the finite Fourier representation of $v(\cdot)$:

$$v(r) = n^{-1} \sum_{j=0}^{n-1} \phi_j \theta^{jr} \tag{10}$$

Verification of all the assertions is direct, and is left to the reader. For example, relation (6) implies that $E(\xi(x)) = E(\xi(x-1)) = \ldots = E(\xi(1))$, whence the first assertion follows. The second follows similarly from (7).

However, what is important is to see the reason for the truth of these assertions. The covariance matrix V has the eigenvectors asserted exactly because these are the eigenvectors of the elementary permutation matrix T, and T commutes with V. The finite Fourier transform $\xi \to \hat{\xi}$ given by

$$\hat{\xi}_j = n^{-1/2} \sum_x \xi(x)\theta^{-jx} \tag{11}$$

is a unitary transformation, i.e. has inverse

$$\xi(x) = n^{-1/2} \sum_j \hat{\xi}_j \theta^{jx} \tag{12}$$

exactly because the matrices $n^{-1/2}(\theta^{jk})$ and $n^{-1/2}(\theta^{-jk})$ have respectively the right eigenvectors of T for columns and the left eigenvectors of T for rows, and so are mutually inverse.

The relation between V and T goes beyond the commutation (7); one could write the representation $V = (v(x - x'))$ as

$$V = \sum_{r=0}^{n-1} v(r)\,T^r \tag{13}$$

The function of displacement $v(r)$ is the *autocovariance function*, the covariance between values of ξ at sites a distance r apart. The quantity ϕ_j is the *spectral density function*. One verifies readily that

$$\phi_j = E|\hat{\xi}_j|^2 \tag{14}$$

where $\hat{\xi}_j$ is the complex Fourier amplitude defined by (11); the real, non-negative character of ϕ_j thus follows.

Exercises and comments

1. A matrix V satisfying the commutation relation (7) is termed a *circulant* matrix. T is itself circulant, and, in virtue of (13), can be termed the elementary circulant matrix.

2. Note that $|V| = \prod_j \phi_j$, so that V is non-singular if and only if no ϕ_j is zero. The matrix $C = V^{-1}$ is then also circulant, with spectral representation

$$C = \sum_j \phi_j^{-1} \zeta_j \zeta_j^+$$

3. Theorem 6.2 determines the properties of a stationary scalar-valued field on P_n. One can also consider a stationary vector-valued field, with $\xi(x) \in \mathbb{R}^p$, say, and Theorem 6.2 generalizes immediately. The $p \times p$ autocovariance matrix $E[\xi(x)\xi(x-r)^{\mathsf{T}}]$ is a function of r alone, $v(r)$ say, of period n in r with $v(-r) = v(r)^{\mathsf{T}}$. If we define ϕ_j, $\hat{\xi}_j$ again by (8), (11) then

$$E[\hat{\xi}_j \hat{\xi}_k^{\dagger}] = \delta_{jk} \phi_j$$

4. There is an obvious d-dimensional version, defined on the direct product of circulant lattices $L = P_{n_1} \otimes P_{n_2} \otimes \ldots \otimes P_{n_d}$; see section 19.2.

CHAPTER 19

Gaussian random fields

1. GAUSSIAN FIELDS ON A FINITE SPACE

Suppose the set of nodes to be the finite set $x = 1, 2, \ldots, n$, and ξ to be the vector with elements $\xi(x)$, as in section 18.6. If all the $\xi(x)$ are jointly normally distributed then one terms ξ a *Gaussian field*.

One uses the convention

$$\text{Law} \ (\xi) = N(\mu, V) \tag{1}$$

to express the fact that ξ is Gaussian with mean vector μ and covariance matrix V. The statistics of a Gaussian field are indeed determined by this second-order specification, and have other properties besides (see Exercises 3 and 5).

If V is non-singular then ξ has probability density

$$f(\xi) = (2\pi)^{-n/2} |V|^{-1/2} \exp\left[-\tfrac{1}{2}(\xi - \mu)^{\mathsf{T}} V^{-1} (\xi - \mu)\right] \tag{2}$$

relative to Lebesgue measure on $\mathbb{R}^n$.

If we work with $\xi - \mu$ rather than ξ as the field variable then we effectively normalize μ to zero and density (2) to

$$f(\xi) \propto e^{-U(\xi)} = e^{-\xi^{\mathsf{T}} C \xi / 2} \tag{3}$$

where

$$C = (c(x, y)) = V^{-1}$$

Here

$$U(\xi) = \tfrac{1}{2} \sum_x \sum_{x'} c(x, x') \xi(x) \xi(x') \tag{4}$$

is an effective potential, whose interactions extend, as we see from (4), only up to second order.

The physical view is that we regard the potential (and so C) as given, and quantities such as V are to be derived from it rather than conversely. We make this explicit.

Definition 1.1 ξ *is a Gaussian field of zero mean if it has the potential* (4), *where the pairwise potential* $c(x, x')$ *is restricted by the condition*

$$U(\xi) \geqslant 0 \tag{5}$$

The definition is one we shall adopt whether the space $\mathscr{X}$ is finite or not. In the finite case condition (5) states that the matrix C should be non-negative definite; this raises issues to which we shall return in a moment.

Recall what we mean by specification of U: the evaluation

$$f(\xi_\Lambda | \xi_{\bar\Lambda}) = \frac{e^{-U(\xi)}}{\int e^{-U(\xi)} d\xi_\Lambda} \tag{6}$$

of conditional densities.

Theorem 1.1 *Definition* 1.1 *implies that*

$$\text{Law}(\xi_\Lambda | \xi_{\bar\Lambda}) = N(-C_{\Lambda\Lambda}^{-1} C_{\Lambda\bar\Lambda} \xi_{\bar\Lambda}, C_{\Lambda\Lambda}^{-1}) \tag{7}$$

Proof We can complete the square in the quadratic form (4) to obtain

$$U(\xi) = (\xi_\Lambda + C_{\Lambda\Lambda}^{-1} C_{\Lambda\bar\Lambda} \xi_{\bar\Lambda})^{\mathsf{T}} C_{\Lambda\Lambda} (\xi_\Lambda + C_{\Lambda\Lambda}^{-1} C_{\Lambda\bar\Lambda} \xi_{\bar\Lambda}) + \dots$$

where $+ \dots$ indicates terms independent of ξ_Λ. Inserting this evaluation in (6) and recalling the definition (1) of the normal density we verify the assertion. ∎

The identification of $C_{\Lambda\Lambda}^{-1}$ as a covariance matrix implies that it, and so also $C_{\Lambda\Lambda}$, is non-negative definite. This is then the reason for the non-negativity condition (5).

The case when Λ consists of a single site x is of interest.

Theorem 1.2 *The optimal interpolation of* $\xi(x)$ *(i.e. estimate of* $\xi(x)$ *in terms of* $\{\xi(y); y \neq x\}$ *is*

$$\breve\xi(x) = -c(x, x)^{-1} \sum_{y \neq x} c(x, y) \xi(y) \tag{8}$$

and the interpolation variance is

$$E[\breve\xi(x) - \xi(x)]^2 = c(x, x)^{-1} \tag{9}$$

The law (1) *is consistent with these one-point conditional distributions if and only if*

$$CV = I \tag{10}$$

$$C\mu = 0 \tag{11}$$

Proof Expressions (8) and (9) are just the conditional mean and variance of formula (7) in the case $\Lambda = x$. This conditional mean is well-known to provide the least square linear estimate of $\xi(x)$ (see Exercises 4 and 5); hence the characterization 'optimal' of the theorem.

To derive the final assertion of the theorem, suppose that ξ has mean μ, covariance matrix V, that the linear least square interpolation is

$$\breve\xi(x) = a - \sum_{y \neq x} b(x, y) \xi(y) \tag{12}$$

and that

$$E[\breve{\xi}(x) - \xi(x)]^2 = D(x) \tag{13}$$

The least square conditions on the coefficients a, b yield the conditions and evaluation of $D(x)$:

$$\sum_y b(x, y)\mu(y) = a \tag{14}$$

$$\sum_y b(x, y)v(y, z) = D(x)\delta(x, z) \tag{15}$$

Here the summations are full summations with the understanding $b(x, x) = 1$.

If we are to identify expressions (12), (13) with (8), (9) respectively then we must have $a = 0$, $b(x, y) = c(x, y)/c(x, x)$, $D(x) = 1/c(x, x)$. Insertion of these evaluations into (14), (15) yields conditions (10), (11). ■

We have required merely that C and so $V = C^{-1}$ should be non-negative definite. It is interesting to ask what happens if either C or V is actually allowed to be singular. These extreme cases are rather exceptional in the case of finite $\mathscr{X}$, but are likely to presage phenomena which can occur more easily in the case of infinite $\mathscr{X}$. We consider them in Exercises 6 and 7.

Note that the unconditional distribution of ξ_Λ is

$$\text{Law}(\xi_\Lambda) = N(0, V_{\Lambda\Lambda}) \tag{16}$$

It is interesting to ask how much this differs from the conditional law (7)—sometimes one is more easily calculated than the other. A partial answer is provided by

Theorem 1.3 *The relation*

$$V_{\Lambda\Lambda}^{-1} = C_{\Lambda\Lambda} - C_{\Lambda\bar{\Lambda}} C_{\bar{\Lambda}\bar{\Lambda}}^{-1} C_{\bar{\Lambda}\Lambda} \tag{17}$$

holds. If the field is Markov then this implies that $(xx')^{\text{th}}$ element of $V_{\Lambda\Lambda}^{-1}$ equals $c(x, x')$ unless x and x' are both neighbours of $\bar{\Lambda}$.

Proof The identity (17) between submatrices of mutually reciprocal matrices is a standard one. For completeness we indicate the derivation in Exercise 8. In the Markov case the $(xx')^{\text{th}}$ element of $C_{\Lambda\bar{\Lambda}}$ is zero unless x is a neighbour of $\bar{\Lambda}$; this fact and formula (17) imply the second assertion. ■

Of course, the mean vector given by (7) is also a function only of $\xi_{\partial\Lambda}$ in the Markov case.

Exercises and comments

1. Expression (1) is a correctly normalized density; we then deduce from it that

$$\int e^{-U(\xi)} d\xi = (2\pi)^{n/2} |C|^{-1/2}$$

where U is the potential (4). This provides an evaluation of the partition function of the field (3), explicit if $|C|$ is regarded as calculable for the values of n of interest.

2. Let ξ be the Gaussian field characterized by (1). Then it follows from (2) that

$$E(e^{\theta^T\xi}) = \exp[\theta^T\mu + \tfrac{1}{2}\theta^T V\theta] \tag{18}$$

Identity (18) may be taken as an alternative expression of (1), meaningful even when V is singular (when (2) is not).

3. We see from (18) that normality of ξ implies that (i) the second-order moments of the field exist, (ii) linear functions of ξ are also Gaussian, and (iii) ξ_Λ and $\xi_{\bar\Lambda}$ are independent if and only if $\operatorname{cov}(\xi_\Lambda, \xi_{\bar\Lambda}) = V_{\Lambda\bar\Lambda}$ is zero. One expresses this last condition by saying that ξ_Λ and $\xi_{\bar\Lambda}$ are *uncorrelated*, otherwise written $\xi_\Lambda \perp \xi_{\bar\Lambda}$.

Conditions (i)–(iii) are in fact also sufficient for joint normality, and so are characterizing. To see this, suppose $\xi_1, \xi_2, \ldots, \xi_s$ independently and identically distributed with mean μ and covariance matrix V, and also 'normal' in that they satisfy (i)–(iii). Then

$$\eta_s := \mu + \frac{1}{\sqrt{s}} \sum_1^s (\xi_j - \mu)$$

is also 'normal' with mean μ and covariance V, and calculations show that $\lim_{s\to\infty} E(e^{\theta^T\eta_s})$ has evaluation (18).

4. Suppose, for simplicity, that ξ has been corrected to zero mean. Consider the quantity

$$\breve{\xi} = V_{\Lambda\bar\Lambda} V_{\bar\Lambda\bar\Lambda}^{-1} \xi_{\bar\Lambda} \tag{19}$$

as an estimate of ξ_Λ based upon $\xi_{\bar\Lambda}$. Its distinguishing feature is that the estimation error $\Delta_\Lambda = \breve{\xi}_\Lambda - \xi_\Lambda$ is uncorrelated with $\xi_{\bar\Lambda}$, i.e. $\Delta_\Lambda \perp \xi_{\bar\Lambda}$. A consequence of this orthogonality is that $\Delta_\Lambda \perp \hat{\xi}_\Lambda - \breve{\xi}_\Lambda$ if $\tilde{\xi}_\Lambda$ is any other estimate of ξ_Λ which is linear in $\xi_{\bar\Lambda}$. From this it follows that

$$\operatorname{cov}(\tilde{\xi}_\Lambda - \xi_\Lambda) = \operatorname{cov}(\breve{\xi}_\Lambda - \xi_\Lambda) + \operatorname{cov}(\tilde{\xi}_\Lambda - \breve{\xi}_\Lambda)$$
$$\geqslant \operatorname{cov}(\Delta_\Lambda)$$
$$\geqslant V_{\Lambda\Lambda} - V_{\Lambda\bar\Lambda} V_{\bar\Lambda\bar\Lambda}^{-1} V_{\bar\Lambda\Lambda} \tag{20}$$

That is, $\breve{\xi}_\Lambda$ is the *linear least square estimate* of ξ_Λ, based on $\xi_{\bar\Lambda}$.

5. The assertions of Exercise 4 do not depend on the assumption of normality. If, however, ξ is Gaussian, then it follows from properties (ii), (iii) of Exercise 3 that Δ_Λ and $\xi_{\bar\Lambda}$ are jointly normal and independent. The distribution of Δ_Λ conditional on $\xi_{\bar\Lambda}$ is then identical with its unconditional distribution: normal with zero mean and covariance matrix $\operatorname{cov}(\Delta_\Lambda)$.

Otherwise expressed, the distribution of ξ_Λ conditional on $\xi_{\bar\Lambda}$ is Gaussian with mean $\breve{\xi}_\Lambda$ and the covariance matrix given in the final member of (20). That is, the conditional expectation $E(\xi_\Lambda | \xi_{\bar{\Lambda}})$ can be identified with the linear least square estimate $\hat{\xi}_\Lambda$.

Expressions (19) and (20) are alternative expressions of the mean and covariance stated in (7); in terms of V rather than of $C = V^{-1}$.

6. Suppose V singular. Then $\operatorname{var}(a^{\mathsf{T}}\xi) = 0$, or

$$a^{\mathsf{T}}\xi = 0$$

in mean square, if a is any element of the null space of V. The vector ξ is then restricted to $\mathscr{V}$, the orthogonal complement of this null space, a Euclidean space of lower dimensionality than n. C is infinite in that $U(\xi) = +\infty$ unless $\xi \in \mathscr{V}$. Essentially, ξ does not have a probability density in $\mathbb{R}^n$, but only in the subspace $\mathscr{V}$.

7. If C is non-singular then $V = C^{-1}$, $\mu = 0$, by (10), (11). If C is singular then μ can take any value in the null space of C, and V is infinite in that

$$\operatorname{var}(a^{\mathsf{T}}\xi) = +\infty$$

for a any element of this null space. In fact, V is also indeterminate to this extent, in that, if V satisfies (10), then so does V^* satisfying

$$C(V - V^*) = 0.$$

That is, $V - V^*$ can be any symmetric matrix whose rows and columns lie in the null space of C.

If we demand that $\operatorname{cov}(\xi) < \infty$ then all these contingencies are excluded. However, in the case of infinite $\mathscr{X}$ it is possible to have C singular and yet for its inverse to be finite, although indeterminate; see section 3.

8. Since $V = C^{-1}$ then the equation systems $Ca = b$ and $a = Vb$ in vectors a and b are identical. If we suppose $b_{\bar{\Lambda}} = 0$ then the second equation system implies that $a_\Lambda = V_{\Lambda\Lambda} b_\Lambda$. Eliminating $a_{\bar{\Lambda}}$ from the first equation system we deduce that $(C_{\Lambda\Lambda} - C_{\Lambda\bar{\Lambda}} C_{\bar{\Lambda}\bar{\Lambda}}^{-1} C_{\bar{\Lambda}\Lambda}) a_\Lambda = b_\Lambda$, and so deduce the identity (17).

2. STATIONARY GAUSSIAN FIELDS ON THE CIRCULANT LATTICE

Suppose that ξ is a Gaussian field on the circulant lattice P_n, stationary in that it is statistically invariant under the circulant shift. By combining the conclusions of section 18.6 with the assumptions of normality we obtain pointers for the formalism of a stationary Gaussian field on an infinite lattice.

We know by Theorem 18.6.2 that $\mu(x)$ is constant and $v(x, x')$ a function $v(x - x')$ of period n. Let us assume μ normalized to zero. It is also convenient to rewrite the Fourier transforms ϕ_j, $\hat{\xi}_j$ defined by (18.6.8), (18.6.11) as

$$\phi(v) = \sum_r v(r) e^{-ivr} \tag{1}$$

$$\hat{\xi}(v) = n^{-1/2} \sum_x \xi(x) e^{-ivx} \tag{2}$$

where $v = 2\pi j/n$, and x, r, j all take values in $(0, 1, 2, \ldots, n-1)$. Summations over v are then understood to be over the set $\{2\pi_j/n; j = 0, 1, 2, \ldots, n-1\}$. In the limit of large n these will become integrals over $[0, 2\pi)$, or, equivalently, over $(-\pi, \pi]$.

Theorem 2.1   *Suppose ξ a Gaussian field on P_n, stationary and with zero mean. Then the probability density of ξ can be expressed*

$$f(\xi) = (2\pi)^{-n/2} \left(\prod_v \phi(v) \right)^{-1/2} \exp\left[-\tfrac{1}{2} \sum_x \sum_y c(x-y) \xi(x) \xi(y) \right]$$

$$= \prod_v (2\pi \phi(v))^{-1/2} \exp\left[-\frac{|\hat{\xi}(v)|^2}{2\phi(v)} \right] \tag{3}$$

where $\phi, \hat{\xi}$ are defined by (1), (2) and x, y, r and $nv/2\pi$ take values in $(0, 1, 2, \ldots, n-1)$. The relation between c and v has the three alternative expressions

$$\sum_y c(x-y) v(y-z) = \delta(x,z) \tag{4}$$

$$c(r) = n^{-1} \sum_v \phi(v)^{-1} e^{ivr} \tag{5}$$

$$\phi(v)^{-1} = \sum_r c(r) e^{-ivr} \tag{6}$$

These assertions can be otherwise expressed: the field has potential

$$U(\xi) = \tfrac{1}{2} \sum_x \sum_y c(x-y) \xi(x) \xi(y)$$

$$= \tfrac{1}{2} \sum_v |\hat{\xi}(v)|^2 / \phi(v) \tag{7}$$

and partition function

$$\int e^{-U(\xi)} d\xi = (2\pi)^{n/2} \left(\prod_v \phi(v) \right)^{1/2} \tag{8}$$

These assertions are all immediate consequences of the evaluations of V^{-1} and $|V|$ implied by Theorem 18.6.2; see Exercise 18.6.2. The final expression of (3) is significant. The quantities $\hat{\xi}(v)$ are the finite Fourier transforms of the field ξ, appropriately scaled. Expression (3) then states that these are independently and normally distributed (if one accepts the factors in this expression as giving the density of a *complex* normal variable). Indeed, the transformation $\xi \to \hat{\xi}$ is unitary, and so of unit Jacobian, so that expression (3) is correct as a $\hat{\xi}$ density.

Singularity of V or of C corresponds to some of the $\phi(v)$ being respectively zero or infinite.

The theorem seems long in statement, but simply starts from the evaluation of the eigenvalues and eigenvectors of V given by Theorem 18.6.2 and states the significant implications.

We should now consider the 'd-dimensional' case

$$L = P_{n_1} \otimes P_{n_2} \otimes \ldots \otimes P_{n_d} \tag{9}$$

For completeness we may also suppose the field vector-valued: $\xi(x) \in \mathbb{R}^m$. We shall take x as a column d-vector whose jth element x_j is the coordinate in P_{n_j} and take the wave-number vector $v = (v_1, v_2, \ldots, v_d)$ as the corresponding row-vector. The elements x_j and $n_j v_j / 2\pi$ then take values in $(0, 1, 2, \ldots, n_j - 1)$ for $j = 1, 2, \ldots, d$. Let us express this by saying that x takes values in L and v takes values in the conjugate lattice $\hat{L}$. Let us also define

$$n = \prod_{j=1}^{d} n_j$$

If we assume the process stationary along all axes (i.e. statistically invariant under all circulant shifts) then $\mu(x)$ is constant (and we again assume it normalized to zero). Also V is d-way circulant in that

$$\mathrm{cov}\,(\xi(x),\,\xi(y)) = v(x - y) \tag{10}$$

where $v(r)$, the autocovariance matrix, is an $m \times m$ matrix with period n_j in r_j ($j = 1, 2, \ldots, d$). Essentially r also takes values in L. If we define the d-dimensional Fourier transforms

$$\hat{\xi}(v) := n^{-1/2} \sum_x \xi(x)\,\mathrm{e}^{-ivx} \tag{11}$$

$$\phi(v) := \sum_r v(r)\,\mathrm{e}^{-ivr} \tag{12}$$

then

$$E\,[\xi(v)\,\xi(v')^\dagger] = \delta(v, v')\,\phi(v) \tag{13}$$

Theorem 2.2   *Suppose ξ an m-vector Gaussian field on the d-dimensional circulant lattice (10), stationary on all axes and with zero mean. Then the probability density of ξ can be expressed*

$$f(\xi) = \left[\prod_v \left[(2\pi)^{-m/2} |\phi(v)|^{-1/2} \right] \right] \mathrm{e}^{-U(\xi)} \tag{14}$$

where

$$U(\xi) = \tfrac{1}{2} \sum_x \sum_y \xi(x)^{\mathsf{T}} c(x - y)\,\xi(y)$$

$$= \tfrac{1}{2} \sum_v \hat{\xi}(v)^\dagger \phi(v)^{-1} \hat{\xi}(v) \tag{15}$$

The matrix potential function $c(r)$ is again related to the matrix autocovariance function $v(r)$ by the equivalent equations (4)–(6).

The path of proof is a clear analogue of that for $d = m = 1$ and we leave verification to the reader. Expressions (14), (15) again indicate the statistical independence of the complex Fourier amplitudes $\hat{\xi}(v)$. Relation (12) of course has the inverse

$$v(r) = n^{-1} \sum_v \phi(v) e^{ivr} \tag{16}$$

The evaluation (8) of the partition function can be carried a little further.

Theorem 2.3 *The partition function has the evaluation*

$$\int e^{-U(\xi)} d\xi = \prod_v [(2\pi)^{m/2} |\phi(v)|^{1/2}] \tag{17}$$

Suppose that $n_1, n_2, \ldots, n_d$ are all increased indefinitely and $v(\cdot)$ defined by relation (16) for fixed $\phi(\cdot)$. One has then

$$\left[\int e^{-U(\xi)} d\xi \right]^{1/n} \to (2\pi)^{m/2} D^{-1/2} \tag{18}$$

where D is the geometric mean of $|\phi(v)|$,

$$D := \exp\left[\frac{1}{(2\pi)^d} \int \log|\phi(v)| \, dv \right] \tag{19}$$

and the integral is over $(-\pi, \pi]^d$, this integral being supposed defined.

As L changes in the way indicated the v-summation over $\hat{L}$ in (16) also changes correspondingly; it is supposed then that one has a fixed spectral density matrix $\phi(v)$ defined for all v in $(-\pi, \pi]^d$.

Proof The first assertion follows from the fact that density (14) is normalized, and the second from the fact that

$$\left[\prod_v |\phi(v)| \right]^{1/n} = \exp\left[n^{-1} \sum_v \log|\phi(v)| \right]. \quad \blacksquare$$

3. STATIONARY GAUSSIAN FIELDS ON $\mathbb{Z}^d$

If we let $n_1, n_2, \ldots, n_d$ all tend to infinity then the lattice (2.9) effectively becomes the cubic lattice $L = \mathbb{Z}^d$, and the conjugate lattice $\hat{L}$ becomes the hypercube $(-\pi, \pi]^d$, summations over $\hat{L}$ becoming integrations with respect to Lebesgue measure over this hypercube.

Suppose $\xi = \{\xi(x)\}$ scalar and stationary on the infinite cubic lattice $\mathbb{Z}^d$. Then the autocovariance function

$$v(r) = \text{cov}(\xi(x), \xi(x - r)) \tag{1}$$

is also defined on $\mathbb{Z}^d$, and the formula (2.12) for the *spectral density function*

$$\phi(v) = \sum_r v(r) e^{-ivr} \tag{2}$$

defines $\phi(v)$ on $(-\pi, \pi]^d$. The inverse relation is the integral evaluation of a Fourier coefficient

$$v(r) = \frac{1}{(2\pi)^d} \int e^{ivr} \phi(v) \, dv \tag{3}$$

where a v-integral is always understood to be over $(-\pi, \pi]^d$, unless otherwise indicated.

There are cases (see Exercise 4) where series (2) does not converge and the spectral density does not exist. In such cases one works with a corresponding measure function $\Phi(\cdot)$, the *spectral measure*, such that (3) is replaced by

$$v(r) = \frac{1}{(2\pi)^d} \int e^{ivr} \Phi(dv) \tag{4}$$

The defining relation (2) would then become

$$\Phi(A) = \sum_r v(r) \int_A e^{-ivr} \, dv \tag{5}$$

However, again we take the physical point of view. That is, we specify the stationary Gaussian field, not by prescription of autocovariance or spectral density, but by prescription of the potential

$$U(\xi) = \tfrac{1}{2} \sum_x \sum_y c(x - y) \xi(x) \xi(y) \tag{6}$$

where each sum is over $\mathbb{Z}^d$. For simplicity we shall keep to the case of a scalar field ($\xi(x) \in \mathbb{R}$) so that the potential function $c(r)$ is itself scalar. Expression (6) implies the specification

$$f(\xi_\Lambda | \xi_{\bar{\Lambda}}) = \frac{e^{-U(\xi)}}{\int e^{-U(\xi)} d\xi_\Lambda}$$

of conditional densities. We must ask what global statistics, and so what values of $\mu(x)$, $v(r)$ and $\phi(v)$, are consistent with this specification.

As ever, the potential function $c(r)$ is constrained by the condition $U(\xi) \geqslant 0$, equivalent to

$$\gamma(v) := \sum_r c(r) e^{-ivr} \geqslant 0 \tag{7}$$

For example, suppose we assume a symmetric Markov field, in that

$$c(r) = \begin{cases} c_0 & r = 0 \\ c_1 & |r| = 1 \\ 0 & \text{otherwise} \end{cases} \tag{8}$$

Then

$$\gamma(v) = c_0 + 2c_1 \sum_j \cos(v_j) \tag{9}$$

and condition (7) implies that we require

$$c_0 \geqslant 2d|c_1| \tag{10}$$

if (8) is to be a valid specification. We may indeed require strict inequality in (10) if the field is to be finite in value; see Theorem 3.3 below.

As in Theorem 1.2 the potential (6) implies that

$$\check{\xi}(x) = -c(0)^{-1} \sum_{r \neq 0} c(r)\,\xi(x-r) \tag{11}$$

is an interpolation of the field at x with interpolation variance

$$E[\check{\xi}(x) - \xi(x)]^2 = c(0)^{-1} \tag{12}$$

in that expressions (11) and (17) are respectively the mean and variance of $\xi(x)$ conditional on $\{\xi(y); y \neq x\}$. From this we derive, again as in Theorem 1.2, an indication of the extent to which the potential determines the global statistics of the field.

Theorem 3.1 *A field with mean $\mu(x)$ and covariance $v(x, y)$ is consistent with the potential (6) only if*

$$\sum_z c(x-z)\mu(z) = 0 \tag{13}$$

$$\sum_z c(x-z)v(z, y) = \delta(x, y) \tag{14}$$

The only stationary fields consistent with the potential are those for which the constant mean value $\mu(x) = m$ is arbitrary if $\gamma(0) = 0$, zero otherwise, and the spectral measure has evaluation

$$\Phi(A) = \int_A \frac{dv}{\gamma(v)} + \Phi_0(A) \tag{15}$$

Here Φ_0 is an arbitrary measure on the set of real v for which $\gamma(v) = 0$.

Proof Relations (13), (14) are just the analogues of (1.10), (1.11) in the present case, and derived in the same way. If we require stationarity then we require that

$\mu(x)$ be constant, equal to m say, and $v(x, y) = v(x - y)$. Equation (13) then becomes

$$\gamma(0)m = 0$$

so that m is zero unless $\gamma(0) = 0$. Equation (14) becomes

$$\frac{1}{(2\pi)^d} \int e^{ivr} \gamma(v) \Phi(dv) = \delta_r, \tag{16}$$

if we appeal to the representation (4). This in turn implies that

$$\gamma(v) \Phi(dv) = dv \tag{17}$$

whence we deduce the characterization (15). ∎

The fact that the potential does not necessarily completely determine the field statistics becomes considerably less mysterious in the next chapter. There we regard the process as developing in time as well as in space, and the question is then whether the spatial process at a given time t tends to an equilibrium as t increases, and whether this equilibrium is independent of initial conditions.

Conditions (13), (14) in general permit many non-stationary solutions; see Exercise 1 for an example and section 20.2 for an indication of their interpretation. However, the theorem indicates that the stationary solution is determined to a high degree by the potential function $c(r)$, and determined uniquely if $\gamma(v) > 0$. We can state

Theorem 3.2   *Let $\hat{L}_+$ be the subset of $(-\pi, \pi]^d$ on which $\gamma(v)$ is strictly positive. Then one has the unique determination*

$$\phi(v) = \gamma(v)^{-1} \tag{18}$$

on $\hat{L}_+$. If the stationary field is to be finite in mean square (i.e. $v(0) < \infty$) then it is necessary that

$$\int_{\hat{L}_+} \gamma(v)^{-1} dv < \infty \tag{19}$$

Proof   The first assertion is a consequence of (15). We have, from (4),

$$v(0) = \frac{1}{(2\pi)^d} \int \Phi(dv) \tag{20}$$

where the integral is over $\hat{L} = (-\pi, \pi]^d$. Since Φ is necessarily a positive measure then finiteness of $v(0)$ will require that

$$\int_A \Phi(dv) < \infty$$

for any subset A of $\hat{L}$. Choosing $A = \hat{L}_+$ and appealing to the evaluation (18) of the spectral density on this set we deduce (19). ∎

The evaluation (18) is of course what we would expect from relation (2.8) of the circulant case. Note that if (18) holds for all v then the interpolation variance (12) has the evaluation

$$E[\breve{\xi}(x) - \xi(x)]^2 = \left[\frac{1}{(2\pi)^d} \int \frac{dv}{\phi(v)} \right]^{-1} \tag{21}$$

the harmonic mean of $\phi(v)$.

To see some of the implications of Theorems 3.1 and 3.2, consider again the Markov case.

Theorem 3.3 *Consider the symmetric Markov potential specified by (8). For $c(r)$ to define a potential it is necessary that $c_0 \geqslant 2d|c_1|$, and the stationary field with this potential function has spectral density function*

$$\phi(v) = \left[c_0 + 2c_1 \sum_j \cos v_j \right]^{-1} \tag{22}$$

on the set of v for which this is finite.

Suppose $c_1 < 0$. Then in the cases $d = 1, 2$ it is necessary and sufficient that $c_0 + 2dc_1 > 0$. In the cases $d \geqslant 3$ it is necessary and sufficient that $c_0 + 2dc_1 \geqslant 0$. If equality holds in this latter case then the field can be represented

$$\xi(x) = \xi^*(x) + \zeta \tag{23}$$

where $\xi^*(x)$ is a Gaussian field with zero mean and spectral density (22) and ζ is an arbitrary Gaussian variable independent of ξ^*.

Proof The first condition is certainly necessary for $U(\xi) \geqslant 0$, and (22) follows from (18). The strict inequality $c_0 > 2d|c_1|$ is certainly sufficient, in that expression (22) then defines an integrable spectral density everywhere. The question is then to find out when equality is permitted in (10); for definiteness we take the case $c_1 < 0$. If $c_0 + 2dc_1 = 0$ then expression for $\gamma(v)$ has a zero just at $v = 0$, implying for $\phi(v)$ a singularity of type $|v|^{-2}$ at the origin. The contribution of this singularity to integral (20) over a set $|v| \leqslant k$ for small k is of order $\int_0^k |v|^{d-3} \, d|v|$, which is finite if and only if $d > 2$.

The criterion $E|\xi(x)|^2 < \infty$ thus requires that equality in (10) be forbidden for $d \leqslant 2$, but allows equality for $d > 2$. Equality in the latter case implies, by (19), that an arbitrary concentration of spectral measure be allowed at $v = 0$. This corresponds to the arbitrary ζ added to the field in (23). The component ζ need not have zero mean, because we know from Theorem 3.1 that m may have any value if $\gamma(0) = 0$. ∎

Exercises and comments

1. Consider the example of the symmetric Markov model (8). Then $\mu(x) = \prod_j \alpha_j^{x_j}$ will satisfy (13) if

$$c_0 + c_1 \sum_j (\alpha_j + \alpha_j^{-1}) = 0$$

Equation (13) thus certainly admits non-constant solutions. These must be oscillatory in some direction if strict inequality holds in (10).

2. How would the assertions of Theorem 3.3 be modified in the case $c_1 > 0$?

3. Note that for the symmetric Markov model (8) one can write the potential of the field into the two alternative forms

$$U(\xi) = \tfrac{1}{2}(c_0 + 2dc_1) \sum_x \xi(x)^2 + \tfrac{1}{2}c_1 \sum_x \sum_j (\xi(x+e_j) - 2\xi(x) + \xi(x-e_j))\,\xi(x)$$

$$= \tfrac{1}{2}(c_0 + 2dc_1) \sum_x \xi(x)^2 - \tfrac{1}{2}\,c_1 \sum_x \sum_j [\xi(x+e_j) - \xi(x)]^2$$

The continuous analogue of these expressions (when the space is $\mathbb{R}^d$ rather than $\mathbb{Z}^d$) is

$$U(\xi) = \alpha \int \xi^2 \, dx - \beta \int \xi \nabla^2 \xi \, dx$$

$$= \alpha \int \xi^2 \, dx + \beta \int |\nabla \xi|^2 \, dx$$

One will require $\alpha \geqslant 0$, $\beta \geqslant 0$, with strict inequality somewhere except in particular circumstances.

4. The purest case of failure of series (2) to converge is that of a pure line spectrum. Consider the one-dimensional case with $\xi(x) = \cos(\alpha x - \theta)$. This is stationary if θ is uniformly distributed on $(-\pi, \pi]$. Then $v(r) = \tfrac{1}{2}\cos(\alpha r)$, so series (2) fails to converge, and one should indeed regard $\phi(v)$ as a sum of δ-function components at $v = \pm\alpha$. Develop the multidimensional analogue.

4. ASYMPTOTIC RESULTS FOR LARGE Λ

We continue to consider the stationary Gaussian field on $\mathbb{Z}^d$ with potential (3.6). In view of assertion (2.14) for the circulant case one is inclined to conjecture that, for large Λ,

$$f(\xi_\Lambda | \xi_{\bar\Lambda}) \sim f(\xi_\Lambda) \sim (2\pi D)^{-|\Lambda|/2} \, e^{-U_\Lambda(\xi)} \tag{1}$$

where

$$D := \exp\left[\frac{1}{(2\pi)^d} \int \log \phi(v)\, dv \right] \tag{2}$$

$$U_\Lambda(\xi) := \tfrac{1}{2} \sum_\Lambda \sum_\Lambda c(x-y)\,\xi(x)\,\xi(y) = \tfrac{1}{2}\xi_\Lambda^{\mathsf{T}} C_{\Lambda\Lambda}\,\xi_\Lambda \tag{3}$$

The approximate equalities of (1) are understood to hold in the sense that the terms of order $|\Lambda|$ in the logarithm of the three quantities are the same.

A result of this type is needed if one is to be able to obtain an evaluation of the partition function for the field: one sees (2.18) as being the expected form of this evaluation. Evaluations of the probability densities in (1) are also needed if one is to make statistical inferences on the structure of the process from observation of ξ_Λ.

Evaluations of type (1) were first deduced by Whittle, initially in the time-series context (Whittle, 1951, 1954a), and then in the spatial context (1954b). The first method used was a simple appeal to the circulant results of section 2; a subsequent method was to appeal to a 'unilateral' representation of ξ:

$$\xi(x) = \sum_{r \in H} a(r)\xi(x-r) + \varepsilon(x) \tag{4}$$

where $\{\varepsilon(x)\}$ is a Gaussian white noise process and H is a 'half-space' to one side of the origin, described precisely in Exercise 1. The quantity D of (2) is exactly the variance of $\varepsilon(x)$. We see then from (3.20), (2.19) and (3.21) that the arithmetic, geometric and harmonic means of $\phi(v)$ can be respectively interpreted as the unconditional variance $\text{var}[\xi(x)]$, the 'extrapolation variance' $\text{var}[\xi(x)|\xi(x-r); r \in H]$ and the interpolation variance $\text{var}[\xi(x)|\xi(x-r); r \neq 0]$.

Both approaches are suggestive, but difficult to rigorize. Here we present a specialized version of an alternative approach due to Künsch (1981) which is both direct and rigorous.

Consider the unconditional density

$$f(\xi_\Lambda) = (2\pi)^{-|\Lambda|/2} |V_{\Lambda\Lambda}|^{-1/2} \exp\left[-\tfrac{1}{2}\xi_\Lambda^T V_{\Lambda\Lambda}^{-1}\xi_\Lambda\right]$$

Theorem 4.1   *Suppose the process p^{th} order Markov, so that $c(r) = 0$ for $|r| > p$ and suppose that Λ grows in such a way that $|\Lambda| \to \infty$, $|\partial^{(p)}\bar{\Lambda}|/|\Lambda| \to 0$. Then the second approximate equality of (1) holds, in that the term of order $|\Lambda|$ in the logarithms of both quantities are the same.*

Proof   We know from (1.17) that

$$\xi_\Lambda^T V_{\Lambda\Lambda}^{-1}\xi_\Lambda - \xi_\Lambda^T C_{\Lambda\Lambda}\xi_\Lambda = o(|\Lambda|)$$

since the $(xy)^{th}$ element of the two matrices is the same unless x and y both belong to $\partial^{(p)}\bar{\Lambda}$. We have thus merely to demonstrate that

$$|V_{\Lambda\Lambda}|^{1/|\Lambda|} \to D \tag{6}$$

as Λ grows in the manner indicated.

Consider a modified version of $\phi(v)$:

$$\phi_\alpha(v) := \left(c(0) + \alpha \sum_{r \neq 0} c(r)e^{ivr}\right)^{-1} \tag{7}$$

If $\phi(v)$ is an integrable spectral density then so is $\phi_\alpha(v)$ for $0 \leqslant \alpha \leqslant 1$. We shall use the obvious notations γ_α, $v_\alpha(r)$, etc. except that we shall denote the α-dependent version of $V_{\Lambda\Lambda}$ by M_α, for simplicity. Then

$$\frac{\partial}{\partial \alpha} \log |M_\alpha| = \mathrm{tr}\left(M_\alpha^{-1} \frac{\partial M_\alpha}{\partial \alpha}\right) = \sum_\Lambda \sum_\Lambda m_\alpha^{xy} \frac{\partial}{\partial \alpha} v_\alpha(y-x) \tag{8}$$

where m_α^{xy} is the $(xy)^{\mathrm{th}}$ element of M_α^{-1}. We have the evaluations

$$m_\alpha^{xy} = c_\alpha(x-y) \tag{9}$$

unless both x and y belong to $\partial^p \bar{\Lambda}$ by (1.17). Further,

$$\frac{\partial}{\partial \alpha} v_\alpha(y-x) = -\frac{1}{(2\pi)^d} \int \frac{S e^{iv(y-x)}}{\gamma_\alpha(v)^2} \, dv \tag{10}$$

where

$$S = \sum_{r \neq 0} c(r) e^{ivr}$$

Inserting (9), (10) into (8) we have then

$$\frac{\partial}{\partial \alpha} \log |M_\alpha| = -\frac{1}{(2\pi)^d} \int \frac{S \sum_\Lambda \sum_\Lambda c_\alpha(x-y) e^{iv(y-x)}}{\gamma_\alpha(v)^2} \, dv$$

$$= -\frac{|\Lambda|}{(2\pi)^d} \int \frac{S}{\gamma_\alpha} \, dv + o(|\Lambda|) \tag{11}$$

The remainder in (11) is $O(|\Lambda|)$ uniformly for $0 \leqslant \alpha \leqslant 1$; integrating (11) with respect to α over this interval we then deduce that

$$|\Lambda|^{-1} \log |V_{\Lambda\Lambda}| = -\frac{1}{(2\pi)^d} \int \log \gamma(v) \, dv + o(1)$$

$$= \frac{1}{(2\pi)^d} \int \log \phi(v) \, dv + o(1) \qquad \blacksquare$$

Künsch applied the calculation (7)–(11) to calculate $\log |C_{\Lambda\Lambda}|$ rather, under assumptions which implied that

$$\mathrm{cov}\,(\xi(x),\, \xi(y)|\xi_{\bar{\Lambda}}) \sim v(x-y)$$

for x, y in Λ and remote from $\bar{\Lambda}$. He also established the first approximate equality of (1) under quite general hypotheses. We have for simplicity restricted ourselves to the more modest assertion of the theorem, but this is useful in that it implies the evaluation of partition function

$$\left[\int e^{-U_\Lambda(\xi)} d\xi_\Lambda\right]^{1/|\Lambda|} \to (2\pi)^{d/2} D^{-1/2}$$

for ξ a Gaussian field which is Markov of any finite order.

Exercises and comments

1. The set H of (2) can be written $\bigcup\limits_{0}^{d-1} H_j$ where H_j is the subset of $\mathscr{X} = \mathbb{Z}^d$ such that $x_1 = x_2 = \ldots = x_{j-1} = 0$, $x_j < 0$ and x_k is unconstrained for $k > j$. One can write

$$\mathscr{X} = H + \{0\} + H^*$$

where $\{0\}$ is the origin and H^* the image of obtained by reversal of all axes. It is the corresponding decomposition of $\gamma(v) = \sum\limits_{\mathscr{X}} c(r)\,e^{-ivr}$ which leads to representation (4).

5. FIELDS GENERATED BY STOCHASTIC DIFFERENCE EQUATIONS

Consider a field generated by a stochastic difference equation

$$\sum_r a(r)\xi(x-r) = \varepsilon(x) \tag{1}$$

where $\{\varepsilon(x)\}$ is a Gaussian white noise process of zero mean and variance v, say. Such models were considered by Whittle (1954b) in an attempt to find the spatial analogue of the temporal autoregressive model.

In the spatial case there is no preferred direction of dependence (whereas in temporal cases one can take a causal point of view, and regard the present as conditioned by the past). Temporal models are always 'causal' or 'realizable' in that $\xi(t)$ is represented as a function of past values $\xi(t-1)$, $\xi(t-2)$, ... and white noise. That is, one regards the present as conditioned by the past. However, in a purely spatial model there is no intrinsically preferred direction of dependence.

Define the generating function

$$A(v) := \sum_r a(r)\,e^{-ivr} \tag{2}$$

Theorem 5.1 *The field generated by model* (1) *has pairwise potential function*

$$c(r) = v^{-1}\sum_s a(s)a(r+s) \tag{3}$$

so that

$$\gamma(v) = v^{-1}A(v)A(-v) \tag{4}$$

Proof The ε have joint probability density proportional to $\exp[-(2v)^{-1}\sum\limits_x \varepsilon(x)^2]$. The linear transformation from the ε to the ξ provided by (1) has constant Jacobian. We thus deduce the potential of the ξ-field as

$$\frac{1}{2v} \sum_x \left[\sum_r a(r) \xi(x-r) \right]^2 = \frac{1}{2v} \sum_x \sum_r \sum_s a(r) a(s) \xi(x-r) \xi(x-s)$$

$$= \tfrac{1}{2} \sum_x \sum_y c(x-y) \xi(x) \xi(y)$$

where $c(r)$ is defined by (3). Remaining assertions follow. ∎

We see from (3) that if $a(r)$ vanishes outside some neighbourhood of the origin then so does $c(r)$. That is, relations (1) of limited range lead to potentials of limited range.

Let us be more specific. We need not assume that the sites $x \pm e_j$ ($j = 1$, $2, \ldots, d$) are the neighbours of x; let us merely assume that there is a neighbour relationship which is translation-invariant, in that

$$\mathcal{N}(x) = x + \mathcal{N}(0) \tag{5}$$

Let us say that scheme (1) has *range p* if p is the smallest value such that $r \notin \mathcal{N}^{(p)}(0)$ implies that $a(r) = 0$. We define the range of the potential similarly.

Corollary 5.1 *A stochastic difference equation* (1) *of range p generates a potential of range up to 2p, and so generates a field which is Markov of order up to 2p.*

Proof The assertions follow from (3). One can see from the one-dimensional case that $2p$ is indeed attainable. ∎

It is not then in general true, as one might briefly have believed, that a model of range p generates a field which is Markov of order p. The obvious case in point is the field specified by (3.8), Markov and symmetric if x does indeed have $x \pm e_j$ ($j = 1, 2, \ldots, d$) as its neighbours. The corresponding $\gamma(v)$, given by (3.9), simply cannot be factorized as in (4) with 'finite-range' factors.

However, we shall see in the next chapter that a *dynamic* linear stochastic differential equation of spatial range p indeed generates, in equilibrium, a Markov process of order p.

6. LITERATURE

The material of sections 1–3 is fairly standard, the main point being that the Fourier transforms of pairwise potential and autovariance, $\gamma(v)$ and $\phi(v)$, are mutually reciprocal. These ideas have probably been developed by several authors in a number of contexts. The formula (2.3) for the Gaussian density in the one-dimensional circulant case was certainly deduced in Whittle (1951) and seen as an approximation to the density of ξ_Λ for large Λ for a process on $\mathbb{Z}$. This approximation was also deduced by unilateral representation of type (4.4) and extended to the multidimensional case in Whittle (1954a, b), where the stochastic difference equation models of section 5 are also proposed and treated. The methods described in section 4 are due to Künsch (1981).

The assertions of Theorem 3.1, essentially determining the stationary Gaussian fields consistent with a given translation-invariant potential, are to be found in Rozanov (1967) and, in a developed form, in Dobrushin (1980a).

CHAPTER 20

Random fields generated by dynamic models

It is in a sense artificial to consider a purely spatial process. In doing so one has lost the notion of a physical mechanism functioning in time, and can no longer appeal to one's intuitions on causality and conditioning.

To have a physical model one must have a dynamic model, and regard the random field as an equilibrium (i.e. as equilibrium distribution) of this dynamic stochastic model. That is, one embeds the spatial process in a spatio-temporal process.

This approach also gives a very natural understanding of the indeterminacies we have encountered; that the potential may not completely determine field statistics. The point is that the equilibrium solution of the dynamic model may under certain circumstances depend upon initial conditions. That is, such indeterminacies reflect a non-ergodicity of the dynamic model. The requirement of spatial stationarity in the limit may require this non-ergodicity to take a somewhat subtle form, but the question is indeed one of non-ergodicity.

One can ring all the changes of discrete or continuous space and time. For the properties we wish to consider it turns out to be natural to take a continuous time formulation, as will indeed be the rule in physical contexts (although all results have a discrete time analogue). The spatial structure will be dictated by convenience; sometimes discrete, sometimes continuous.

The field variable, now being also time-dependent, will be written $\xi(x, t)$. We shall revert to the notation $\xi(x)$ when we come to consider purely equilibrium distributions.

1. REVERSIBILITY AND MARKOV FIELDS

We consider a process in continuous time, and on a general graph Γ, so that a neighbour structure is specified.

Let $\eta = \eta(x)$ and $\eta' = \eta'(x)$ denote two possible numerical values of the field $\xi(x, t)$ at a given time. As in section 18.2 we shall use the notation $C_x(\eta')\eta$ to describe the field which is equal to η except at the single site x, where it takes the value $\eta'(x)$. That is, the transition

$$\eta \to C_x(\eta')\eta \tag{1}$$

is the transition

$$\eta(y) \rightarrow \begin{cases} \eta(y) & y \neq x \\ \eta'(y) & y = x \end{cases}$$

Let us make the following assumptions about the spatial-temporal process.

(i) The process is homogeneous Markov in time.
(ii) The only possible transitions are the one-point transitions of type (1).
(iii) These transitions are *locally* conditioned, in that the intensity for transition (1) is a function only of $\eta'(x)$ and of $\eta(y)$ for $y \in \mathcal{N}(x)$.
(iv) The process is time-reversible.

Theorem 1.1 *Assume that the process satisfies conditions (i)–(iv) above and reaches equilibrium. Then the equilibrium field is spatially Markov.*

Proof It follows from assumption (iii) that we can write

$$\lambda(\eta, C_x(\eta')\eta) = \kappa(\eta'(x), \eta_{\mathcal{N}(x)}) \tag{2}$$

for κ a specific function of the arguments indicated. If $P(\xi)$ is the equilibrium distribution of the purely spatial field ξ then time-reversibility implies the detailed balance relation

$$P(\xi)\kappa(\eta(x), \xi_{\mathcal{N}(x)}) = P(\eta)\kappa(\xi(x), \eta_{\mathcal{N}(x)}) \tag{3}$$

for η such that

$$\eta(y) = \xi(y) \qquad (y \neq x)$$

For given x and $\eta(x)$ relation (3) implies a factorization of $P(\xi)$ in which $\xi(x)$ is separated from $\xi(y)$ if y is not a neighbour of x. Since there is such a factorization for any x then, by Brooks's theorem, there is a factorization of $P(\xi)$ in which all non-neighbours are separated. That is, the field is Markov, by Theorem 18.3.1. ∎

Correspondingly, the equilibrium field is Markov of order p if $\mathcal{N}(x)$ is replaced by $\mathcal{N}^{(p)}(x)$ in (2). As noted in Exercise 18.3.2, this follows by simple redefinition of the neighbour relationship.

It is assumption (iii) which expresses the property of 'local' interaction in space. The theorem (due to Spitzer, 1971) leads to the conclusion sought but not found in section 19.5; that a local character in stochastic relationships between variables should lead to a similarly local character of equilibrium statistics. The specification must be dynamic before this is true. There is also the strong assumption of reversibility, to be substituted in the next section by a totally different assumption: of linear dynamics driven by Gaussian noise.

Exercises and comments

1. Consider a version of Theorem 1.1 assuming dynamic reversibility rather than reversibility.

2. FIELDS GENERATED BY LINEAR DYNAMICS

Consider the scalar field $\{\xi(x, t); \ x \in \mathbb{Z}^d, \ t \in \mathbb{R}\}$ satisfying the linear dynamic equation

$$\frac{\partial \xi(x, t)}{\partial t} + \sum_r b(r)\xi(x - r, t) = \varepsilon(x, t) \tag{1}$$

where the driving process $\{\varepsilon(x, t)\}$ is white both in x and t. That is, $\varepsilon(x, t)$ has zero mean and

$$E\left[\sum_x \int \theta(x, t)\varepsilon(x, t)\,dt\right]^2 = v\sum_x \int \theta(x, t)^2\,dt \tag{2}$$

for all $\theta(x, t)$ for which the right-hand number of (2) is defined. If we postulate no more, then we have essentially determined the second-order properties of ξ. If we postulate that the injected white noise ε is Gaussian, then ξ will remain Gaussian if it begins so (since $\xi(\cdot, t)$ is a linear function of initial value $\xi(\cdot, 0)$ and of $\varepsilon(\cdot, \tau)$ for $0 \leqslant \tau < t$).

Model (1) implies that there is a field, described by the process variable ξ, whose intrinsic dynamics are linear and deterministic, but which is driven by a random noise input ε. One might, for example, be representing the strings of an Aeolian harp, stirred by the wind (although wind-forces will be white neither in space nor time, and (1) will then have to be supplemented by spatial boundary conditions). For another example, one might be considering the variation of nutrient density in the soil, nutrient being evened out by diffusion and cultivation, and new nutrient being injected in some random fashion. (Again, for realism one would require non-linearity of diffusion and a non-uniformity with depth.) Yet another example that has been considered is the effect on aeroplane runways or roads of the alternating scuffing by traffic and smoothing by maintenance engineers. We return to this example in section 4.

Let us define the moments

$$\mu(x, t) = E(\xi(x, t))$$

$$v(x, y, t) = \text{cov}\,(\xi(x, t), \ \xi(y, t)).$$

One could also define covariances between field values at different times as well as at different sites, but we consider only the 'cross-sectional' statistics of the spatial field at particular times.

Theorem 2.1 *The first and second moments obey the differential equations*

$$\frac{\partial \mu(x, t)}{\partial t} + \sum_r b(r)\mu(x - r, t) = 0 \tag{3}$$

$$\frac{\partial v(x, y, t)}{\partial t} + \sum_r b(r)(v(x, y - r, t) + v(x - r, y, t) = v\delta(x, y) \tag{4}$$

These results follow from (1) by standard formal calculations; see Exercise 1.

Before saying anything about these equations, let us consider the particular case when the field is initially stationary; i.e. $\{\xi(x, 0)\}$ is stationary as a spatial process. The same is then plainly true for all t, so that

$$\mu(x, t) = \mu(t)$$

$$v(x, y, t) = v(x - y, t)$$

say. We can define the time-dependent spectral generating function

$$\phi(v, t) = \sum_r v(r, t) e^{-ivr}$$

Theorem 2.2 *If the field is spatially stationary initially then it remains so, and the mean and spectral density satisfy*

$$\frac{\partial \mu(t)}{\partial t} + B(0)\mu(t) = 0 \tag{5}$$

$$\frac{\partial \phi(v, t)}{\partial t} + [B(v) + B(-v)]\phi(v, t) = v \tag{6}$$

where

$$B(v) := \sum_r b(r) e^{-ivr}. \tag{7}$$

These equations have solution

$$\mu(t) = e^{tB(0)} \mu(0) \tag{8}$$

$$\phi(v, t) = \gamma(v)^{-1} + [\phi(v) - \gamma(v)^{-1}] e^{-tv\gamma(v)} \tag{9}$$

where

$$\gamma(v) := v^{-1}[B(v) + B(-v)] = 2v^{-1} \operatorname{Re} B(v). \tag{10}$$

The field thus converges to a statistical equilibrium in which it has finite mean and variance if and only if

$$\gamma(v) \geqslant 0 \tag{11}$$

$$\int \gamma(v)^{-1} \, dv < \infty \tag{12}$$

where $v \in (-\pi, \pi]^d$. *Inequality* (11) *must be strict if the equilibrium is to be independent of initial conditions.*

The sequence of assertions makes the proofs evident. If spectral density functions do not exist then one would write (9) more circumspectly as

$$\Phi(dv, t) = \gamma(v)^{-1} \, dv + [\Phi(dv, 0) - \gamma(v)^{-1} \, dv] e^{-tv\gamma(v)}.$$

If (11), (12) hold and the field is initially zero we see from (9) that

$$\gamma(v) = \phi(v, +\infty)^{-1}$$

We can thus (cf. (19.3.17)) identify $\gamma(v)$ with the $\gamma(v)$ of (19.3.7); the Fourier transform of the pairwise potential function. In view of this, the limit form of (8), (9) as $t \to \infty$ illuminates the assertions of Theorem 19.3.1. The equilibrium field is undetermined by the potential in just exactly those aspects in which it is sensitive to initial conditions.

The strict version of (11)

$$\text{Re } B(v) > 0 \tag{13}$$

now appears as the necessary and sufficient stability condition. That is, let $\mathscr{F}$ be the set of spatial processes on $\mathbb{R}$ which are stationary and Gaussian. We know that a process in $\mathscr{F}$ remains in $\mathscr{F}$ under the dynamics expressed by (1). It will converge to a limit in $\mathscr{F}$ only if (11), (12) hold, and will converge to a limit independent of initial conditions if and only if (13) holds.

However, condition (13) by no means implies that the spatial process will converge to its unique limit in $\mathscr{F}$ from an initial point outside $\mathscr{F}$. In particular, the process may well not converge to its stationary limit if it is initially non-stationary. For example, there are cases (see Exercise 2) where one can, consistently with (13), find solutions v of

$$\alpha + B(v) = 0 \tag{14}$$

for real positive α. The trajectory

$$\mu(x, t) = e^{\alpha t + ivx} \tag{15}$$

then satisfies (3). It follows from (13) that the solution v of (14) cannot be real. Solution (15) then corresponds to an initial spatial non-stationarity in ξ which grows with time, despite fulfilment of the stability condition (13).

However, it may be said that to add any such component of non-stationarity, however small, to a stationary process is not indeed a small perturbation of the process. The fact that v in (15) is not real implies that expression (15) grows exponentially fast in some x-direction. The perturbation is thus *not* small for indefinitely large x in that direction.

One assumes that condition (13) probably assures that the property of spatial stationarity is locally stable in some sense.

The identification of expression (10) with the potential transform $\gamma(v) = \sum_r c(r) e^{-ivr}$ implies that

$$c(r) = v^{-1} [b(r) + b(-r)]. \tag{16}$$

We thus deduce

Theorem 2.3 *Suppose that a translation-invariant neighbour relation is defined on the lattice, as in (19.5.5). If $b(r)$ and $b(-r)$ are both zero for r outside $\mathcal{N}^{(p)}(0)$ then the equilibrium field generated by (1) is Markov of order p.*

The assertion is an immediate consequence of (16). It supplies the conclusion we would have hoped for in section 19.5.

Exercises and comments

1. Consider a version of (1) in which the 'field' at time t is a vector ξ of finite dimension satisfying

$$\dot{\xi} + B\xi = \varepsilon.$$

Here B is then a square matrix and ε is a vector white noise process satisfying

$$E\left[\int \theta(t)^{\mathsf{T}} \varepsilon(t)\, dt\right]^2 = \int \theta(t)^{\mathsf{T}} N\theta(t)\, dt.$$

Show that $\mu(t) := E(\xi(t))$ and $V(t) := \operatorname{cov}(\xi(t))$ satisfy the equations

$$\dot{\mu} + B\mu = 0$$

$$\dot{V} + BV + VB^{\mathsf{T}} = N$$

2. Consider the one-dimensional version of (1) with $b(0) = 1$, $b(1) = b(-1) = \frac{1}{3}$ and all other $b(r)$ zero. Then

$$B(v) = 1 + \tfrac{2}{3}\cos v > 0.$$

However, the equation

$$\alpha + 1 + \tfrac{1}{3}(z + z^{-1}) = 0$$

has real (negative) solutions z for any $\alpha > 0$; the mean value $\mu(x, t) = z^x e^{\alpha t}$ then satisfies (3).

3. Note that model (1) is time-reversible if $B(v) = B(-v)$. In this case the conclusion of Theorem 2.3 follows of course also from Theorem 1.1.

4. Suppose example (1) modified to the discrete-time version

$$\xi(t) = B\xi(t-1) + \varepsilon(t)$$

Show that V satisfies the recursion

$$V(t+1) = BV(t)B^{\mathsf{T}} + N$$

5. Theorem 2.2 implies that model (1) has, in equilibrium, a spatial spectral density function

$$\phi(v, +\infty) = \frac{v}{B(v) + B(-v)} \tag{17}$$

Suppose the model replaced by the discrete-time version

$$\xi(x, t) = \sum_r b(r)\xi(x - r, t - 1) + \varepsilon(x, t) \tag{18}$$

Show, from example (4), that for this model (17) is replaced by

$$\phi(v, +\infty) = \frac{v}{1 - B(v)B(-v)} \tag{19}$$

This can be true only under a stability condition, which will be that $|B(v)| < 1$ for real v. Equality is permitted in the inequality provided ϕ remains integrable, but the spectral measure is then again indeterminate at points of equality.

3. FIELDS IN $\mathbb{R}^d$; THE DIFFUSION-INJECTION MODEL

The continuous space analogue of (2.1) would be a model

$$\dot{\xi}(x, t) + B\xi(x, t) = \varepsilon(x, t) \tag{1}$$

where B is a linear operator over the x argument, translation-invariant in that there is a function $B(v)$ such that

$$Be^{ivx} = B(v)e^{ivx} \tag{2}$$

For example, for the purely integral operator

$$B\xi(x, t) = \int b(r)\xi(x - r, t)\, dr \tag{3}$$

we have

$$B(v) = \int b(r)e^{-ivr}\, dr \tag{4}$$

and for the purely differential operator

$$B = \prod_j \left(\frac{\partial}{\partial x_j}\right)^{n_j} \tag{5}$$

we have

$$B(v) = \prod_j (iv_j)^{n_j} \tag{6}$$

At least to begin with we suppose ε to be white noise in $\mathbb{R}^{d+1}$, so that

$$E[\int\int \theta(x, t)\varepsilon(x, t)\, dx\, dt]^2 = v\int\int \theta(x, t)^2\, dx\, dt$$

although generalizations are easily accommodated (see equation (10)).

Under these conditions the treatment and assertions of Theorem 2.2 transfer bodily, with only the modifications that $B(v)$ is now defined by (2) rather than by (2.7) and that v now takes values in $\mathbb{R}^d$.

Theorem 2.3 will also transfer, and is natural if we regard B as an integral operator of type (3), with the support of $b(r)$ defining the 'spatial range' of the operator. However, what will often happen is that B is certainly local in that it is a linear combination of differential operators of type (5). In this case an important character is its order: the degree of the polynomial $B(v)$. We see from (2.9) that $\gamma(v)$ is a polynomial of the same degree.

A particular model of interest is that in which the scalar field ξ obeys the linear stochastic differential equation

$$\frac{\partial \xi}{\partial t} = \nabla^2 \xi - \alpha^2 \xi + \varepsilon \tag{7}$$

where

$$\nabla^2 = \sum_j \left(\frac{\partial}{\partial x_j} \right)^2$$

We shall term this the *diffusion-injection* model. It was first considered by Whittle (1962) and makes several of the points on dimension-sensitive effects and scale-invariance which have been seen as more generally significant later.

Model (7) represents a field into which new variation is being injected continuously by the driving noise ε. This variation is spread out in space (and so smoothed) by the presence of diffusion term $\nabla^2 \xi$. The strength of the field tends to be damped by the term $-\alpha^2 \xi$.

For example, one might regard $\xi(x, t)$ as describing concentration of some nutrient in the soil at point x, and at a time t. (For this purpose one might regard the soil as two-dimensional—variation of nutrient density with depth is plainly going to be non-stationary.) The term ε then represents additions of nutrient, in a manner totally random in space and in time (and so irregular to an extreme, but the extreme case is of interest in itself, and can be moderated later). The term $\nabla^2 \xi$ then represents a physical diffusion of nutrient laterally in the soil, tending to equalize concentration. The term $-\alpha^2 \xi$ represents an actual loss of nutrient, due, for example, to oxidation, leaching, or extraction by plants.

An alternative application is to a situation first noted and partially analysed by Akaike (1960). Consider an aeroplane runway. Its surface is constantly scuffed by aircraft wheels, particularly at landing. Attempts are made to restore the smoothness of the surface by regularly grading the runway, i.e. by using a grader to scrape the surface smooth. The effect of these two processes is observed to be that, in the course of time, the runway develops long smooth undulations. These become stronger and stronger unless there is a conscious attempt to relevel the runway globally (rather than smooth it locally).

One could imagine the mechanism as being essentially represented by model (7). The field variable ξ is now the variable height of the runway, the new variation ε is the scuffing produced by aircraft, the diffusion term represents the local smoothing produced by grading, and the term $-\alpha^2 \xi$ is absent. Grading is, of course, intermittent rather than continuous, and its scraping action is a nonlinear smoothing, but the smoothing provided by a diffusion captures the essentials.

For this model

$$B(v) = |v|^2 + \alpha^2 \tag{8}$$

so that the field has equilibrium spectral density

$$\phi(v) = \frac{v}{2(|v|^2 + \alpha^2)} \tag{9}$$

This model can be regarded as the simplest which is spatially Markov and non-degenerate (i.e. which shows dependence in all directions; see Exercise 4). It is not derivable from a purely spatial stochastic differential equation, but, as we see, is indeed generated by a simple spatio-temporal model. The model also demonstrates very naturally the sensitivity to dimension which we have already observed in Theorem 19.3.3.

Theorem 3.1 *Consider the spectral-density function* (9) *of the diffusion-injection model.*

 (i) *If one chooses* $\alpha = 0$ *then the integral* $\int \phi \, dv$ *diverges at* $v = 0$ *if* $d = 1, 2$ *but not for* $d \geqslant 3$. *This corresponds to the fact that model* (7) *does not then have a stationary limit if* $d < 3$, *but does if* $d \geqslant 3$.

 (ii) *The integral* $\int \phi dv$ *diverges at* $|v| = \infty$ *if* $d \geqslant 2$. *This corresponds, not to non-existence of a stationary limit, but just to the fact that the limit field is 'tinted noise', and belongs to the same generalized class of processes as does white noise itself.*

The assertions of the theorem summarize a number of points that can be appreciated intuitively. We shall give an informal proof that will indeed indicate the course a formal proof would follow, but will also amplify the qualitatively important points.

If we set $\alpha = 0$ then we remove damping from the process; $B(v)$ becomes of order $|v|^2$ for small v and $\phi(v, t)$ develops a singularity of order $|v|^{-2}$ with increasing time. This is enough to destroy the possibility of a stationary limit in dimensions one and two.

This is in fact just the phenomenon which Akaike observed in the one-dimensional case, first empirically, and then as the necessary consequence of an appropriate model. He noted that, for the repeatedly scuffed and scraped runway, long undulations built up. The long undulations correspond just to periodic components of small wave-number, and so to a major component of spectral density at small v. However, the amplitude of undulations also builds up indefinitely in the course of time if matters are left to themselves, corresponding to the impossibility of stationary limit behaviour and a $|v|^{-2}$ singularity in the spectral density.

In three and more dimensions one can have a stationary limit even in the absence of damping. The dimensionality is high enough that injected variation can escape to infinity and avoid a build-up. On the other hand, as one now knows from Theorem 2.2, the undamped model shows a susceptibility to initial conditions, in that model (7) will not change the mean level of ξ which existed initially.

Coming to assertion (ii) we seem to have another type of divergence for $d \geqslant 2$. This is not at all serious, however, and certainly does not correspond to a non-stationarity. Suppose we modify the driving noise ε so that it is white in time, but has spectral density $\phi_\varepsilon(v)$ in space. One finds readily that (9) generalizes to

$$\phi(v) = \frac{\phi_\varepsilon(v)}{2(|v|^2 + \alpha^2)} \tag{10}$$

If ϕ_ε is bounded at its origin and integrable then ϕ is integrable, and there is no irregularity. One might for example choose

$$\phi_\varepsilon(v) = \begin{cases} v & |v| \leqslant w \\ 0 & |v| > w \end{cases}$$

so that ε is white noise with a high wave-number cut-off. By choosing w large enough we can make ε approximate white noise, but the finiteness of cut-off gives ε finite power—i.e. finite variance.

The divergence of $\int \phi \, dv$ for expression (9) then corresponds solely to divergence of $\int \phi_\varepsilon \, dv$; i.e. to irregularities of the input. This is removed if one regularizes the input sufficiently.

However, there is no need to do this if one regards the white noise process itself as well-defined, and there is now a theory of 'generalized processes' which makes it so (see e.g. Gikhman and Skorokhod, 1979). A process with constant ϕ is termed 'white', and a process with non-constant and integrable ϕ often termed 'coloured'. A process with non-constant ϕ for which $\phi(v)$ converges to zero too slowly at infinity that $\int \phi \, dv$ should converge there might be termed 'tinted', and belongs to the same generalized class of processes as does white noise itself.

Akaike (1960) and Cox (1984) presented one-dimensional models for the particular cases of runway wear and yarn-spinning respectively which explained the development with time of a singularity of the spectral density at $v = 0$. The statement and analysis of the more general model (1) are due to Whittle (1962); this seems to have provided the first instance (9) of the formula (2.17).

Exercises and comments

1. Note that the spectral density function (9) is a function only of $|v|$. This is an indication of the stochastic rotation-invariance of the spatial process in equilibrium, and so of its isotropy.

2. The spectral density function (9) corresponds to a potential

$$U(\xi) = 2v^{-1} \int [|\nabla \xi|^2 + \alpha^2 \xi^2] \, dx$$

for the spatial field $\{\xi(x)\}$. The fact that this cannot be written

$$U(\xi) = \int (A\xi)^2 \, dx$$

for A a differential operator means, again, that the spatial field cannot be seen as the solution of a purely spatial stochastic differential equation $A\xi = \varepsilon$.

3. Cox (1984) considered a model of yarn-spinning similar to the Akaike runway model, but with spatial stretching replacing the operation of spatial smoothing. So, the thickness of the yarn constitutes a one-dimensional field; this suffers a continuing dilation of x, injection of white noise, and a necessary rescaling of ξ. Show that the time-dependent spatial spectral density will then obey an equation

$$\dot{\phi} + \alpha v \phi = v$$

It will thus develop a v^{-1} singularity rather than a v^{-2} singularity as t increases.

4. The simplest spatial model in $\mathbb{R}^d$ might be considered to be a first-order stochastic differential equation

$$\left[a_0 + \sum_{1}^{d} a_j \frac{\partial}{\partial x_j} \right] \xi(x) = \varepsilon(x)$$

where ε is spatial white noise. Show that a change of co-ordinates reduces this to

$$\left[\frac{\partial}{\partial x_1} + \alpha \right] \xi(x) = \varepsilon(x)$$

The model is thus seen to specify essentially a collection of independent one-dimensional Markov processes, in that the field is independent at points at which $(x_2, x_3, \ldots, x_d)$ differs, Markov for constant $(x_2, x_3, \ldots, x_d)$ and variable x_1. In this sense the model is strongly degenerate.

4. POWER LAW AUTOCOVARIANCES, SELF-SIMILAR PROCESSES AND FRACTALS

We continue with the diffusion-injection model (3.7). The autocovariance function corresponding to the spectral density function (3.9) can be calculated from standard integrals; we find

$$v(r) = \begin{cases} \dfrac{v e^{-\alpha|r|}}{2\alpha} & (d = 1) \\[2ex] \dfrac{v}{2\pi} K(\alpha|r|) & (d = 2) \\[2ex] \dfrac{v e^{-\alpha|r|}}{4\pi r} & (d = 3) \end{cases} \tag{1}$$

This is of course a function of $|r|$ alone, because of the isotropy of the model.

The expressions for $v(r)$ at a given r do not have a finite limit as $\alpha \downarrow 0$ for $d = 1, 2$. This reflects the non-existence of a stationary limit in these cases, as we have already noted. However, in the case $d = 3$ we can indeed let α tend to zero, and obtain

$$v(r) \propto |r|^{-1} \tag{2}$$

More generally, a spectral density $\phi \propto |v|^{-2}$ corresponds to an autocovariance

$$v(r) \propto |r|^{2-d} \qquad (d > 2) \qquad (3)$$

The divergence $v(0) = +\infty$ can either be accepted as a characteristic of tinted noise, or can be removed by truncating or tapering ϕ at infinity, cf. (3.10), (3.11). Even if one adopts the latter course, the $|v|^{-2}$ behaviour of ϕ at the origin will still induce the power-law behaviour (3) in the autocovariance for large $|r|$.

Autocovariance with power-law tails are certainly of common physical occurrence; they occur in turbulence, and statistical and continuum mechanics. Fairfield Smith (1938) made a classic collection of observations on agricultural uniformity trials which implied, as Whittle (1962) demonstrated, the almost universal phenomenon of a power law spatial autocovariance $|r|^{-p}$ in yield density. We give the histogram (the graph of observed frequencies) of the exponent p in Fig. 20.4.1. One sees that p falls in the range $\frac{1}{2} \leqslant p \leqslant \frac{3}{2}$, with strong peaking in frequency at the value $p = 1$, and lesser peakings at the extreme values $p = \frac{1}{2}, \frac{3}{2}$. A point of the 1962 paper was that it demonstrated the $|r|^{-1}$ behaviour as being explicable by an undamped diffusion-injection model in three dimensions; see equation (2).

It is just the power-law autocovariances which are *self-similar* in that they are invariant in form under a dilation of space:

$$v(kr) \propto v(r)$$

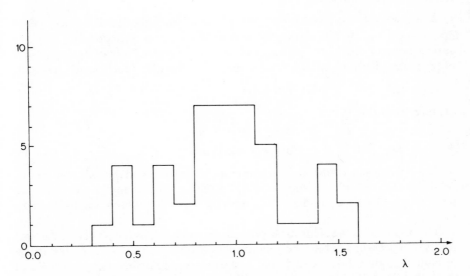

Fig. 20.4.1 A histogram of the exponents λ in the power-law autocovariances $s^{-\lambda}$ deduced from Fairfield Smith's uniformity trial data: 46 trials of different crops.

for arbitrary given positive scalar k (see Exercise 1). We may say then that the processes with autocovariance $|r|^{-p}$ are not only stationary and isotropic, but also stochastically invariant under dilations (actually, semi-invariant in the sense explained in Exercises 1, 2).

Statistical invariance under dilations is just the concept behind Mandelbroit's theory of *fractals* (1977) in which the particular autocovariance $|r|^{-1}$ plays a special role.

More generally, there is considerable interest now in processes which are *self-similar* (see e.g. Taqqu, 1982) or *auto-models* (Dobrushin, 1980b) whose entire statistics are semi-invariant under dilations (see Exercise 2). These play a considerable role in the rescaling theory of critical phenomena (see, for example, Reichl (1980), p. 318). According to this the detailed geometry of lattices, etc. should be irrelevant to critical effects, and these effects should remain even after space dilations. Exploitation of this observation has led to the extensive renormalization theory and at least a partial determination of critical exponents, etc.

Of course, it is physically unrealistic to expect invariance under all dilations. A large enough dilation would bring one to a scale at which the microstructure becomes evident, and other factors enter. This corresponds to the fact that, for example, we do not really expect the power-law behaviour of the autocovariance function to persist right down to indefinitely small distances r.

Exercises and comments

1. A uniform dilation of space by a factor k about an origin x_0 will replace $\xi(x)$ by $\xi(x_0 + k(x - x_0))$. If ξ is stationary with autocovariance function $v(r)$ then the autovariance of the dilated process will be $v(kr)$, independently of x_0. Invariance of $\{\xi\}$ under dilations is understood to mean that there is a function $a(k)$ such that

$$v(kr) = a(k)v(r) \qquad (k > 0, r \in \mathbb{R}^d) \qquad (4)$$

Deduce from this that

$$v(r) = a(|r|)v(r/|r|) \qquad (5)$$

$$a(k_1 k_2) = a(k_1)a(k_2) \qquad (6)$$

An autocovariance function is continuous, so then is a, and the only continuous solutions of (6) are of the form $a(k) = k^{-p}$. Relation (5) then implies a dependence $|r|^{-p}$ on $|r|$. The direction-dependent term $v(r/|r|)$ will be absent for an isotropic process.

2. The origin x_0 of the dilation plays much the same role as the pivot a of the time reversal of section 4.1; both become irrelevant only if the process is stationary. Assume then that the process $\{\xi(x)\}$ is stationary, and define the dilation $K\xi$ of the full realization ξ by

$$K\xi[x] = \xi(kx)$$

One can say that the process is *self-similar* or *dilation-invariant* if there is a function $a(k, l)$ such that

$$EH(K\xi) = a(k, l)EH(\xi)$$

for all homogeneous functionals H of realization ξ for which $EH(\xi)$ exists. Here l is the degree of the homogeneous function H; i.e.

$$H(\lambda\xi) = \lambda^l H(\xi)$$

for positive scalar λ.

APPENDIX 1

Stochastic invariance and equivalence

Let $x(t)$ be a random function of t, and so a random mapping $\mathcal{T} \to \mathcal{X}$ if t takes values in $\mathcal{T}$ and x takes values in $\mathcal{X}$ for each t. The argument t is often termed the *parameter* (again one of those overworked words) and $\mathcal{T}$ the *parameter space*. In Chapters 1 to 17 we interpret t as time, taking values in $\mathbb{R}$ or $\mathbb{Z}$ according as time is continuous or discrete. In Chapters 18 to 20 we consider random functions of space, of space-time or of position on a discrete graph. In such cases the analogue of what we now term t labels a space coordinate, a space-time coordinate or the node of a graph respectively.

We shall use X to label the complete course of the function $x(t)$ as t varies. The quantity X is itself random, giving the *realization* of the function, the actual rule for the $\mathcal{T} \to \mathcal{X}$ mapping in a given case. The realization takes values in $\mathcal{X}^{\mathcal{T}}$, for which we shall adopt the more convenient notation $\mathcal{R}$. We shall assume the stochastic behaviour of the model fully enough specified that one can in principle calculate $E\phi(X)$ for all functions $\phi: \mathcal{R} \to \mathbb{R}$ in some class of interest, $\mathcal{F}$.

One often wishes to express the idea that the statistical characteristics of the process $\{x(t)\}$ remain unchanged under some transformation of the process. In this volume we have occasion to consider such invariance under the transformations of time translation (section 1.5), time reversal (Chapter 4), state conjugation (section 4.6), spatial translations (section 18.6 and Chapter 19), spatial rotations (section 20.3) and spatial dilations (section 20.4).

So, suppose that S is a function from $\mathcal{R}$ to $\mathcal{R}$, defining a transformation $X \to SX$ of realizations. We shall say that the random mapping X is *stochastically invariant under S*, written $S \rightsquigarrow SX$, if for any ϕ in $\mathcal{F}$ the function ϕS also belongs to $\mathcal{F}$ and

$$E\phi(SX) = E\phi(X) \tag{1}$$

If X is stochastically invariant under a number of such transformations, $S_1, S_2, \ldots$ then it is also invariant under products of such transformations. $S_j S_k, S_j S_k S_l$, etc. That is, it is stochastically invariant under the *semi-group $\mathcal{S}$ of transformations* generated from $\{S_1, S_2, \ldots\}$.

Suppose that the semi-group thus generated is finite. Then it must in fact be a *group* of transformations, in that it also contains the identity element I and the inverse S^{-1} of any of its elements S. The simplest examples of this situation is the

439

group of size 2 generated from the time reversal operator R of section 4.2 for which $R^2 = I$, or from the state conjugation operator C of section 4.6, for which also $C^2 = I$. The discrete circulant shift T of section 18:5 satisfies $T^n = I$, and generates the cyclic group of order n.

If X is stochastically invariant under both S and S^{-1} we shall say that X and SX are *stochastically equivalent*, written $SX \sim X$. In this case, existence of *either* side of (1) implies existence of the other plus the equality (1). For, suppose that ϕS belongs to $\mathscr{F}$. Then, by stochastic invariance under S^{-1} the function $\phi SS^{-1} = \phi$ also belongs to $\mathscr{F}$ and

$$E\phi(SX) = E\phi(SS^{-1}X) = E\phi(X)$$

In general, we can say that if X is stochastically invariant under the elements of a group $\mathscr{S}$ of transformations then the quantities SX are stochastically equivalent, for all elements S of $\mathscr{S}$.

In many cases (indeed, all of those mentioned except state conjugation) the transformations $X \rightarrow SX$ are induced by a transformation $t \rightarrow \sigma t$ of the parameter. That is

$$SX[t] = x(\sigma t) \tag{2}$$

where σ is something like a translation, rotation, dilation, etc. of $\mathscr{T}$. Several such transformations $\sigma_1, \sigma_2, \sigma_3, \ldots$ constitute the generators of a semi-group Σ, the corresponding semi-group generated from the induced transformations $S_1, S_2, S_3, \ldots$ is just $\mathscr{S}$. If σ^{-1} exists then so does S^{-1}, and is the transformation induced by σ^{-1}. Thus, if Σ is a group, so is $\mathscr{S}$.

Suppose $\mathscr{T}$ is finite. If we regard X as a vector with elements $x(t)$ then the S induced by (2) is a matrix if the transformations σ are one-to-one in that distinct elements of $\mathscr{T}$ are mapped on to distinct elements then σ is a *permutation* and S a *permutation matrix*. Inverses then exist, and S is orthogonal:

$$S^{-1} = S^{\mathsf{T}}.$$

The simplest non-trivial example is the 'circulant stationary' process of section 18.5. Here $\mathscr{T}$ is the set $t = 1, 2, \ldots, n$ and the parameter transformation considered is the one-step cyclic permutation

$$\sigma t = t + 1 \qquad (t = 1, 2, \ldots, n - 1)$$
$$\sigma n = 1$$

The corresponding matrix operating on X is the circulant matrix defined in (18.5.2).

The operation σ when iterated generates the cyclic group of order n. It is indeed a group, because $\sigma^n = 1$, and σ^s can be identified with $\sigma^{n-s}(s = 0, 1, 2, \ldots, n)$. Stochastic invariance of X under T thus implies stochastic equivalence of all cyclic translates $T^s X$.

Suppose now that $\mathscr{T}$ is the set $\mathbb{Z}_+ = \{0, 1, 2, \ldots\}$. We can consider the forward translation $\sigma t = t + 1$ with the corresponding operator T on X. Then statistical invariance of X under T implies statistical invariance of X under all the semi-group of transformations $\{T^s; s = 0, 1, 2, \ldots\}$. One cannot, however, infer stochastic equivalence of the random functions $T^s X$ because σ (and so T) does not possess an inverse.

However, if $\mathcal{T} = \mathbb{Z}$ then σ has the inverse specified by

$$\sigma^{-1} t = t - 1$$

and T correspondingly has an inverse. In this case, stochastic invariance of X under T and T^{-1} implies stochastic equivalence of all translates $T^s X$ ($s \in \mathbb{Z}$).

In the case of $\mathcal{T}$ equal to $\mathbb{R}_+$ or $\mathbb{R}$ then there will be analogous statements. That is, if $\mathcal{T} = \mathbb{R}_+$ then X is stochastically invariant under T^s for all $s \geqslant 0$ if it is so for all s in some positive interval $[0, \delta]$. If $\mathcal{T} = \mathbb{R}$ then the translates $T^s X$ are stochastically equivalent for all s if X is stochastically invariant under T^s for all s in some non-empty interval $(-\delta, \delta)$. However, there is not a 'smallest possible' translation corresponding to T in the cases $\mathcal{T} = \mathbb{Z}_+$ or $\mathbb{Z}$. That is, rather than generators T and T^{-1} one will have an infinitesimal generator

$$\frac{\mathrm{d}}{\mathrm{d}t} = \lim_{s \to 0} \frac{T^s - I}{s}$$

where the limit must be through positive values of s in the case $\mathcal{T} = \mathbb{R}_+$, but is not thus restricted in the case $\mathcal{T} = \mathbb{R}$. One then has, formally

$$T^s = \exp\left(s \frac{\mathrm{d}}{\mathrm{d}t}\right)$$

The considerations outlined above are the most general and superficial. If one takes account of the specific properties of the group Σ (or the equivalent group $\mathscr{S}$) then one can derive similarly specific properties of the statistics of X. This programme has been carried out for second-order statistics in particular by MacLaren (1963) and Hannan (1965). The simplest non-trivial case is again that of the cyclic group, analysed in section 18.6.

APPENDIX 2

Hamiltonian structure

A classic Hamiltonian system of dimension d has a $2d$-dimensional state variable $(p_1, p_2, \ldots, p_d; q_1, q_2, \ldots, q_d)$, where p_j and q_j are the jth components of momentum and of position respectively. If chosen in the right canonical form these satisify the system of dynamical equations

$$\dot{p}_j = -\frac{\partial H}{\partial q_j}$$

$$\dot{q}_j = \frac{\partial H}{\partial p_j}$$

(1)

where $H = H(p, q)$ is the *Hamiltonian*, an expression of the energy of the system as a function of (p, q).

The transition operator Λ of (2.6.5) (specializing to (2.8.3) in the deterministic case) thus has the action

$$\Lambda \phi = [H, \phi] := \sum_j \left[\frac{\partial H}{\partial p_j} \frac{\partial \phi}{\partial q_j} - \frac{\partial H}{\partial q_j} \frac{\partial \phi}{\partial p_j} \right]$$

(2)

on a scalar function $\phi(p, q)$, where the *Poisson bracket* $[H, \phi]$ is also defined by expression (2).

If we consider a function $\psi(p, q, t)$ then the total rate of change of $\psi(p(t), q(t), t)$ with t (i.e. the rate of change in the value of ψ as (p, q) actually moves along its trajectory) is

$$\frac{d\psi}{dt} = \frac{\partial \psi}{\partial t} + [H, \psi]$$

(3)

The condition that a function $\phi(p, q)$ should be an invariant of the motion is then that

$$[H, \phi] = 0$$

(4)

Relation (4) is satisfied if ϕ is a function of $H(p, q)$, although there may well be other invariants; see the discussion at the end of the appendix.

The operator Λ is skew-symmetric in that its adjoint relative to Lebesgue

442

measure on $\mathbb{R}^{2d}$ formally has the action (cf. (2.8.5))

$$\Lambda^{\mathsf{T}}\phi = -\sum_j \left[\frac{\partial}{\partial q_j}\left(\phi \frac{\partial H}{\partial p_j}\right) - \frac{\partial}{\partial p_j}\left(\phi \frac{\partial H}{\partial q_j}\right)\right]$$

$$= -[H, \phi] = -\Lambda\phi \tag{5}$$

Otherwise expressed: $\frac{1}{i}\Lambda$ is Hermitian.

The equation satisfied by the probability density $\pi(p, q, t)$ relative to Lebesgue measure is then

$$\dot{\pi} + [H, \pi] = 0 \tag{6}$$

by Theorem 2.8.1. Thus, any probability density $\pi(p, q)$ satisfying

$$[H, \pi] = 0 \tag{7}$$

is a possible equilibrium density. That is, π can be any function of the invariants of the motion and, in particular, any function of energy $H(p, q)$.

We deduce from (3), (6) that

$$\frac{d\pi}{dt} = 0 \tag{8}$$

That is, the local density does not change in time as one follows a representative point along its trajectory. This is effectively Liouville's theorem.

There must certainly be invariants of the motion other than energy H. Indeed, the fact that the $2d$ quantities p_j, q_j ($j = 1, 2, \ldots, d$) can be expressed in terms of the single quantity t implies that there are $2d - 1$ functionally independent invariants of the motion. However, most of these constrain the path on a scale of detail which becomes finer and finer with increasing t. Those that do not (the 'separating integrals') correspond to invariance of $H(p, q)$ under some transformation of its arguments. For example, suppose that there exists a d-vector b such that $H(p, q + \alpha b)$ is independent of the value of the scalar parameter α. It follows then from (1) that $b^{\mathsf{T}}p$ has zero rate of change, and so is an invariant of the motion; this is a generalized statement of the conservation of linear momentum.

If $H(p, q)$ has no such symmetries then H itself is the only significant invariant. Suppose, for example, that the system consists of a number of molecules constrained to lie inside a fixed vessel of irregular shape. The walls of the vessel will correspond to the walls of a potential well, so that the molecules conserve their total energy but change their total linear momentum in collisions with the wall. The existence of the containing vessel means that H is not invariant under translations of physical space, and linear momentum of the molecules cannot then be conserved. The irregular shape of the vessel means that H is not invariant under rotations of physical space so that angular momentum of the molecules cannot either be conserved. It is plausible that H itself is indeed the only significant constant of the motion.

APPENDIX 3

Constrained maximization

We sketch the elements of the subject now known as *convex programming*, since the text makes constant appeals to this material, especially in Part II. At the end of the Appendix we indicate also the local Lagrangian conditions (Kuhn–Tucker conditions) which hold under weaker assumptions, and are required for the non-convex optimization problem of section 9.7. For standard and extensive treatments of these topics the reader is referred to texts such as those by Luenberger (1969), Mangasarian (1969), Rockafellar (1969), Stoer and Witzgall (1970) or Whittle (1971).

Suppose $g(x)$ a scalar function of an n-vector variable x whose behaviour we discuss in some convex set $\mathscr{X} \in \mathbb{R}^n$. Then a row vector α is said to constitute a *subgradient* to $g(x)$ at a value $\bar{x}$ if

$$g(x) - g(\bar{x}) \geqslant \alpha(x - \bar{x}) \qquad (x \in \mathscr{X}) \tag{1}$$

If g possesses a gradient g_x at $\bar{x}$ then α must coincide with this. The class of convex functions on $\mathscr{X}$ is exactly that class of functions possessing subgradients at all points of $\mathscr{X}$.

If the inequality reverse to (1) holds then α is a *supergradient*, and supergradients are to concave functions as subgradients are to convex functions.

Consider now the problem of maximizing a function $f(x)$: $\mathscr{X} \to \mathbb{R}$ in $\mathscr{X}$ subject to linear constraints

$$Ax = b \tag{2}$$

where b is a vector of finite dimension. $b \in \mathbb{R}^m$, say. This constrained optimization problem is referred to as the *primal problem*. Let $\mathscr{B}$ be the set of b for which equations (2) possess a solution in $\mathscr{X}$, i.e. for which the primal problem is *feasible*. Let $U(b)$ be the supremum of $f(x)$ in $\mathscr{X}$ under constraints (2), a function defined in $\mathscr{B}$. One then has

Theorem A3.1 *Suppose $\mathscr{X}$ convex and f concave. Then $\mathscr{B}$ is convex and $U(b)$ is concave in its domain of definition $\mathscr{B}$.*

The really important assertions follow immediately; these give a much stronger form of the classic Lagrange multiplier treatment of constrained maximization.

Theorem A3.2 *Suppose $\mathscr{X}$ convex, f concave and $U(b)$ finite, at least in the interior of $\mathscr{B}$. Then, at least for b in the interior of $\mathscr{B}$, it can be asserted that*
 (i) *There exists an m-vector y such that a value of x freely maximizing the Lagrangian form*

$$L(x, y) := f(x) + y^{\mathsf{T}}(b - Ax) \tag{3}$$

   solves the constrained optimization problem.
 (ii) *Any such vector y has the property that y^{T} is a supergradient to $U(\cdot)$ at the prescribed value of b.*
(iii) *Any such vector y minimizes the maximized Lagrangian form*

$$\sup_{x \in \mathscr{X}} [f(x) + y^{\mathsf{T}}(b - Ax)] \tag{4}$$

(iv) *Determination of y by appeal to the dual minimization problem (iii) is equivalent to the requirement that the x maximizing L should satisfy the constraints (2).*

Statement (i) is a statement of the Lagrangian principle stronger than the classic one in that the optimal x actually maximizes L in $\mathscr{X}$ rather than simply constituting a stationary point. Statement (ii) is the familiar 'marginal cost' interpretation of a Lagrangian multiplier. If $U(\cdot)$ possesses a gradient at b then y^{T} agrees with it, and we have

$$y_j = \frac{\partial U(b)}{\partial b_j} \tag{5}$$

That is, y_j measures the sensitivity of the maximum attainable to a change in b_j.

Statement (iii) formulates the *dual problem*; it states that a relevant value of the Lagrangian multiplier vector y solves the dual problem. Otherwise expressed, if x and y solve primal and dual problems respectively, then (x, y) is a saddle-point of the Lagrangian form $L(x, y)$.

Theorem A3.2 summarizes the results appealed to repeatedly in Chapters 5, 7 and 8. Cases of failure of these assertions can be nicely codified.

Theorem A3.3 *Suppose $\mathscr{X}$ convex and f concave. Then*

 (i) *If the primal problem is infeasible (i.e. $b \notin \mathscr{B}$) then the minimal value of the form (4) characterizing the dual problem is $-\infty$.*
(ii) *If $b \in \mathscr{B}$ but $U(b) = +\infty$ then the dual problem is infeasible, in that the form (4) equals $+\infty$ for all y.*

In this theory a natural relaxation of constraint (2) is to require that $b - Ax$ belong to a convex cone $\mathscr{C}$. The only such relaxation in fact needed in the text is to the inequality constraint

$$Ax \leq b \tag{6}$$

corresponding to $\mathscr{C} = \mathbb{R}^m_+$. This can be rewritten as an equality constraint

$$Ax + z = b \tag{7}$$

with $(x, z) \in (\mathscr{X}, \mathbb{R}^m_+)$. The vector of *slack variables* measures the margin of inequality in (6), (and so measures the numbers of free lines in (4.7.11), the numbers of free units of various types in (5.8.14) and the amounts of unused resources in (8.3.9)). The feasible set $\mathscr{B}$ will now be defined as the set of b compatible with (6) for some x in $\mathscr{X}$, and $U(b)$ will be defined on this larger set.

By applying Theorem A3.2 to the problem in which constraint (6) is written as the equality constraint (7) in the augmented variable (x, z) one derives

Theorem A3.4 *Consider the maximization of concave $f(x)$ in convex $\mathscr{X}$ subject to the inequality constraints (6). Then the assertions of Theorem A3.2 continue to hold with the additions that*

(i) *a relevant multiplier vector y will always satisfy $y \geqslant 0$ and $y^{\mathsf{T}} z = 0$.*
(ii) *The dual problem stated in Theorem A3.2(iii) is qualified by the constraint $y \geqslant 0$.*

The fact that $y \geqslant 0$ follows from the fact that $b' \geqslant b$ now implies that $U(b') \geqslant U(b)$; the function $U(\cdot)$ is non-decreasing in all arguments. The relation $y^{\mathsf{T}} z = 0$ has the consequence that if $z_j > 0$ in the optimal solution then $y_j = 0$. That is, 'resources present in excess have zero marginal value', or constraints which are not *active* for the optimal solution can be discarded from the problem.

The most celebrated special case of these results is that in which one maximizes a linear form $f(x) = c^{\mathsf{T}} x$ in $x \geqslant 0$ subject to (6); one classic form of a *linear programme*. The dual problem is then the minimization of $y^{\mathsf{T}} b$ in $y \geqslant 0$ subject to

$$y^{\mathsf{T}} A \geqslant c^{\mathsf{T}} \tag{8}$$

and the dual of this dual (i.e. that derived if one takes account of constraints (8) by Lagrangian multipliers x) is just the primal problem.

Among the optimal solutions of this primal problem there is at least one which is *basic*, i.e. such that the number of non-zero elements in the vector (x, z) solving the equality form (7) of the constraints does not exceed the number m of constraints. This is, the number of non-zero elements of x does not exceed the number of active constraints. This is the assertion appealed to in Theorem 8.3.1(iii).

If the number of constraints is allowed to become infinite then one enters the field of *semi-infinite programming*, which may present new features. This is not a generalization we have occasion for in the text, since the number of types of constituent in restricted abundance (elements, resources) is always assumed finite. However, the analogue of x is a vector $n = \{n_r\}$ representing the numbers of molecules of various types r. The set $\mathscr{R}$ of values of r may well be infinite. In Part IV this was indeed the essence of the analysis, in that it was this infinity which led to critical effects.

If x is infinite-dimensional then $U(b)$ is still a function of m arguments only, and so one may expect the assertions of Theorems A3.2 and A3.4 still to hold at values of b in whose neighbourhood $U(\cdot)$ is finite. However, the difficulty may be that there are values of b at which the maximizing x is not such that its components x_j are arbitrarily small outside some sufficiently large but finite set $\mathscr{J}$ of j. This is the very effect that leads to critical transitions, and is revealed simply by the fact that one cannot find a multiplier value y such that the x maximizing the Lagrangian form (3) will satisfy the constraints (2). As an example we can just formulate the situation considered in section 6.3 in the notation of this appendix. Consider the maximization of

$$f(x) = \sum_{j=1}^{\infty} [x_j(1 + \log \gamma_j) - x_j \log x_j] \tag{9}$$

in $x \geqslant 0$ subject to the scalar constraint

$$\sum_j j x_j = b \tag{10}$$

Taking a Lagrangian multiplier y for constraint (10) we deduce a solution

$$x_j = \gamma_j e^{-jy} \tag{11}$$

We determine y from the constraint (10), i.e. from

$$h(y) := \sum_j j \gamma_j e^{-jy} = b \tag{12}$$

If all γ_j are non-negative then it is plain that y decreases with increasing b. However, there may simply not be values of y which will satisfy (12) for given b. Consider the case

$$\gamma_j = y_c^{-j}/j^3$$

and define

$$b_c = h(y_c) = \sum_j j^{-2}$$

Then for $b \leqslant b_c$ we can find a value $y \geqslant y_c$ such that (12) is satisfied. For $b > b_c$ one would require $y < y_c$, and then $h(y)$ diverges.

If the functions and sets involved do not obey the convexity and linearity assumptions of Theorem A3.2 then the conclusions of that theorem will not hold in general. This is the situation for the network optimization problem of section 9.7, for which the criterion function C of (9.7.2) is in general not convex in the variables (λ, w) and the constraint (9.7.5) is not linear in these variables). However, Lagrangian assertions of a local nature may still be made.

Suppose that $f(x)$ is to be maximized in $\mathscr{X}$ subject to a vector constraint

$$g(x) = b \tag{13}$$

where g is not necessarily linear. As before, let $U(b)$ be the supremum of $f(x)$ under this constraint, and suppose that it is in fact attained by an optimal solution $x = x(b)$. Then $U(\cdot)$ is characterized by the statement that the inequality

$$f(x) \leqslant U(g(x)) \tag{14}$$

holds for all x in $\mathscr{X}$, with equality for some x for every possible value of the vector g. That is, if the value b is realizable in that, for that value of b, equations (13) can be satisfied for some x in $\mathscr{X}$, there is then an $x(b)$ such that

$$g(x(b)) = b$$
$$f(x(b)) = U(b) \tag{15}$$

where $x(b)$ is of course a solution of the constrained optimization problem.

Theorem A3.5 The Kuhn–Tucker conditions. *Suppose that the maximization of $f(x)$ in $\mathscr{X}$ subject to (13) for a given value of b is solved by $x = x(b)$, and that f and g possess first derivatives at $x(b)$ and U possesses first derivatives at b. Then the Lagrangian form (3) is non-increasing as x is varied locally from $x(b)$ within $\mathscr{X}$, if y^{T} is given the value U_b.*

Proof A permitted local direction of variation s would be one for which

$$x = x(b) + \varepsilon s$$

belongs to $\mathscr{X}$ for all sufficiently small positive ε. By comparing the inequality (14) at this value of x with the equality (15) in the limit $\varepsilon \downarrow 0$ one deduces the inequality

$$(f_x - U_b g_x)s \leqslant 0$$

at $x = x(b)$ for the prescribed b. This is just the assertion of the theorem. ∎

The case when the constraint (13) is replaced by the relaxed version

$$b - g(x) \in \mathscr{C} \tag{16}$$

can again be reduced to the previous case by the introduction of slack variables. We can write (16) as an equality constraint

$$g(x) + z = b$$

and work in terms of the augmented variable $(x, z) \in (\mathscr{X}, \mathscr{C})$.

There are various *constraint qualifications* which ensure that the gradient U_b does indeed exist at the prescribed b-value; we refer to the texts quoted for details.

References

AKAIKE, H. (1960) On a limiting process which asymptotically produces f^{-2} spectral density. *Ann. Inst. Statist. Math. Tokyo*, **12**, 7–11.

BARBOUR, A. D. (1976) Networks of queues and the method of stages. *Adv. Appl. Prob.*, **8**, 584–591.

BARTHOLOMAY, A. F. (1958) Stochastic models for chemical reactions. I. Theory of the unimolecular reaction process. *Bull. Math. Biophys.*, **20**, 175–190.

BARTLETT, M. S. (1949) Some evolutionary stochastic processes. *J. Roy. Statist. Soc. B*, **11**, 211–229.

BASKETT, F., CHANDY, K. M., MUNTZ, R. R., and PALACCOS, F. C. (1979) Open, closed and mixed networks of queues with different classes of customers. *J.A.C.M.*, **22**, 248–260.

BELLMAN, R. (1960) *Introduction to Matrix Analysis*. McGraw-Hill.

BETHE, H. A. (1935) Statistical theory of superlattices. *Proc. Roy. Soc.*, **A 150**, 552–575.

BIGGS, N. L. (1977) *Interaction Models*. Cambridge U.P.

BOLLOBAS, B. (1985) *Random Graphs*. Academic Press.

BOLTZMANN, L. (1896) *Vorlesungen über Gastheorie*. English translation (1974), University of California Press, Berkeley.

BOXMA, O. J. and KONHEIM, A. G. (1981) Approximate analysis of exponential queueing systems with blocking. *Acta Informatica*, **15**, 19–66.

BRANFORD, A. J. (1983) Self-excited random processes. Ph.D dissertation, Cambridge, UK.

BRANFORD, A. J. (1985) A self-excited migration process. *J. Appl. Prob.*, **22**, 58–67.

BROOK, D. (1964) On the distinction between the conditional and the joint probability approaches in the specification of nearest-neighbour systems. *Biometrika*, **51**, 481–483.

BURMAN, D. Y. (1981) Insensitivity in queueing systems. *Adv. Appl. Prob.*, **13**, 846–859.

BURMAN, D. Y., LEHOCZKY, J. P., and LIM, Y. (1984) Insensitivity of blocking probabilities in a circuit-switched network. *J. Appl. Prob.*, **21**, 853–859.

CHANDY, K. M., HOWARD, J. H., and TOWNSLEY, D. F. (1977) Product form and local balance in queueing networks. *J.A.C.M.*, **24**, 250–263.

CHAY, S. C. (1972) On quasi-Markov random fields. *J. Multi. Analysis*, **2**, 14–76.

COX, D. R. (1984) Long-range dependence, a review in *Statistics—an appraisal* (eds H. T. David and H. A. David), 55–76. Iowa State University Press.

DARVEY, I. G., NINHAM, B. W., and STAFF, P. J. (1966) Stochastic models for second-order chemical reaction kinetics. The equilibrium state, *J. Chem. Phys.*, **45**, 2145–2155.

DAVIS, M. H. A. (1984) Piecewise-deterministic Markov processes; a general class of non-diffusion stochastic models. *J. Roy. Statist. Soc. B*, **46**, 353–388.

DOBRUSHIN, R. L. (1980a) Gaussian random fields—Gibbsian point of view. In *Multicomponent Random Systems* (eds R. L. Dobrushin and Ya. G. Sinai). M. Dekker, New York, pp. 119–151.

DOBRUSHIN, R. L. (1980b) Automodel generalized random fields and their renorm group. In *Multicomponent Random Systems* (eds R. L. Dobrushin and Ya. G. Sinai). M. Dekker, New York, pp. 153–198.

DUBIN, D. A. (1974) *Solvable Models in Algebraic Statistical Mechanics*. Oxford UP.

EHRENFEST, P. and EHRENFEST, T. (1907) *Physik. Z.*, **8**, 311–

ERDÖS, P. and RÉNYI, A. (1960) On the evolution of random graphs. *Mat. Kutató. Int. Közl.*, **5**, 17–60.

ERDÖS, P. and SPENCER, J. (1974) *Probabilistic Methods in Combinatorics*. Academic Press.

FAIRFIELD SMITH, H. (1938) An empirical law describing heterogeneity in the yields of agricultural crops. *J. Agric. Sci.*, **28**, 1–23.

FRANKEN, P., KÖNIG, D., ARNDT, U., and SCHMIDT, V. (1982) *Queues and Point Processes*. Wiley.

FLORY, P. J. (1953) *Principles of Polymer Chemistry*. Cornell University Press.

GIHMAN, T. I. and SKOROHOD, A. V. (1971) *The Theory of Stochastic Processes, Vol. III*. Springer.

GOLDBERG, R. J. J. (1952) A theory of antigen-antibody reactions. I. *J. Amer. Chem. Soc.*, **74**, 5715.

GOLDBERG, R. J. J. (1953) A theory of antigen-antibody reactions. II. *J. Amer. Chem. Soc.*, **75**, 3127.

GOOD, I. J. (1960) Generalisations to several variables of Lagrange's expansion, with applications to stochastic processes. *Proc. Camb. Phil. Soc.*, **56**, 366–380.

GOOD, I. J. (1963) Cascade theory and the molecular wieght averages of the sol fraction. *Proc. Roy. Soc. A.*, **272**, 54–59.

GORDON, M. (1962) Good's theory of cascade processes applied to the statistics of polymer distributions. *Proc. Roy. Soc. A*, **268**, 240–259.

GORDON, M. and ROSS-MURPHY, S. B. (1978) *J. Phys. A. Math. Gen.*, **11**, L155.

GORDON, M. and TEMPLE, W. B. (1976) *Chemical Applications of Graph Theory*, 300–332. Ed. A. T. Balaban. Academic Press.

GORDON, W. J. and NEWELL, G. F. (1967) Closed queueing systems with exponential servers. *Operations Research*, **15**, 254–265.

GRIFFEATH, D. and LIGGETT, T. M. (1982) Critical phenomena for Spitzer's reversible nearest particle systems. *Ann. Prob.*, **10**, 881–895.

GRIMMETT, C. R. and STIRZAKER, D. R. (1982) *Probability and Random Processes*. Oxford University Press.

HALL, P. (1983a) On the roles of the Bessel and Poisson distributions in chemical kinetics. *J. Appl. Prob.*, **20**, 585–599.

HALL, P. (1983b) On the interpretation of random fluctuations in competing chemical systems. *J. Appl. Prob.*, **20**, 877–883.

HANNAN, E. J. (1965) Group representations and applied probability. *J. Appl. Prob.*, **2**, 1–68.

HORDIJK, A. (1984) Insensitivity for stochastic networks. In *Mathematical computer performance and reliability* (Eds. G. Iazeolla, P. J. Courtois and A. Hordijk) Elsevier (North Holland).

HORDIJK, A. and VAN DIJK, N. (1983a) Networks of queues. *Proc. International Seminar on Modelling and Performance Evaluation Methodology, INRIA, Vol. 1*, 79–135.

HORDIJK, A. and VAN DIJK, N. (1983b) Adjoint processes, job local balance and insensitivity for stochastic networks. *Bull. 44th Session Int. Stat. Inst., Vol. 50*, 776–788.

ISHAM, V. (1981) An introduction to spatial point processes and Markov random fields. *Int. Stat. Rev.*, **49**, 21–44.

JACKSON, J. R. (1957) Networks of waiting lines. *Operations Res.*, **5**, 518–521.

JACKSON, J. R. (1963) Jobshop-like queueing systems. *Management Science*, **10**, 131–142.

JANSEN, U. and KÖNIG, D. (1980) Insensitivity and steady-state probabilities in product form for queueing networks. *Elektronische Informationsverarbeitung und Kybernetik*, **16**, 385–397.

KAC, M. (1947) On the notion of recurrence in discrete stochastic processes. *Bull. Amer. Math. Soc.*, **53**, 1002–1010.

KAC, M. (1959) *Probability and Related Topics in Physical Sciences.* Interscience, New York.

KELLY, F. P. (1975a) Networks of queues with customers of different types. *J. Appl. Prob.,* **12**, 542–554.

KELLY, F. P. (1975b) Markov processes and Markov random fields. *Bull. Int. Inst. Statist.,* **46**, 397–404.

KELLY, F. P. (1976) Networks of queues. *Adv. Appl. Prob.,* **8**, 416–432.

KELLY, F. P. (1979) *Reversibility and Stochastic networks.* Wiley.

KELLY, F. P. (1982a) Networks of quasi-reversible nodes. *Applied Probability—Computer Science, the Interface: Proc. of the ORSA-TIMS Boca Raton Symposium* (ed. R. Disney), Birkhauser, Boston, Cambridge, Mass.

KELLY, F. P. (1982b) The throughput of a series of buffers. *Adv. Appl. Prob.,* **14**, 633–653.

KELLY, F. P. (1985) Stochastic models of computer communication systems. *J. Roy. Statist. Soc. B.,*

KELLY, F. P. (1986) Blocking probabilities in large circuit-switched networks. *Adv. Appl. Prob.,* **18**,

KINDERMANN, R. and SNELL, J. L. (1980) *Markov Random Fields and their Applications.* American Mathematical Society.

KINGMAN, J. F. C. (1969) Markov population processes. *J. Appl. Prob.,* **6**, 1–18.

KOENIG, D., MATTHES, K., and NAWROTZKI, K. (1967) *Verallgemeinerungen der Erlangschen und Engsetschen Formeln (Eine Methode der Bedienungstheorie).* Akademi-Verlag, Berlin.

KOLMOGOROV, A. N. (1936) Zur theorie det Markoffschen ketten. *Mathematische Annalen,* **112**, 155–160.

KÜNSCH, H. (1981) Thermodynamics and statistical analysis of Gaussian random fields. *Z. Wahrsch. vorw. Geb.,* **58**, 407–421.

LUENBERGER, D. G. (1969) *Optimization by Vector Space Methods.* Wiley.

LUENBERGER, D. G. (1979) *Introduction to Dynamic Systems.* Wiley.

McLAREN, A. D. (1963) On group representations and invariant stochastic processes. *Proc. Camb. Phil. Soc.,* **59**, 431–450.

McQUARRIE, D. A. (1967) Stochastic approach to chemical kinetics. *J. Appl. Prob.,* **3**, 413–478.

MANDELBROIT, B. B. (1977) *Fractals, Form, Chance and Dimension.* W. H. Freeman, San Francisco.

MANGASARIAN, D. L. (1969) *Nonlinear Programming.* McGraw-Hill.

MATTHES, K. (1962) Zur Theorie der Bedienungsprozesse. *Trans. 3rd Prague Conf. Int. Theory.*

MOORE, T. and SNELL, J. L. (1979) A branching process showing a phase transition. *J. Appl. Prob.,* **16**, 252–260.

MORAN, P. A. P. (1961) Entropy, Markov processes and Boltzmann's H theorem. *Proc. Camb. Phil. Soc.,* **57**, 833–842.

MORGAN, B. J. T. (1976) Stochastic models of grouping changes. *Adv. Appl. Prob.,* **8**, 30–57.

MUNTZ, R. R. (1972) Poisson departure processes and queueing networks. *IBM Research Report RC 4145.* IBM Thomas J. Watson Research Centre, Yorktown Heights, New York.

ONSAGER, L. (1944) Crystal statistics, I. A two-dimensional model with an order-disorder transition. *Phys. Res.,* **65**, 117–149.

ORRISS, J. (1969) Equilibrium distributions for systems of chemical reactions with applications to the theory of molecular absorption. *J. Appl. Prob.,* **6**, 505–515.

PENROSE, O. (1970) *Foundations of Statistical Mechanics.* Pergamon Press.

PERCUS, J. K. (1971) *Combinatorial Methods*. Springer, Berlin.

PHILLIPS, O. M. (1973) The equilibrium and stability of simple marine biological systems, I. *The American Naturalist*, **107**, 73–93.

PHILLIPS, O. M. (1974) The equilibrium and stability of simple marine biological systems, II. *Arch. Hydrobiol.*, **73**, 310–333.

POLLETT, P. K. (1982) *Distributional approximations to networks of queues*. Ph.D. dissertation. Cambridge, U.K.

POSTON, T. and STEWART, I. (1978) *Catastrophe Theory and its Applications*. Pitman.

PRESTON, C. J. (1974) *Gibbs States on Countable Sets*. Cambridge University Press.

REICH, E. (1957) Waiting times when queues are in tandem. *Ann. Math. Statist.*, **28**, 768–773.

REICHL, L. E. (1980) *A Modern Course in Statistical Physics*. University of Texas Press.

REIMAN, M. I. (1984) Open queueing networks in heavy traffic. *Math. Oper. Res.*, **9**, 441–458.

RÉNYI, A. (1970) *Foundations of Probability*. Holden-Day, San Francisco.

ROCKAFELLAR, R. T. (1969) *Convex Analysis*. Princeton University Press.

ROSS, S. M. (1980) *Stochastic Processes*. Wiley.

ROZANOV, Y. A. (1967) On Gaussian random fields with given conditional distributions. *Theory Prob. Appl.*, **12**, 381–391.

SCHASSBERGER, R. (1977) Insensitivity of steady-state distributions of generalised semi-Markov processes. Part I. *Ann. Prob.*, **5**, 87–99.

SCHASSBERGER, R. (1978a) Insensitivity of steady-state distributions of generalised semi-Markov processes. Part II. *Ann. Prob.*, **6**, 85–93.

SCHASSBERGER, R. (1978b) Insensitivity of steady-state distributions of generalised semi-Markov processes with speeds. *Adv. Appl. Prob.*, **10**, 836–851.

SCHASSBERGER, R. (1978c) The insensitivity of stationary probabilities in networks of queues. *Adv. Appl. Prob.*, **10**, 906–912.

SCHASSBERGER, R. (1979) A definition of discrete product-form distributions. *Zeitschrift für Op. Res.*, **23**, 189–195.

SMOLUCHOWSKI, M. von (1916) *Phys. Z.*, **17**, 557–585.

SMOLUCHOWSKI, M. von (1917) *Z. Phys. Chem.*, **92**, 129.

SCHUSTER, P., SIGMUND, K., and WOLFF, R. (1980) Mass action kinetics of self replication in flow reactors. *J. Math. Anal. Appl.*, **78**, 88–112.

SPITZER, F. (1971) *Random Fields and Interacting Particle Systems*. Mathematical Association of America.

SPITZER, F. (1979) Markov random fields on an infinite tree. *Ann. Prob.*, **3**, 387–398.

SPOUGE, J. L. (1984) Polymers and random graphs: asymptotic equivalence to branching processes. *J. Stat. Phys.* (to appear).

STEPANOV, V. E. (1970a) On the probability of connectedness of a random graph. *Teoriya Veroyatnostei i ee Prim.*, **15**, 55–67.

STEPANOV, V. E. (1970b) Phase transitions in random graphs. *Teoriya Veroyatnostei i ee Prim.*, **15**, 187–203.

STOCKMAYER, W. H. (1943) Theory of molecular size distribution and gel formation in branched chain polymers. *J. Chem. Phys.*, **11**, 45–55.

STOCKMAYER, W. H. (1944) Theory of molecular size distribution and gel formation in branched polymers. II. General cross-linking. *J. Chem. Phys.*, **12**, 125–131.

STOER, J. and WITZGALL, C. (1970) *Convexity and Optimization in Finite Dimensions*. Springer.

TALLIS, G. M. and LESLIE, R. T. (1969) General models for r-molecular reactions. *J. Appl. Prob.*, **6**. 74–87.

TAQQU, M. S. (1982) Self-similar processes and related ultraviolet and infrared catastrophes. In *Random Fields* (eds Fritz, Lebowitz and Szasz). North-Holland, pp. 1057–1096.

WALRAND, J. and VARAIYA, P. (1980) Interconnections of Markov chains and quasi-reversible queueing networks. *Stochastic Process. Appl.*, **10**, 209–219.

WASSERMAN, S. P. (1978) Models for binary directed graphs and their applications. *Adv. Appl. Prob.*, **10**, 803–818.

WATSON, G. S. (1958) On Goldberg's theory of the precipitin reaction. *J. Immunology*, **80**, 182–185.

WHITTLE, P. (1951) *Hypothesis Testing in Time Series Analysis*. Almqvist and Wicksell, Uppsala.

WHITTLE, P. (1952) Some results in time series analysis. *Skand. Aktuarietidskr.*, **35**, 48–60.

WHITTLE, P. (1954a) Some recent contributions to the theory of stationary processes. Appendix (pp. 196–228) to the second edition of H. Wold: *A Study in the Analysis of Stationary Time Series*. Almqvist and Wicksell, Uppsala.

WHITTLE, P. (1954b) On stationary processes in the plane. *Biometrika*, **41**, 434–449.

WHITTLE, P. (1955) Reversibility in Markov processes. Unpublished paper.

WHITTLE, P. (1962) Topographic correlation, power-law covariance functions, and diffusion. *Biometrika*, **49**, 305–314.

WHITTLE, P. (1965a) Statistical processes of aggregation and polymerisation. *Proc. Camb. Phil. Soc.*, **61**, 475–495.

WHITTLE, P. (1965b) The equilibrium statistics of a clustering process in the uncondensed phase. *Proc. Roy. Soc. Lond. A*, **285**, 501–519.

WHITTLE, P. (1967) Nonlinear migration processes. *Bull. Int. Inst. Statist.*, **42**, 642–647.

WHITTLE, P. (1968) Equilibrium distributions for an open migration process. *J. Appl. Prob.*, **5**, 567–571.

WHITTLE, P. (1971) *Optimisation under Constraints*. Wiley.

WHITTLE, P. (1972) Statistics and critical points of polymerisation processes. *Proc. of Symp. on Statistical and Probabilistic Problems in Metallurgy*. Supplement to *Adv. Appl. Prob.*, 199–215.

WHITTLE, P. (1975) Reversibility and acyclicity. *Perspectives in Probability and Statistics*, 217–224. Applied Probability Trust.

WHITTLE, P. (1977a) Cooperative effects in assemblies of stochastic automata. *Proc. Symp. to Honour Jerzy Neyman*, 335–343. Polish Scientific Publishers, Warsaw.

WHITTLE, P. (1977b) An extremal characterisation of the equilibrium state for a spatial competition process. *Bull. Int. Inst. Stat.*, **40**, No. 2, 392–396.

WHITTLE, P. (1980a) Polymerisation processes with intrapolymer bonding. I. One type of unit. *Adv. Appl. Prob.*, **12**, 94–115.

WHITTLE, P. (1980b) Polymerisation processes with intrapolymer bonding. II. Stratified processes. *Adv. Appl. Prob.*, **12**, 116–134.

WHITTLE, P. (1980c) Polymerisation processes with intrapolymer bonding. III. Several types of unit. *Adv. Appl. Prob.*, **12**, 135–153.

WHITTLE, P. (1981) A direct derivation of the equilibrium distribution for a polymerisation process. *Teoriya Veryatnostei*, **26**, 350–361.

WHITTLE, P. (1982a) Criticality and the emergence of structure. *Evolution of Order and Chaos* (ed. H. Haken), 264–269. Springer.

WHITTLE, P. (1982b) Semi-spatial models of socio-economic transition. *Statistics in Theory and Practice* (ed. B. Ranneby), 299–304. Swedish University of Agricultural Sciences.

WHITTLE, P. (1982c) *Optimisation Over Time*, Vol. 1. Wiley.

WHITTLE, P. (1983a) *Optimisation Over Time*, Vol. 2. Wiley Interscience.

WHITTLE, P. (1983b) Relaxed Markov processes. *Adv. Appl. Prob.*, **15**, 769–782.

WHITTLE, P. (1983c) Competition and bottlenecks. *Probability, Statistics and Analysis* (eds J. Kingman and G. Reuter), 277–284. L. M. S. Lecture Note Series 79, Cambridge UP.

WHITTLE, P. (1984a) Weak coupling in stochastic systems. *Proc. Roy. Soc. Ser. A*, **395**, 141–151.

WHITTLE, P. (1984b) Optimal routing in Jackson networks. *Asia-Pacific J. Op. Res.*, **1**, 32–37.

WHITTLE, P. (1985a) Partial balance and insensitivity. *J. Appl. Prob.*, **22**, 168–176.

WHITTLE, P. (1985b) Scheduling and characterisation problems for stochastic networks. *J. Roy. Statist. Soc.*, *B*.

WHITTLE, P. (1985c) Random graphs and polymerisation processes. *Discrete Mathematics*.

WHITTLE, P. (1986) Partial balance, insensitivity and weak coupling. *J. Appl. Prob.*

ZACHARY, S. (1983) Countable state space Markov random fields and Markov chains on trees. *Ann. Probab.*, **11**, 894–903.

ZERMELO, E. (1896) *Ann. Physik*, **57**, 485.

Index

(*continued from front*)